ENVIRONMENTAL IMPACT ASSESSMENT

ENVIRONMENTAL IMPACT ASSESSMENT
Practical Solutions to Recurrent Problems

DAVID P. LAWRENCE
Lawrence Environmental

WILEY-INTERSCIENCE

A JOHN WILEY & SONS, INC., PUBLICATION

For general information on our other products and services please contact our Customer Care Department within the U.S. at 877-762-2974, outside the U.S. at 317-572-3993 or fax 317-572-4002.

Wiley also publishes its books in a variety of electronic formats. Some content that appears in print, however, may not be available in electronic format.

Library of Congress Cataloging-in-Publication Data:

Lawrence, David P. (David Phillip), 1947–
  Environmental impact assessment : practical solutions to recurrent
  problems / David P. Lawrence.
       p.   cm.
  ISBN 0-471-45722-1 (cloth)
  1. Environmental impact analysis.   I. Title.
  TD194.6.L39 2003
  333.7′14–dc21

                                                              2003003582

Printed in the United States of America

10 9 8 7 6 5 4 3 2 1

To Barbara

CONTENTS

PREFACE

This book was born from a nagging concern about how the environmental impact assessment (EIA) process is commonly depicted in EIA literature and applied in practice. It also stems from a perception that EIA practitioners need more help to cope with the many competing demands and recurrent problems encountered in their day-to-day work. More effective EIA process design and management can, I believe, help practitioners in their efforts to balance competing demands and to ameliorate recurrent problems.

My uneasiness about the EIA process has arisen over the past three decades. Over that period I have become increasingly convinced that there are far more process design and management choices available in practice than are customarily conveyed in EIA literature. It also seemed to me that EIA regulatory analyses usually began from and sought to refine current requirements rather than exploring, at a more fundamental level, the full range of potential regulatory choices.

This uneasiness was reinforced through my ongoing interest in EIA and planning processes. I have maintained a joint interest in planning and EIA for many years. I have practiced and taught in both fields and have addressed the interrelationships between the two fields through graduate and undergraduate papers, a doctoral dissertation, a series of journal articles, and considerable EIA process management experience. A central feature of planning theory is the plurality of overlapping and competing prescriptive planning theories. A central feature of EIA is the largely unitary approach to process design. Planning theory literature can be extremely frustrating! It is plagued by hyperbole, jargon, and, until very recently, a huge gulf between theory and practice. Still the claims, counterclaims, debates, and critiques alert the reader to the dangers of hidden assumptions and to the value of multiple perspectives. Such debates exist in EIA literature, but they are more muted. They also seem to take more for granted regarding shared assumptions and perspectives. Depictions of the EIA process, in particular, are, from my perspective, less diverse than they should be.

My concern about the multiple demands and recurrent problems faced by EIA practitioners is a product of both direct experience and interchanges with other practitioners. EIA practitioners must counterbalance multiple, often conflicting internal and external demands. Frequently, it is expected that EIA requirements, procedures, and documents should be rigorous, rational, practical, substantive, democratic, collaborative, ethical, and adaptable simultaneously. These demands commonly reflect fundamentally different perspectives on the environment and on the appropriate role of EIA in decision making. Almost invariably, perspective differences are translated into varying interpretations of critical issues, the nature and significance of potential effects, and most centrally (in terms of the purpose of this book), how best to proceed from proposal inception to final proposal decision making and implementation. Difficulties encountered in dealing with multiple demands and perspectives often coalesce as recurrent problems that hamper effective EIA process design and management. EIA practitioners need additional assistance in navigating through this minefield.

This book is intended to help EIA participants (regulators, managers, EIA specialists, other study team specialists, nongovernment organizations) and observers (commentators, instructors, students) to contribute jointly to more effective EIA processes. Effective processes can help refine and achieve EIA regulatory objectives and further the goals of EIA as a form of environmental management. The book challenges the prevailing assumption that EIA should be structured around a unitary EIA process. It begins by identifying, through a scenario, eight recurrent problems encountered in EIA practice. The characteristics of multiple variations of conventional EIA processes, at both the regulatory and applied levels, are then presented. These analyses open up consideration of available regulatory and applied EIA process design and management choices. But they address the recurrent problems only partially. The residual problems that remain provide the springboard for a description and analysis of eight EIA processes for coming to grips with recurrent problems. The description of each of these EIA processes provides examples from practice, defines the problem, and identifies a direction for improvement. For each we then detail major relevant conceptual distinctions, describe how a process to reduce the problem would operate at the regulatory level (based on an overview of EIA requirements in the United States, Canada, Europe, and Australia), and explain how a process to reduce the problem would operate at the applied level. We next assess how well each process satisfies ideal EIA process characteristics. Each analysis ends with a summary overview and the identification of links between the conceptual analysis and the practice examples. In the final chapter we address how to link and combine EIA processes to operate in situations characterized by multiple, overlapping problems. EIA literature and literature from such related fields as planning, environmental and resource management, risk assessment and management, site selection and evaluation, and public participation are drawn upon to characterize and assess each EIA process.

The analyses and solutions offered in this book are far from definitive. Hopefully, they are practical. I believe that sufficient knowledge and experience now exist regarding the recurrent problems such that major pitfalls can be identified and possible improvements suggested. I am not sufficiently naive to suggest that we are on the brink of delineating that elusive core body of common knowledge that is supposed to characterize "mature" fields. I have serious doubts as whether such a quest is even desirable. I also appreciate that there are immense impediments to significant improvements to EIA practice, many of which lie beyond the control or influence of EIA practitioners. But I still believe that sufficient operating room remains within which EIA practice enhancements are possible. I also maintain that the EIA process is at the core of many such improvements. Hopefully, this book will contribute to such efforts.

I wish to thank the following people for contributing their thoughtful and insightful stories for inclusion in this book: Dave Abbott, Ralf Aschemann, Jo Anne Beckwith, Alan Bond, Roger Creasey, Alan Diduck, Patricia Fitzpatrick, Bob Gibson, Dave Hardy, Nick Harvey, Annie Holden, Peter Homenuck, Leslie Matthews, Bruce Mitchell, Robin Saunders, Darryl Shoemaker, and John Sinclair. I also wish to thank the anonymous reviewers and the staff of John Wiley & Sons (most notably Bob Esposito and Jonathan Rose) for their constructive suggestions and guidance.

A great many colleagues have provided encouraging comments and/or have influenced my thinking and writing regarding EIA process management over the past several years. A very partial list includes Rabel Burdge, Dave Cressman, Peter Croal, Bob Dorney, George Francis, Bob Gibson, Peter Homenuck, Eric Hunter, Larry Martin, Jim Micak, Greg Michelenco, John Page, Donna Pawlowski, Barry Sadler, Paul Scott, John Sinclair, Graham Smith, Margaret Smith, Roger Suffling, Richard Szudy, and Tom Wlodarczyk.

My thanks to the following for permission to reproduce, without charge, the following copyrighted material (full details are provided in the reference list at the back of the book):

- Beech Tree Publishing (Figure 6.5)
- Elsevier Science, Inc. (Figure 3.1)
- Imperial College Press (Figures 2.3 to 2.6, 2.10, Tables 2.4, 2.6, and 2.7, and selective text from Lawrence, 2001)
- National Association of Environmental Professionals (Figures 2.7 to 2.9, 2.11 to 2.18)
- Springer-Verlag New York, Inc. (Figures 5.2 and 5.3)

Finally and most important, I am especially grateful to my wife, Barbara, for her patience, encouragement, and support throughout this lengthy process.

Lawrence Environmental   DAVID P. LAWRENCE
dlawren@telus.net

CHAPTER 1

INTRODUCTION

1.1 HIGHLIGHTS

This book is intended to enhance environmental impact assessment (EIA) practice. It provides practical solutions to EIA practitioners for major, recurrent problems encountered in daily EIA practice.

- The scenario presented in Section 1.2 highlights the problems. It illustrates how a failure to anticipate and respond to varying perspectives can contribute to the collapse of a seemingly well-designed and well-managed EIA process.
- In Section 1.3 we use insights from the scenario to identify key prerequisites to formulating a strategy that can cope with the problems that may arise from multiple perspectives.
- In Section 1.4 we go back to the fundamentals. We use an EIA definition and an overview of EIA characteristics to identify implications for overall EIA process management and for accommodation of the perspectives displayed in the scenario.
- In Section 1.5 we address the current state-of-the-art of EIA process management. We test the need for a strategy in light of current and emerging EIA practice. The analysis is based on an overview of EIA patterns and trends and a review of recurrent shortcomings in EIA practice.

Environmental Impact Assessment: Practical Solutions to Recurrent Problems, By David P. Lawrence
ISBN 0-471-45722-1   Copyright © 2003 John Wiley & Sons, Inc.

- In Section 1.6 we identify the EIA process as the organizing framework around which a strategy should be built. We explain why the EIA process in general and alternative EIA processes in particular are essential to the effort.
- In Section 1.7 we present a strategy to facilitate more effective EIA process management.
- In Section 1.8 we suggest how EIA stakeholders may use this book.
- In Section 1.9 we highlight major themes and conclusions.

1.2 A NOT-SO-HYPOTHETICAL SCENARIO

1.2.1 Brave Beginnings

A private proponent decides to establish a new hazardous waste treatment facility. Management realizes that there will be numerous licensing requirements, including the preparation and approval of an EIA. Accordingly, a consulting team is hired to prepare the EIA documentation and to ensure that all approval requirements are satisfied. A preliminary design is prepared for a state-of-the-art facility. An overview of available properties is conducted. A site is selected in a general industrial park a couple of miles outside a medium-sized community. An option is taken out on the property. Local community officials express a willingness to accept the facility because of the tax revenue to be generated and a promise to share a portion of the facility revenues with the local community. Two municipal councilors express reservations because of a fear that the facility might stigmatize the community.

The EIA process has a promising beginning. A core study team is assembled with ample EIA and regulatory approval experience. A variety of engineering and environmental specialists, together with an expert in public participation, are added to the team. A preliminary study design is prepared. Initial scoping sessions are conducted with government officials to identify regulatory requirements, concerns, and priorities. An initial set of public meetings and open houses are convened to identify public concerns and preferences. The study program is modified to accommodate public and agency concerns. The EIA is divided into a clearly defined sequence of steps. Provision is made for public and agency input into each step.

In the early months of the process, the focus is on establishing a sound environmental baseline and on refining facility characteristics. Several mitigation options are screened and compared in the ongoing effort to prevent and ameliorate adverse impacts. Initial background papers are prepared documenting baseline conditions, study methodology, the analysis of alternatives, and preliminary impact predictions. Impact predictions are then refined and impact significance determined for both individual and cumulative impacts. A concerted effort is made to mitigate potentially significant adverse impacts. In a few cases, this necessitates comparing mitigation options. These various analyses are consolidated first in working and background papers and then in a draft impact statement. Summary reports are prepared for each document. Documents are circulated for initial agency comment and are used as the basis for discussions and presentations at public meetings and open houses. All comments and suggestions are recorded. Responses are provided to each comment received, including a detailing of how and where the comments are addressed in the EIA documentation.

1.2.2 Cracks in the Foundation

Public opposition begins to mount during this period. Initially, this opposition comes from individuals. It is not long before a local opposition group is formed. Local and then regional environmental organizations quickly join the fray. The local community groups are concerned about potential human health effects, possible declining property values, and community stigma. They strongly criticize the limited, closed, and informal procedure adopted for selecting the preferred site. The environmental groups question the need for the facility, arguing that it is old technology that should be superseded by waste reduction, reuse, and recycling initiatives. They challenge the "growth ethic" inherent in the predicted use of the facility and argue that the proposed facility undermines the cause of environmental sustainability.

Several faculty members from the local university also voice their opposition. They focus their comments on the scientific validity of the impact predictions. They especially point to the failure to use control communities, the lack of peer review, the excessively descriptive analysis, the questionable statistical analyses, the crude models employed, and the short duration of the baseline studies. In addition, they stress that the studies fail to adequately address uncertainties, low probability–high consequence risks, and perceived risks. The opposition to the facility culminates in a raucous public meeting. Many members of the public attending the meeting stress that public involvement in the process has been at best tokenism and at worst manipulation. Considerable frustration is expressed about what is seen as a loss of community control. Many participants argue that the process is neither open nor fair. They also suggest that it is unfair to locate such a facility in an area that generates such a small proportion of the waste, has several similar facilities, and is social and economically disadvantaged. Frequent reference is made to the mixed track record of the proponent in other communities. Several municipal councilors soon reconsider their initial support for the facility.

Initial agency reactions to the documents are mixed at best. As they work their way through the lengthy documents, some reviewers have difficulty in determining whether specific regulatory requirements and policies have been addressed explicitly. Other reviewers question the clarity of the methodology, challenge the methods or data sources used, argue that the methods have been misapplied, or suggest that conclusions are insufficiently substantiated. The alternatives analysis becomes a focal point of criticism. Several reviewers argue that a wider range of alternatives should have been considered, criteria are not explicitly defined or consistently applied, criteria are not ranked, and sensitivity analyses have not been undertaken to explore the implications of alternative criteria rankings and varying interpretations of mitigation potential and the implications of uncertainty. Substantial document modifications are made to address public and agency concerns and preferences. However, it is apparent that document modifications alone will not be sufficient to quell the tide of opposition that is building against the facility.

1.2.3 Hasty Repairs

In the face of this mounting opposition, the proponent decides to retrench and reconsider how best to proceed. A community advisory committee is established to ensure the ongoing involvement of all interests affected. A community conciliator, acceptable to all parties, is hired to chair the committee. Funding is provided to the committee to hire specialists to peer review all the major technical analyses. A separate subcommittee is established to formulate an impact management and local benefits strategy. The strategy is to ensure a greater level of local participation and control in facility operations, management, monitoring, and contingency planning. It also is to formulate local benefits and compensation policies and procedures for both local residents and the overall community. A parallel government advisory committee is established to better coordinate regulatory interactions.

1.2.4 Too Little, Too Late

The costs and the duration of the process have greatly increased—to the considerable exasperation of the proponent. The reformulated approach has some success in addressing many of the technical, scientific, and community control concerns. Broader environmental sustainability and social equity concerns are largely beyond the committee's mandate. Several options advanced by facility opponents are not addressed, on the grounds that they are impractical or beyond the control of the proponent. The negative perceptions of the proponent, the facility, and the EIA process are ameliorated only slightly by these efforts. Some environmental and community groups either refuse to participate in the modified process or opt out when it becomes evident that the committee agenda will be confined largely to refinements to technical analyses and to impact management. Several municipal councilors come to the conclusion that the likelihood of a satisfactory middle ground is remote and decide to add their voices to those of the facility opponents. More parties withdraw from the community advisory committee under a barrage of criticism from the groups they ostensibly represent.

It is increasingly evident that it is virtually impossible to reverse the momentum that has built up against the facility. Faced with the prospect of continued intense local opposition and protracted legal battles, the proponent decides that the costs of proceeding are simply too great and the likelihood of project approval too low. The application is withdrawn and the proponent decides that it will concentrate instead on upgrading and expanding existing facilities in other communities.

1.3 FIRST PRINCIPLES

1.3.1 An Open Mind

The preceding scenario is all too common in EIA practice. Admittedly, criticisms directed at any one process tend to be narrower. It is an overstatement and oversimplification to suggest that problems such as those cited above can always be avoided or resolved. It is equally inappropriate to conclude that a negative outcome is inevitable. Certainly, some criticisms of EIA practice are overstated, unfair, and unreasonable. Sometimes, conflicting perspectives cannot be reconciled to the point where accommodations are possible. Sometimes the environmental consequences of proposed projects are simply unacceptable, regardless of process-related considerations. But that is not always the case. Just as often, arguably more so, the process fails because it is inadequately designed and managed. Many process-related problems can be avoided or reduced significantly. There is a substantial knowledge base, in both EIA and related fields, to draw upon. The task is not easy. It begins with an appreciation that many of the criticisms of EIA practice are valid. It is furthered by openness to alternative ways of "getting the job done."

1.3.2   Starting from Perspectives

How, then, to move from the types of problems cited in the scenario to better EIA process management? A reasonable place to start is with the perspectives and messages contained in the scenario. In brief, the major perspectives are as follows:

- The local university faculty members and some peer reviewers argued that the EIA process, documents, and methods should have been more scientifically rigorous.
- The environmental groups and some government reviewers made the case that alternatives were too narrowly defined and were not evaluated systematically and consistently.
- The environmental groups concluded that the EIA process and documents failed to adequately advance long-term environmental quality and sustainability principles and goals.
- The proponent and some reviewers felt that the process and documents were too lengthy and costly. They were also concerned that the EIA documents were insufficiently linked to specific regulatory approval requirements, policies, and guidelines.
- Local community groups and some politicians expressed the view that they were losing control over their lives and their community. They did not trust the proponent and had little faith in the government.
- Local community groups and individuals took the position that the EIA process was largely closed and that their views and positions were not considered seriously.
- Local community groups and politicians argued that the EIA process was unfair and that the benefits and costs from the proposed facility were unfairly distributed.
- Environmental groups, some local residents, and some government reviewers felt that the risks and uncertainties associated with the proposed facility were not adequately anticipated or managed.

1.3.3   Widening the Envelope

The positions adopted by the stakeholders regarding the failures of the EIA process in the scenario are far from novel, either in EIA practice or more broadly in public and private decision making. They crop up repeatedly in EIA literature and practice. They are evident, for example, in the debates between advocates of a scientific EIA process and proponents of a more streamlined, practical EIA process; between adherents of an apolitical, collaborative EIA process and advocates of a political, conflict-based democratic EIA process; between those characterizing EIA as essentially procedural requirements and those stressing that EIA should advance environmental quality and sustainability objectives; between advocates of a technical, rational EIA process and those arguing that EIA is a form of adaptive environmental management and among supporters of the rational–technical, community control, and social equity site selection approaches.

These debates are not confined to EIA. Similar, arguably more intense and more fully articulated, debates have been occurring for many years in related fields of practice, such as urban and regional planning and environmental management and in related disciplines such as sociology and philosophy. No consensus has emerged regarding the preferred approach or mix of approaches in these other fields. The debates have, however, proven highly instructive and insightful. The breadth and depth of the perspectives, the intensity of the debates, and the failure to agree on a middle ground (both within and outside EIA) suggest that accommodating the perspectives cited in the scenario will not be easy. The perspective differences are deeply rooted. They cannot simply be integrated by minor additions and adjustments to a conventional EIA process. More fundamental changes are required.

1.3.4   Reading the Underlying Messages

The scenario provides some initial clues as to how such changes might take place. The EIA practitioner's perspective is critical. The EIA practitioner is very familiar with the subtle issues and problems that arise in practice. She or he has attempted, with varying success rates, to integrate multiple perspectives. The EIA practitioner will evaluate any proposed solutions in terms of whether or not they will contribute to more effective management of recurrent, practice-based problems, and if so, to what extent. Problem analyses and proposed solutions, which stray from the practitioner's perspective, run the risk of irrelevance. Proposed solutions will need to be grounded in and derived from practice. The stories and perspectives of other process participants must, of course, supplement the practitioner's story. Perceptions of both problems and the effectiveness of solutions will undoubtedly vary among stakeholders.

Sometimes there will be limited differences in stakeholder perspectives. In such cases, concerns and positions can readily be identified during scoping sessions and easily accommodated in EIA documents. As long as there is an ongoing dialogue with stakeholders to ensure that the initial consensus is maintained, the focus of the

EIA process can be the expeditious satisfaction of regulatory requirements. This book provides advice regarding EIA management in such situations. However, as the scenario suggests, often there are multiple stakeholders and substantial differences in positions. These positions tend to be rooted in fundamental value and perspective differences. Such situations can be especially problematic for EIA practitioners. This book is largely oriented toward situations where there are multiple, often conflicting perspectives regarding the conduct of and products from the EIA process. The scenario suggests that a firm understanding of the nature and implications of perspective differences must be an essential building block to more effective EIA process management.

To some, EIA practice is all about document preparation. To others, the focus should be compliance with regulatory requirements. Document preparation and meeting regulatory requirements are clearly important, sometimes critical. But the process breakdowns in the scenario are more related to such matters as the scope of the analysis, core values, principles and priorities, basic assumptions, public involvement provisions, impact-significance interpretations, the choice and application of methods and impact, and risk and uncertainty management procedures. All these concerns pertain to EIA process planning and management. *Planning* is used here in its broadest sense (i.e., the guidance of future action) (Forester, 1989). Management is concerned with guiding, controlling, and administering the process. The scenario seems to suggest that planning and managing the EIA process is pivotal when addressing multiple, often conflicting perspectives.

1.4   THE BASICS

In this section we use an EIA definition and an overview of EIA characteristics to identify implications for the overall EIA process management and for efforts to accommodate the perspectives displayed in the scenario.

1.4.1   Implications of an EIA Definition

EIA is defined as a systematic process of:

- Determining and managing (identifying, describing, measuring, predicting, interpreting, integrating, communicating, involving, and controlling) the
- Potential (or real) impacts (direct and indirect, individual and cumulative, likelihood of occurrence) of
- Proposed (or existing) human actions (projects, plans, programs, legislation, activities) and their alternatives on the
- Environment (physical, chemical, biological, ecological, human health, cultural, social, economic, built, and interrelations)

This definition demonstrates that EIA is a process that blends numerous activities. It shows that EIA encompasses a broad definition of the environment, of alternatives, and of proposed actions. The systematic exploration of interrelationships (e.g., among environmental components, between the proposal and the environment, among alternatives, among impacts) is clearly crucial in EIA. Any EIA process must be designed and managed to accommodate these attributes. The definition hints at some of the perspectives present in the scenario. It underscores, for example, the need for the process to be systematic, adaptive, open, and focused on environmental change. The need to measure, predict, interpret, and manage environmental change, with and without a proposed action, points toward the rigorous application of methods.

1.4.2 Implications of Other EIA Characteristics

EIA is a purposeful, normative activity. It seeks to further specific environmental values and to operate with specific ethical limits. The objectives advanced for EIA are many and varied. Most directly, it attempts to ensure that proposed actions are environmentally sound. This objective is usually facilitated by action-forcing EIA requirements and institutional arrangements. EIA requirements address such matters as documentation, the nature and scope of analyses, and public and agency involvement. Increasingly, EIA is viewed as an instrument (one among many) for furthering specific environmental and societal objectives, especially sustainability. EIA is sometimes seen as a means (again one among many) of achieving more open, informed, coordinated, unbiased, and systematic planning and decision making. EIA objectives and requirements imply and, to some degree, demand a systematic, open, and collective process, a process that draws upon an interdisciplinary and interprofessional knowledge base and that informs public and private decision making.

These EIA characteristics provide a further sense of what practitioners should consider in designing and managing EIA processes. The EIA process should, for example, be guided by specific environmental values. These values can assist in determining and evaluating both alternatives and proposed actions. Value-based determinations of the acceptability and significance of actions, alternatives, criteria, procedures, and impacts also will be required. The emphasis on values raises the issue of whose values should guide the process and, in turn, the roles that should be assumed by all parties potentially interested in and affected by a proposed action. The recognition of planning and decision-making roles within the EIA process points to the need to link the process to EIA institutional arrangements and more broadly to public and private planning and decision making.

More specific connections to the perspectives illustrated in the scenario are evident from these characteristics. The interdisciplinary nature of EIA and the need to apply methods to identify, predict, interpret, and manage environmental change, for example, could favor the integration of scientific methods. The recognition that the EIA process should be systematic and should incorporate such activities as alternatives and impact analyses is conducive to integrating rational planning process attributes and to applying rational planning methods. The emphasis on tangible environmental improvement underscores the need to fully integrate environmental "substance" into EIA procedures and methods. The regulatory requirements, administrative procedures, and ties to decision making demonstrate that EIA is inherently political. Also implied are the practical considerations associated with operating efficiently and effectively within the review and approval apparatus surrounding EIA. The open and collective nature of the EIA process raises questions regarding which stakeholders are to be involved, when they are to be involved, and how they are to be involved. One way of responding to these questions is partial or complete stakeholder control of the process. The value-full nature of EIA inevitably leads to questions regarding how ethical issues such as procedural and substantive fairness are to be addressed in the EIA process. It is logical to design an adaptable EIA process and to manage risks and uncertainties once it is acknowledged that the process is nonlinear, focuses on complex relationships, and anticipates and manages future conditions.

Even the most basic description of EIA characteristics encompasses, or at least implies, elements of the perspectives displayed in the scenario. Any EIA process, therefore, should, to the extent practical and appropriate, integrate such perspectives. The EIA characteristics provide only broad hints regarding how the perspectives should be characterized and how they should be integrated into the process. No guidance is offered regarding how much emphasis should be accorded to each perspective or how multiple, sometimes conflicting perspectives should be balanced and reconciled. A closer scrutiny of the EIA process is required.

1.5 THE BASELINE

In this section we address the current and emerging state of the art of EIA process management. Historical and emerging trends are described briefly. Examples of major recurrent EIA practice shortcomings that remain, notwithstanding advances in EIA practice, are identified. The shortcomings are used to determine if the problems cited in the scenario are indeed real and recurrent. The trends provide the basis for assessing whether contemporary EIA practice is likely to offer ready solutions to the problems identified.

1.5.1 Patterns and Trends

The roots of EIA, as a formalized (i.e., action-forcing) procedure for assessing the potential environmental effects of proposed actions, are not deep. First introduced in the United States in 1969 [the National Environmental Policy Act (NEPA)], the field has expanded rapidly, especially over the past two decades. EIA is now applied in almost 100 countries. The definition of the environment has broadened from an early emphasis on physical and biological effects to an increased concern with social, cultural, human health, and ecological effects. In the early years, direct impacts, usually from large capital projects, tended to be the focus. Much greater emphasis is now placed on indirect and cumulative effects and on interconnections with other projects and activities. EIA requirements have been applied, to varying degrees, to policies, plans, programs, legislative proposals, technologies, development assistance, products, and trade agreements. EIA also has been adapted to different jurisdictional types and settings (e.g., third-world and transitional economies, indigenous decision-making regimes).

Boundary spanning, both within EIA and to related fields, has received considerable attention. Frameworks and strategies have been formulated for addressing interrelationships among disciplines and EIA types. EIA has been linked to resource management, urban and regional planning, risk assessment, environmental management systems (EMS), and efforts to protect threatened and endangered species, communities, and ecosystems. From an initial focus on procedural requirements, much greater stress is now placed on the integration of substantive environmental concerns such as biodiversity, environmental justice and sustainability, on the adaptation of the precautionary and pollution prevention principles, and on the application of traditional knowledge.

The grounding of EIA in practice has been another priority. EIA documents and review procedures have been streamlined. Greater use is made of EIA document standards [e.g., document format and style, electronic publishing, use of geographic information systems (GIS)]. An increased effort has been made to integrate practice-based knowledge through greater use of EIA quality and effectiveness analyses, monitoring and auditing procedures, and applied research. EIA capacity building, training, and the pooling of information, knowledge, and experience have received more attention. Methods and procedures for facilitating earlier and more collaborative public participation have been refined.

The spotlight in EIA literature and practice is now being turned on new challenges. Transboundary impacts and the integration of macroenvironmental issues, such as global warming and biodiversity, are particular concerns. EIA is being adapted to protect global environmental resources (e.g., the oceans, arctic, and antarctic), to address the implications of global economic activity, to assess the repercussions of international bodies and agreements (such as the European Union and NAFTA), and to explore the consequences of economic trends such as privatization.

1.5.2 Recurrent Shortcomings

These internal advances and external challenges have furthered the rapid evolution of EIA. The realization of EIA objectives has been, at best, a mixed endeavor. EIA has undergone a stocktaking in recent years. Various commentaries and effectiveness studies conclude that EIA has generally contributed to more environmentally sound projects and to more environmentally sound and open decision making (Andrews, 1997; Caldwell, 1997; US CEQ, 1997a). EIA, however, has fallen well short of its aspirations as a major policy instrument for environmental protection and enhancement and as a significant contributor to the cause of environmental sustainability (Clark, 1997; Ortolano and Sheppard, 1995; Sadler, 1996; US CEQ, 1997a; World Bank, 1997).

Some argue that EIA procedures and methods commonly lack scientific rigor. Precise, verifiable predictions are often not made. Data confidence limits do not tend to be established. The significance of residual impacts is frequently not specified. The consideration of cumulative and transboundary effects tends to be weak. These recurrent limitations have contributed to a call for the greater use of scientific methods, especially in baseline data development, in predicting cumulative environmental changes, and in impact management.

The rationality of EIA procedures and methods continues to be questioned. Procedures for both generating and evaluating alternatives often could be more comprehensive, transparent, systematic, and explicit. Similar concerns have been raised regarding the formulation and application of mitigation and monitoring measures, both prior and subsequent to decision making. These shortcomings have contributed to inconsistencies in EIA quality and to an EIA process that can be difficult to understand or reproduce.

EIA practice continues to be criticized for failing to adequately advance environmental objectives. EIA requirements, it is suggested, are often avoided and are rarely applied to policies and programs. Even when EIA requirements are applied, EIA documents tend to be treated as ends rather than as means. As a result, impacts are not always minimized, environmentally unsound projects are sometimes approved, irreversible impacts are often not avoided, and development is seldom placed on a sustainable basis. More emphasis needs to be placed on, it is argued, place-based EIA approaches and on measuring substantive environmental contributions.

The practicality of the EIA process is a concern of many proponents and decision makers. Misgivings remain regarding the efficiency, cost, duration, and level of detail of the EIA process and documents. EIA is often poorly integrated into proponent decision making. Interagency coordination problems persist. More fundamentally, the efficacy of EIA, relative to other environmental policy instruments, is questioned increasingly. Community groups and individual members of the public often believe that they are peripheral to EIA-related decision making. Sometimes, they argue, they are treated as adversaries or are manipulated by the process. EIA processes can diminish rather than enhance local control of matters that affect the community directly. Frequently, public involvement occurs late in project planning. There is considerable variation in the quality of public participation programs. More creative and proactive public involvement programs are needed but are applied much too rarely.

Many EIA processes are still faulted for their failure to address process and outcome fairness adequately. Social impacts continue to receive insufficient attention. The process can be especially unfair to the least advantaged. Inequities in the distribution of benefits and adverse impacts are often not adequately offset. Sometimes, projects compound rather than ameliorate inequities. Finally, too often EIA processes are inflexible. They tend not to anticipate potential changes and sometimes react poorly when changes do occur. Frequently, uncertainties and human health and ecological risks receive insufficient attention in EIA documents and procedures.

1.5.3 Implications of Shortcomings

The recurrent shortcomings closely parallel the perspectives displayed in the scenario. Both the perspectives and the shortcomings suggest that the EIA process should be more scientifically rigorous, should be rational and consistent, especially regarding the treatment and evaluation of alternatives, should lead to more substantive environmental improvements, should be more efficient and more closely linked to decision making, should ensure that local communities have more democratic control over matters that affect them, should provide for earlier, more ongoing, and more collaborative public involvement, should devote more attention to fairness and equity concerns, especially for the least advantaged, and should be more flexible and make a greater effort to integrate risk and uncertainty considerations. The shortcomings convey the impression that the problems arising from perspective differences, as cited in the scenario, are indeed real and recurrent. They also suggest that EIA practice continues to fall short of the mark in addressing these concerns.

1.5.4 Grounds for Optimism

The patterns and trends offer some basis for optimism in addressing the recurrent shortcomings. The increased emphasis on an environmental management EIA approach, coupled with the stress on EIA effectiveness and systematic monitoring and auditing, are consistent with a more rigorous and scientific EIA process. Ongoing efforts to identify, screen, and compare alternatives more systematically, especially in strategic environmental assessment (SEA), site selection, and impact management, could help make the EIA process more rational. The major push to integrate biodiversity, sustainability, and other global environmental concerns into EIA theory and practice supports the effort to make EIA practice more environmentally substantive.

Numerous initiatives to streamline documents and review procedures and to better integrate practice-based knowledge auger well for more practical EIA practice. EIA capacity building and training efforts, the greater application of community control site selection approaches, the integration of traditional knowledge, and the adaptation of EIA to indigenous decision-making regimes all further the cause of local control. Earlier public involvement and greater use of continuous and collaborative public involvement procedures are conducive to more open and participative EIA practice. Initiatives to integrate environmental justice into EIA requirements and practice and the application of social equity principles, criteria, and measures in site selection and in impact management agreements facilitate EIA process and outcome fairness. The numerous efforts to span boundaries within EIA, to link EIA and related fields, to integrate ecological and human health risk concerns and to better manage uncertainties, and to contribute to a more adaptive EIA process.

1.5.5 Closing the Circle

Optimism caused by positive initiatives does not mean that the problems depicted in the scenario and identified as recurrent shortcomings are solved or will be solved.

The positive initiatives are scattered across EIA literature and practice and through-out the literature and practice of related fields. Inevitable gaps remain between EIA theory and practice and between problems and solutions. Relevant resource materials may not always be available to EIA practitioners. They may not be in forms that can readily be adapted for application in EIA practice. These materials are not consistently presented, analyzed, or compared. Greater headway would have been made to ameliorating the recurrent problems if all the potentially relevant resources had been consolidated, refined, and testing for application in practice.

It would be highly imprudent to assume that the recurrent problems will simply disappear because of several scattered, albeit impressive initiatives. It would be wiser to assume that they will not. The relevant materials need to be collated, reviewed, and analyzed. Systematic procedures and methods for addressing the problems, both individually and collectively and at both the regulatory and applied levels, need to be formulated and tested. These procedures and methods need to be in forms that can readily be applied in EIA practice. Preferably they should emerge from EIA practice.

1.6 A STRUCTURE

The preceding sections indicate that there is a genuine need to come to grips with the problems identified in the scenario. They illustrate that there is a considerable knowledge base and many promising initiatives to draw upon. They also make it clear that these resources, although necessary, are not likely to resolve recurrent problems in EIA practice. The shortest path to improvement does not lie simply with tossing an array of potentially relevant methods and procedures at each problem, either individually or collectively. A coherent structure is required. The EIA process provides that structure. In this section we describe the EIA process briefly. We then explain why the EIA process is important and why multiple EIA processes are necessary to any strategy to address the problems.

1.6.1 The EIA Process

At a most basic level a process is a series of actions directed toward an end. The end in this case is more effective management of the problems that often result from multiple perspectives in EIA practice. The primary instrument, adopted to achieve that end, is the EIA process. As illustrated in Figure 1.1, some important distinctions must be made before the choice and arrangement of EIA process actions can be considered. EIA operates within a framework established by institutional arrangements. The institutional arrangements provide the basis for determining whether EIA requirements are applied (i.e., screening). They also offer guidance regarding administrative procedures, documentation, and various planning activities. The planning process activities interact with proposal and alternative characteristics and with environmental conditions. The EIA administrative, planning, and documentation processes are all reviewed and contribute to decision making.

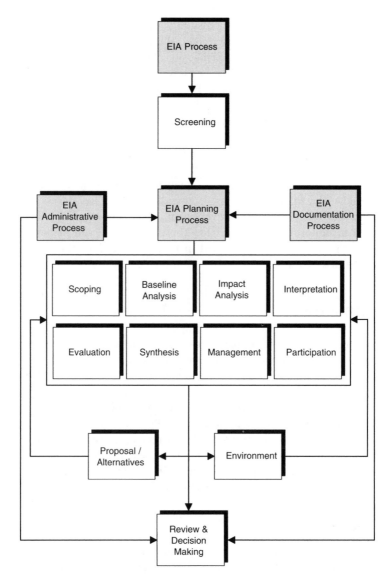

Figure 1.1 EIA process.

EIA requirements usually provide only general guidance regarding the conduct of each process type. This is especially the case for EIA planning process activities. Major decisions may be identified. Early public involvement is usually encouraged. Specific guidance is commonly offered for conducting individual planning process activities. More detailed requirements and guidelines are often provided for administrative (such as document circulation and agency review) procedures and the content of EIA documents. EIA practitioners then design and manage the EIA process within the framework established by EIA requirements.

Figure 1.2 is an example of how EIA process activities might be arranged. As Figure 1.2 demonstrates, some activities are discrete and others are continuous. There are numerous interactions among activities. As explored more fully in Chapter 2, there are many ways in which the EIA process can be managed. Some management procedures are better able than others to address the problems that arise from diverging perspectives.

1.6.2   Why Is the EIA Process Important?

EIA practitioners have a great deal of discretion, in the sequence in which EIA activities are arranged, in the extent to which EIA activities are subdivided, in the choice of activity inputs and outputs, and in the choice and nature of interconnections among EIA activities. EIA practitioners, in concert with other stakeholders, determine if and to what extent potential problems that result from varying perspectives will be resolved. The varying perspectives, as cited in the scenario, generally focus on how the EIA is or is not conducted. Considerable potential for preventing and ameliorating the problems stemming from varying perspectives lies with better EIA process management. A great many methods, an extensive knowledge base, a vast pool of experience, and a diverse array of values and perspectives are available to EIA practitioners. The EIA process integrates these inputs. A poorly managed process will more than offset any benefits resulting from, for example, rigorous application of the latest methods. The EIA process also is the bridge between EIA regulatory requirements and EIA practice. Because of the breadth of their application, EIA requirements must necessarily be general. At best, EIA requirements will reduce the incidence and severity of bad EIA practice. Highly effective EIA practice and high-quality EIA documents will occur only when effectively framed by EIA processes.

The EIA process varies greatly depending on the type of proposal, the type of EIA, the local and regional environmental conditions, and the types of impacts anticipated. The effectiveness of an EIA is often highly dependent on how well the EIA process fits the context. The roles assumed by stakeholders, both planned and unplanned, are often crucial to project success or failure. The EIA process determines which stakeholders are involved and when and how they are involved. It determines how stakeholder concerns and suggestions are solicited and addressed. The EIA process is the bridge between EIA practice and decision making. It determines the decision points and establishes how EIA technical and procedural activities and decision making are linked and integrated. Decision makers often reject EIA analyses and documentation when those interconnections are established and managed inappropriately.

EIA texts, manuals, and guides usually provide only very general EIA process characterizations. These characterizations are rarely tested and refined in practice. When the EIA process is discussed, it tends to be in terms of document preparation and the satisfaction of government requirements. EIA resource materials generally

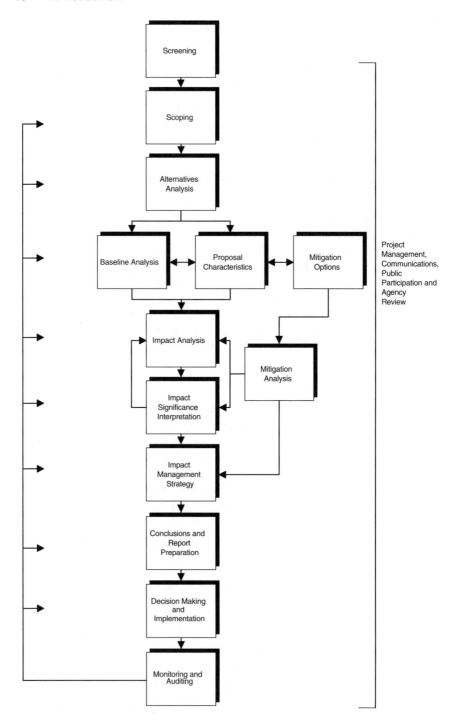

Figure 1.2   Example of a conventional EIA process.

concentrate on EIA institutional arrangements, EIA methods, and the adaptations required for predicting different types of impacts and for assessing various types of proposals. This deficiency in the literature (notwithstanding the central role of the EIA process) underscores the need to formulate and apply more effective and adaptable EIA processes.

1.6.3   Why Are Multiple EIA Processes Important?

EIA processes should be designed to encompass stakeholder perspectives. A multi-pronged approach can narrow the gap between process characteristics and stake-holder values. Different processes will be appropriate in different situations. As illustrated in Figure 1.3, EIA practitioners could benefit from being able to

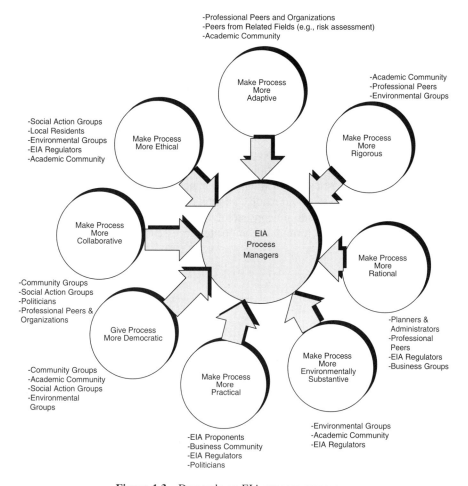

Figure 1.3   Demands on EIA process managers.

pick and choose from a suite of EIA processes as they seek to meet multiple demands and to match process and context. The procedures for addressing any one demand are many and varied. A coherent understanding of how each demand can be addressed is essential before strategies for counterbalancing overlapping, emerging, and conflicting demands can be explored and applied.

The EIA process has been scrutinized before. Numerous variations have been suggested. There are many references to alternative approaches scattered across EIA literature. Process differences between project-based EIA and SEA processes have received particular attention. There is an ongoing debate between advocates of a more scientific as compared with a more practical EIA process. A similar debate has occurred, especially within social impact assessment (SIA), among proponents of technical systems, consensus-based processes, and political, conflict-based processes. Much has been written regarding study team roles and responsibilities within the EIA process. Alternative processes, such as adaptive environmental assessment and management, have been advanced. The application of alternative models (e.g., advocacy planning) from related fields, such as planning theory, has been suggested. What has not occurred is the consolidated presentation, analysis, comparison, and integration of the available concepts and insights into EIA process forms readily applicable to EIA practice.

Related fields of practice (such as urban and regional planning and environmental management) and related disciplines (such as sociology and philosophy) have demonstrated the many insights that can be acquired when multiple models or frameworks are applied to a complex problem or situation. As a field of theory and practice, EIA is very much the exception in its common assumption that a single process model (with minor adaptations) can be applied to any and all situations. A rich array of applied research, methods, and conceptual frameworks (both within EIA and in related fields) could readily be integrated into multiple EIA processes. Once multiple EIA processes are consistently formulated, presented, and analyzed, it becomes much easier to formulate composite processes that balance and integrate a range of perspectives.

1.7   A STRATEGY

In this section we present a strategy for facilitating more effective EIA process management. The EIA process provides an organizing structure for both regulatory and applied level analyses. Both the conventional EIA process (in a diversity of manifestations) and multiple EIA processes (one for each perspective) are presented and assessed. The same evaluation criteria are applied to each of the EIA processes presented in Chapters 3 to 10. Composite EIA processes are described. Applied examples, EIA literature, and selective literature from related fields contribute to the major conceptual distinctions, models, and frameworks. The regulatory analysis is based on a review of readily available (i.e., largely on the Internet) federal or senior-level EIA legislation, regulations, and guidelines from four jurisdictions: the United States, Canada, Australia, and the European Union.

An iterative, analysis–synthesis relationship is inherent to EIA practice. It also is intrinsic to this book. Interactions between EIA requirements and practice are explored in each of Chapters 2 to 10. Anecdotal examples from EIA practice are integrated into each of Chapters 3 to 11. Major conclusions and lessons are integrated for conventional EIA processes (at the end of Chapter 2) and for individual EIA processes (at the end of each of Chapters 3 to 10). In Chapter 11 we present multiple approaches for synthesizing the individual EIA processes and identify residual challenges and priorities. A comprehensive treatment of the subject matter addressed by this book, at a uniform level of detail, is neither practical nor appropriate. The analysis is selective, sometimes arbitrarily so. It was necessary to focus on some subjects, treat other subjects more selectively and at a broader level of detail, and not consider other potentially relevant topics.

This analysis focuses on managing the EIA process. It seeks to ameliorate the negative consequences that sometimes stem from perspective differences. The regulatory analysis is based on senior-level EIA requirements and guidelines in the United States, Canada, Australia, and the European Union that bear directly on EIA process management. The applied analysis focuses on the roles and responsibilities of EIA process managers. It assesses and reformulates conventional EIA processes, formulates, analyzes, integrates, refines multiple alternative EIA, and processes, and formulates composite EIA processes. Both the regulatory and applied analyses are based largely on pertinent EIA literature and practice. A broader and more selective analysis is undertaken of other EIA aspects connected to EIA process management and literature and practice in related fields.

EIA requirements outside the four jurisdictions are not considered. It is possible that insights and lessons from this analysis may be of value to other jurisdictions. Adaptations to address, for example, EIA practice in developing and transitional countries have not been included. No effort has been made to trace the long and complex intellectual traditions that underlie the various EIA processes. These traditions span hundreds of years and encompass tens of thousands of references. Although intriguing (albeit a major undertaking), it would be impossible to provide a just and accurate treatment of those traditions without detracting from the central thrust of the book, which is to provide a succinct and practical reference to EIA practitioners.

Finally, any treatment of the EIA process, no matter how practical, will never fully match the subtle complexities encountered in EIA practice. Each proposal and each setting are, in important, often not readily apparent respects, unique. Hopefully, much can be learned from EIA references such as this book. Nevertheless, the EIA process and methods must always be designed and adapted, jointly with stakeholders, to fit unique project and setting related circumstances. Both the approach and the circumstances will change and evolve, preferably in tandem.

1.8   A ROAD MAP

Table 1.1 identifies examples of how EIA process managers, EIA study team specialists, EIA regulators, nongovernmental organizations (NGOs), and EIA instructors might wish to use this book.

Table 1.1   Suggested Approaches for Using This Book

EIA process managers	Read the book for an in-depth review of alternative EIA process management approaches and approach combinations.
	Use Chapters 2 to 10, as appropriate, to avoid and minimize the recurrent problems.
	Read the highlights and summing up subsections for an initial impression of each chapter.
	Use the figures and tables, with text support, for process design and management.
	Draw upon individual chapters as issues are identified.
	Use the checklist tables to evaluate, monitor, and adjust the EIA process.
	Use the criteria application text to ensure that the processes are applied with a full appreciation of process strengths and limitations.
	Take a particular look at Chapter 6 for study team management strategies.
	Use Chapter 11 for composite planning process management.
	Use the references for follow-up.
EIA study team specialists	Review the highlights and summing up subsections to obtain an initial impression of available processes.
	Review the descriptions of process activities, inputs, and outputs in Chapter 2 to understand specialist integration procedures.
	Review Chapter 6 to understand decision-making links and roles within the study team.
	Natural and social scientists may be especially interested in Chapters 3, 5, and 10.
	Urban and regional planners and resource managers may be especially interested in Chapter 3.
	Social scientists may wish to take a close look at Chapter 9.
	Public participation specialists may find Chapters 7 and 8 especially relevant.
	Any specialist concerned with risk and uncertainty management may wish to take a closer look at Chapter 10.
EIA regulators	Review Chapter 2 and the regulatory sections of each of Chapters 3 to 10.
	Peruse the highlights, summing up subsections, figures, and criteria application text in determining possible regulatory reform approaches.
	Use the flowcharts and checklist tables for evaluating EIA documents.
	Review the definition-of-problem sections to alert proponents and project managers to potential and emerging problems.
	Take a closer look at Chapter 6 for examples of procedures for expeditiously satisfying regulatory requirements.
	Review Chapter 11 for an overview of composite process approaches.

Table 1.1 (*Continued*)

Nongovernmental organizations	Take an especially close look at Chapters 7 and 8.
	Use the criteria application and checklist tables to evaluate EIA processes and documents.
	NGOs are likely to find Chapters 5, 7, 8, 9, and 10 especially pertinent to their concerns.
	Peruse the highlights, definition-of-problem, direction, and summing-up sections and Chapter 11.
EIA instructors	Use the highlights, the scenario, and the definition of the problem to introduce each process type.
	Structure lectures around the scenario, the figures, the tables, and the summing-up sections.
	Use the anecdotes to identify insights and lessons from practice.
	Use the process characterizations to structure the presentation of methods.
	Use the checklist tables as a means for structuring class discussion.
	Use the alternative approaches as a basis for debates and discussions regarding which process should receive more emphasis, and why, under which situations.
	Use the regulatory analysis and checklist tables in workshops and class exercises to evaluate EIA requirements.
	Use the process evaluation text in workshops and class exercises to evaluate each process type.
	Use Chapter 11 to structure debates and discussions regarding alternative integration strategies.
	Use checklists to evaluate EIA documents.
	Use the concept tables and reference list to facilitate student research efforts.

1.9 SUMMING UP

This book is intended to enhance EIA practice. It is directed especially toward the needs of EIA practitioners who manage EIA processes. It provides practical solutions to major, recurrent problems encountered in daily EIA practice. A scenario is used to identify several recurrent EIA process-related problems. These problems stem largely from a failure to anticipate and respond adequately to major stakeholder perspective differences. An open mind to alternative approaches is an essential prerequisite to effective process management. The perspectives identified in and the messages derived from the scenario should also be considered carefully. The perspectives crop up repeatedly in the literature and practice in EIA and related fields. Accommodating multiple, often conflicting perspectives is a difficult task. Adopting and maintaining an EIA practitioner's perspective, in combination with the perspectives of other stakeholders, will facilitate the analysis of problems and potential solutions.

The EIA definition and other EIA characteristics provide an initial sense of the perspectives and the requisite characteristics of an EIA process. Recurrent shortcomings of EIA practice closely parallel the problems portrayed in the scenario. An overview of EIA patterns and trends demonstrates that the shortcomings are unlikely to resolve themselves through the normal evolution of EIA practice. The EIA process is pivotal to managing many recurrent shortcomings of EIA practice effectively. Formulating, analyzing, and synthesizing multiple EIA processes can further enhance EIA process management. A strategy is presented for more effectively managing EIA processes. Initially, the strategy assesses how well various versions of the conventional EIA process, with appropriate modifications, respond to problems arising from perspective differences. It is then used to formulate, assess and combine EIA processes that respond to each perspective. Pertinent literature and reviews of EIA requirements in four jurisdictions support the analyses. A focused and selective approach is required because of the breadth of the subject matter.

CHAPTER 2

CONVENTIONAL EIA PROCESSES

2.1 HIGHLIGHTS

In this chapter we describe and assess conventional choices for controlling and shaping the EIA process through EIA legislation, regulations, and guidelines. We also describe and assess conventional choices for designing and managing the EIA process. The EIA process is addressed from the practitioner's perspective (i.e., EIA regulators and individuals responsible for preparing EIA documents and for managing EIA processes within the framework established by EIA requirements and guidelines).

- The analysis begins in Section 2.2 with the problem, which is the need for better EIA process management. The desire is to reduce the incidence and severity of recurrent process-related problems, through both regulatory guidance and applied practice. The approach taken to address the problem is described and substantiated.
- In Section 2.3 we describe, assess, and suggest improvements to conventional approaches to regulating and guiding the EIA process. The analysis is based on an overview of how four jurisdictions (the United States, Canada, Australia, and the European Union) seek to guide the EIA process. The analysis encompasses screening guidance, guidance for individual activities, and integration and coordination guidance.

Environmental Impact Assessment: Practical Solutions to Recurrent Problems, By David P. Lawrence
ISBN 0-471-45722-1   Copyright © 2003 John Wiley & Sons, Inc.

- In Section 2.4 we describe, integrate, and enhance the EIA process as conventionally portrayed in EIA literature. The analysis encompasses general process management, EIA process inputs, outputs and links, and EIA process adaptations.
- In Section 2.5 we describe the major insights and lessons derived from the analysis, including the potential for conventional EIA processes to address recurrent problems adequately. A summary checklist is presented.

2.2   DEFINING THE PROBLEM AND DECIDING ON A DIRECTION

The problem for this chapter is twofold. First, there is the question of whether conventional characterizations of the EIA process, as typically portrayed in EIA literature, adequately reflect the choices available. Second, there is the question of whether conventional EIA process guidance and practice, even if reformed substantially, can respond adequately to the recurrent problems identified in Chapter 1. The direction is first, to present a range of available choices at the regulatory level; second, to present a range of available choices at the applied level; and third, to assess whether those choices can provide a comprehensive response to the recurrent problems.

2.2.1   Regulatory Level

The point of departure for the regulatory analysis is the EIA process control and guidance provided by existing EIA legislation, regulations, and guidelines. The focus is on regulatory guidance of the EIA process as it is, in terms of broad patterns, and as it could be, in terms of good practice and if reconstructed and reformed. The analysis is based on an overview and assessment of publicly available (largely through the Internet) EIA legislation, regulations, and guidelines, at senior government levels in the United States, Canada, Australia, and the European Union. A detailed comparative analysis of EIA requirements is not undertaken. Instead, the focus is on helping EIA regulators to guide EIA process management more effectively. Good practice criteria structure the analysis.

There is a tendency in EIA literature to focus on the details of individual EIA regulatory systems. EIA regulators must do so if they are to administer the system effectively. EIA practitioners must do so if they are to operate within the system effectively and efficiently. There are numerous texts that seek to aid practitioners in understanding and operating within individual EIA systems. All of this is well and good, but sometimes a broader perspective is required.

EIA regulatory systems are not static. Reforms, refinements, and modifications are commonplace. Sometimes change takes the form of a fundamental restructuring of or replacement of EIA and related legislation. This was the case over the past decade in Canada, Australia, and the European Union. While administering the system as it is, EIA regulators, must determine how the system can be improved either by modifications and refinements or by more fundamental changes. If the benchmark is simply the system as it is, the basic questions that lead to more fundamental changes may never or rarely be asked. Individual modifications and refinements may, moreover, when aggregated, result in an inefficient and ineffective patchwork of reforms and existing practices. Individual elements may operate at cross-purposes. Significant flaws, gaps, and inconsistencies may (and based on Chapter 1 do) remain. Even if the existing system is not replaced, the repairs should be based on basic principles and objectives, an appreciation of practical alternative approaches and experiences elsewhere, a systematic evaluation of the existing system, and a coherent strategy that builds on strengths and ameliorates weaknesses.

Whether the intent is to replace, reform, or just fine-tune EIA requirements and procedures, the benchmark should be *good regulatory practice*. The good regulatory practice criteria are derived from good practice examples. Patterns in requirements and procedures across the four jurisdictions provide a sense of the gaps between what is and what could be.

EIA legislation and regulations should spell out both aspirations (e.g., goals, objectives, principles, policies, priorities) and minimum requirements (e.g., thresholds, standards, criteria, areas of application, roles and responsibilities). EIA legislation and regulations should result in a consistent and acceptable level of EIA practice. Ideally, the gap between EIA aspirations and requirements should be narrow. EIA guidelines can further diminish that gap by facilitating compliance and by contributing to the quality of EIA documents and to the effectiveness of the EIA process. EIA guidelines have the additional benefit of flexibility. They can be adapted as the state of the art and practice of EIA evolves. They also can be adjusted for different setting and proposal types and for individual applications. A delicate balancing act is required. Too general EIA legislation and regulations will contribute to a low, or at least highly inconsistent, level of EA practice. EIA legislation and requirements which micromanage every aspect of EIA practice, are likely to stifle innovation and inhibit necessary adaptations.

2.2.2 Applied Level

Identifying EIA regulatory choices is not enough. There is considerable discretion, within the framework established by EIA requirements and guidelines, to manage the EIA process effectively and efficiently or not to do so. Moreover, EIA requirements and guidelines do not and cannot adequately convey the many process management choices potentially available to EIA practitioners. Once EIA requirements are satisfied, it is also not enough simply to apply the EIA process as presented in any one of several EIA texts. Although these process depictions can be very helpful, numerous versions of the EIA process are presented. Most EIA process descriptions are not identified as one among many contributions to an ongoing debate and discussion. Alternative processes and process variations are rarely described, compared, or critically evaluated. Instead, it is commonly assumed that there is only one EIA process. Occasionally, allowance is made for individual process variations. Usually, little importance is attached to these variations. The EIA practitioner is left in the difficult position of designing and managing the EIA process without the benefit of an array of readily accessible process guidance choices that can be integrated, combined, adapted, and applied to suit the circumstances.

Applied analysis describes the broad patterns in which the EIA process is conventionally portrayed in EIA literature. It then integrates and enhances the EIA process characterizations, with particular emphasis on identifying choices available to EIA process managers. Finally, it identifies residual gaps and priorities (with special reference to the recurrent problems) to be pursued in subsequent chapters.

Many perspectives on how the EIA process should be designed and managed are integrated into the analysis. Various ways of identifying and structuring EIA process activities and activity components are considered. Procedures for integrating key inputs into the EIA process, such as EIA requirements, public and agency concerns, substantive environmental priorities, methods and pertinent values, and knowledge and experiences, are considered. Key EIA process outputs are identified. Consideration is given regarding how to establish links to proposal planning, decision making, and related actions and fields. Process management choices are consolidated in tabular form. Several tables and figures are presented that extend beyond conventional EIA process portrayals.

The applied analysis is based on an overview of major EIA texts and mainstream EIA journals. It focuses on prescriptive portrayals of the EIA process. A broader range of EIA literature and literature in related fields is considered in the analyses in Chapters 3 to 11. The emphasis is less on comparing the EIA process depictions than on identifying and illustrating the range of conventional process management choices available to EIA practitioners. Integration and refinement of conventional EIA process portrayals provide a baseline for assessing whether or not the recurrent problems cited in Chapter 1 require further consideration. The specific nature of the recurrent problems and how they might best be addressed are explored in greater detail in Chapters 3 to 11.

2.3 CONVENTIONAL REGULATORY EIA APPROACHES

2.3.1 Screening

The first step in designing or adapting any EIA regulatory system is determining (1) what should trigger an application of EIA requirements and (2) which particular set of EIA requirements should be applied. These two screening steps, highlighted in Figure 2.1, focus on various actions (what), proponents (who), and environments (where). Often, screening decisions are based on action, proponent, and environmental combinations. Each decision involves a significance determination (i.e., whether or not a level of importance has been reached where EIA requirements should be instituted). Added to the mix is an effort to achieve two not always complementary purposes: (1) building environmental considerations into proponent and action-related decision making and (2) protecting and enhancing the environment.

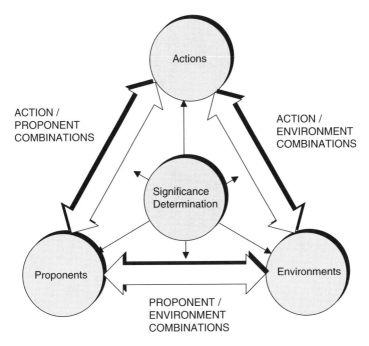

Figure 2.1   Screening combinations.

Table 2.1 lists examples of good practice screening characteristics for each of the three elements (individually and collectively) and for significance determination. The EIA systems in the four jurisdictions each grapple with the screening elements, seek to achieve the purposes, and exemplify the good practice screening characteristics in different ways and to varying degrees. Each jurisdiction exhibits an uneven mix of positive and negative attributes.

Proponents   Proponent-driven EIA systems (e.g., in the United States and Canada) focus largely on integrating environmental concerns into public decision making. Such systems have the potential to facilitate more environmentally sound public decision making. Refinements can be introduced by individual agency requirements and guidelines. Provisions can be made for multiproponent proposals, both within the same government level or involving intergovernmental undertakings. Allowances can be made for voluntary EIA preparation. Proponent-driven EIA systems do not tend to apply to private undertakings, except selectively through approval, land, and funding triggers. EIA guidance can be provided to the private sector and to other government levels. Consideration can also be given to the EIA implications of privatization (as has occurred in the United States) and to the proponent's environmental record (as occurs in Australia).

Proponent-driven EIA screening systems vary greatly in terms of efficiency and effectiveness. Under such systems, unless a concerted effort is made to focus on the

Table 2.1 Examples of Good Practice Screening Characteristics

Proponents	Actions	Environments	Interactions	Significance Determination
Applies to public and private proponents	Automatic triggers for major actions that almost always have significant environmental effects	Automatic rejection if unacceptable environments	(A-E)—action thresholds adjusted to significance/sensitivity of environment	Significance defined
Exclusion of proponents that do not perform environmentally significant actions	Triggers for proposal characteristics likely to induce significant effects (e.g., emissions, waste disposal)	Requirements adjusted depending on environmental and resource carrying and regenerative capability	(A-E)—action thresholds adjusted where cumulative effects a concern	Provision for early exclusion of unacceptable effects
Funding and approval triggers early in process and consistent	Exclusion of actions with minimal adverse effects	Broad definition of environment and effects	(A-E)—provisions to group proposals geographically (e.g., mineral exploration)	Well-defined significance thresholds
Scope of funding and approval triggers consistent with scope of likely effects	Triggers for policies, programs, and plans (strategic environmental assessment) likely to induce significant effects	Provisions for extraterritorial application	(A-E)—focuses requirements on action characteristics likely to significantly affect sensitive/significant environmental components, processes, interactions, species, and habitats	Provisions for degrees of significance (scaling systems)
Triggers adapted to proponent characteristics (e.g., agency characteristics, indigenous EA regimes)	Triggers for legislative changes, regulatory changes, and trade agreements likely to induce significant effects	Provisions to apply to global commons (e.g., antarctic, oceans)	(P-E) requirements and guidelines focus on proponent characteristics most likely to affect environment significantly	Explicit significance criteria
Provisions to group agencies (e.g., co-proponents)	Adjustment of requirements for different action types and characteristics	Effect requirements and guidelines (e.g., biodiversity, social, health, environmental justice)		Explicit and consistent significance determination process
Provisions to coordinate levels of government (e.g., federal, state/provincial/territorial, municipal, indigenous)		Transboundary effect requirements		Significance determinations adjusted based on consideration of; magnitude/intensity, mitigation and enhancement potential, availability of reasonable alternatives, risks and uncertainty, public policies, requirements and priorities, degree of controversy, reversibility, carrying capacity, sensitivity, and significance of potentially affected environment and context
		Provisions for areawide assessments (e.g., watersheds, SEAs of land-use plans)		

Provisions to prevent and minimize duplication and overlap (e.g., delegation, harmonization)
Voluntary participation provisions
Consideration of proponent's environmental record
Individual agency requirements/guidelines
Assessment of privatization implications
EIA guidance for business (e.g., links to environmental management systems)

Adjustment of SEA requirements for different sectors
SEA-project EIA links (e.g., tiering)
Action guidelines (e.g., best practice)
Individual proposal requirements/guidelines (e.g., terms of reference)
Provisions to group actions by type (e.g., site remediation, weapons decommissioning)
Provisions to combine related actions (e.g., precedent setting)
Selective application for expansions to existing projects where significant effects likely
Provisions to apply to existing action (with consent of proponent)
Small project guidance

Setting requirements/ guidelines
Adjustments in requirements to match regional characteristics
Regional guidelines
Special requirements for specific sensitive/ significant/hazardous environments
Endangered/threatened/rare species and habitat triggers
Contaminated site guidance
Special provisions for culturally sensitive lands (e.g., archaeologically significance, indigenous peoples' reserves, land claims)

(P-E) requirements and guidelines focus on locations and location types favored by proponent
(P-A) requirements and guidelines focus on actions undertaken by proponents that are most likely to result in significant effects
(P-E) requirements and guidelines seek to integrate environmental considerations into all proponents' actions

Recognition that significance can be positive or negative
Recognition that individually insignificant actions can be cumulatively significant
Provision for stakeholder involvement in significance determination
Significance determination integrated into each EIA process step
Provision for integration of significance determination methods

agencies and agency actions that have the potential to generate significant environmental effects, substantial resources may be devoted to screening large numbers of government actions of very limited environmental significance. Well-defined, focused, consistently applied, open, and transparent screening procedures are critical under such systems. If the proponent-oriented systems are perceived as wasteful and ineffective, pressure may mount to consider other means of building environmental values and perspectives into public decision making. It also may make it very difficult to extend EIA requirements more broadly to the actions of others (i.e., other government levels and private proponents), even when environmental impacts are highly significant.

A more fundamental concern with a proponent-driven screening approach is that protecting and enhancing the environment is addressed indirectly. The point of departure is the proponent. Links to the environment occur through document circulation procedures, significance interpretations, and other environmental requirements that may be layered on or combined with the EIA review and approval process. This may not be a major problem if substantive environmental requirements are explicit and consistently applied and integrated with EIA procedural requirements. The danger is that a patchwork of evolving, formal, and informal environmental requirements will be applied inconsistently at different points in the EIA process. Environmental protection and enhancement will not, as a consequence, be addressed coherently and may be difficult to discern or measure.

Reliance on (sometimes late and inconsistently applied) approval and funding triggers also can be problematic. Procedural modifications may reduce this problem by, for example, ensuring early and consistent triggers. The larger issue is whether it is appropriate for a narrowly defined approval or funding requirement to trigger a broadly defined set of EIA requirements. Should the focus, instead, be on sensitive and significant environmental components and systems (as is the case in Australia and California) and/or on major proposals likely to induce significant adverse environmental effects (as is the case in Europe and in most Canadian provinces)? The Australian approach largely avoids the late and inconsistent trigger issue (by excluding approval and funding triggers and by focusing on matters of national environmental significance) but raises questions regarding how well and how systematically environmental concerns are being integrated into commonwealth decision making.

An additional problem with a proponent-driven screening approach is the potential for duplication and overlap with the EIA requirements of other government levels. Other government levels (e.g., states, provinces, territories, municipalities, indigenous peoples) often establish separate EIA requirements. Approval and funding triggers can mean that the same proposal is subject to more than one set of EIA requirements. Formal or informal senior government participation in the EIA reviews of other governments can lead to different parties addressing the same substantive concerns. There also may be issues related to timing coordination and to conflicting requirements and perspectives.

These potential problems can be ameliorated by formal (e.g., harmonization agreements and accords, multiparty EIA systems) and informal (e.g., joint project committees, combined requirements) coordination mechanisms. The duplication and overlap problem can largely be eliminated by dropping the approval and funding triggers and by introducing the types of accreditation procedures included in the Australian EIA system. Industry and lower government levels usually favor such approaches. Senior government officials and environmental organizations, in contrast, often take the position that duplication and overlap is largely a myth. The bigger issue, from their perspective, concerns gaps in application and the tendency of lower government levels to align themselves too closely with the interests of industry. They argue that measures to minimize duplication and overlap will undermine environmental protection and enhancement efforts.

The many and often conflicting perspectives regarding proponent-driven EIA systems are not readily resolvable. Independent auditing procedures could help test conflicting claims and, at least tentatively, identify opportunities for improvements. Cross-jurisdictional comparisons can help test alternative approaches to responding to those opportunities. The larger, and still outstanding, question is whether a single public proponent is the appropriate focus of an EIA screening system.

Actions   An action-driven EIA screening system begins with proposal types and proposal characteristics deemed likely to induce significant environmental effects. The EIA systems in Europe and in several Canadian provinces are largely action driven, although location, impacts, and other considerations influence screening decisions. Projects (often with project-characteristics thresholds) that require an EIA are listed. Projects that may require an EIA (subject to a case-by-case examination or thresholds or criteria) also may be identified. Selection criteria for assessing projects that may be subject to an EIA are sometimes provided. These criteria are usually refined and interpreted in guidance documents. Sometimes an action-driven component is grafted onto an essentially proponent-driven EIA system. Canadian federal EIA requirements, although triggered by federal agency actions, funding, lands, and approvals include an excluded project list, a list of major projects that require a comprehensive study, and a list of projects that may be subject to a screening procedure. The U.S. EIA system applies to all U.S. agency actions except those excluded through categorical exclusions or other exemptions.

An action-driven screening system can be broadened and adapted to encompass different proposal types (e.g., legislation, policies, plans, programs, trade agreements). It focuses EIA efforts on those proposals most likely to induce significant environmental effects, regardless of whether proponents are public or private. Actions can be grouped by type, by geographic area, for different sectors, and by interrelationship (as when one action represents a precedent for others). EIA requirements for different proposal types can be linked (e.g., by tiering programs and project-level EIAs). Requirements can be instituted and guidelines prepared to provide more specific controls and guidance for various proposal types and for specific proposals. Proposal guidelines can describe common effects, types of alternatives and mitigation measures, and "good practices." Additional guidance can be provided for existing projects and for various types of small proposals, even when such proposals are not subject to formal EIA requirements.

An action-driven EIA screening system is not without drawbacks. Actions and/or identified action characteristics are, by definition, surrogates for impacts. Sometimes the match is a poor one. The range of proposals can be too narrow or too wide. Thresholds can be too high (especially when environmental settings are especially sensitive or significant) or too low. Cumulative effects can receive insufficient attention. The broader goal of integrating environmental concerns into decision making is addressed only selectively. Measures are available for partially offsetting these disadvantages. The discretionary project lists [such as those contained in the European Directive (Annex II) and in the Canadian Inclusion List Regulations] make it possible to apply EIA requirements to more project types and characteristics and with more sensitivity to local circumstances than with systems that automatically classify projects as in or out. Including categories of project locations and potential effects (see, e.g., Annex III to the European Union's Project Directive) partially address the issues of cumulative effects and sensitive and significant environmental settings. Further refinements are possible through generic screening and with project-type and project-specific guidelines. Cumulative effects potential can be added as a screening consideration (as, e.g., in proposed amendments to the Canadian Environmental Assessment Act) and addressed through cumulative effects requirements and guidelines.

Environments An environment or effects-based EIA screening system starts from the premise that beginning with the environment is the best and most direct way to protect and enhance the environment. Environment can refer to generic environment types (e.g., wetlands, floodplains), specific environmental locations (e.g., national trails and parks), species or habitat types (e.g., rare and endangered species), effect types (e.g., climate change, environmental injustice, biodiversity) and extraterritorial environments (e.g., antarctic) or effects (e.g., transboundary air pollution).

An environment or effects screening approach necessitates a matching of jurisdictional level/mandate (e.g., national) and environmental or effect significance. Ideally, a multitiered environment-based EIA system would differentiate among globally/internationally, nationally, state/provincially/territorially, regionally, and locally significant environments/effects. The globally, internationally, and nationally significant environments and effects would be addressed through international agreements and national EIA requirements. States, provinces, and territories could address the intermediate level, presumably in consultation with the senior level. Regionally and locally significant environments and effects could be addressed at the municipal level, generally with the advice and guidance of upper levels. Each successive level down the hierarchy would be nested within the upper levels. This screening approach works best if there is a strong connection between EIA and spatial planning (through, e.g., SEAs and areawide assessments) and if there are complementary EIA systems at each level.

An environment/effects-based screening approach focuses directly on protecting and enhancing the environment rather than on improving projects and decision-making procedures. Distinctions are explicitly drawn among spatial significance levels (e.g., international, national, regional, local). Clear thresholds of acceptability and significance are defined. Unsuitable proposals can be excluded early in the decision-making process. An environmental screening system is conducive to a place-based environmental management approach. Sensitive and significant areas, resources, species, and habitats can be protected. Cumulative and carrying capacity environmental issues and effects can be considered directly using natural boundaries. EIA requirements can be extended to address transboundary and global environmental issues. Requirements can be adjusted to match regional and setting characteristics. Guidelines can be prepared to address the EIA treatment of different environments and effects. Dialogue with other government levels concerning joint and coordinated environmental protection and enhancement efforts can be facilitated.

An exclusively environment/effects-based screening approach, however, tends to be selective (i.e., only the most important environments and species are protected). Systems-level environmental concerns, which transcend sensitive and significant environmental components, may not be adequately considered. There is an implicit assumption that clear thresholds of significance can readily be identified. Subtle interrelationships, degrees of significance, and cumulative effects concerns in non-sensitive environments may receive insufficient attention. Urbanized and degraded environments may not be adequately considered. The environmental benefits of a broadly based effort to integrate environmental concerns in proponent decision making and project planning may be neglected. The approach may be especially problematic if complementary EIA systems are not established at the subnational level and/or if only partial and scattered regional environmental databases are available to support the system.

The Australian and California EIA requirements come closest to an environments/effects-driven EIA system. Each focuses on explicitly identifying and applying significance thresholds. Other, more proponent- and action-driven EIA screening approaches contain environmental, effects, and place-based elements. There is the danger that the effort to protect and enhance the environment will fall short of a coherent and consistent strategy, when such elements are added or linked to proponent and/or action-driven EIA screening systems. The place of the environment in such EIA systems is more likely to be the product of a disjointed, partial, and sometimes inconsistent historical process.

Proponent, Action, and Environmental Interactions   The EIA systems in all four jurisdictions blend, albeit in different ways, action, proponent, and environmental elements. As is evident from the preceding subsections, each system exhibits a different mix of the good practice screening characteristics listed in Table 2.1. There is ample room for improvement with each system. The road to an optimal system (if such exists) does not lie with a primary emphasis on any one element. Such an approach is bound to undermine efforts to enhance both decision making and the environment. It does not follow that the answer then lies in a balanced EIA system that gives equal emphasis to all three elements. Not all the elements are equally important. Arguably, the ultimate goal is protecting and enhancing the environment, presumably in a manner consistent with sustainable development principles and precepts. Although laudable, integrating environmental concerns back into proponent and proposal decision making is at best either a means or an intermediate end toward the sustainability goal. Accordingly, greater emphasis should be placed on the environmental element and to links between the proponent and action elements to and from the environment.

It should not be assumed that the only approaches available for devising an EIA screening system involve beginning with one element and then folding in the other elements. Perhaps enhancing decision making, proposals and the environment can be approached by different but interrelated EIA screening systems. Perhaps, other environmental management tools are more appropriate for addressing one or more of the goals. It is easy to become complacent with an EIA screening system, especially if there is tangible, often anecdotal, evidence of success. The tendency may be to search for ways of building on those successes by refining and improving the current system. What may be required (and may become increasingly evident to the extent that EIA and sustainability are merged) are more fundamental changes, perhaps involving a basic rethinking and reconstruction of the EIA screening system.

The proponent and action elements of an EIA screening system can either contribute to or undermine (either directly or by diverting necessary resources) sustainability. It is especially important that the action and proponent elements of an EIA system be efficient (to minimize wasted resources) and effective (in the sense of advancing, not inhibiting sustainability). A positive contribution to sustainability is not sufficient if other, less resource-consumptive, environmental management tools can make equivalent or greater sustainability contributions. It is important that the various aspects of the environmental element are part of a coherent strategy, a strategy that operates both within EIA and embraces other public and private environmental management and planning requirements and initiatives.

The action–proponent connections help focus EIA screening reform on actions and proponents (both individually and collectively) most likely to induce significant adverse and positive environmental effects. The magnitude of positive contributions to sustainability should always be considered. The action–environment and proponent–environment interconnections are crucial. They make it possible to adjust EIA requirements based on environmental significance and sensitivity, adopt areawide and place-based approaches to project and activity assessment, place project-related evaluation within the context of broaden environmental management and planning efforts, focus EIA requirements on actions and proponents most likely to induce significant environmental effects, provide a coherent approach to cumulative effects, and build back environmental concerns and priorities into action and proponent-related decision making.

Significance Determination   As is evident from Figure 2.1, significance determination is pivotal in any EIA screening system. It is the tool for determining which actions, proponents, environments, and action–proponent–environment combinations are subject to which (if any) EIA requirements. An ill-defined, ad hoc, and closed significance determination procedure can rapidly reduce an otherwise exemplary EIA screening system to little more than simply paper-shuffling. Significance determination criteria and procedures vary greatly in their breadth and level of detail. Considerable progress has been made (most notably in Australia and in California) in defining significance thresholds and criteria, in identifying the major elements of the significance determination process, and in linking significance determinations to effect magnitude, mitigation potential, environmental characteristics, uncertainties, cumulative effects potential, and public policies and requirements. There also is a broad recognition that significance determination is a subjective, value-full activity that necessitates stakeholder involvement.

More attention could be devoted to such matters as the basis for determining the significance of actions and proponents, the integration of significance into all EIA activities (not just screening), the question of degrees of significance (significance does not have to be viewed as an absolute threshold), specific stakeholder roles in significance determinations, and the available methods to aid significance determinations. Refinements to and alternative formulations of the significance determination process also are needed.

2.3.2 Individual Activities

Regulating and guiding the EIA process does not end with screening and significance determination. It is also necessary to control and guide individual EIA activities. Table 2.2 lists examples of good regulatory practice general and specific (for each EIA activity) characteristics. There is a considerable variation across the four jurisdictions in the controls and guidance provided. The patterns that emerge appear to have evolved in an ad hoc manner, often in response to deficiencies in practice. The net result is a highly uneven regulatory management approach.

General   Overall guidance, across EIA process activities, necessitates a delicate balancing act. Ideally, EIA requirements (legislation and regulations) should identify objectives, spell out minimum requirements, and include general performance standards or criteria. Guidelines can then provide more specific guidance. Successively more specific guidance can be provided first, through environment and proposal guidelines, and second, through proposal and environment-specific requirements and guidelines. Guidance for individual activities can be offered through both general guides (that address all EIA activities but at a broad level of detail) and specific guides (that provide more detailed guidance for individual activities). Guidelines that are too superficial are likely to be of little value in facilitating good EIA practice. Overly specific requirements and guidelines can inhibit good practice innovations and adaptations, especially when there are multiple and changing perspectives regarding good practice standards and methods. General EIA process guides are helpful because they provide an overview of all EIA activities at a consistent level of detail. They also can give the reader a sense of how the individual activities fit together into an overall process.

Table 2.2 Examples of Good Regulatory Practice: Individual Process Activities

General Characteristics

Objectives, minimum requirements and performance standards/criteria included in legislation or regulations

Identifies activity objectives and principles

Defines key terms

Addresses methodological issues (e.g., level of detail)

Describes the process for undertaking the activity, including possibility of alternative approaches

Provides examples of the role of methods within the activity

Describes potential stakeholder roles and responsibilities within the activity

Identifies links to EIA regulatory requirements and to related activities

Requires consideration of uncertainties and associated implications

Provides good-practice examples and case studies

Identifies potential pitfalls and obstacles

Identifies follow-up references and sources

Provides more specific guidance (e.g., environment types, proposal types, effects type, area-specific, proposal-specific)

Maintains a balance between good practice control/guidance and ensuring sufficient flexibility to apply alternative approaches, innovate, and make necessary adaptations to suit local and proposal specific conditions

Scoping

Provides for scoping as a formal decision-making step in the EIA process (e.g., approval of terms of reference, potential for proposal rejection)

Identifies the role of scoping (e.g., focusing) in each EIA process activity

Provides for the scoping of significant environmental components and processes, data sources, effects, issues, alternatives, proposal characteristics, stakeholders, uncertainties, and proponent characteristics

Ensures sufficient flexibility to adjust process after scoping

Proposal Characteristics

Focuses on proposal characteristics most likely to induce significant environmental effects

Identifies minimum information requirements for proposal characteristics (e.g., status, location, scale, stages, service, land and resource requirements, components, processes, design, emissions, effluents, residuals, and interactions among proposal characteristics)

Provides for links to alternatives, mitigation, land-use planning, and related proposals

Provides for early and ongoing links between proposal planning and EIA process

Recognizes that proposal characteristics will evolve and change

Baseline Analysis

Broad definition of environment

Requires justified boundaries for analysis (e.g., temporal, spatial, ecological, administrative)

Provides for the consideration of patterns over space and time (e.g., existing environmental degradation and hazards, environmental carrying capacity)

Table 2.2   (*Continued*)

Facilitates focusing on sensitive and significant environmental components and processes most likely to be affected

Identifies potentially significant environmental components and processes

Facilitates consideration of links among environmental elements (e.g., physical, biological, ecological, social, economic)

Provides for links to impact prediction, monitoring and state-of-the-environment reporting

Impact Analysis and Synthesis

Broad definition of effects

Provisions for characterizing impact dimensions (e.g., intensity, duration, frequency, reversibility, direct, indirect, and cumulative)

Impact identification and prediction guidance (including examples of methods)

Refined guidance for effect (e.g., biodiversity, social, cultural, noise, environmental quality, health) types

Explicit consideration of transboundary effects

Provisions for considering interactions among activities and effects

Linked to alternatives and mitigation analyses

Alternatives Analysis

Requires identification of purpose and need

Provides guidance for alternatives identification

Provides overview of alternatives generation and evaluation process (including possible stakeholder roles)

Identifies and defines alternatives that must be considered (e.g., no action, environmentally preferred, alternatives to proposal, alternative means, alternatives outside jurisdiction)

Indicates when alternatives must be considered (e.g., when potentially significant effects)

Identifies types of alternatives that could be considered depending on circumstances (e.g., siting)

Identifies possible approaches to screening alternatives

Provides criteria examples

Points out need to consider differences in criteria importance

Identifies possible approaches and methods for comparing alternatives

Links alternatives analysis to scoping, significance interpretation and mitigation

Mitigation and Enhancement

Broad definition (e.g., prevention, amelioration, rehabilitation, restoration, compensation, enhancement, local benefits)

Provides examples of typical methods

Requires consideration and documentation of mitigation measures when potentially significant effects

Requires consideration of feasibility, effectiveness and consequences of methods

Linked to proposal characteristics, significance determination, monitoring (e.g., mitigation effectiveness), and legal requirements (e.g., compliance and enforcement)

Provisions to integrate individual measures into action plan

(*Continued*)

Table 2.2 (*Continued*)

Methods

Guidance and examples for each process activity and for major environmental components and types of effects

Identifies characteristics, strengths, and limits of methods

Provisions for integration of traditional knowledge

Sponsoring of methods research, methods symposiums, and research institutions

Links to technical guides in related areas of jurisdiction and related fields

Documentation

Identifies documentation requirements for each EIA decision-making step

Provides for interim documentation; encourages documentation that traces EIA process

Guidance: style, format, level of detail, length, cover sheet, contents, list of preparers, rationale for interpretations, conclusions and recommendations, treatment of uncertainties, summaries, use of graphics and mapping, cross-references to other documents, source and reference list, use of appendices, indexes, and keywords, and electronic publishing standards

Guidance: document circulation procedures

Guidance: documentation of agency and public involvement (including treatment of comments and suggestions)

Guidance: appendices, draft and final reports; supplemental studies

Contents guidance: notifications, project registry/referral forms, decision-making record, approval requirements, hearings record, and post-approval documents

Management

Identification of potential triggers for follow-up work

Monitoring guidance (e.g., compliance, environmental changes, effects, mitigation effectiveness, public concerns)

Identification of types of parameters that may require monitoring

Enforcement requirements

Guidance: contingency provisions

Liability provisions

Requirements and guidance: terms and conditions

Provisions for links to joint areawide monitoring and management

Guidance: integration of individual measures within overall program

Legal authority specified (e.g., enforcement, power to remedy damages, offences, penalties, responsibilities)

Identification of links between monitoring and baseline analysis, impact prediction, significance determination, proposal characteristics, and mitigation

Cumulative effects monitoring provisions

Monitoring record-keeping provisions

Identifies links between proposal monitoring and adaptive environmental management/ecosystem management

Auditing

Provisions to audit EIA document quality, methods use and effectiveness, prediction accuracy, influence on sensitive and significant environmental components and processes, mitigation measure effectiveness and monitoring, and enforcement effectiveness

Table 2.2   *(Continued)*

Provisions to review relevant research

Provisions to audit efficiency and effectiveness of EIA legislation, regulations, guidelines, and procedures

Provisions to audit responsiveness to agency and public comments and suggestions

Provisions to audit sustainability contribution

Links to auditing of the state of the environment

Participation

Identification of public participation rights and principles

Requirement for public and agency involvement prior to each decision point

Public and agency involvement provisions for each EIA activity; emphasis on early and ongoing involvement (includes post-approval involvement)

Guidance: public and agency involvement procedures

Requirements and guidance regarding notification procedures, public registry, electronic registry, and preparation of summary documents

Requirements for public and local government access to, dissemination of, and involvement in each EIA document

Requirement to document public comments and suggestions; and proponent responses to each

Provisions for additional measures (e.g., participant funding) for traditionally underrepresented groups and organizations

Legal standing provisions; access to justice

Stakeholder lists

Guidance: interdisciplinary teams

Mediation and independent inquiry provisions

Special provisions for indigenous peoples' involvement

Public participation effectiveness reviews

Provisions for consultation with neighboring states regarding transboundary effects

Timelines conducive to public involvement

Review and Decision Making

Integrates regulatory requirements into all guidance documents

Provisions for voluntary preparation of EIA documents

Includes measures to ensure that decision-making transparent, open, and accountable

Explicit criteria, policies, and standards for application at each decision-making stage

Guidance: project review team

Mediation and public hearings provisions; including criteria for when enacted and appeal provisions

Explicit review policies and requirements consistently applied by each review agency

Specifies roles and responsibilities of each review participant

Interjurisdictional review, harmonization, and involvement provisions

Explicit links to related decisions and requirements

Procedural guidance

Each decision justified in writing

Record of decision provisions

Limits on actions during review

Provisions to evaluate EIA decision-making effectiveness

A useful general tool for EIA process guidance in the United States is an EIS checklist (US DOE, 1997). The checklist identifies, using a series of questions, the content requirements of various EIS report sections (e.g., cover sheet, summary, purpose and need, proposed action and alternatives, affected environment, environmental effects, overall considerations). It also addresses document format quality and style, regulatory requirements, procedural considerations, draft and final EIS considerations, and the treatment of specific environmental effects. The checklist is cross-referenced to specific regulatory requirements. The U.S. Environmental Protection Agency (US EPA, 1998b) has prepared a training course, "Principles of Environmental Impact Assessment." The training course goes somewhat further in providing guidance for individual EIA activities and in addressing such matters as forecasting and assessment methods, report writing, reviewing and evaluating, and study team management.

The European Commission has also prepared a review checklist (ERM, 2001c) to assist EIA reviewers in assessing information submitted by developers. This review checklist addresses such matters as project description; alternatives analysis; descriptions of the environment, mitigation measures, and effects; nontechnical summaries; difficulties preparing reports; and general approach. General Canadian EIA guidelines (CEAA, 1994, 1998) provide an overview of the typical scope of several EIA activities. More detailed guidance for selective activities (e.g., significance determination, project descriptions, purpose, need and alternatives analysis, cumulative effects assessment) is provided in separate guides. Regulations to the Australian EPBC Act (2000) identify minimum content requirements for each EIA document. These requirements pertain to such matters as a description of the proposal, the treatment of alternatives, descriptions of the area likely to be affected, identification and prediction of impacts, proposed safeguards and mitigation measures, other approvals and conditions, the proponent's environmental record, and information sources.

EIA legislation and regulations in the four jurisdictions fall short of the general performance standards identified (under "General Characteristics") in Table 2.2. Minimum legislative and regulatory requirements, checklists, and general guides provide a measure of consistency in treating EIA process activities. However, requirements and guidelines tend to be more detailed for some activities and less so for others. No rationale is provided for these differences. More attention is devoted to administrative procedures and general document contents than to the consistent and systematic treatment of process activities, interrelationships among process activities, and EIA methods.

Specific Activities The four jurisdictions have numerous requirements and have prepared a large number of guideline documents pertaining to individual EIA activities. There are many examples of innovative approaches. Collectively, the four jurisdictions come close to meeting most of the criteria listed in Table 2.2. Additional and more detailed guidance could be provided, especially for baseline analysis, alternatives evaluation, impact prediction and mitigation, compensation, and management. There is considerable variation within each jurisdiction in the level of control and guidance provided for individual EIA activities. A greater pooling of knowledge and experience across jurisdictions would be highly beneficial.

It is not sufficient to implement EIA requirements and prepare EIA guidelines in a manner consistent with the criteria listed in Table 2.2. While EIA continues to evolve rapidly, as a field of theory and practice, sufficient knowledge and experience have been acquired to be able to distinguish between good and inadequate practice. EIA guidelines that systematically draw upon applied research and case studies can identify reasonable minimum standards coupled with good practice performance standards. Most EIA jurisdictions, for example, have prepared public involvement guidelines. Such guidelines tend to summarize regulatory requirements, identify a few general principles and provide an overview of the characteristics, strengths, and limitations of a standard set of consultation procedures. This often dated and largely descriptive approach fails to convey a state of practice that is much further along in providing for earlier and more extended public involvement, for sharing decision making, for consensus building and conflict resolution, and for facilitating the involvement of traditionally underrepresented groups and organizations. A more concerted effort could be made to formulate and refine EIA requirements and guidelines that are conducive to EIA practice as it could and should be (and sometimes is) rather than as it too often is or was 10 to 20 years ago.

Occasionally, EIA requirements and guidelines are too precise. In the past, selective evaluation and site selection (for waste management projects) guidelines in Ontario, Canada, maintained that EIA practice should be limited to a defined range of specific methods, criteria, and procedures. Such an approach is problematic. There are many evaluation and site-selection procedures available, each with a different mix of advantages and disadvantages. EIA literature contains many debates among advocates of alternative siting and evaluation procedures. Proponents need the flexibility to be able to select, integrate, and adapt methods, jointly with stakeholders, which are appropriate to local circumstances. Maintaining such flexibility does not preclude EIA requirements and guidelines spelling out objectives, principles, and performance standards. Examples of methods, including their strengths and limitations, and good and bad practice examples also can be provided.

A similar danger exists with EIA document requirements. Document requirements can be extremely helpful in facilitating consistency and in ensuring that minimum content requirements are satisfied. At some point, however, very detailed document format and content requirements can reduce EIA to a fill-in-the-blank "cookbook" exercise. The focus should not be exclusively on preparing EIA documents in accordance with requirements. The primary emphasis should remain on protecting and enhancing the environment and on facilitating more environmentally sound decision making and undertakings. There also is the question of whether the EIA process suffers when too much stress is placed on document preparation requirements. EIA documents should be outputs from and should reflect the EIA process structure. EIA document requirements are worthwhile, but only if they reinforce EIA objectives and do not inhibit innovative process design and management approaches that seek to better achieve EIA objectives.

2.3.3 Integration and Coordination

Integration and coordination are central attributes of regulatory EIA process management. Figure 2.2 illustrates interconnections among various integration and coordination categories. Table 2.3 lists examples of good regulatory practices.

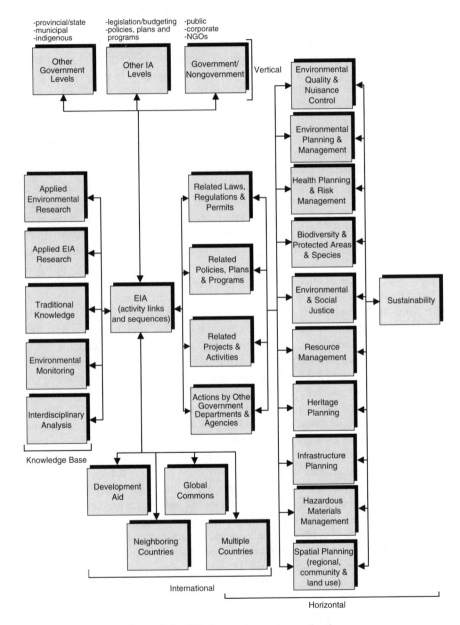

Figure 2.2 EIA integration and coordination.

Table 2.3 Examples of Good Regulatory Practice: EIA Integration and Coordination

International—Global Commons	Vertical—Other Government Levels	Horizontal—Related Policies, Programs, and Plans	Knowledge Base—Applied EIA Research
Extends EIA requirements to actions by nations that might affect global commons	Minimizes EIA duplication and overlap among government levels (e.g., accreditation)	Provides for and encourages areawide assessments	Establishes and supports EIA research centers
Extends EIA requirements to boundary areas of commons	Integrates approval requirements of all government levels in EIAs	Undertakes SEAs of regional and sectoral plans	Sponsors EIA research by universities, research institutions, and consultants
Coordinates EIA requirements and strategies for global commons	Harmonizes EIA requirements and procedures (to ensure a consistently high level of EIA practice and to address potential conflicts)	Establishes EIA systems based on natural boundaries (e.g., watersheds)	Coordinates EIA research among jurisdictions
Integrates global commons concerns (e.g., climate change, rare and endangered species)	Provides for joint or cooperative EIAs	Integration of EIA within multilevel sectoral and spatial planning	Provides for case studies of good EIA practice
Integrates international conventions for protection of commons into EIA requirements	Coordinates EIA procedures and activities	Integrates EIA requirements into regional and land-use planning	
	Provides for indigenous EIA regimes (establishment and coordination)	Systematically and formally integrates into EIA sustainability and other substantive environmental concerns	
	EIA guidelines and assistance for lower government levels		

(Continued)

Table 2.3 (Continued)

International—Multiple Countries	Vertical—Other IA Types	Horizontal—Related Projects and Activities	Knowledge Base—Traditional Knowledge
Harmonizes EIA requirements among countries	SEAs (policies, program) frame project-level EIA	Includes cumulative effects requirements	Formally recognizes in EIA legislation potential value of traditional knowledge to EIA practice
Harmonizes environmental legislation, strategies, and plans that provide context for and substantive content to EIA	Tiering provisions (e.g., project-level EIAs summarize and incorporate by reference higher-level SEA documents)	Provides cumulative effects guidance	Integrates traditional knowledge into EIA practice
Facilitates multijurisdictional EIA coordination, consultation, and planning	Legislative requirements for both SEAs and project levels (recognizing differences)	Provides for regional environmental databases to support cumulative effects assessment	
Extends applications to actions by nationals in other countries	Provides for EIAs of legislative proposals		
Integrates international EIA and other environmental conventions and agreements			
Assesses environmental impacts of trade agreements			

International—Neighboring Countries	Government/Nongovernment	Horizontal—Actions by Other Government Department and Agencies	Knowledge Base—Environmental Monitoring
Transboundary EIA agreements and coordination	Links and integrates EIA and EMS (environmental management systems)	Provides for interagency coordination agreements (to address timetables, disputes, etc.)	Integrates EIA monitoring into areawide monitoring
Adherence to transboundary EIA conventions	EIA application to public and private projects (recognizing differences)	Provides for forums (e.g., task forces, committees) to facilitate coordination	Integrates results from areawide monitoring into SEA and EIA practice
Conducts EIAs for projects in neighboring jurisdictions	Provides forums and mechanisms (e.g., roundtables, committees) for multistakeholder EIA dialogue	Integrates (where appropriate) and coordinates EIA and permitting	Provides for cumulative effects monitoring
Assesses environmental impacts of trade agreements	Provides for voluntary EIA preparation	Includes well-defined interagency EIA circulation, review and consultation requirements, procedures, and mechanisms (e.g., review teams)	Links EIA outputs to state-of-the-environment and environmental pressure indicators, where practical
	Assessment of EIA implications of privatization		
	Provides for research, coordination, and auditing roles by independent nongovernment organizations		
	Considers environmental record of proponent		

(Continued)

45

Table 2.3 (*Continued*)

International—Development Aid	Horizontal—Related Laws, Regulations, and Permits	Knowledge Base—Applied Environmental Research	Knowledge Base—Interdisciplinary Analysis
Applies, with adjustments, EIA requirements to development assistance undertakings Assists in establishing and reforming EIA systems, including capacity building	Merges EIA and other environmental regulatory requirements (where appropriate); formally linked and coordinated; guidance provided Formally incorporates substantive environmental requirements into baseline and significance determination EIA activities Analyzes related requirements to ensure complementary Brings EIA requirements under the umbrella of sustainability requirements	Provides mechanisms (e.g., networks, committees) to integrate relevant applied research into EIA requirements and guidelines Initiatives to sponsor applied environmental research of relevance to EIA practice	Emphasizes EIA requirements and guidelines on importance of transcending disciplinary boundaries Sponsors interdisciplinary EIA research

Interrelationships among EIA Activities   The individual EIA process activities, described in Section 2.3.2, are highly interrelated. Several criteria listed in Table 2.2 refer to connections between pairs of EIA activities. EIA requirements and guidelines in the four jurisdictions refer in passing to many such interconnections. However, no systematic effort has been made to trace or to provide guidance regarding patterns of interconnections among EIA activities. The implicit assumption tends to be made that EIA activities are discrete, nonrecurrent stages in the EIA process. In practice, EIA process activities often occur multiple times, in different forms and in different sequences. Some ways of arranging EIA activities are more conducive than others to good EIA practice. The frequency, sequence, and interrelationships among EIA activities are pursued in greater depth in Section 2.4.

Integration and Coordination with International EIA Activities   There are several ways in which EIA requirements can address international matters. EIA requirements can be extended directly to the actions of nationals that might affect the global commons (e.g., arctic and antarctic regions). International EIA strategies can be adopted (such as the U.S. Arctic Environmental Protection Strategy). Global environmental issues (climate change, rare and endangered species) and international environmental conventions can be integrated into EIA requirements (as occurs in Australia and Canada). EIA requirements can be extended to and applied within boundary waters, airspace, and the seabed. There have been numerous efforts to coordinate and harmonize EIA requirements among (as in the European Union) and between (as between the United States and Canada) countries. Transboundary effects and the environmental implications of trade agreements have received particular attention (George et al., 2001). These coordinative efforts often extend to other environmental management areas (e.g., airshed management, watershed management, coastal zone management). All four jurisdictions have extended EIA requirements, in modified forms, to development aid initiatives. Efforts also have been made to facilitate EIA capacity building in third-world nations.

Although these selective initiatives are a start, there would appear to be considerable room for improvement in protecting the global commons (especially the world's oceans) and in addressing international environmental issues. The pattern of international environmental strategies, plans, and requirements falls well short of what will be required to adequately manage global and international environmental problems and issues. The role of EIA process management within such efforts is far from well defined. Recent sustainability assessment and SEA initiatives may be indicative of what may be required. Substantial changes in the EIA process as conventionally portrayed are likely to be necessary. In Chapter 5 we explore these issues in much greater detail.

Vertical Integration and Coordination   Vertical EIA coordination encompasses interactions among government levels (e.g., federal, state/provincial/territorial, municipal, indigenous peoples), among EIA forms (e.g., legislative, policy, program, plan, and project EIA) and between government- and nongovernment-initiated forms of EIA. Considerable attention has been devoted by all four jurisdictions to interactions among the EIA requirements of various government levels. Many formal (e.g., agreements, multilevel systems, accreditation) and informal (e.g., integration of approvals, committees) mechanisms facilitate EIA harmonization and coordination among levels. Senior governments provide EIA advice, guidelines, and expertise to other government levels. Special provisions are often made for establishing and coordinating with indigenous EIA regimes. These mechanisms can facilitate joint action, reduce duplication and overlap, and contribute to more open, responsive, and effective EIA systems at all levels.

Whether or not such measures achieve their objectives, and if so, to what extent, is open to debate. Stakeholders have widely differing perspectives regarding the need for and efficacy of EIA harmonization and coordination measures. Clearly, there is a need for coordination. Coordination is likely to be easier if the EIA systems complement one another. Complementary EIA systems are not necessarily the same. As is evident from the previous sections, there are many options for constructing an EIA system. It would be premature to suggest that one system type is preferable in all cases. It is possible that different system types are more suited to one government level than another. Further adjustments are likely to be necessary and desirable, depending on area characteristics and preferences. The water tends to be further muddied by different perspectives regarding the appropriate roles of government, jurisdictional "turf wars," different modes of operation, varying levels of available resources, and the inevitable value-system and personality differences.

Notwithstanding these differences and complications, cooperation is more frequently the norm than is conflict and competition. Even if it is formalized in agreements and accords, evidence of cooperation should not preclude further efforts to improve coordination. EIA reforms to date have tended to concentrate on a single level at a time, albeit with some consideration of links to other levels. EIA regulatory experience has probably advanced to the point that it would now be appropriate to evaluate design and management options systematically for multilevel EIA systems. Part of this evaluation would entail an assessment of the effectiveness of existing harmonization and coordination measures.

Vertical integration also refers to interconnections among EIA types. EIA literature and many EIA regulators recognize that project-level EIA is more effective when framed by and cross-referenced to the strategic environmental assessment (SEA) of legislative proposals, policies, programs, and plans. It is also broadly acknowledged that SEA, and consequently SEA requirements, are and should be different from project-level EIA. All four jurisdictions have SEA provisions, although the requirements vary considerably. The United States (e.g., through tiering) and Europe (through the SEA and Project Directives and in multilevel EIA systems in countries such as the Netherlands) have explored the fit between SEAs and project-level EIAs.

The more immediate issue is the limited extent to which SEA is practiced. There remains a debate between SEA advocates and those that believe that legislation, policies, programs, and plans should frame project-level EIA but question the need for SEA. The middle ground tends to be occupied by those who suggest that SEA should be integrated, formally or informally, into existing legislative, policy, program, and plan formulation and review mechanisms. The debate is compounded by varying perspectives regarding whether SEA requirements should be combined with EIA requirements (as occurs in the United States), should be an option under EIA requirements (as occurs in Australia), should involve a separate set of legal requirements (as occurs in Europe), or should be addressed through a more informal set of procedures (as occurs in Canada). Opinions also vary concerning the appropriate features of any requirements and how best to interconnect the levels. Requirements may, for example, need to be adjusted, depending on whether a policy, plan, or program SEA is involved (Fischer, 2002). A consensus position is unlikely to emerge any time soon, although there seems to be a gradual drift toward embedding EIA in a broader public decision-making context. Audits, case studies, and comparative reviews would help advance the discussion beyond abstract, potentially intractable debates.

The relationship between government and nongovernment EIA systems has been explored selectively only in the four jurisdictions. Many private projects are subject to EIA requirements in action-driven EIA systems. Funding and approval triggers, under proponent-driven EIA systems, sometimes apply to private projects. Environment-driven systems apply to private projects that may result in particular effects on designated environments. The United States provides for voluntary EIA preparation and has explored the EIA implications of privatization. Canada has considered the possible links between EIA and corporate environmental management systems (EMSs). Occasionally, independent nongovernmental organizations (e.g., the Australian and New Zealand Environment and Conservation Council) have helped facilitate EIA reform. Much more could be done in addressing how best to adapt EIA requirements for application to public–private and private proposals, in facilitating the integration of EIA principles and practices and corporate management (through, e.g., the merging of EIA and EMS requirements and systems) and in defining a more proactive role for nongovernmental organizations in EIA regulatory practice.

Horizontal Integration and Coordination Horizontal integration and co-ordination includes links between EIA requirements and related laws, regulations and permits, related policies, plans and programs, related projects and activities, and actions by other government departments and agencies. As illustrated in Figure 2.2, this pattern of interconnections encompasses links to both environmental management (in a diversity of sectors) and spatial planning. Sustainability could represent an umbrella concept that facilitates coordination and integration.

Several approaches are available for linking EIA requirements and related laws, regulations, and permits. EIA requirements can be embodied in separate legislation (the approach adopted at the federal level in Canada and in the United States) but linked to related laws, regulations, and permits through approval triggers, significance determination requirements, and requirements added on during the review and approval process. Alternatively, EIA requirements can be integrated with more wide-ranging environmental legislation (the approach adopted in Australia and in some European countries and Canadian provinces). Sometimes, separate EIA and environmental requirements are merged selectively, as has occurred for some U.S. coastal zone management, transportation, and resource management plans.

It does not necessarily follow that components are better coordinated and integrated when legislation or requirements are merged. However, merged legislation or requirements appear to offer greater potential for shared objectives, a clear division of responsibility, a coherent system of links, and the flexible application of a range of implementation and management measures. The more fundamental issue is not so much whether EIA requirements should or should not be in separate legislation but rather, how best to coordinate and integrate EIA requirements with other related laws, regulations, and permits. A pattern of links is in place in all EIA systems. What is largely undetermined and which could be addressed through applied research is how well those links perform in ensuring that the various sets of requirements do not leave important gaps, do not work at cross-purposes, and are as efficient and as effective as they could be.

The links between EIA and related policies, plans, and programs are addressed most directly through the comments and requirements of review agencies and departments. It is generally preferable if review agencies explicitly identify and document in advance (to the extent practical) the policies, standards, criteria, and objectives that will be used to evaluate EIA documents. This approach can facilitate greater review consistency. It also can help reveal policy gaps and contribute to bridging the EIA review and the policy, program formulation, and administration functions of line agencies. SEAs, areawide assessments, EIA systems based on natural boundaries, the grouping of activities over space, and the integration of EIA procedures into regional and land use planning can all further the integration of EIA with sectoral and spatial policies, plans, and programs. The net result of such initiatives over time could be formal or informal sectoral and spatial environmental management regulatory hierarchies. Such systems could be partially or fully integrated, possibly under the general umbrella of a sustainability strategy. Regardless of the eventual outcome (planned or unplanned) of such efforts, a continuing need remains to explore, assess, and reform (where needed) the pattern of relationships between EIA and related policies, plans, and programs.

The links between EIA and related projects and activities has received increasing attention in all jurisdictions, largely because of concerns with cumulative effects. The United States, Canada, and the European Union all require the consideration of cumulative effects and all provide cumulative effects assessment guidance. It is generally easier to address cumulative effects through SEA. Gaps in regional environmental databases can severely hamper cumulative effects assessment. All jurisdictions group interrelated projects and activities geographically and by type. It is unclear how effective EIA requirements have been in addressing related projects and activities.

The interrelationships between EIA and the actions of other government departments and agencies are many and varied. All four EIA jurisdictions spell out in regulations and guidelines document circulation procedures, agency roles and responsibilities, and review requirements and procedures. Document circulation alone can be a slow and cumbersome process. It also is not conducive to facilitating joint action, resolving disputes, or clarifying potential misunderstandings. Greater use is therefore being made of review committees, teams, and task forces. Sometimes, coordination is formalized in interagency agreements and in merged requirements. A review of EIA requirements and procedures should include an assessment of the effectiveness of the links between EIA requirements and procedures and other government actions.

Knowledge Base The adequacy of the knowledge base for and from EIA requirements and procedures is a crucial determinant of good regulatory EIA practice. Knowledge to support EIA requirements can come from applied EIA and environmental research, traditional knowledge, environmental monitoring, and interdisciplinary analysis. Part of the EIA knowledge base comprises the lessons and insights acquired through formulating and applying EIA requirements and procedures.

Limited and sporadic links between EIA requirements and the EIA knowledge base are likely to result in outdated, ineffective, and inefficient requirements. All jurisdictions sponsor targeted research, increasingly in the form of case studies and good practice examples. Greater recognition is being given to the role of traditional knowledge in the EIA process and to the need to transcend individual disciplines. EIA regulatory practice is highly dependent on sound environmental data, especially when addressing transboundary, regional, and cumulative effects. Close liaison with those agencies and governments responsible for managing such databases is, therefore, essential. Once again the relationships between EIA and sectoral and spatial environmental management and planning are likely to be critical. In many instances, carefully scoped supplementary analyses will be needed to adequately support SEA and EIA analysis and review. The division of responsibility between public agencies and public proponents in establishing and maintaining reasonable, regional environmental databases will often require particular attention.

2.4 CONVENTIONAL APPLIED PROCESSES

2.4.1 General EIA Process Management

There are many process management choices available to EIA practitioners. There is no magical formula for navigating through the available choices. Figure 2.3 categorizes some possible choices. Decisions can be made regarding the appropriate activities and activity components (including choices regarding activity sequence, form, frequency, duration, and interactions), inputs to the process, outputs from the process, and links between the EIA process and decision-making and related

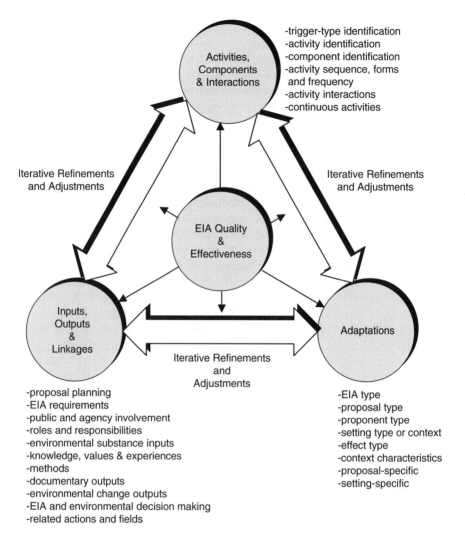

-trigger-type identification
-activity identification
-component identification
-activity sequence, forms
 and frequency
-activity interactions
-continuous activities

Activities, Components & Interactions

Iterative Refinements and Adjustments

Iterative Refinements and Adjustments

EIA Quality & Effectiveness

Inputs, Outputs & Linkages

Adaptations

Iterative Refinements and Adjustments

-proposal planning
-EIA requirements
-public and agency involvement
-roles and responsibilities
-environmental substance inputs
-knowledge, values & experiences
-methods
-documentary outputs
-environmental change outputs
-EIA and environmental decision making
-related actions and fields

-EIA type
-proposal type
-proponent type
-setting type or context
-effect type
-context characteristics
-proposal-specific
-setting-specific

Figure 2.3 EIA process management elements. (From Lawrence, 2001.)

activities and fields, adapting the process for different EIA, proposal, proponent, effect and setting types and to match proposal and effect specific characteristics, and how to build into process management insights and lessons from EIA quality and effectiveness analyses. Support figures (Figures 2.4 to 2.18) are provided at the end of the chapter to assist practitioners in making EIA process management decisions. The choices presented in these tables and figures are far from definitive. However, they do offer a cross section of possible choices. Table 2.4 identifies examples of general process management choices.

Table 2.4 Examples of General EIA Process Design Choices

Choice Types	Examples of Choices				
Trigger type	Proposal (Figures 2.4 and 2.5)	Ends (Figure 2.6)	Setting (Figure 2.7)	Contingency model (Figure 2.8)	Class or categorical assessment (Figure 2.17)
Range of activities and components	Narrow (Figure 2.4)	Front-end elaborations (e.g., screening, scoping, ends) (Figures 2.5, 2.6, and 2.12)	Midprocess elaborations (e.g., impact identification, prediction, interpretation) (Figure 2.6)	Process extension (e.g., monitoring, management, auditing) (Figures 2.5 and 2.6)	Combinations (Figures 2.5 and 2.6)
Treatment of alternatives	No alternatives (Figure 2.11)	Alternatives in general (Figures 2.4 and 2.5)	Alternatives by type (e.g., alternative means, alternatives to, alternative locations) (Figures 2.6 and 2.12 to 2.14)	Alternative combinations (Figures 2.15 and 2.16)	
Treatment of proposal characteristics	Determined from outset (Figure 2.4)	Concept and refinement (Figure 2.5)	Progressive refinement (Figures 2.6 and 2.9)	Integrated with proposal planning (Figure 2.10)	
Sequence and frequency (Table 2.5)	Linear (Figure 2.4)	Linear and activities in different forms (Figure 2.6)	Nonlinear and activities in single form (Figures 2.5 and 2.9)	Nonlinear and activities in different forms (Figure 2.10)	Continuous activities (Figures 2.6 and 2.10)
Range of interactions	None	Limited interactions (Figure 2.4)	Extensive interactions (Figures 2.5 and 2.6; Table 2.5)	Iterative (Figure 2.9)	Restructured (Figure 2.10)

Source: Adapted from Lawrence (2001).

Trigger Types A logical departure point for EIA process design is the action or actions that trigger the process. Figures 2.4 and 2.5 are examples of processes initiated by a proposal. Once the process is triggered, the focus shifts to predicting and managing the effects associated with the action proposed. Figure 2.6 begins with ends or goals (e.g., purpose to be satisfied, need to be met, opportunity to be taken advantage of, problem to be solved). Alternative ways of achieving the goals are then generated and evaluated. The proposal emerges from this procedure.

Figure 2.7 is an example of an EIA process triggered by an environmental setting (e.g., an environmentally sensitive area). Decisions regarding the acceptability of proposed actions focus on avoiding and minimizing adverse environmental effects on the environmental feature. Figure 2.8 envisions EIA process design as contingent on who is proposing to undertake the activity, the type of activity being proposed, the characteristics of the activity, the characteristics of the environment potentially affected, and the major anticipated effects. Various classes of EIA process designs can be formulated to match different combinations of these characteristics. Figure 2.18 illustrates a class, proposal type, or categorical EIA process. A proposal type initiates the procedure of preparing a class EIA document. Individual proposals within the class require no further approvals if they comply with the provisions of the class EIA document.

Range of Activities and Components There is no standard set of EIA activities. Table 2.5 describes briefly the frequency and interactions associated with 12 examples of major EIA process activities. There are many choices regarding process startup (i.e., determining the process and the proposal), the main body of EIA activities (i.e., the analysis and interpretation of environmental condition changes, with and without a proposal and with and without mitigation), and the termination of the process (i.e., proposal approval or rejection and associated measures to monitor and control effects). Figure 2.5 refines the process startup by including a two-step procedure for determining proposal characteristics and by incorporating screening and scoping steps. Figure 2.6 includes need and justification inputs. It also adds purpose, study design, EIA requirements, screening of alternatives, and the preliminary assessment of the environment and of the proposal activities. Figures 2.10 and 2.12 integrate ends or goals into the process.

Many elaborations and refinements also are possible in the middle EIA process stages. Figure 2.5 distinguishes among impact identification, impact prediction, and significance determination. It also refers to direct, indirect, and cumulative effects. Figure 2.6 adds data collection and compilation, criteria, indicator and parameter formulation, review of comparable proposals, and risk and uncertainty analysis activities. Examples of other candidates for midprocess activities include estimating the probability of impacts (Westman, 1985), determining database adequacy, refining study methodologies (Marriott, 1997), and describing and evaluating environmental systems (Morris and Thérivel, 1995).

Very broad or very fine distinctions also are possible in the closing stages of the EIA process. Figure 2.4 simply ends with preparing and reviewing EIA documents. Figure 2.5 adds decision making, implementation, monitoring, and management.

Figure 2.6 integrates individual impact management measures within an impact management strategy or a monitoring program (Canter, 1993a). It also provides for attaching terms and conditions, for monitoring proposal, effects, and mitigation/compensation effectiveness, and for auditing the process and its outcomes. EIA quality and effectiveness analyses can evaluate direct and indirect outcomes.

Table 2.5 Examples of EIA Activity: Frequency and Sequence

Screening

Screening is most commonly placed at or near the outset of the process, when it is used to determine whether and which EIA requirements will or will not be applied. Screening can occur at several points in the process if it is treated as the application of one or more exclusionary criteria. Alternatives can be screened. Insignificant environmental components and impacts can be excluded. The proposal, at the end of the process, can be determined to be unacceptable.

Scoping

Scoping is generally placed near the beginning of the process, usually after screening. Sometimes it is preceded by an analysis of need and occasionally by the identification and evaluation of alternatives to the proposal. Scoping can be treated as synonymous with focusing or bounding. If such is the case, a form of scoping takes place near the outset of every activity. Scoping procedures can, for example, be applied to determine which alternatives, issues, environmental components, and impacts should be considered, to decide on which interests should be represented, to bound the consideration of indirect and cumulative effects, to decide on which proposal characteristics are relevant to the impact analysis, to determine which areas of uncertainty should be considered further, and to decide on the scope or coverage of EIA documents.

Proposal Characteristics

The establishment of proposal characteristics is a progressive process. One level of detail is needed for screening and progressively greater levels of detail are required to undertake scoping, to assess alternatives (often, the alternatives analysis is at several levels of detail), to identify, predict, and interpret impacts, to determine mitigation potential, to make a final determination of proposal acceptability, and to monitor and manage impacts. Often, the refinements to proposal characteristics at one stage will lead to a reconsideration of previous stages—a form of sensitivity analysis.

Baseline Analysis

Baseline analysis is commonly divided into two stages: an initial environmental overview for screening and scoping purposes and a more detailed environmental evaluation to provide a basis for impact prediction and interpretation. Baseline analysis can be viewed as occurring whenever supplementary environment data are incorporated into the analysis. Baseline analysis, therefore, also occurs in alternatives evaluation, in the determination of mitigation measures, and in conjunction with monitoring and auditing.

(Continued)

Table 2.5 (*Continued*)

Impact Analysis

Impact analysis is most commonly treated as occurring once in the process—the detailed identification and prediction of impacts. However, an initial determination of impacts sometimes occurs as part of screening and is intrinsic to scoping. Impacts also are identified and predicted in alternatives analyses, in considering mitigation measures, and in determining and then applying monitoring measures.

Interpretation and Evaluation

Interpretation and evaluation both refer to the act of judging significance, importance, desirability, or acceptability. Interpretation more commonly refers to determining environmental component (e.g., valued ecosystem components) and impact significance. However, significance determinations also must be made during screening (need for EIA), scoping (major issues), proposal characteristics (those most likely to induce significant adverse impacts), in alternatives analysis (to screen options and rank criteria), in public participation (major concerns), and in impact management (impacts warranting mitigation and monitoring). Evaluation is more commonly associated with need, alternatives, and proposal acceptability. However, as noted below, alternatives consideration, and therefore evaluation, is a recurrent activity.

Synthesis

Synthesis or integration is rarely an explicit EIA activity. It is implied in references to determining environmental systems characteristics, to summarizing impacts, to predicting cumulative effects, to integrating individual impact management measures within broader strategies, and to consolidating findings, conclusions, and recommendations. Arguably, if analysis (whole divided into parts) is recurrent in EIA, the same can be said of synthesis (parts formed into a whole).

Alternatives

The placement and frequency of alternatives analyses varies dramatically among depictions of the EIA process. Occasionally, it is excluded altogether. Sometimes it is placed at the outset of the process and in other cases it is near the end—as a means to prevent and reduce impacts associated with the proposal. Some process descriptions place the identification of alternatives near the beginning. They then may draw such additional distinctions as alternatives to the proposal (assessed near the beginning often in scoping) and alternative methods (to be undertaken in parallel with or subsequent to impact analysis) and management options (near the end or post approval). An alternative perspective on the subject is to view alternatives analyses as a means of identifying and assessing the available choices at every decision point in the process. Applying this perspective would also mean tthe explicit assessment of participation, baseline, synthesis, and documentation choices.

Management

A distinction is generally drawn between mitigation (usually undertaken in parallel with impact analysis) and monitoring (decided before approval and implemented after). Sometimes, reference is made to mitigation as part of the alternatives analysis.

Table 2.5   (*Continued*)

Occasionally, a distinction is drawn between monitoring and auditing and between individual management measures and the integration of measures within an impact management strategy or plan. Management is sometimes used in a quite different sense—project management as an ongoing process function. If management is defined more broadly to encompass any action that influences, organizes, or controls something else, management can also be seen as a recurrent activity present, albeit in different forms, in every other activity. There are management aspects to, for example, screening, scoping, the design and execution of public and agency participation procedures, the conduct of the baseline analysis, the assessment of alternatives, and the documentation of the process.

Participation

Early versions of the EIA process tended to limit public and agency participation to the review of EIA documents and, in some instances, to scoping. More recent depictions provide for public and agency involvement in other stages, such as the alternatives assessment, the determination of sensitive and significant environmental components, the determination of impact significance, the selection of the appropriate mix of mitigation and monitoring measures, and the determination of proposal acceptability. Occasionally, provision is made for front-end public involvement in, for example, the determination of need and in post-approval monitoring. There is considerable variation in the activities identified as meriting public and agency involvement. Arguably, public and agency involvement should occur throughout the process, especially whenever information is required (they are an important source) and judgments must be made. Public involvement is particularly important prior to decisions.

Documentation

Early process descriptions tended to focus on a single document—the EIA report. The acknowledgment of documentation requirements has been extended progressively to front-end decision making (screening and scoping), to the review of the EIA report (draft and final report), and to post-approval documentation requirements (monitoring and auditing reports). Increasingly (in parallel with opening up of the process to greater public and agency participation), the need for more interim documents (to record each process stage), more summary documents (to facilitate public participation), and more technical support reports and appendices (for technical review purposes) has been recognized.

Decision Making

There has been a tendency to assume that the EIA process consists of no more than three decisions: screening (is an EIA required?), scoping (what should the EIA encompass?), and EIA review (Is the EIA acceptable?). It has increasingly been recognized, especially for large and controversial projects, that there are multiple decision points in an EIA process. Large gaps between formal decisions can lead to a lengthy review process and costly revisions if extensive analyses undertaken over a lengthy period must be considered all at once. The introduction of additional decision-making stages (e.g., adequacy of baseline analysis, adequacy of alternatives analysis, appropriateness of significance interpretations, suitability of proposed management measures) can reduce the likelihood of such problems. This orientation shift parallels the move toward a more incremental documentation approach.

Treatment of Alternatives Alternatives analysis, although often acknowledged as pivotal to effective EIA practice, tends to be too narrowly defined (Armour, 1990a; Steinemann, 2001). There are numerous ways to integrate alternative analyses into the EIA process. Occasionally, there are no reasonable alternatives, as illustrated in Figure 2.10. Alternatives analysis may be a means of avoiding or managing impacts (Figures 2.4, 2.14, and 2.15) (Smith, 1993; Westman, 1985), of selecting and refining a proposal (Figure 2.12) (Glasson et al., 1999; Jain et al., 1993), of locating a facility (Figure 2.13) (Burdge, 1994; Sadler, 1996) or of a combination of these functions (Figure 2.15). The alternatives analysis can contribute to scoping (Figure 2.5), to impact analysis (Figures 2.4 and 2.5), and to impact management (Figure 2.6). It assumes different forms (e.g., screening and comparison) and is conducted at varying levels of detail. The analysis of alternatives can be treated as a recurrent activity (Figure 2.9) or as intrinsic to all EIA decision making (i.e., the identification and evaluation of the available choices bearing on each EIA decision) (Figure 2.10).

Treatment of Proposal Characteristics The proposed action, in its most basic form, is defined at (or even before) the outset of the EIA process (Figure 2.4). The balance of the process then prevents and ameliorates the negative proposal effects (through mitigation and/or alternatives analyses) and determines the acceptability of the proposal, after negative effects are minimized. A more refined approach broadly outlines the proposal at the outset (sufficient to undertake screening and scoping) and then defines it more precisely (for and from impact, mitigation, and management analyses) (Figure 2.5). The proposal characteristics analysis can begin with a need or a problem to be addressed rather than a preliminary proposal. It can be refined further through terms and conditions and in response to monitoring outcomes (Figure 2.6). In practice, proposal characteristics tend to be refined progressively through the EIA process (Figures 2.9 and 2.10), partly in response to other EIA activities but also as proponents (and facility designers and operators) modify proposal characteristics to eliminate and ameliorate potential risk and impact sources.

Sequence and Frequency EIA activities can be arranged in many ways. More basic EIA process characterizations (as exhibited in Figures 2.4 and 2.5) view each activity as a discrete process stage. As detailed in Table 2.5, more refined processes demonstrate that each activity can assume multiple forms and either occur at several points or occur throughout the EIA process (Figure 2.6). EIA process portrayals become increasingly complex (as is readily evident from Figure 2.6) when demonstrating the recurrent (albeit in different forms) and continuous nature of many EIA process activities. Figure 2.9 copes with this problem by depicting the EIA process as a series of iterations, each involving the same basic set of activities. Figure 2.10 pursues this theme further by designing the process around a series of decisions. Actions (the equivalent of activities), ends, means, and participants contribute to each decision. The process, surrounding each decision, is adjusted to contextual factors and results in documentary and environmental products or outputs.

Range of Interactions Even the most basic EIA process portrayals (e.g., Figure 2.4) recognize that EIA activities are interrelated beyond incorporating inputs from previous stages and connecting them to succeeding stages. EIA process depictions commonly identify key interconnections and provide feedback loops, especially from monitoring and management (Figure 2.5). More complex EIA process characterizations (e.g., Figure 2.6) provide for multiple interconnections. All EIA activities are potentially interconnected, as illustrated in Figure 2.10. Certain interconnections are inevitably more important than others. It would be imprudent to assume that the same small number of interconnections will always be the major links among EIA process activities. EIA process managers should be highly sensitive to potential interconnections, at both the outset of the process and as the process unfolds.

2.4.2 EIA Process Inputs, Outputs, and Links

There are numerous potential inputs to and outputs from the EIA process. There also are external links to decision making and to related decisions, fields, and activities. Each interconnection shapes and is influenced by the EIA process. There are many choices available regarding interconnection roles in the EIA process. Table 2.6 provides examples, cross-referenced to the tables and figures, of some possible choices.

Proposal Planning Links The relationship between the EIA process and the proposal life cycle has been addressed in several ways. One approach views the proposal as a single point (at the outset) input to the EIA process (Figure 2.4). Another sees EIA as a periodic input to the project life cycle (Figures 2.5 and 2.6) (McDonald and Brown, 1990). EIA activities or stages feed into or between proposal cycle stages. A third portrays EIA and the proposal cycle as parallel but sporadically linked processes (Figure 2.10) (Sadler, 1996; World Bank, 1997). A forth blends EIA and the proposal life cycle into a single process (Barrow, 1997; Glasson et al., 1999; Mayda, 1996; Noorbakhsh and Ranjan, 1999).

The first three approaches treat EIA and proposal planning as separate but connected procedures. Seeing EIA as an input to the proposal life cycle ignores feedback from EIA to the project life cycle. Viewing EIA and the proposal life cycle as connected sporadically parallel processes fails to fully reflect the highly iterative relationship between EIA and proposal planning. Presentations of both approaches vary, regarding which EIA stages occur prior to, in parallel with, or subsequent to which proposal life-cycle stages.

Proposal planning should input to EIA. Proposal information is required for numerous EIA activities. Sometimes this entails scanning ahead for detailed information. Frequently, selective probes into preliminary and final design are necessary when predicting impacts on sensitive and significant environmental components and to address mitigation and monitoring potential and effectiveness. Often, several proposals are compared well into the process to ensure that social and environmental benefits are maximized.

Table 2.6 Examples of Input, Output, and Linkage EIA Process Design Choices

Choice Types	Examples of Choices				
Proposal planning links	No links to proposal planning (Figure 2.4)	Periodic inputs from proposal planning (Figures 2.5 and 2.6)	General recognition of ongoing two-way links (Figure 2.10)	Specific periodic two-way links	EIA and proposal planning fully integrated
EIA requirements inputs	No connections specified	Identified as a general input (Figure 2.6)	Linked at major decision points (Figure 2.10)	All connections specified	Fully integrated
Public and agency involvement provisions	None	Limited and late (Figure 2.4)	Major decision points (Figure 2.5)	Continuous involvement provisions (Figure 2.6)	Continuous involvement with consensus building and conflict resolution provisions (Figure 2.9)
Roles and responsibility inputs	Roles and responsibilities unspecified	Study team roles specified (Figure 2.6)	Study team and reviewers roles specified	Roles for all participants specified (Figure 2.10)	Peer review provisions (Figures 2.6 and 2.10)
Environmental substance inputs	No connection acknowledged (Figures 2.4 and 2.5)	Identified as a general input	Specific categories of substantive environmental concerns identified (Figure 2.6)	Environmental concerns fully integrated (Figure 2.10)	—
Knowledge, values, and experience inputs	Not acknowledged (Figures 2.4 and 2.5)	Need to integrate values acknowledged (Figure 2.6)	Values integrated as ends (Figure 2.10)	Need for applied research recognized (Figures 2.6 and 2.9)	Need to integrate knowledge, values and experiences into each EIA activity recognized

Methods	Not acknowledged (Figures 2.4 and 2.5)	Identified as a general input (Figure 2.10)	General roles of methods in process specified (Table 2.5)	Selective role of methods identified (e.g., applied research, peer review) (Figure 2.6)	Role of methods in each process activity specified (Table 2.5)
Documentary inputs and outputs	Final report only (Figure 2.4)	Draft and final reports (Figure 2.5)	Draft, final, and multiple interim reports (Figure 2.6)	Documentation inputs and outputs as a continuous activity (Figure 2.10)	—
Environmental outputs	Documents assumed to be only outputs (Figure 2.4)	Recognized through impact management (Figure 2.5)	Recognized through impact management and auditing (Figure 2.6)	Explicit links to environmental quality (Figures 2.10 and 2.19)	—
Decision-making links	No connections (Figure 2.3)	Approvals at end (Figure 2.4)	Approvals at end with conditions and follow-up (Figure 2.5)	Staged approval (Figures 2.16 and 2.5.)	Multiple decision points specified (Figure 2.10)
Related environmental decisions	No connections (Figures 2.4 and 2.5)	General recognition of need to link to other decision-making levels (Figure 2.6)	General recognition of need to link to related decisions (Figures 2.6 and 2.10)	Specific links to other decision-making levels (Figures 2.9 and 2.17)	EIA integrated into related decisions
Related fields	No connections (Figures 2.4 and 2.5)	General recognition of need to link to related fields (Figures 2.6 and 2.7)	Specific links between process and related fields (Figure 2.20)	EIA and related fields integrated within broader systems framework	Sustainability assessment
Related activities	No connections (Figure 2.5)	Addressed through cumulative effects assessment (Figure 2.5)	Addressed through comparable activity review (Figures 2.6 and 2.7)	Addressed through comparable setting review (Figure 2.10)	Combinations (Figure 2.10)

Source: Adapted from Lawrence (2001).

EIA and proposal planning interact continuously in practice. Both evolve through the process. There are no clear dividing lines among design refinements, alternatives analyses, and mitigation. The process should connect and, where practical, integrate proposal planning and EIA, consistent with the oft-stated EIA goal of integrating environmental concerns into planning (including presumably proposal planning) and decision making. Integration is more than merging activity components. The whole exceeds the "sum of the parts." This necessitates (1) identifying and exploring interrelationships between EIA and proposal planning, (2) determining how each changes under a merged process, and (3) defining integrated process characteristics that transcend both EIA and proposal planning.

EIA Requirement Inputs   EIA process depictions often blend generic process steps with EIA requirements. It is not sufficient to make the general point that EIA requirements should be integrated into the EIA process, as in Figures 2.6 and 2.10. Screening requirements, notification procedures, significance criteria, document types, agency, public review and comment requirements, and links to related environmental standards, laws, rules, goals, and guidelines can be referenced (Bass and Herson, 1993; Canter, 1996; Erickson, 1994; Marriott, 1997). Occasionally, citations link process activities and components directly to EIA legislative and regulatory requirements (March, 1998; Jain et al., 1993).

It can be problematic to separate generic EIA characterizations from EIA requirements. Simply adhering to a generic process runs the risk of not satisfying EIA requirements. Designing an EIA process to meet EIA requirements, although necessary, fails to benefit from EIA as a well-developed and rapidly expanding field of study and practice. The EIA process should be consistent with both EIA requirements and good practice standards. EIA processes also should evolve in parallel with changing regulatory requirements and advances in EIA theory and practice.

Public and Agency Involvement Provisions   The EIA process is rarely a closed technical process, with no provision for public or agency involvement. Agencies sometimes chose not to involve the public in routine screenings or restrict public involvement to draft or final EIA document review (Figure 2.4). A more open approach provides for public involvement prior to major decisions (Figure 2.5). Increasingly, provisions are made for ongoing public involvement (e.g., through public advisory committees) (Figure 2.9) and for consensus building and conflict management initiatives (Figures 2.6) (Manring et al., 1990). Involvement includes both public participation and communications (Figures 2.6 and 2.9) (Morgan, 1998). The public, in conjunction with other stakeholders, can be a full participant in EIA planning and decision making (Figure 2.10). Sometimes a proactive effort is needed to involve traditionally unrepresented and underrepresented groups and interests. Public participation can be extended into post-approval facility and environmental management (Morrison-Saunders et al., 2001).

Environmental Substance Inputs   Specific environmental priorities have been stressed in recent years in EIA theory and practice. This orientation shift is reflected in EIA subfields, such as social, ecological, human health, climate, and gender impact assessment. It is exemplified by the effort to integrate biodiversity, environmental justice, pollution prevention, sustainability, and the precautionary principle into EIA process activities (Figure 2.6). Substantive environmental concerns and priorities can and arguably should be integrated into each EIA activity (Figure 2.10). The treatment of environment as a process input presupposes that the basic EIA process will not be altered by the nature of inputs and outputs. Alternative process characteristics might emerge if the process were to begin from specific environmental priorities, requirements, and limits.

Role and Responsibility Inputs The EIA process involves multiple and ongoing interactions among numerous participants (Figure 2.10). Participants can include, for example, a proponent(s), a core team, an interdisciplinary study team (Figure 2.6), an evaluation team, peer reviewers (Figure 2.6), sponsors, funding agencies, user groups, elected representatives, governments, (supranational, national, regional, and local governments and statutory agencies), nongovernment organizations (international, national, regional, and local), national, regional, and local communities, interested and affected members of the public, and the media (Morgan, 1998). Occasionally, roles are ascribed to individual EIA process activities (Kreske, 1996).

The major parties should participate in designing and executing the EIA process. Roles should be determined jointly. Care should be taken to avoid inappropriate role definitions. Often, "technical" matters are assumed to be the exclusive province of engineers, scientists, and other specialists. The public is seen as participating in judgmental activities such as criteria ranking, alternatives comparison, and proposal acceptability. What such artificial distinctions ignore are the considerable knowledge and experience possessed by the public and the numerous interpretations and judgments that fall within the technical realm. Each party will inevitably assume multiple and shifting roles and responsibilities. Many roles will be shared. Role sharing will alter both role definitions and participants.

Any exploration of EIA process role definition will necessitate grappling with such issues as: How open and democratic should the EIA process be? How are the roles of generalists and specialists to be reconciled? What is the appropriate balance between rigor and practicality? Given these complexities, role definition should be approached cautiously, openly, and collectively. For each activity, roles should be identified and substantiated. Roles will be blended. They also will evolve and change through the process.

Knowledge, Values, and Experience Inputs EIA is not a value-free technical procedure. Relevant knowledge and experiences are not limited to study team specialists and government reviewers. EIA process characterizations sometimes recognize the need to integrate multiple values and sources of knowledge and experience (Lahlou and Canter, 1993). Occasionally, the need to undertake applied research, to incorporate traditional knowledge, and to provide for peer review is acknowledged (Figure 2.6) (Wiles et al., 1999). Sometimes, professional and institutional standards and criteria are considered. Often, multiple values are considered directly through EIA goals and objectives (allowing for conflicting goals or goal rankings) or indirectly by acknowledging a diversity of stakeholders (who bring to the process numerous perspectives, values, interests, and experiences) (Figure 2.10). The increasing stress on consensus building and conflict resolution also reflects an appreciation that values and perspectives will often be at odds. EIA process characterizations rarely explicitly integrate a diversity of values, interests, and experiences into each EIA process activity. Yet to be determined is whether the EIA process itself might have to be reconfigured to fully accommodate such diverse and often conflicting inputs.

Methods Inputs   Methods selection and application are central to EIA literature and practice. Some process descriptions either view the EIA process and methods as unrelated, or, more likely, that the interconnections are so self-evident as not to warrant explicit recognition (Figures 2.4 and 2.5). Occasionally, general references are made to the need to integrate methods (Figures 2.6 and 2.10). The more common practice is to present generic methods that can be applied to several EIA activities (e.g., checklists, matrices, networks, and models) and to describe methods applicable to major EIA activities (e.g., scoping, impact prediction, cumulative effects assessment, impact management, public participation, alternatives analysis) and/or to major environmental components or effects (e.g., air, soil, water, noise, biological, social, cultural, economic) (Barrow, 1997; Canter, 1996; Gilpin, 1995; Harrop and Nixon, 1999; Morgan, 1998; Morris and Thérivel, 1995).

Documentary Outputs   The EIA process, very narrowly defined, is concerned primarily with document preparation, review, and approval. Figure 2.4 shows the EIA process as culminating in document preparation and review. Figure 2.5 provides for draft and final EIA report preparation. Draft reports allow for additional agency and public involvement in document review and refinement. Figure 2.6 depicts a process with provisions for interim report preparation and review. The EIA process, presented in Figure 2.6, shifts the orientation away from documentation as a process end result and toward documentation as inputs to and outputs from a phased planning and decision-making process. The latter perspective is even more evident in Figures 2.9 and 2.10.

Environmental Outputs   Sometimes the EIA process is envisioned as ending either with EIA documents (Figure 2.4) or with EIA document approval, rejection, or amendment (Figures 2.7 to 2.9 and 2.11 to 2.18). This approach fails to determine whether the primary purposes of EIA, such as environmental protection and enhancement, are being achieved. The addition of a monitoring and management step partially addresses this question. Effects are monitored and managed and mitigation effectiveness is determined (Figure 2.5). The inclusion of an auditing step makes it possible to address broader environmental output questions, such as the accuracy of impact forecasts and the environmental effectiveness of the EIA process and of EIA institutional arrangements (Figure 2.6). The EIA process can be broadened and extended still further to assess direct and indirect EIA environmental quality contributions relative to other environmental management instruments and from a variety of perspectives (Figure 2.10).

Decision-making Links   EIA seeks to facilitate more environmentally sound decision making. Often, decision making is treated as an event that follows the EIA process (Figure 2.4) (Morris and Thérivel, 1995; Spaling et al., 1993). More commonly, it is portrayed as a single stage or event, usually preceded by EIA review and followed by implementation, monitoring, and sometimes auditing (Glasson et al., 1999; Morgan, 1998; Wiesner, 1995; Wood, 1995) (Figures 2.7 to 2.9 and 2.10 to 2.18). Decision making is generally defined, in such cases, as reaching a decision on (1) whether the EIA provides a sound decision-making basis and (2) whether to approve (with or without conditions) the proposal or to permit an appeal (Lee, 2000).

Some process depictions identify screening, scoping (Barrow, 1997; Harrop and Nixon, 1999), public participation, preferred alternative selection (Canter, 1996), and postproject review (Gilpin, 1995) as additional decision-making steps. Few and widely separated decisions can make it difficult and costly to alter or reconsider choices. An iterative and incremental EIA planning and decision-making process may ameliorate such difficulties (Figure 2.9), especially if the types and sequence of decisions are identified explicitly (Figure 2.10).

Related Decisions   Organizational and EIA decision making can be linked (Offringa, 1997) potentially to the point of a fully integrated organizational/EIA planning/decision-making process. EIA can be connected to public policy making (Caldwell, 1988; Sadler, 1996), to program development (Devuyst, 1999), to planning (macro and micro) (Mayda, 1996), to regulatory requirements (international law, legislation, regulations, guidelines) (Morgan, 1998; Ortolano, 1997), to institutional arrangements, to interest representation (Smith, 1993), and to implementation control mechanisms (Vanclay and Bronstein, 1995). EIA also can be connected to state-of-the-environment reports, environmental audits, and national environmental accounts systems (Barrow, 1997).

Tiering can link EIA levels (policy SEA, plan SEA, program SEA, and project EIA) to equivalent decision-making levels (Glasson et al., 1999) and to government levels (e.g., national/federal, regional/state, subregional, local) (Barrow, 1997; Harrop and Nixon, 1999; Lee, 2000). The connections (as they are and as they could or should be) between specific EIA activities and decision-making forms and levels also can be explored. Interconnections can be the first step toward integration, possibly within integrated assessment frameworks (Fischer, 2002; Ravetz, 1998).

Related Fields   EIA is one of many related environmental management fields of practice. Other examples include environmental and resource planning and management, risk assessment and management, urban and regional planning, life-cycle assessment, public participation, communications and conflict resolution, and corporate environmental management. EIA process managers can obtain insights and lessons from these related fields (Elling, 2000; Morgan, 1998; Ridgeway, 1999; Sánchez and Hacking, 2002; Smith, 1993). They can refer to links among fields and make use of integrative frameworks, possibly under the general umbrella of sustainability (Gilpin, 1995; Sadler, 1996; Smith, 1993; Van Der Vorst et al., 1999).

Most EIA process characterizations include no explicit links to related fields of practice (Figures 2.4 and 2.5). Occasionally, the general need to link EIA to other fields is acknowledged (Figures 2.6 and 2.10). Many integrative frameworks are available from EIA literature. These frameworks suggest, for example, the need to link EIA to related disciplines (e.g., within the earth, life, and social sciences), to interdisciplinary studies, and to frameworks that transcend individual disciplines (especially as they relate to sustainability) (Glasson et al., 1999; Morgan, 1998).

Related Activities The EIA process generally revolves around a proposed set of actions. Proposed actions are commonly placed within the context of other historical, current, and probable future actions. Basic EIA process descriptions either fail to acknowledge such links (Figure 2.4) or only make the general point that related activities should be addressed through cumulative effects assessment (CEA) (Figure 2.5). Sometimes specific CEA activities are integrated into the EIA process (Figure 2.6). Occasionally, cumulative effects are integrated into each EIA process activity (Lawrence, 1994). In some instances, the EIA process is modified to address cumulative effects more effectively (e.g., context scoping, more emphasis on follow-up, project-regional CEA links) (Baxter et al., 2001).

Related activities also can be addressed through comparable activity, proposal, and environmental reviews (Figures 2.6 and 2.10). These reviews provide data and knowledge to facilitate, for example, impact prediction. Sometimes such analyses are formalized, with control communities or environments, consistent with natural and social science experimental design procedures (Beanlands and Duinker, 1983; Burdge, 1994).

2.4.3 EIA Process Adaptations

EIA assumes many forms and includes numerous subfields. One size does not fit all for EIA process management. Adaptations are likely to be necessary. Examples of EIA process design adaptation choices are listed in Table 2.7.

EIA and Effect Types EIA can be subdivided based on, for example, ecological, social, human health, economic, and cumulative environmental components or effects. Although sharing many characteristics, there are significant differences among EIA subfields in both perspectives and methods. They also vary in the extent to which conventional EIA process stages, steps, or activities are assumed.

Ecological impact assessment, economic, and human health impact assessment process characterizations generally integrate substantive concerns into conventional

Table 2.7 Examples of Adaptation EIA Process Design Choices

Choice Types			Examples of Choices		
EIA and effects type	No recognition of need to adapt process for different EIA types (Figure 2.4)	General recognition of need to adapt for different IA types (Figures 2.5 and 2.7)	Recognition that EIA types could shape EIA process (Figure 2.10)	Recognition of specific EIA types	Specific EIA process modifications for different EIA and effect types
Proposal type	No recognition of need to adapt EIA process for different proposal types (Figures 2.4 to 2.6)	General recognition of need to adapt EIA process for different proposal types (Figure 2.8)	Recognition that proposal types could shape EIA process (Figure 2.10)	Recognition of specific proposal types	Specific EIA process modifications for different proposal types
Setting and context type	No recognition of need to adapt EIA process for different setting types (Figures 2.4 to 2.6)	General recognition of need to adapt EIA process for different settings and contexts (Figure 2.7)	Recognition that settings and contexts could shape EIA process (Figure 2.10)	Recognition of specific settings and contexts	Specific EIA process modifications for different settings and contexts

(Continued)

Table 2.7 (Continued)

Choice Types	Examples of Choices				
Proponent type	No recognition of need to adapt EIA process for different proponent types (Figures 2.4 to 2.6)	General recognition of need to adapt EIA process for different proponent types	Recognition that proponent type differences could shape EIA process (Figure 2.10)	Recognition of specific proponent types (e.g., private sector, indigenous communities)	Specific EIA process modifications for different proponent types
No provisions	General auditing provisions (Figure 2.6)	Specific document quality provisions	Specific process effectiveness provisions	Specific quality and effectiveness provisions (Figure 2.6)	Quality and effectiveness provisions
No provisions	Need to consider overlaps and interconnections acknowledged (Figure 2.6)	Specific interconnections and areas of overlap identified	Integrative frameworks within EIA (including process implications)	Integrative frameworks include EIA and related forms of environmental management (process implications identified)	Overlaps and interconnections

Source: Adapted from Lawrence (2001).

EIA activities (Arquiaga et al., 1994; Canter, 1996; Health Canada, 2000; Treweek, 1999; Westman, 1985). SIA, gender assessment, and CEA process descriptions also largely integrate content and methods into conventional EIA activities or stages (Branch et al., 1993; Interorganizational Committee, 1994; Lawrence, 1994; Verloo and Roggeband, 1996). SIA process descriptions tend to point to process-related social costs, stress the need for public scoping at the process outset, emphasize the importance of public participation and shared decision making, and note that impact prediction and interpretation includes determining the probable responses of affected publics (Interorganizational Committee, 1994). Opinions vary as to whether prediction should remain the central function with SIA or whether the focus should be on restructuring the SIA process into a form of communicative rational deliberation, co-learning, and participatory decision making (Lockie, 2001). CEA is sometimes a separate stage, but more often, methods, models, and perspectives are integrated into conventional EIA activities.

Figures 2.5 and 2.8 make the general point that the EIA process may need to be adapted for different impact assessment types. Figure 2.10 allows for the possibility that impact assessment and effect type could be factors that might shape the EIA process. The EIA process design implications of differences and similarities among environmental disciplines need to be considered further. Frameworks and methods for integrating EIA types can facilitate such efforts (Ravetz, 1998).

Proposal Type   EIA can and is applied (to varying degrees) to many action types: legislation, regulations, policies, plans, programs, projects, technologies, products, and development assistance and trade agreements. Again, there is the question of how the EIA process might change for different proposal types (Fischer, 2002). SEA and technology assessment process characterizations often parallel EIA stages (Barrow, 1997; Dalal-Clayton and Sadler, 1998; Wood and Dejeddour, 1992). Greater stress, however, is placed on such front-end activities as prescreening, goal, objective, and target setting, visioning, problem definition, bounding, need, and justification (Glasson et al., 1999; Noble, 2000a; Vanclay and Bronstein, 1995) and on policy and decision-making integration processes (Partidário, 1996).

SEA and technology assessment portrayals emphasize procedural and methodological differences from EIA. They also highlight the dangers of uncritically transferring and applying EIA practices to policies, plans, programs, and technologies. Their planning processes are less structured, formal, technical, and bounded and more uncertain, proactive, continuous, and political (Partidário, 1996; Porter, 1995; Smith, 1993; Thérivel, 1993). They are conducted at a broader level of detail and encompass a wider range of issues, choices, decision makers, and publics (Thérivel, 1993; Tywoniuk, 1990).

Alternatives often overlap and emerge as the process unfolds. The no-change alternative tends to be considered rather than the no-action alternative. The no-action alternative is usually impractical (Coates, 1990). Most process characteristics, attributed to SEA and technology assessment, also apply to life-cycle assessment, legislative and regulatory assessment, and area-wide assessment.

Setting or Context The EIA process necessarily varies by location. The need to consider local, regional, national, and international ecological, social, cultural, economic, political, legal, administrative and institutional systems, interests, trends, patterns, policies, constraints, and prospects is broadly recognized (Erickson, 1994; Morgan, 1998; Smith, 1993). Increasing consideration is being given to interconnections among contextual variables, integrating insights from pertinent knowledge sources (e.g., traditional knowledge), placing EIA within broader sustainability frameworks, using EIA to test for sustainability and placing EIA within a global context (George, 1999; Gilpin, 1995; Glasson et al., 1999; Morgan, 1998; Sadler, 1996).

EIA process management implications have received less attention. The subject has been addressed selectively, most notably concerning procedural and methodological adaptations in developing and transitional countries (Lee, 2000). The EIA process could be adapted to classes of contextual characteristics. It can then be refined to suit individual site, area, and local conditions. The EIA process should positively influence contextual variables, consistent with environmental and sustainability objectives, imperatives, and limits.

Proponent Type A question, which has received less attention, is whether proponent characteristics should influence EIA process management. Public- and private-sector proponents, for example, vary in their perspectives and in their mandates. Private-sector proponents usually, for example, provide only a limited range of products or services and are unable to expropriate land. They also need to obtain an economic return on investment, are limited in their ability and willingness to take financial risks, must be able to respond flexibly to market conditions, have a limited ability to predict or manage the actions of others, and will seek to maintain and enhance their competitive positions. These differences will affect the range of reasonable alternatives available to private proponents and can limit the extent to which information and decision-making authority can be shared. EIA regulators and EIA managers need to consider the implications of these differences.

Proponent differences will also be relevant for EIA systems established by indigenous peoples and in third-world countries (Lee, 2000; Ross, 1990). Perspective, decision making, and cultural characteristics may need to shape the EIA process rather than being treated simply as inputs. The communities involved could determine the required EIA process changes, with advice (if necessary) from EIA specialists. These specialists should be experienced in working with such communities and should be open to alternative EIA process perspectives and assumptions.

EIA Quality and Effectiveness EIA process and management is very much a work in progress. It should evolve in response to the lessons and insights provided though EIA quality and effectiveness assessments. These assessments can aid in the evaluation of EIA institutional arrangements, processes, methods, documents, and direct and indirect outcomes (Lawrence, 1997a).

Overlaps and Interconnections Overlaps and interconnections among EIA subfields have been explored through, for example, tiering (Morgan, 1998; Noble,

2000a; Nooteboom, 2000), integrative impact assessment, environmental planning frameworks, land suitability analysis, and sustainability assessment (Devuyst, 1999; Noorbakhsh and Ranjan, 1999; Ravetz, 1998; Sadler, 1996; Warner, 1996). Implications of these tools and frameworks for EIA process management require more attention. Valid differences should be respected in any integrative efforts.

2.5 SUMMING UP

In this chapter we have addressed the question of whether conventional EIA regulatory and process characterizations adequately convey the available choices. We also considered whether conventional EIA process guidance and practice, even if substantially reformed, can respond adequately to the recurrent problems. These questions were addressed through an overview of a range of EIA regulatory and process design and management choices. The regulatory analysis is based on a review of EIA requirements and guidelines at the senior government levels in the United States, Canada, Australia, and the European Union. The analysis formulates and elaborates on good practice criteria for addressing screening, individual EIA process and integration, and coordination activities. The good-practice criteria, coupled with the commentary provided in the text, are intended to facilitate enhanced EIA regulatory process control and guidance.

The applied analysis seeks to enhance EIA process design and management by integrating and then extending from a variety of EIA process characterizations portrayed in EIA literature. Tables identify examples of possible choices. Supporting tables and figures indicate relevant distinctions and illustrate process management approaches. The text summarizes process management implications. The checklist in Table 2.8 is to help EIA regulators structure their reviews of the adequacy of

Table 2.8   Checklist: Conventional Regulatory and Applied EIA Processes

The response to each question can be yes, partially, no, or uncertain. If yes, the adequacy of the approach taken should be considered. If partially, the adequacy of the approach and the implications of the areas not covered should be considered. If no, the implications of not addressing the issue raised by the question should be considered. If uncertain, the reasons why it is uncertain and any associated implications should be considered.

GENERAL: REGULATORY

1. Do the EIA requirements (legislation and regulations) clearly spell out (both overall and for each section of the requirements):
    a. Goals/objectives?
    b. Principles?
    c. Policies/priorities?
    d. General performance standards/criteria?
2. Do the EIA requirements clearly detail minimum requirements for:
    a. Screening?
    b. Significance determination?
    c. Each individual EIA activity?
    d. Each integration and coordination mechanism?

(Continued)

Table 2.8 (*Continued*)

    e. Addressing each recurrent problem area?
3. Is there a direct and traceable link between the EIA requirements and EIA aspirations (as represented by goals, objectives, principles, and policies)?
4. Does each EIA guideline:
    a. Facilitate compliance with EIA requirements?
    b. Contribute to EIA document quality?
    c. Contribute to EIA process effectiveness?
    d. Contribute to environmental protection, enhancement, and sustainability?
    e. Provide for adaptations as EIA evolves and changes?
    f. Provide for adaptations for different proposal and setting types?
    g. Provide for refinements to suit individual proposals and settings?
5. Are EIA guidelines:
    a. Consistent with the leading edge of good practice?
    b. Sufficiently specific to facilitate good EIA practice?
    c. Not so narrow as to inhibit good practice innovations and adaptations?
    d. Not so narrow as to fail to allow for multiple and changing good practice standards and perspectives?
    e. In a form that makes it possible to distinguish between good and inadequate practice?
    f. Based on the systematic use of applied research and case studies?
6. Are EIA requirements and guidelines regularly reviewed and modified based on independent effectiveness audits?

SCREENING

7. Is each of the good practice screening characteristics (see Table 2.1) addressed?
8. How effectively does the screening system address the objectives of:
    a. Building environmental considerations into proponent decision making?
    b. Building environmental considerations into action-related decision making?
    c. Protecting and enhancing the environment?
    d. Focusing on significant proponents, actions, and environments?
    e. Adapting to different proposal, action, and environment types?
9. Is environmental sustainability the preeminent objective of the screening system?
10. How effectively does the screening system avoid or minimize such screening-related concerns as:
    a. Inefficiency and lack of focus?
    b. Ill-defined requirements?
    c. Inconsistent procedures and requirements?
    d. Largely closed procedures with limited provisions for stakeholder involvement?
    e. Late triggers?
    f. Basis for triggers inconsistent with scope of requirements?
    g. Narrowly applied?
    h. Environment only addressed partially and indirectly?
    i. Incompatible with EIA systems of other government levels (e.g., gaps, duplication, overlaps)?
    j. Poor match between proponents/actions and the potential to induce significant adverse effects?
11. Does the screening system blend proponent, action, and environmental elements coherently in a manner that facilitates sustainability?
12. Does the screening system provide explicit significance thresholds and criteria?

Table 2.8 (*Continued*)

13. Does the screening system provide a systematic significance determination process, including the definition of stakeholder roles?

INDIVIDUAL ACTIVITIES

14. Are each of the good practice individual activity characteristics (see Table 2.2) addressed?
15. Do review checklists and EIA process guides provide adequate guidance for each EIA process activity at a consistent level of detail?

INTEGRATION AND COORDINATION

16. Are each of the good practice integration and coordination characteristics (see Table 2.3) addressed?
17. Is systematic guidance provided for addressing interrelationships among EIA process activities?
18. Do EIA requirements make adequate provisions for protecting the global commons and for addressing international and transboundary issues and effects?
19. Are EIA systems among government levels complementary and coordinated effectively and efficiently?
20. Are EIA types (e.g., legislative, SEA, project-level) integrated and coordinated effectively and efficiently?
21. Are SEA requirements (or their equivalent) sufficiently stringent, and are they applied as widely as they should be?
22. Is adequate consideration given to public- and private-sector EIA links?
23. Are EIA requirements integrated and coordinated effectively and efficiently with related:
    a. Laws, regulations, and permits?
    b. Policies, plans, and programs?
    c. Projects and activities?
    d. Actions by other government departments and agencies?
24. Is the environmental knowledge base for EIA integrated systematically and regularly into EIA requirements and guidelines?

GENERAL EIA PROCESS DESIGN AND MANAGEMENT

25. Are the trigger(s) for the EIA process identified, appropriate, and substantiated?
26. Are the EIA process activities appropriate:
    a. At the front end of the process?
    b. In the middle stages of the process?
    c. In the closing stages of the process?
27. Are the EIA process activity components appropriate?
28. Does the process address systematically, at the appropriate stages, all potentially reasonable alternatives?
29. Are proposal characteristics integrated progressively and systematically into the EIA process?
30. Are the EIA process activities in the appropriate sequence?
31. Are all the relevant forms of each activity integrated into the process in the appropriate way?
32. Does the EIA process identify and consider the implications of all the relevant interconnections among the EIA process activities?

EIA PROCESS INPUTS, OUTPUTS, AND LINKAGES

33. Is the EIA process linked to and integrated systematically with proposal planning?

(*Continued*)

Table 2.8 (*Continued*)

34. Are EIA requirements fully integrated into the EIA process?
35. Is appropriate provision made for public involvement in the EIA process?
36. Is appropriate provision made for agency involvement in the EIA process?
37. Are substantive environmental concerns and priorities adequately integrated into the EIA process?
38. Does the EIA process make full provision for the integration of all relevant knowledge, values and experiences?
39. Are appropriately applied and fully substantiated methods systematically integrated into the EIA process?
40. Are there appropriate documentary outputs from the EIA process?
41. Are the environmental implications of the EIA process considered fully and adequately?
42. Are the links between the EIA process and EIA-related decision making addressed adequately?
43. Are the links between the EIA process and environmental decisions addressed adequately?
44. Does the EIA process address adequately links between the proposed action(s) and related past, current, and potential future activities?
45. Is adequate consideration given to comparable activities and settings?
46. Are relevant connections to related fields of practice considered adequately?

EIA PROCESS ADAPTATIONS

47. Are appropriate adaptations made to the EIA process for different EIA types and effects?
48. Are appropriate adaptations made to the EIA process for the proposal type?
49. Are appropriate adaptations made to the EIA process for the proponent type?
50. Are appropriate adaptations made to the EIA process for:
    a. Proposal-specific characteristics?
    b. Setting-specific characteristics?
    c. Proponent-specific characteristics?
    d. Effect-specific characteristics?
51. Are EIA quality and effectiveness analyses integrated adequately into EIA process design and management?
52. Is adequate consideration given to the overlaps and interconnections among the process adaptation areas listed above?

RECURRENT PROBLEM AREAS

53. Does the EIA process apply soundly and rigorously, as appropriate, scientific standards, knowledge, methods, and procedures?
54. Does the EIA process provide a sound, consistent, and traceable decision-making basis?
55. Does the EIA process contribute adequately to substantive environmental management and sustainability improvements?
56. Is the EIA process practical and efficient, and does it facilitate implementation?
57. Is the EIA process conducive to local community control?
58. Does the EIA process facilitate collaboration with all interested and potentially affected stakeholders?
59. Does the EIA process adequately consider fairness and equity concerns?
60. Does the EIA process adequately consider risk and uncertainty concerns?
61. Can the EIA process rapidly and adequately anticipate and respond to new information, knowledge, and perspectives?

existing and proposed EIA legislation, regulations, and guidelines. It also can help EIA practitioners evaluate, refine, and reform current EIA process management approaches.

The EIA choices described in this chapter could be supplemented by further inter jurisdictional comparisons of EIA requirements, guidelines, and practices. Information, knowledge, and experience sharing would be highly beneficial. Workshops, joint studies, and collaborative efforts (such as the joint preparation of EIA proposal and setting type guidelines) are likely to lead to regulatory and applied enhancements well beyond what is practical within individual jurisdictions.

More frequent and comprehensive effectiveness analyses (from multiple stakeholder perspectives) of EIA requirements, guidelines, and practices also are conducive to enhanced EIA process management. Such reviews need to ask basic questions regarding what is and is not working and why, and to assess the options available for enhancing the levels of regulatory and applied practice. Often, it is far from clear whether and to what extent the control and guidance provided and the level of practice is adequate, appropriate, or has unintended secondary consequences. Frequently, only a narrow range of choices for elevating regulatory and applied practice is considered systematically. The search for potential approaches can be advanced by case study analyses and by applied research. The scope of potential improvements should not be limited to refinements. Basic regulatory restructuring and a redefinition of what is considered adequate and good EIA practice should always be a possibility.

The EIA regulatory approaches in the four jurisdictions address the recurrent problems in a variety of ways, as detailed in Chapters 2 to 10. In Chapters 2 to 10 we also explore the potential for additional regulatory enhancements. Conventional EIA process portrayals (within the framework of EIA requirements and guidelines), as described in this chapter, only partially address the recurrent problems identified in Chapter 1. For example, provision is often made for integrating scientific knowledge, using peer and comparable proposals and settings reviews selectively, and linking baseline and impact prediction to monitoring and management. However, such measures fall well short of a truly scientific EIA process, as advanced by many EIA critics. Similarly, conventional EIA process characterizations can provide a sound, consistent, systematic, and traceable decision-making basis, especially in the generation and evaluation of alternatives. But they generally fail to adequately reflect the many subtle distinctions, the numerous competing models, and the rich debate surrounding the issue of rationality in planning and decision making.

Substantive environmental and sustainability concerns can be partially integrated into conventional EIA processes. But conventional processes are not conducive to the fundamental perspective changes required to address such matters fully. They also fail to fully integrate fairness and equity concerns, risk and uncertainty considerations, and other values, forms of knowledge, perspectives, and ideals that fall outside the traditional purview of EIA. Conventional processes may make ample provision for public and agency involvement, but they do not incorporate the basic perspective and procedural reforms required for shared or delegated decision making. These shortcomings are partially attributable to a tendency to tightly circumscribe ends, potentially available alternatives, and the grounds for rejecting proposals. These limited perspectives also inhibit creative problem and opportunity identification and exploration.

Conventional EIA processes are potentially adaptable. But they are not generally well suited to complex and uncertain situations, where rapid adaptations to new information, knowledge, and perspectives are required. Considerable progress has made in focusing the EIA process (especially through scoping) and in more effectively linking EIA and decision making. It would seem unlikely that these recurrent problems can be fully addressed by further adaptations and refinements to conventional EIA requirements and guidelines and to conventional EIA processes. More fundamental reorientations, as described in Chapters 3 to 11, are needed. The practice-based anecdotes, presented in Chapters 3 to 11, also may provide insights of value for selecting among the regulatory and applied choices presented in this chapter.

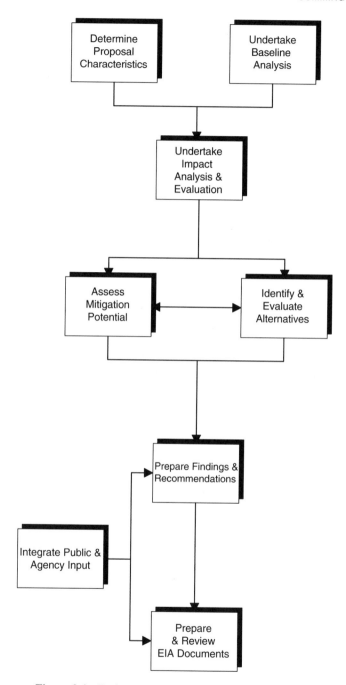

Figure 2.4 Basic EIA process. (From Lawrence, 2001.)

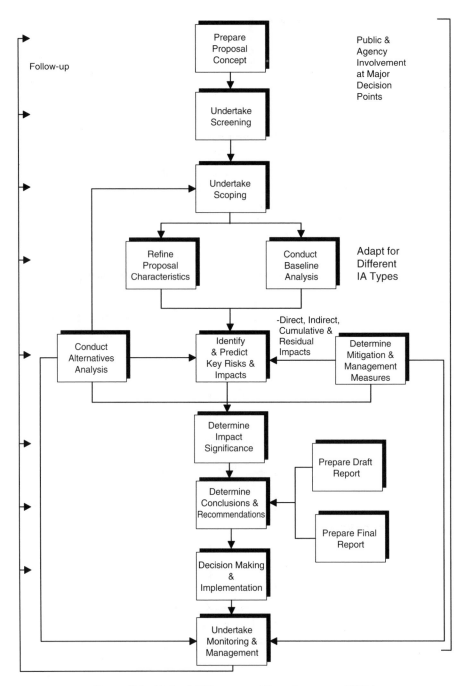

Figure 2.5   Refined EIA process. (From Lawrence, 2001.)

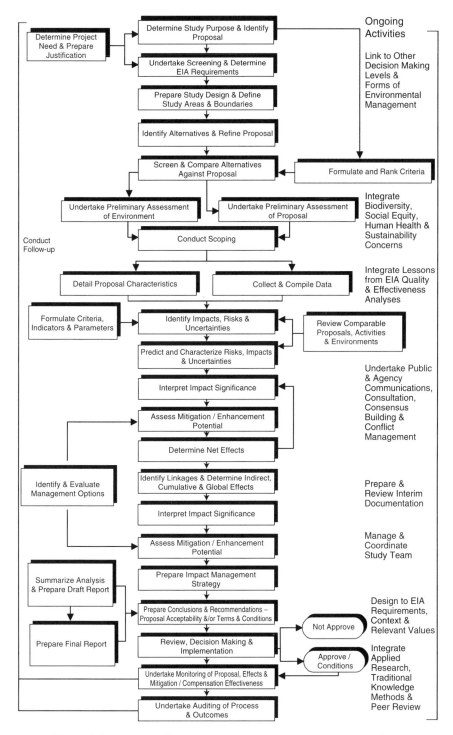

Figure 2.6 Example of a complex EIA process. (From Lawrence, 2001.)

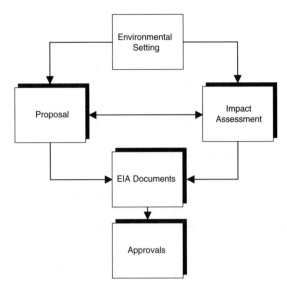

Figure 2.7 EIA process triggered by the environmental setting. (Adapted from Lawrence, 1994.)

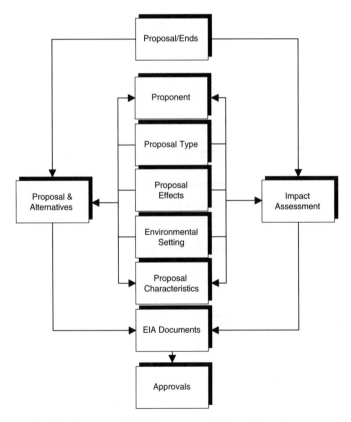

Figure 2.8 Contingency model EIA process. (Adapted from Lawrence, 1994.)

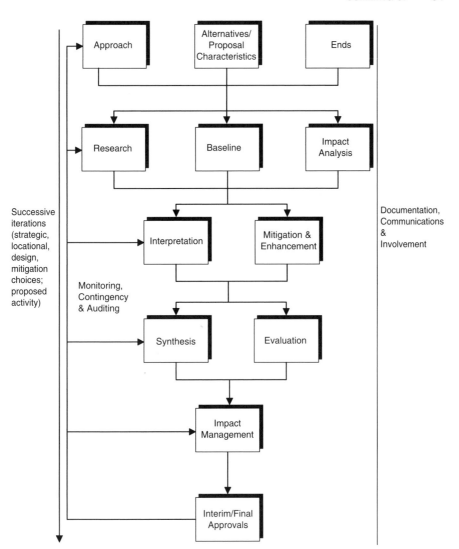

Figure 2.9   Iterative EIA process. (Adapted from Lawrence, 1994.)

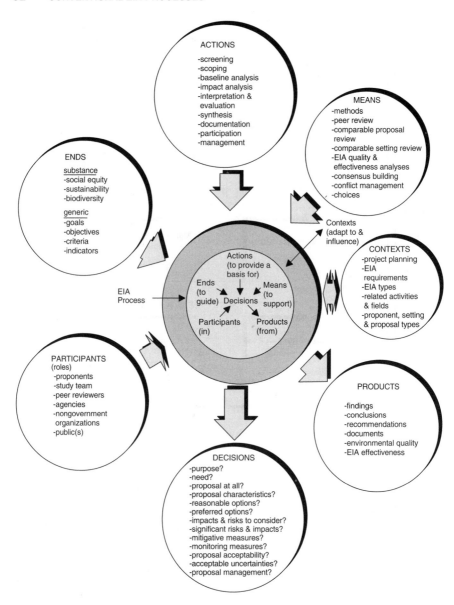

Figure 2.10 Reconstructed EIA process. (Adapted from Lawrence, 2001.)

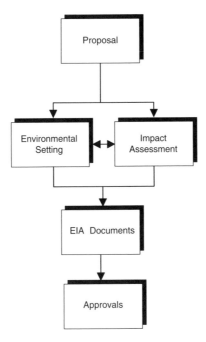

Figure 2.11   No alternatives. (Adapted from Lawrence, 1994.)

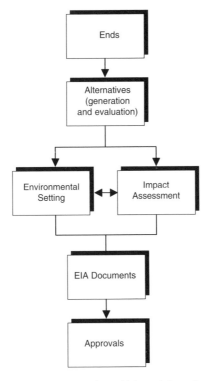

Figure 2.12   Alternatives to project. (Adapted from Lawrence, 1994.)

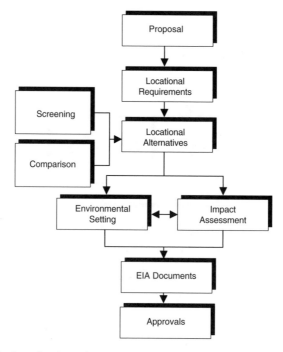

Figure 2.13 Locational requirements/alternatives (Adapted from Lawrence, 1994.)

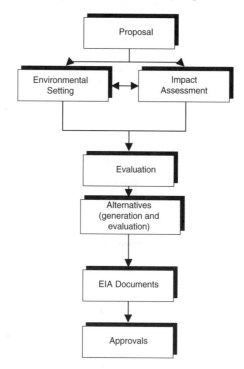

Figure 2.14 Alternatives to avoid or manage impacts. (Adapted from Lawrence, 1994.)

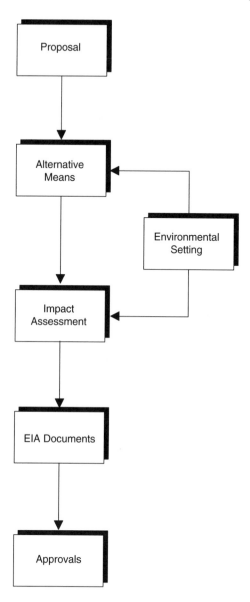

Figure 2.15 Alternative means to carry out proposal. (Adapted from Lawrence, 1994.)

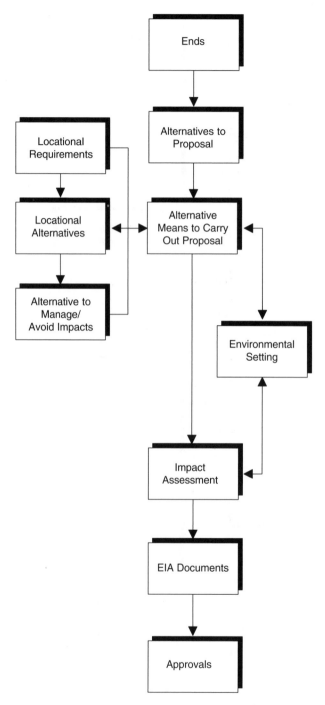

Figure 2.16 Alternatives to proposal and alternative means to carry out proposal. (Adapted from Lawrence, 1994.)

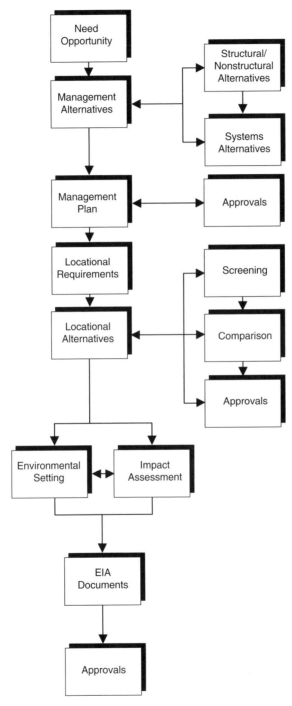

Figure 2.17 Staged approvals. (Adapted from Lawrence, 1994.)

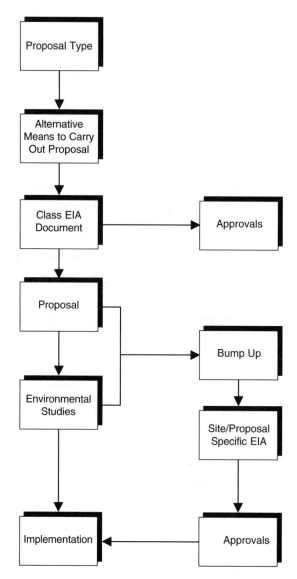

Figure 2.18 Class EIAs. (Adapted from Lawrence, 1994.)

CHAPTER 3

HOW TO MAKE EIAs MORE RIGOROUS

3.1 HIGHLIGHTS

In this chapter we test the premise that EIA processes, documents, and methods should be more scientifically rigorous. In Chapter 1 we identified lack of scientific rigor as a recurrent shortcoming of EIA practice. This shortcoming was addressed partially in Chapter 2 through choices that can contribute to more systematic, explicit, and interdisciplinary EIA analysis. However, the reforms and refinements introduced in Chapter 2 fall well short of a truly scientific EIA process, as called for by many EIA critics. Also lacking is an exploration of alternative positions regarding the appropriate role of science in the EIA process. In this chapter we seek to remedy those deficiencies.

- The analysis begins in Section 3.2 with two applied anecdotes. The stories describe applied experiences associated with efforts to make EIA practice more rigorous.
- The analysis in Section 3.3 then defines the problem, which is the inadequate and ineffective use of scientific principles, knowledge, and methods in the EIA process. We provide practical advice for making EIA documents and processes more rigorous while allowing for the limits of science in EIA practice.

Environmental Impact Assessment: Practical Solutions to Recurrent Problems, By David P. Lawrence
ISBN 0-471-45722-1   Copyright © 2003 John Wiley & Sons, Inc.

- In Section 3.4 we identify relevant principles and assumptions commonly ascribed to analytical science. Debates concerning analytical science are explored both in general and for applied fields such as EIA. These analyses provide the basis for defining a scientific EIA process.
- In Section 3.5 we detail how a scientific EIA process could be implemented at the regulatory and applied levels. In Section 3.5.1 we infuse a "scientific" perspective into EIA regulatory requirements and guidelines, and in Section 3.5.2 integrate a scientific perspective into applied processes.
- In Section 3.6 we assess how well the scientific EIA process presented in Section 3.5 satisfies ideal EIA process characteristics.
- In Section 3.7 we highlight the major insights and lessons derived from the analysis. A summary checklist is provided.

3.2 INSIGHTS FROM PRACTICE

3.2.1 The Role of Administrative Interpretation in Making EIA Processes More Rigorous

An EIA was carried out for a major hospital redevelopment in London. The project involved the demolition of listed buildings. An environmental statement was submitted, but the decision-making body concluded that more information was required. A supplementary environmental statement was requested. The supplementary statement, which included a justification for the demolition of the listed buildings, was submitted after a period of three months. This additional work resulted in an expensive delay for the project.

In this example the decision-making body required that an extra assessment be carried out on cultural heritage issues, despite the fact that the consultants had basically followed common practice. Consideration of cultural heritage is required under the EIA regulations in force in England and Wales. But the text of those regulations simply states that environmental statements must contain a "description of the aspects of the environment likely to be affected by the development, including, in particular, population, fauna, flora, soil, water, air, climatic factors, material assets, including the architectural and archaeological heritage, landscape and the inter-relationship between the above factors." There is no clarification of what an assessment of "material assets, including the architectural and archaeological heritage" might mean. It tends to be interpreted as just listing the number of designated sites affected. In London in particular, which has a long and generally recorded history, such a view is too narrow. It does not take into account important aspects such as open space and the urban character described as *townscape effects*.

The initial environmental statement was thus found wanting in the assessment of cultural heritage effects. The decision-making body that took this position had a statutory right to request supplementary information. It might have been possible to avoid this problem if cultural heritage had been assessed more rigorously rather than being presented in purely descriptive terms. Such an analysis would have been easier to undertake had additional generic guidance been provided. Also, better scoping, particularly involving the decision-making body, may have identified the degree of rigor required in the cultural heritage assessment. Any decision-making body has some discretion in interpreting and applying requirements. A failure to adequately anticipate and jointly agree upon the range and depth of analyses required for each issue identified in the regulatory requirements will almost inevitably lead to a requirement for supplementary investigations and often a potential for project delays.

The final decision regarding whether this hospital project can proceed has yet to be made. There are complications because of other, unrelated proposed developments. The decision makers have recognized the need to exercise caution and to consider cumulative and interactive effects. This latest development underscores the importance of taking into account interconnections with other decisions and projects.

This story demonstrates that simply meeting regulatory requirements, as conventionally interpreted and applied, can be problematic. There is often considerable administrative discretion in interpreting and applying regulatory requirements. This suggests the need to maintain close contact with regulatory officials to minimize the likelihood of surprises late in the review and approval process. More broadly, it points to the dynamic nature of regulatory interpretations, especially when such requirements are defined broadly and imprecisely. Interpretations will evolve in part because environmental practice lags behind but still draws upon the changing state of social and natural scientific knowledge, methods, and practices. This does not mean that full application of the protocols of social and natural sciences is necessary or even practical. It does suggest the need to employ specialists with the requisite applied social scientific and natural scientific skills and knowledge and to be aware of changing requirements and expectations in other jurisdictions in comparable situations. It should not be assumed that regulators are unaware of the changing state of the art and practice of their fields and are unwilling to adjust their interpretations accordingly. It also implies, especially when delays are likely to be costly, that a more rigorous treatment of the subject than has occurred in the past will often be a wise investment. It is increasingly apparent that in the future it will commonly be necessary to draw even more heavily on social and natural scientific methods and protocols to allow for the possibility of interactive and cumulative effects and to take into account potential sustainability implications.

ALAN BOND
School of Environmental Sciences
University of East Anglia

3.2.2 Salvaging Rigor in an EIA Process

The Jubilee Point private marina proposal at Glenelg, South Australia, called for the construction of a private marina, a residential village on land reclaimed from the sea, more housing units on reclaimed land within the adjacent inlet, a tourist hotel across part of South Australia's most popular beach, and other facilities, such as a sailing club, a relocated boat ramp, car parking, a pier, and associated commercial and tourist activities.

The draft environmental impact statement (EIS) was placed on public exhibition in early 1986. Numerous concerns were raised by public submissions. In addition, there were extensive government comments and concerns. A major concern was with coastal processes and sand management since the project bisected the metropolitan beaches and the existing sand management scheme. There were also major concerns about financial implications for government, visual and amenity effects, alienation of public land and beach, potential impact on the marine environment and archaeology, problems with water quality, issues of traffic and parking, drainage and flooding, social impact, proximity to the airport, and effects of construction and other noise impacts.

The proponent attempted to tackle the major problem of sand management through extensive engineering calculations and a novel Crawl Cat approach to sand dredging and pumping. Notwithstanding the engineering studies, the Coast Protection Board had reservations about the sand volume calculations and the costs of sand management. In response, the proponent produced a greatly modified scheme in its supplement, released in 1986. The supplement was more detailed than the draft EIS. Extensive supplementary material to the main text was also provided, particularly for sand management. The extent of modifications prompted calls for public comments on the revised scheme. The Minister for Environment and Planning took the unprecedented step of requesting public comment on the supplement. Further public submissions were received. Given the sensitive nature of the project, the proponents were given an opportunity to comment on the assessment report before the minister received it. The proponent's vehement opposition to parts of the assessment report raised questions about the role of the assessment report since it then had no status in the legislation. The opportunity for testing these questions in court never arose because the report was modified by some senior government officers in consultation with the proponents before the EIS was given official recognition three months later in 1987.

By this time, the environmental backlash and local community opposition to the project was having an effect on local politics. The local authority, originally in favor of the project, withdrew its support. The council realized that apart from local opposition, the sand management costs could be almost as high as the anticipated extra tax revenue from the project. The withdrawn council support left the state government in an invidious position. In order for the state government to break the deadlock and save face, a separate Jubilee Point review committee was set up in late 1987, headed by a local Queen's counsel. After considerable reworking of information already contained in the assessment report, the committee's final report did not make any substantial alterations to the findings of the assessment report, but attempted to differentiate between some "facts" and "opinions" in the report. With an "independent" report described as "thorough and factual," the premier of the state was able to announce just prior to Christmas 1987 that the project had been rejected.

This story is largely a depiction of an EIA process that became compromised—a process that failed to systematically identify, predict, and interpret key effects and reasonable options or to constructively involve stakeholders in either building consensus or resolving conflicts. The story does, however, suggest that sometimes a systematic and rigorous external and independent evaluation can provide a sound decision-making basis, when mutual distrust, conflicts, and flawed analyses and coordination efforts preclude interested and affected parties from coming to mutually acceptable and beneficial accommodations. It would probably be preferable if such independent, external advice were provided near the outset of the process rather than when the planning and decision-making process has largely unraveled.

NICK HARVEY
Department of Geographical and Environmental Studies
University of Adelaide

3.3 DEFINING THE PROBLEM AND DECIDING ON A DIRECTION

The preceding stories demonstrate that there is a potential role for the sciences in the EIA process. But through references to independent advice and administrative discretion, they offer only a preliminary hint of what that role might and should be. Critics of the prevailing and, in their judgment, too limited role of science in EIA practice argue that EIA processes, documents, and methods are too often ill-defined, biased, subjective, and excessively descriptive (Whitney, 1986). They generally point to the limited foundation of sound scientific theory and knowledge (Greer-Wooten, 1997). They note that objectives are usually poorly stated. Study designs and standards of inquiry, they suggest, tend to be weak to nonexistent (Whitney, 1986). Commonly, they argue, spatial and temporal boundaries are either not defined or are defined too narrowly (Greer-Wooten, 1997). They indicate that methods for characterizing environmental conditions, predicting changes with and without the proposal, and for managing effects are frequently vague, overly descriptive, poorly substantiated, and inconsistent with scientific standards and protocols (Beanlands and Duinker, 1983; Greer-Wooten, 1997).

The critics tend to express particular concern with the limited attention devoted to the variability of natural phenomena, to environmental and impact interactions that transcend disciplinary boundaries, to comparable proposals and environments, and to post-approval monitoring and auditing of environmental conditions, the accuracy of impact prediction, and the effectiveness of management measures (Morgan, 1998; Whitney, 1986). They further contend that qualified scientists are insufficiently involved in conducting and peer reviewing EIA documents, methods, and procedures (Brown, 1986). These deficiencies result in, the critics conclude, unreliable predictions, avoidable uncertainties, an unsound decision-making basis, the diminished credibility of science and scientists, a negligible contribution to the accumulation of knowledge, and a degraded environment (Morgan, 1998; Whitney, 1986).

These views are not shared uniformly. Many argue that there are numerous dangers associated with a more scientific EIA process. Some even suggest that a scientific EIA process is inherently inappropriate. Between these two poles is a considerable middle ground occupied by those who would selectively apply, adapt, temper, and modify analytical scientific methods and principles. Further complicating the issue is a plethora of alternative conceptions of the nature and role of science as it is and as it could be applied for planning and decision-making purposes.

The journey from a desire for a more scientific EIA process to its application therefore involves intermediate steps. First, an overview of the principles and assumptions commonly ascribed to analytical science needs to be provided as a point of departure. Then debates regarding if and how analytical science might be modified in general and for applied fields such as EIA need to be explored. The guidance for implementing a scientific EIA process addresses management at the regulatory and applied levels. Ideal EIA process criteria are used to identify the positive and negative tendencies of a scientific EIA process.

3.4   SELECTING THE MOST APPROPRIATE ROUTE

The debates surrounding the applied role of science tend to treat classical analytical science as the touchstone. Analytical science is viewed by some as a role model for applied fields such as EIA. Others focus on its characteristics in making a case either against the use of science or in favor of substantial adaptations. Table 3.1 lists examples of terms that commonly crop up in the debate. Table 3.2 identifies characteristics often ascribed to analytical science. These characteristics sometimes operate in dramatic tension. Almost all analytical science characteristics are debated intensely. Any exploration of science in EIA must inevitably touch on some of these "science wars" debates (Giere, 1999). Table 3.3 highlights contrary positions for 10 of these debates. It also identifies potential middle ground. The table presents only a superficial and partial description of an immense, evolving, and complex series of discussions. It would be highly presumptuous to seek to characterize, much less resolve, any of these debates. However, as described in the following subsections, some general EIA process management lessons and insights can be identified.

3.4.1   Absolute Truth versus Relativism

The absolute truth versus relativism debate suggests that EIA practitioners need not be constrained to a choice between the quest for absolute truth and the "anything goes" perspective of relativism. EIA practitioners can draw upon the insights and methods of the natural and social sciences to enhance understanding, facilitate explanations, and contribute to improved decision making. As illustrated in Figure 3.1, EIA practice can make a small contribution to middle-range theory building and successively greater contributions to micro theory building and pre-theory. These contributions will undoubtedly vary greatly in their quality, accuracy, scope, simplicity, and fruitfulness (Gower, 1997). They may not fully satisfy the standards of analytical science. A tidy hierarchy of mutually consistent and supportive theories is unlikely to emerge. A plurality of overlapping and often competing theories is the more likely result. But these efforts can, when due allowance is

Table 3.1   A Few Key Terms in Analytical Science

Applied theory	Findings applied to the solution of problems
	Focus on facilitating decision making
Deduction	Logical rules determine general premises, hypotheses, or theories
	Conclusions about particulars follow from general premises
	Testing approach often referred to as hypothetico-deductive method
Empiricism	Research orientation that emphasizes facts, observations, and experiences over theory and conceptual reflection
Grounded theory	Grounded in data obtained by research (contrasts with formal abstract theory)
Induction	Process leads from particular facts and observations to general conclusions
Knowledge	True belief acquired by a reliable method
Normative theory	Hypotheses or other statements about what is right or wrong, desirable or undesirable, just or unjust in society
Positivism	Rejects metaphysical speculation in favor of observation and experimentation as the preferred source of knowledge about the world
	Aims to construct general laws or theories that express relationships between phenomena
Principles	General rules for constructing models
Reductionism	Belief that all phenomena can be reduced to a few laws and principles
Rigor	Strict precision; exactness
Science	Systematic series of empirical activities (methods) for constructing, representing, and analyzing knowledge about phenomena being studied
	Set of normative commitments shared by a community of scholars
	Occupation (scientists seeking to establish a body of knowledge)
	Knowledge (tested facts and theories)
	Applied to human needs and purposes (when applied rather than basic or pure science)
Scientism	Claim that the positivist method is the only true method of obtaining knowledge
Theory	Collection of hypotheses and predictions amenable to experimental testing
	Organizes our concepts of and understating of the empirical world in a systematic way
	Guide for defining what types of observations need to be made to understand phenomena
	Guide for interpreting observations

Sources: Bird (1998), Giere (1999), Gower (1997), Patterson and Williams (1998), Rothman and Sudarshan (1998), Wilson (1998).

Table 3.2   Characteristics Commonly Ascribed to Analytical Science

Objective	Values separable from facts
	Science should not be subject to preconceptions (unbiased, impartial)
	Science and scientists can and should be value-free
Independent	Independent from subject, values, moral and political commitments, and environment (disinterested observer)
	Judged on academic grounds; no reflection on individual
Reducible	Deducible from the smallest number of possible axioms
	Search for laws and lawlike generalizations (ideal universal)
	Largest amount of information with least effort (elegant)
Heuristic	Builds on knowledge base (each addition contributes to enhanced understanding)
	Continually testing and improving
	Joined by theory
Methodological	Accepted procedures for observing phenomena
	Importance of observation, data, and evidence
	Rigorous (precise), reliable, and standardized methods of investigation
	Preference for quantitative results and experimental methods (most satisfactory form of evidence)
	Reliability enhanced when multiple methods applied
Technological	Reciprocal relationship between science and technology
	Great advances in sciences often related to new tools
	Omnipresence in science of machines, instruments, and experimental setups
Prone to positivism and scientism	Science preferred source of knowledge (sometimes argued only valid source of knowledge)
Natural science model	Physics as model for natural sciences
	Natural sciences as model for social sciences
	Scientific model for humanities
	Scientific (normative) model for planning and decision making (e.g., EIA, policy science, scientific management, scientific planning, scientific politics)
Explanatory	Proper roles for science: measurement, observation, explanation, and prediction
	No "ought" in science
	Facilitates understanding
	Conducive to decision making (knowledge base)
Verifiable	Explicit assumptions and procedures
	Traceable and replicable procedures
	Possible to determine if correct (verifiable) or not (falsification) (possible to be weakly or strongly verifiable)
	Prizes observation and measurement as primary means of explaining phenomena in comparable situations
Real	Sometimes succeeds in stating the truth or a good approximation of
	Knowledge is experienced based (meaning grounded in observation)
	Objective world; can be observed and recorded in an objective manner

Table 3.2   (*Continued*)

Collective	Collective activity of discovery
	Knowledge is shared (open, iterative)
	Advanced by constructive discussion, analysis, and criticism (organized skepticism)
	Must satisfy standards of peers
Pluralistic	Multiple paradigms, theories, and concepts
	Heterogeneous: many divisions within field
	Multiple perspectives on definition, practice, and application of science
Consilient	Underlying methodological unity
	Fundamental laws and principles underlie every branch of learning (unity of knowledge)
	Trend toward bridging of fields (interconnections, consistencies, middle ground, transcending concepts)
Certain	Sufficient knowledge to measure and predict with accuracy
	Manageable uncertainty
Causal	Events have determinate causes
	Causes precede events
Complex	World viewed as a set of complex problems
	Large number of variables and interrelationships
	Still amenable to scientific methods
Beneficial or benign	Can provide sound basis for decision making
	Major contributions to society and environment

Sources: Barrow (1998), Curtis and Epp (1999), Dawkins (1998), Erckmann (1986), Gower (1997), Greene (1999), Lemons and Brown (1990), Miller (1993), Patterson and Williams (1998), Pickering (1995), Porritt (2000), Rothman and Sudarshan (1998), Wilson (1998).

Table 3.3   EIA and the Science Wars

Positions	Middle Ground	Positions
Absolute	Build beyond individual circumstances but adjust to context (restricted generalizations)	*Relative*
Seek objective truth or good approximation		Phenomena too complex and fluid to be reduced to small number of laws, theories, models, and principles
Reduce phenomena to a small number of laws, theories, models, and principles	Science as model building rather than search for laws or theory building; principles as model-building rules (similar to real world)	Phenomena inseparable from observer or context
Build from tested and accepted body of knowledge (foundational)	Versimitude—nearness to the truth	All points of view valid
Scientific methods as only true methods of obtaining knowledge; a model for all fields	Good practice, broadly defined, rather than foundational truth	Multiple, competing, and overlapping theories, concepts, and perspectives (antifoundational)

(*Continued*)

Table 3.3   (*Continued*)

Positions	Middle Ground	Positions
Rigor	Natural and social sciences a resource	*Relevance*
Focus on precision; achieved through the strict application of scientific principles and methods	Be rigorous within practical limits	Focus on problem-solving value; minimize adverse environmental effects and contribute to substantive environmental improvements
Predictable phenomena	Boundaries, assumptions, and models can be explicit and substantiated	
Boundaries are established, competing hypotheses are formulated and tested (measurable verification)	Use control, comparable, and pilot studies	Goal to provide a sound decision-making basis; verification a secondary objective
Significance based on statistical analysis	Monitoring can test predictions	
Methods and results peer reviewed	Target applied research to problems	Science standards considered unrealistic and inappropriate
Only way to ensure valid and reliable predictions	Degrees of rigor; higher degree of rigor possible with simple, small, and clearly structured problems; lower degree of rigor with complex, large-scale problems with temporal variability and limited control	Work within limited resource and time constraints
Goal to advance science	Selective peer review is appropriate	Phenomena too complex and uncertain to be precisely predicted
Objective	Still value in independence and transparency	*Subjective*
Strive to be impartial, unbiased, and free from preconceptions and value positions	Strive to minimize bias and to advance environmental values explicitly	Science and EIA are inherently value-full; value judgments in every EIA activity
Values and facts are separable; focus on objective facts	Make assumptions, constraints, and value positions explicit; also can substantiate	Should be guided by values; EIA should be directed toward environmental enhancement
Only way to avoid distortions and maintain transparency	Reduce bias with peer review and good practice; also can make process more open to involvement of nonscientists	Subjective judgments (e.g., genius, imagination, and invention) essential
Analysis must be independent to be credible	Still strive for accurate predictions and effective management	If value judgments not acknowledged will be implicit and distorting

Table 3.3 (*Continued*)

Positions	Middle Ground	Positions
Beneficial to Environment	Science has applied different models of nature in different circumstances	*Detrimental to Environment*
Environmentally benign; value neutral		Science separates humans from rest of nature
Better environmental knowledge leads to better management	Ecological science does not separate humans from nature; also true for holistic science	Science integral to dominant world view (aggressive, exploitive, reductionist, arrogant)
Evidence of environmental benefits from scientific research	Science can advance environmental values if integrates environmental ethical principles, imperatives, and standards	Reinforces a nonsustainable economic and social culture
Scientists at forefront of environmental causes		
Beneficial to Democracy	Ameliorate antidemocratic tendencies (more partici- pative and compassionate and less arrogant)	*Detrimental to Democracy*
Science politically neutral or beneficial to democracy		Tendency of science and scientists to be arrogant
Scientific knowledge facilitates sound, democratic decision making	Recognize knowledge limits Encourage community- initiated research and interactions between scientists and community (civic science)	Overreliance on expert knowledge and interpretations inhibits democratic debate
Scientists generally exhibit moral values consistent with democratic principles (persuasion over force, shared knowledge)	Integrate other knowledge sources such as traditional knowledge	Science used for political persuasion purposes
	Blend science with ethical democratic principles, perspectives, and imperatives	Political interests distort scientific priorities and findings
Scientific methods (traceable, open, testing of evidence, reasoned argu- ment) conducive to open political dialogue		Limited democratic controls over science or technology
	Approximations of truth and good practice still ideals to strive for	
Espoused	Although not steadily cumulative (many wrong turns, backtracks, and lulls), still progressive over time	*Applied*
Descriptions of analytic science an ideal to strive for		Standards of science not realistic and do not reflect how science is practiced
Hypotheses are formulated and confirmed or falsified	Much to learn from good practice and good practitioners	Should seek to better understand and learn from applied scientific practice (nonlinear, evolving, interdependent with machines, discursive, creative, subjective, disjointed); much dialogue and negotiation
Knowledge is progressively accumulated	Move iteratively back and forth between espoused and applied	
Theories and models are largely consistent	EIA can contribute at the level of pre-theory and sometimes micro theory building	Proceeds in "fits and starts" in a zigzag path; many gaps and uncertainties
		Models often inconsistent and incomplete

(Continued)

Table 3.3 (*Continued*)

Positions	Middle Ground	Positions
Predict and Control	Draw on insights of chaos and complexity theory	*Manage and Adapt*
Ability to predict and control a success criterion	Adopt a cautious (precautionary where warranted) approach; hedge away from large risks	Many uncertainties associated with prediction
Causation assumed; ability to identify and verify determinate causes assumed	Identify and explore uncertainties and implications	Causal connections difficult to identify and trace; much less predict and control
Cautious and skeptical predictions	Place more emphasis on false negatives (type II errors)	Difficult to match comparable situations; too many unique factors
Seeks to minimize both false positives (no effects when effects predicted) and false negatives (effects when none predicted)	Appreciate limits of holistic approaches and of adaptive management; ensure applied in appropriate situations	Complex systems not predictable (emergent properties, discontinuous shifts)
Identified effects can be avoided or reduced to acceptable levels		Shift orientation to more adaptive and holistic management approach
Analysis	Analysis and synthesis are both inherent to EIA; can move iteratively between the two; one does not have to precede the other	*Synthesis*
Reductionism the cutting edge of science; many successes	Analysis and synthesis are complementary; maintain in a dramatic tension	Synthesis has been neglected in both science and EIA
Analysis serves to identify essential elements and foundational principles	Holistic science could have something to offer with "wicked," trans-scientific, and "messy" problems; value of application elsewhere less clear	Countertrend toward theories that integrate existing theories and span natural and social science disciplines
Necessary to address individual component impacts before can address interrelationships and cumulative effects (build from particular to general)	Appreciate the limits of holistic science; advantages may be more than offset by limits in some circumstances	Systems knowledge necessary to address role of components; whole rarely simply the sum of the parts
Inevitable trend toward greater specialization as knowledge accumulates		Shift in orientation in EIA toward synthesis (cumulative effects) and synthesis frameworks (such as sustainability)
		Value of holistic scientific approaches

Table 3.3 (*Continued*)

Positions	Middle Ground	Positions
Explanation	Possible to blend explanatory (measurement and prediction) with prescriptive (interpretation and management); need to explore which blends effective and appropriate and which are not	*Prescription*
The focus of science should be description and explanation (no ought in science)		Much scientific research is applied
Should pursue knowledge for own sake		Scientists often advocates for how scientific knowledge should or should not be applied
Science tainted when values and social objectives introduced	EIA (although more tightly circumscribed) can be treated as a form of applied research while still serving prescriptive ends	Many social science theories are normative
Values not capable of verification; when applied tendency to degenerate into social engineering	Closer matching of society's needs and thrust of scientific research (already occurring)	Applied fields, such as management, EIA, and planning have adapted applied scientific methods and theories
No role for science in EIA	Recognize problems not amenable to scientific methods	Scientific community (open and civil knowledge acquisition, sharing, and testing) a model for society
Unified	Search for underlying connections and construct macro integrative frameworks	*Plurality*
Argument for one governing model of science and scientific methods	Respect valid differences; insights from multiple perspectives and tools	Plurality of scientific paradigms scattered across natural and social sciences; partially intruding into applied fields
Scientific knowledge cumulative; new theories incorporate existing theories	View integration and differentiation as complementary strategies in the quest for knowledge	Divisions reflect fundamental value, methods, and perspective differences; also result of subject matter and context differences
Increasingly theories transcending disciplinary boundaries	Accept necessity of multiple theories, models, and methods but apply performance standards to separate reliable/desirable from unreliable/undesirable	Multiple divisions both within and among disciplines; especially within social sciences and between natural and social sciences
Successful natural science methods and habits of thought should be extended to social sciences and beyond (including to applied fields such as EIA)	Adopt open, tolerant, but critical posture to new theories (including those on science fringes)	Line between science and nonscience blurred
		Even competing and overlapping integrative frameworks and concepts

Sources: Barrow (1998), Beanlands and Duinker (1983), Beder (1993), Bird (1998), Bowler (1992), Caldwell (1988), Curtis and Epp (1999), Giere (1999), Gower (1997), Greene (1999), Greer-Wooten (1997), Healey (1997), Heinman (1997), IEMTC (1995), Lee et al. (1992), Lemons and Brown (1990), Miller (1993), Morgan (1998), Mostert (1996), O'Riordan (1995), Ozawa (1991), Parkin (1996), Patterson and Williams (1998), Pickering (1995), Porritt (2000), Powers and Adams (1997), Raiffa (1982), Ritzer (1996), Rorty (1991), Rothman and Sudarshan (1998), Rowson (1997), Wilson (1998).

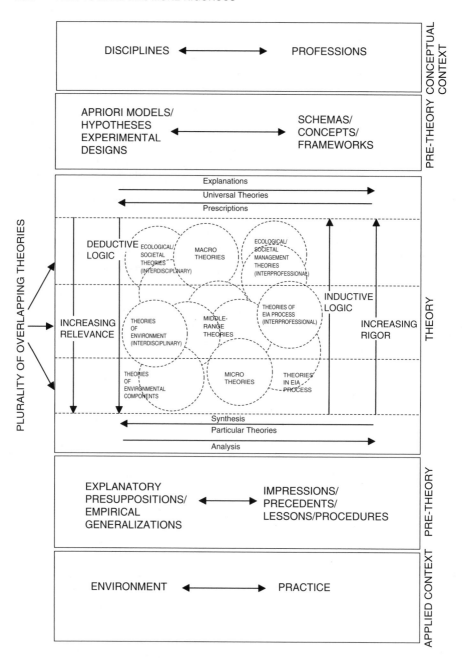

Figure 3.1   EIA theory levels. (From Lawrence, 1997c.)

made for contextual variations, improve our understanding of and our role within the environment.

Much EIA practice is either atheoretical or at the pre-theory level. Nevertheless, carefully formulated, applied, shared, and refined EIA models, designs, concepts, frameworks, precedents, lessons, and tightly circumscribed generalizations still can lift EIA practice beyond an exclusive focus on individual proposals and settings. A core body of good-practice EIA knowledge is emerging. This process has accelerated in recent years with the proliferation of EIA quality and effectiveness analyses. The construction of an EIA knowledge base cannot and should not be limited to the application and adaptation of a narrowly defined range of scientific perspectives and procedures. The claims and contributions of other scientific and nonscientific perspectives and of other modes of reasoning must also be recognized and accommodated (Healey, 1997). It will often be helpful to view EIA-related issues from multiple perspectives (sometimes this means alternative world views) and to apply multiple scientific and nonscientific paradigms and methods (Patterson and Williams, 1998). It is especially important that nonscientists be given a voice, expressed, if possible, in well-structured public interest arguments (Parkin, 1996).

3.4.2 Rigor versus Relevance

The rigor versus relevance debate in EIA is largely a false dichotomy. Natural and social scientific knowledge is a valuable resource for EIA practitioners. It contributes to understanding and facilitates both significance interpretations and management actions (Healey, 1997). Appreciating the practical limits of EIA practice, empirically adequate impact predictions are still conducive to better decision making and to better post-approval environmental management. Boundaries, assumptions, and models can and should be explicit and substantiated. Pilot studies and reviews of comparable proposals and environments can contribute to more accurate predictions and to more effective management actions. Monitoring can test the accuracy of impact predictions. Peer review can test methods and methods application. The substantive and methodological knowledge acquired through EIA practice can be shared more broadly. Applied science, conducted under the auspices of EIA practice, can be targeted to problems, opportunities, knowledge gaps, and uncertainties (IEMTC, 1995).

There can, moreover, be different degrees, standards, and forms of rigor. There are alternatives to analytical science (such as new or holistic science, chaos and complexity theory, regulatory or applied science), which, to varying degrees, relax and adapt the standards of analytical science. Rigor can be applied selectively. The issue then becomes which scientific standards are appropriate and reasonable given the constraints and objectives of EIA practice, rather than whether EIA can or should be either rigorous or relevant. There is, however, a danger (some would say a "slippery slope") with selectively abandoning and adapting scientific standards. At some point one has ventured so far beyond the realm of applied science that notwithstanding scientific "labels," what remains is no more than unsupported speculative thinking (Miller, 1993). Appreciating such risks, it still seems possible for EIA practice to effectively blend both rigor and relevance. But any mixing of rigor and relevance will need to be defined and substantiated carefully, systematically, and explicitly.

3.4.3 Objectivity versus Subjectivity

EIA practice inevitably combines the objective and the subjective. Clearly, EIA practitioners are not value-free. Much of EIA practice is subjective. This does not mean that those advocating objectivity in science and in EIA do not have valid points. Underlying the appeal for objectivity is a concern that if independence and transparency are simply abandoned, EIA practitioners, documents, and procedures will be biased (often, implicitly) against other stakeholders and against the environment whenever either conflicts with proponent interests. Some commentators will naturally conclude that such biases are inherent in EIA. This is undoubtedly true to some degree, but bias can be reduced and environmental values can be applied.

Bias can be ameliorated if subjective judgments regarding assumptions, constraints, the choice and application of methods and procedures, interpretations, and uncertainties are explicit, unambiguous, and substantiated (Beder, 1993; Mostert, 1996; Rothman and Sudarshan, 1998). Reports and methods can be peer reviewed. Professional codes of practice can be adhered to strictly. Sometimes it is preferable if the proponent does not prepare some (e.g., interest groups generate their own data) or all (e.g., EIA reports and/or peer review reports are prepared for a third party) EIA documents (Beder, 1993). A tradition of openness and honesty in reporting can be fostered (Lee et al., 1992). Raw data and input reports can be made available (Beder, 1993). "Whistle blowers" can be encouraged and protected (Beder, 1993). All interested and affected parties can be consulted freely and openly and involved in interpretations and decisions (Mostert, 1996; Rorty, 1991).

Norms, values, and the interests of all parties can be systematically integrated into the EIA process (Mostert, 1996; Parkin, 1996). Explicit environmental values, objectives, criteria, and ethical standards can be identified and applied (Mostert, 1996). Differences and trade-offs among value-based positions can be explored and substantiated systematically and explicitly. The role of scientists and nonscientists in contributing to judgments will vary by EIA process activity. Scientists could assume a greater role in impact analysis and monitoring. Nonscientists could take the lead in identifying issues and in impact evaluation and decision making (Morgan, 1998). This blending and blurring of the objective and the subjective does not mean that realism in science and in EIA need be abandoned (Rorty, 1991). Although scientists and scientific methods are subjective (especially as applied in EIA practice), this does not mean that there is no objective truth. Predictions can be more or less accurate. Environmental management and EIA practice can be more or less effective (Giere, 1999; Gower, 1997). What is required is that the objective, subjective, and objective–subjective elements of EIA practice be transparent, substantiated, jointly determined (with interested and affected parties), and conducive to environmental enhancement.

3.4.4 Beneficial versus Detrimental to Environment

EIA is intended to benefit the environment or at least minimize detrimental effects on the environment. Scientific EIA approaches, in common with science, have been criticized for contributing to environmental degradation by reinforcing an aggressive, exploitive, reductionist, and arrogant worldview (Bowler, 1992; Porritt, 2000). The net result, it is argued, is a nonsustainable economic and social culture (O'Riodan, 1995).

EIA practitioners need to guard against these negative tendencies. They can moderate or replace subtle assumptions regarding such matters as the levels of certainty and control, the beneficial links between science and technology, and the preeminence of scientists and scientific knowledge. The claims of holistic scientific theories regarding greater respect for the environment also can be scrutinized and tested (Bowler, 1992). Scientific principles and methods can be guided by and integrated with environmental ethical principles, imperatives, and standards (see Chapter 9).

3.4.5 Beneficial versus Detrimental to Democracy

A common benefit attributed to EIA is more open, transparent, informed, and democratic decision making. However, an overreliance on and uncritical acceptance of expert knowledge and interpretations (even in inherently subjective areas) can inhibit democratic debate (Bowler, 1992). Political values, perspectives, and interests can be subsumed implicitly within objective expert opinions. Science as expressed through EIA can become a tool for political persuasion (Ozawa, 1991). The interests that sponsor applied EIA research can distort priorities and findings.

These negative tendencies can be offset and ameliorated. The contributions of scientists can be acknowledged without scientists acting like or being treated as all-knowing seers (Rorty, 1991). Community-initiated research and interactions between scientists and the community can be encouraged (Heinman, 1997). Other knowledge sources, such as traditional knowledge, can be integrated into decision making. Greater consideration can be given to subfields such as civic science (Porritt, 2000). An increased effort can be made to blend scientific EIA practice with ethical democratic principles, perspectives, and imperatives (Bowler, 1992).

3.4.6 Espoused versus Applied Science

In an applied field such as EIA, the major discrepancy between the theory and practice of science does not mean that the espoused model of science has nothing to offer EIA practitioners. Except to its fiercest critics, EIA practice is acknowledged to have generally improved over the past two decades. Arguably, improvements would have been greater had there been more systematic effort to define and build on a core body of good practice. A systematic exploration of EIA as applied (an inductive analysis) is likely to reveal patterns of good and bad EIA process management. Although these patterns are largely at the level of pre-theory (as illustrated in Figure 3.1), they could lead over time (if tightly circumscribed within contextual limits) to micro-theory building. Similarly, EIA concepts and frameworks for EIA process management can be refined and tested in practice (i.e., a deductive analysis). Instead of choosing between espoused and applied science and EIA, practitioners could iteratively move between the two in a progressive (albeit disjointed) process of knowledge accumulation, derivation, and application.

3.4.7 Predict and Control versus Manage and Adapt

It is an overstatement to suggest that science as applied in EIA is ill equipped to deal with risks and uncertainties or that complexity is the "order (or perhaps more appropriately the "disorder") of the day" in EIA practice. Uncertainty can be partially addressed in EIA by integrating techniques such as human health and ecological risk assessment. Gaps, uncertainties, and value assumptions, together with their implications, can be explicitly identified and explored (Lemons and Brown, 1990). Conservative assumptions can be applied. Greater emphasis can be placed on minimizing type II errors (effects when none are predicted) (Reckhow, 1994). There could be situations where applying the precautionary principle is warranted (e.g., new technologies, processes or chemicals, catastrophic potential), where adaptive management is appropriate, or where elements of holistic science are helpful.

Care should be taken not to overstate predictive and control limits or to abandon potentially valid and useful elements of analytical science. Holistic science, moreover, may produce intriguing concepts and frameworks. But these concepts and frameworks may bear only a passing resemblance to reality (i.e., patterns imposed on rather than tested against or derived from the surrounding world) (Miller, 1993).

3.4.8 Analysis versus Synthesis

Analysis is a central attribute of EIA practice. Potentially affected components and functions of the environment must be determined. Impacts must be identified, predicted, interpreted, and managed. However, neither science nor EIA end or should end with analysis. EIA has always been integrative. Overall conclusions regarding both preferred alternatives and proposal acceptability must be reached to establish a sound decision-making basis. More attention is now devoted in EIA theory and practice to interrelationships among environmental components, as reflected, for example, in ecological and socioeconomic models. Practical approaches have been formulated and applied for addressing cumulative effects; for considering economic, social, and biophysical interconnections (interdisciplinary rather than multidisciplinary analysis and synthesis increasingly under the umbrella of sustainability); and for integrating individual measures within management strategies (Caldwell, 1988). Also, as described in Chapter 2, greater attention is being devoted to interrelationships between EIA and other forms of environmental management.

EIA practice constraints necessitate professional judgment, adaptation, and creativity. Holistic science could be especially useful when "wicked," trans-scientific, and "messy" problems must be managed (Miller, 1993). Analytical science could be better suited to situations where problems can readily be circumscribed, where much is known, and where available methods appear adequate (Miller, 1993). For most EIA problems it could be better to maintain the analytic and holistic components in a "dramatic tension." Analytical methods and perspectives can be tempered by judgment, systems thinking, and a willingness to adapt and create. Holistic approaches need to be derived from or tested by empirical evidence obtained, where practical, by the judicious use of scientific methods and protocols.

3.4.9   Explanation versus Prescription

EIA is both explanatory (what effects are likely to occur?) and prescriptive (how are negative effects to be avoided and managed and positive effects to be enhanced?). The prescriptive EIA role (i.e., to advance environmental values) is consistent with a management orientation. EIA practice is not dissimilar to applied scientific research. Both are sponsored by government and industry and are directed toward social, environmental, and business purposes. EIA-related research, although perhaps more tightly circumscribed, is still a form of scientific research. It can still, with appropriate qualifications, apply scientific principles and methods.

Many sociological, economic, and political science theories are normative. They extend beyond explanation. They also prescribe how institutions and society at large could more efficiently and effectively operate (Rorty, 1991). Applied fields such as management and planning also have combined the prescriptive and the explanatory under the umbrella of scientific planning and management, albeit with mixed results. EIA can draw on both the positive and negative lessons acquired in these fields.

3.4.10   Unified Science versus a Plurality of Sciences

The divisions within science are considerable, notwithstanding many integrative efforts. The same heterogeneous pattern of competing and overlapping theories and frameworks is repeated in applied fields such as planning, environmental management, and EIA. A core body of EIA knowledge and methods has yet to emerge more than tentatively. The debates surrounding whether science is, or should be, unified or pluralistic (including the middle-ground positions) indicate that defining good EIA practice will not be a simple task. Being aware of the divisions within and among the natural and social sciences, as well as efforts to transcend divisions, could help interpret and place in context EIA divisions and integrative initiatives.

An open and tolerant, albeit critical posture to new theories, concepts, and frameworks is likely to be more conducive to insights and applications of value to EIA practice than a strict application of analytical scientific protocols. Often, it will be necessary to select a mix of tools from both science and EIA, appreciating their characteristics, strengths, limitations, and interconnections. These tools can then be adapted and applied to match proposal and environmental characteristics. No easy task—but perhaps one that is becoming less difficult as lessons and insights are increasingly emerging from EIA quality and effectiveness analyses.

3.5 INSTITUTING A SCIENTIFIC EIA PROCESS

3.5.1 Management at the Regulatory Level

The four jurisdictions (the United States, Canada, the European Union, and Australia) identify numerous general science-related principles for application in EIA practice. Reference is made to applying an interdisciplinary approach (natural and social sciences) to assuring scientific integrity, to describing and applying credible methods accepted by the scientific community, and to the use of peer review to assess scientific methods and results. Stress is placed on the desirability of objective analyses, unbiased language, explicit, substantiated, and conservative assumptions, and on the use of scientific criteria and methods for determining uncertainty and error (e.g., confidence limits, the determination of type I and type II statistical errors). It is suggested that scientific notations should be used and explained and that scientific references should be identified.

The four jurisdictions, to varying degrees, sponsor research to address knowledge and methodological gaps, undertake quality and effectiveness analyses of EIA institutional arrangements, and EIA documents to help build and test a core body of good practice and promote science-based and flexible management approaches to extend science-based approaches beyond pre-approvals and to link EIA to related environmental management instruments. The development and extension of environmental databases has been promoted for ensuring a scientifically sound information base for EIA practice. Some consideration has been given to accrediting environmental professionals to ensure that qualified social and natural scientists are involved in preparing EIA documents. Provision is sometimes made for using independent scientific panels and advisors to ensure that sound and independent scientific advice is available, as needed, for screening, scoping, report review, and monitoring. Independent EIA centers have been established in Europe to undertake and coordinate EIA-related research. There are scattered examples of science-related methodological guidance for specific EIA activities (e.g., forecasting), types of environments (e.g., ecological processes), and types of effects (e.g., cumulative effects). There also are examples of scientific-research guidelines and limits (e.g., in environmentally sensitive areas, involving indigenous peoples). The consistent identification and application of these principles and measures at the regulatory level could establish a foundation for a greater emphasis on sound science in EIA practice.

Table 3.4 provides four examples of studies initiated to better integrate science and EIA practice. Each extends beyond general principles and provisions. The Interagency Report (IEMTC, 1995) explains the benefits of science and describes how "sound and right" science can be integrated into decision making, the ecosystem

Table 3.4   Examples of Suggested Regulatory Approaches to the Scientific Treatment of EIA

ECOSYSTEM APPROACH: HEALTHY ECOSYSTEMS AND SUSTAINABLE ECONOMICS (IEMTC, 1995)

U.S. interagency task force established for implementing an ecosystem approach to environmental management

Stresses the need for both sound and right science

Identifies the benefits of a sound scientific basis for decision making

Identifies roles of sound science in the ecosystem approach

Identifies integration deficiencies and barriers to scientific support for ecosystem management decisions

Describes need for and constraints to collaboration among scientists, managers, and the public

Identifies scientific knowledge, methods, technology, and other substantive science and information management gaps

Identifies research priorities

Describes roles of science in adaptive management

Provides examples of what works and what does not

Addresses the role of science and information in a series of case studies

GUIDELINES AND PRINCIPLES FOR SOCIAL IMPACT ASSESSMENT (INTERORGANIZATIONAL
COMMITTEE, 1994)

A group of social scientists formed a committee to prepare a set of guidelines and principles to assist agencies and private interests in fulfilling their obligations under the U.S. National Environmental Policy Act (NEPA)

Defines social impacts

Identifies ways in which social and biophysical impacts are similar and are different

Describes how a comparative SIA method can facilitate impact prediction and the determination of alternatives and mitigation measures

Identifies examples of SIA variables and project/policy stages

Describes steps in the SIA process (including baseline conditions variables, projection methods, and significance criteria)

Identifies SIA principles (including advance identification of methods and assumptions, the use of trained social scientists employing social science methods, the management of uncertainty with monitoring and mitigation, the use of published scientific literature, secondary data and primary data, and plans for data gaps)

ECOLOGICAL FRAMEWORK FOR EIA IN CANADA (BEANLANDS AND DUINKER, 1983)

Object to determine the extent to which the science of ecology could improve EIA practice in Canada

Identifies shortcomings in the scientific quality of assessment studies

Presents a basis for improvement, including;

    Links among science, values, and decisions

    The role of peer review

    The recognition of scientific requirements (boundaries, quantification, modeling, prediction, and study design)

    The development of an ecological perspective (lessons from experience, the need to conceptualize, ecological scoping, developing a study strategy)

    Organizing the approach (for initial understanding, in support of prediction, for testing hypotheses)

(Continued)

Table 3.4   (*Continued*)

Requirements for organizing and conducting impact studies (facilitating implementation, the identification of valued ecosystem components, defining a context for impact significance, developing and implementing a study strategy, specifying the nature of predictions, undertaking monitoring)

Report recommends that the requirements be adopted, that scientific advisory committees be established, that monitoring be formally recognized as integral to the assessment process, and that organizations and institutions that employ research scientists and natural resource experts actively encourage their involvement in EIA

HANDBOOKS ON EFFECTS PREDICTION (JELTES AND HERMENS, 1990)

Handbooks prepared to enhance EIA practice in the Netherlands

Prepared by research institutes on behalf of government (assessed by a review committee and edited by science writers)

Handbooks for interested and involved laypersons and as a reference for experts

Introductory volume (impact prediction) and volumes for environmental components (air, surface water, soil, groundwater, flora, fauna and ecosystems, landscape), receptors, and polluting agents (risk assessment, noise, health assessment, interactions among environmental components)

Introductory volume provides an overview of prediction (models, stages, methods)

Methods analysis includes methods descriptions, inputs, outputs, effects, assumptions and limits, reliability and sensitivity, links to other models, examples, background information, and literature

Series to be updated regularly

Used in educational courses

Could initiate new research and result in impact prediction methods improvements

Considerable international interest

---

approach, and adaptive management. It acknowledges the limits of science, identifies deficiencies and priorities, and provides specific advice regarding what does and does not work when attempting to marry science and environmental management. The Interorganizational Committee guidelines and principles (1994) provide specific advice concerning where and how social scientific knowledge, methods, and variables can be integrated into social impact assessment (SIA) practice. The Beanlands and Duinker study (1983) identifies shortcomings in applying ecological science in EIA practice. It provides specific recommendations for improving regulatory and applied EIA practice. Specialists (subject to independent review) prepared the Netherland's handbooks on effects prediction. The handbooks focus on sound methodology for application by experts. They are to be undated to reflect a changing knowledge base. They also are to be circulated for review and educational purposes (Jeltes and Hermens, 1990). These studies demonstrate that it is possible, at the regulatory level, to institute additional science-related requirements and to suggest specific roles for natural and social scientific knowledge and methods in EIA practice.

There are dangers when extending beyond general science-related principles and requirements, as demonstrated in the overview of the science wars presented in

Section 3.4. Even the general scientific principles already incorporated into EIA requirements are debatable and can be interpreted in multiple ways. It is essential to proceed with caution when instituting science-based EIA requirements. It would be easy to fall into the trap of insisting on science-based requirements, requirements that are hotly debated in science and which could constrain as much as benefit EIA practice. A more flexible, performance-based approach could capitalize on the benefits of science while appreciating its limits for applied fields such as EIA.

3.5.2 Management at the Applied Level

Figure 3.2 shows an example of a scientific EIA process. Figure 3.2 and the process description that follows incorporate many scientific EIA elements. EIA process managers and participants can pick and choose relevant or appropriate elements. A scientific EIA process treats EIA as an experiment. Predictions are hypotheses to be tested. The process is driven by independent, skilled, and qualified natural and social scientists. The scientists strive to ensure a rigorous, objective, and open process consistent with scientific principles and protocols. Scientific findings and interpretations are assumed to provide a sound decision-making basis.

Startup The startup to a scientific EIA process involves several highly interrelated activities. The problem or opportunity to be addressed must be determined. Objectives, which represent the purpose of the experiment, must be formulated. Constraints and assumptions for bounding the experiment need to be established. It is necessary to identify the initial proposal characteristics that trigger the process. Preliminary methods for collecting, analyzing, integrating, and interpreting data must be identified. The need, an extension of the problem or opportunity, must be determined. The appropriate aspects of the context must be identified. Initial hypotheses, the preliminary alternative explanations suitable for testing, must be formulated. The experimental design, a research program for testing the hypotheses, must be prepared.

Experimental design principles are incorporated, where practical, into the scientific EIA process. Initial hypotheses of interest are formulated. The hypotheses are explanations of the likelihood and magnitude of potential impacts. Hypotheses are formulated for alternative proposal characteristics (options including the null hypothesis) and for alternative impact predictions for the proposal. The experiment is designed to suit the context, to achieve the objectives, and to operate within the constraints and assumptions. Graphical and statistical testing procedures using appropriate sampling frames support the analyses and the interpretations. The experiment is designed to minimize both type I (concluding that there are effects when none exist) and type II (concluding that there are no effects where effects have occurred) errors.

Constraints (e.g., limits to knowledge) and assumptions (e.g., temporal and spatial boundaries, values) are explicit and substantiated. The study design addresses the implications of uncertainties and of contextual factors. A high level of rigor (the experimental methods and test protocols of analytical science) is applied when

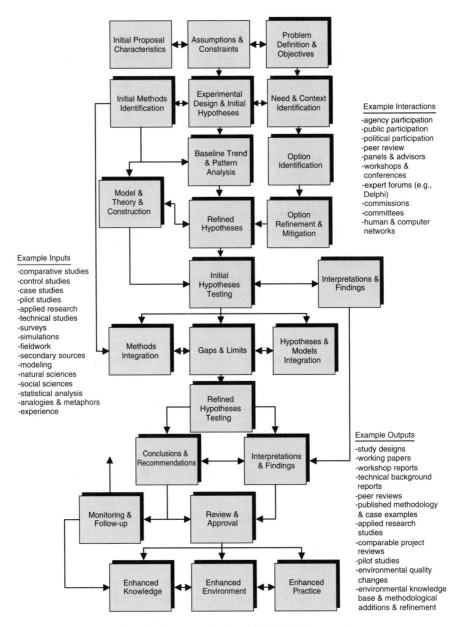

Figure 3.2 Example of a scientific EIA process.

cause–effect relationships are simple and clearly structured. A lower level of rigor (more selective and superficial quasiexperimental and judgmental activities) is applied for more complex, larger-scale, longer time horizons characterized by high levels of variability and low levels of predictability and control (Beanlands and Duinker, 1983; Patterson and Williams, 1998).

Multiple scientifically defensible methods structure the data collection and analysis, link and integrate impact predictions, and facilitate interactions among scientists (Bird, 1998). The methods are reliable and consistent (over a range of spatial and temporal scales). The startup activities are progressively refined as new data are incorporated into the EIA process.

Analysis   The analysis is highly structured. Multiple, reliable, and preferable precise environmental and social criteria and indicators are selected—criteria and indicators that can reveal (ideally, statistically significant) changes in ecological and socioeconomic conditions. The level of detail (e.g., individual organisms, species, populations, communities) at which change occurs and can reliably be detected and predicted is selected carefully (Beanlands and Duinker, 1983). The analysis identifies and focuses on sensitive and significant social, economic, and ecological components, processes, and functions. Historical and emerging cumulative effects are identified.

Data, which can readily be aggregated or disaggregated, are collected over a range of temporal and spatial scales. Ideally, the EIA database naturally extends from regional social, economic, and ecological indicators. The regional and greater statistics establish a context for subsequent impact interpretations. The baseline analysis is dynamic (trends and patterns over time and space) and conducive to systems level understanding. The data are suitable for testing impact prediction accuracy.

Hypotheses are refined based on insights obtained from the baseline analyses. Predictions or implications are deduced for each competing hypothesis (Curtis and Epp, 1999). A premium is placed on predictive precision (especially predictions that are easiest to test by observation and experiment) and causal hypotheses (correlations between two kinds of events) (Bird, 1998; Wilson, 1998). Predictions are first likely to be broad approximations. They are refined as more detailed data are integrated into the analysis (Greene, 1999). The object or process being studied is separated from its context to control confounding variables (Miller, 1993). Predictions address the magnitude, frequency, extent, and likelihood of potential effects. The bases for predictions are explicit.

Synthesis   Interrelationships among environmental components and among potential impacts are addressed through models and occasionally, preliminary theories. Multiple model types (conceptual, mathematical, physical, biological, social, economic, human health, and ecological risk) are applied. The models convey a systems level perspective on natural and socioeconomic conditions and provide a framework for predicting changes, including cumulative effects. They also are useful for testing options and mitigation methods (alternative hypotheses).

The models are derived from baseline spatial and temporal patterns and trends and from reviews of comparable, control, and pilot studies. They also draw upon natural and social science literature. They focus on critical environmental components and interrelationships. The refined hypotheses, emerging from the model and theory construction exercises, trace webs of causal connections beginning from the proposal, extending through various levels of direct and indirect effects, and ending with systems level effects. The models and theories are described and justified.

Models, theories, and hypotheses initially tend to be narrowly defined, generally within individual disciplines. A second round of model and theory building is usually necessary to address interconnections among the models and theories. Ideally, integrated assessment models or overarching theories can be constructed (Bird, 1998; Greene, 1999; Ravetz, 1998). Interconnections across consistent and mutually supportive theories and models are preferred. More commonly, interconnections and inconsistencies are addressed only partially and subjectively (Rothman and Sudarshan, 1998).

Hypothesis Testing   The real measure of a scientific EIA process occurs when the validities of alternative hypotheses are rigorously tested (Curtis and Epp, 1999). Hypothesis testing involves collecting empirical evidence and then modeling environmental conditions. An interrupted time series design for hypothesis testing tends to be preferred. This entails periodic tests, measurements, and observations of relevant variables at equally spaced intervals. The proposal is introduced at a predetermined interval. Hypotheses are tested (falsified) both prior (using comparative, control, case, and pilot studies) and subsequent (through monitoring) to approval (Burdge, 1994).

Hypothesis testing applies to the impacts predicted for the proposed action and to the impacts that could result from options and mitigation measures. Options tend to be excluded where severe impacts are probably based on reliable impact predictions and where there is a high degree of uncertainty and potentially severe consequences. Option comparison relies on a combination of social and natural scientific (preferably quantitative) indicators and substantiated scientific interpretations.

Gaps, limits, and uncertainties are explicitly identified together with potential implications. Uncertainties can be considerable given the complexity of the phenomena and the time and resource limits usually inherent in EIA practice. Consequently, there is a tendency to incorporate procedures and assumptions that minimize the likelihood of underestimating the incidence and severity of adverse impacts. These procedures are explored more fully in Chapter 10.

Interpretations   Extensive use is made of statistical tests of significance to facilitate interpretations in scientific EIA processes. Interpretations are often influenced by uncertainties regarding data reliability, the potential for systemic bias, and nonlinear relationships. Both quantitative and logical techniques are used to analyze the evidence and to reach judgments based on the weight of evidence. Interpretation is the creative component of science. Scientific interpretations encompass both espoused (reliance on statistical and other quantitative analyses) and applied (a systematic, creative, and collective endeavor) dimensions. In the latter case the boundaries between scientific and nonscientific EIA processes overlap.

Approvals and Post Approvals   Scientific findings and interpretations provide the primary decision-making basis. A scientific experimental design guides and structures the monitoring and follow-up program. Monitoring tests the accuracy of impact predictions and the effectiveness of mitigation measures. Monitoring data can lead to new and refined hypotheses that are, in turn, tested through further monitoring. Follow-up assesses the validity and effectiveness of experimental design features, data collection and analysis procedures, model design and application, interpretations, and the overall scientific EIA process. Monitoring results can be incorporated into regional environmental databases.

Inputs, Outputs, and Interactions   A scientific EIA process is far from closed. It extends from existing natural and social scientific knowledge (Brown, 1986). It draws on interdisciplinary knowledge and on the experiences and insights of applied research scientists and EIA practitioners (Caldwell, 1988). Model construction and hypotheses testing (prior to approvals) requires the systematic use of comparative, control, case, and pilot studies. Knowledge gaps are addressed through targeted research. Baseline analyses apply such scientific tools as surveys, field investigations, modeling, and computer simulations (Barrow, 1998). Statistical analyses aid significance interpretations. Analogies and metaphors often help structure the analysis but need to be confirmed through experimentation (Rothman and Sudarshan, 1998).

A scientific EIA process is a collective endeavor. A team of natural and social scientists manage the process and provide specialist environmental knowledge and methods. Other scientists are involved as government reviewers and as specialist advisors and peer reviewers. Peer reviewers can assess the correctness of procedures and the plausibility of findings and conclusions (Hirschmann, 1994). Interactions among scientists occur through committees, workshops, commissions, panels, and expert forums. Links to the broader scientific community are maintained by circulating and publishing (where practical) research findings and by means of human and computer networks (Barrow, 1998). Broader agency, political, and public participation occurs prior to major decisions and in the review of documentary outputs. The public and politicians assume a more prominent role in defining the problem and objectives, in contributing to significance interpretations, in suggesting options, and in tempering and testing conclusions and recommendations.

Scientific EIA processes tend to have numerous interim documentary outputs (e.g., study designs, working papers, applied research studies, pilot studies, technical reports, comparable project reviews, workshop reports, background studies, peer reviews). The findings from scientific and technical support studies are fully integrated into core documentation. EIA process documentation extends into post approval with the preparation and circulation of monitoring results. Consistent with good scientific practice, assumptions, methods, and findings are fully and fairly represented. They also are independently evaluated and subjected to rigorous criticism. A scientific EIA process is expected to add to the environmental and EIA knowledge base and to contribute to environmental protection and enhancement.

Adaptations and Variations   There are multiple perspectives in the scientific community (as highlighted in Table 3.3) regarding what constitutes "the" scientific method or indeed whether a plurality of methods is both necessary and desirable. The same is the case for a scientific EIA process. The process, described in the preceding subsections, largely conforms to the tenets of analytical science. Several important modifications have been made, most notably the greater emphasis on integration, the infusion of environmental values and objectives, and the direct links to decision making.

This process could be tempered or replaced by, for example, a holistic, a management, a complexity, or a civic scientific EIA process. These other scientific EIA processes are treated here as tempering considerations or at best variations rather than as alternatives to the analytical model. This is partly because they are not nearly as fully developed and have been applied to a much more limited extent. Also, if viewed as a replacement to analytical science, there is some question as to whether the process that emerges is still primarily scientific in orientation.

Viewed solely in a tempering capacity, holistic or new science points to the value of creativity, intuition, imagination, judgment, and flexibility and to the potential roles of multiple trans-scientific and systems perspectives and frameworks (Miller, 1993; Porritt, 2000). Management science demonstrates the need to place EIA-related science more firmly in the context of decision-making priorities, requirements, and limits. Complexity science underscores the central position of uncertainty, especially in applied fields such as EIA. It also systematically addresses the implications of uncertainty for the ability to predict and manage and for the design and adaptation of planning processes and organizations (Patterson and Williams, 1998; Rothman and Sudarshan, 1998). Civic science points to the need for and value of having interested and affected members of the public assume a less peripheral role in the EIA process. Civic science argues that the public should assume a valuable role in shaping the process and in contributing to such tasks as data collection, analysis, and interpretation (Heinman, 1997; Ozawa, 1991; Porritt, 2000).

The multiparadigm view of science is intriguing. It forms part of the foundation for the structure of this book. It offers many insights regarding potential ways of linking, integrating, and transcending multiple overlapping models within and among disciplines and fields of practice. It also clearly demonstrates the many pitfalls and potential losses associated with pushing the integration process too far. To avoid repetition, the themes and insights derived from an exploration of multiparadigm approaches to science have been incorporated into Chapter 11, where we address interrelationships among the EIA processes.

3.6 ASSESSING PROCESS EFFECTIVENESS

Table 3.5 presents criteria and indicators for assessing the positive and negative tendencies of the EIA processes presented in this and in the subsequent seven chapters. In this section we summarize examples of positive and negative tendencies of the scientific EIA process. A scientific EIA process is clearly aimed toward injecting more *rigor* into EIA practice. Scientific knowledge, methods, and standards can

Table 3.5   Criteria for Assessing EIA Processes

Criteria	Example Indicators
Rigorous	Sound, independent, and unbiased application of scientific standards and protocols
	Conducive to integration and application of scientific and technical knowledge and methods
	Conducive to participation by scientists
	Explicit and substantiated assumptions, findings, interpretations, conclusions, and recommendations
	Facilitates contribution to scientific knowledge base
Comprehensive	Thorough treatment of relevant physical, biological, social, cultural, and economic effects
	Conducive to addressing interrelationships and cumulative effects
	Conducive to a broad definition of problems and opportunities
	Conducive to a holistic perspective
Systematic	Provides an explicit and traceable decision-making basis
	Systematically identifies and assesses potential objectives
	Systematically identifies, assesses, and applies methods
	Systematically identifies and assesses options and impact management methods
	Systematically identifies, predicts, and manages potential effects
Substantive	Guided by environmental values and ideals
	Conducive to integration of environmental knowledge and perspectives
	Facilitates substantive contribution to enhanced environmental quality
	Conducive to realization of sustainability
Practical	Efficient and effective use of available resources
	Proven and credible methods and procedures, consistent with good practice
	Clearly defined, appropriate, and realistic roles and responsibilities
	Focuses on major issues and trade-offs
	Conducive to decision making and implementation
	Realistic expectations and standards
Democratic	Conducive to maintenance and enhancement of stakeholder influence
	Accommodates and applies traditional knowledge
	Conducive to delegation of authority to stakeholders and local communities
	Sensitive to political implications
	Provides for and potentially conducive to stakeholder acceptance
Collaborative	Conducive to stakeholder understanding and involvement
	EIA process jointly defined and undertaken with stakeholders
	Facilitates consensus building
	Facilitates conflict resolution
	Roles and responsibilities jointly defined with participants

(Continued)

Table 3.5 (*Continued*)

Criteria	Example Indicators
Ethical	Facilitates procedural and distributional fairness
	Process guided and shaped by ethical imperatives and standards
	Conducive to recognizing rights and meeting responsibilities of interested and affected parties
	Explicitly addresses ethical issues, implications, trade-offs, and dilemmas
	Conducive to addressing social and environmental fairness, equity, and justice concerns from multiple perspectives
Adaptive	Conducive to anticipation of and rapid adaptation to changing circumstances
	Facilitates creative identification and exploration of problems and opportunities
	Designed to match and evolve with context
	Conducive to systematic consideration of risks and uncertainties
Integrative	Conducive to the integration of diverse values, forms of knowledge, perspectives, and ideals
	Considers implications for and from related decisions
	Facilitates integration with proposal planning
	Adapts, integrates, and transcends individual disciplines, professions, and EIA types
	Links and integrates (where appropriate) EIA with related environmental management forms and levels

be better integrated. Participation by scientists can be facilitated. Assumptions, methods, and findings are likely to be explicitly presented. A scientific EIA process can be more independent and less prone to bias. But it will not be objective and value-free. Implicit biases can be hidden beneath the mask of objectivity. EIA practice could make a modest contribution to scientific knowledge, appreciating the limits imposed by available time, resources, and decision-making requirements.

A scientific EIA process could, by building on the natural and social knowledge base, facilitate a thorough analysis of many types of physical, biological, social, cultural, and economic effects. The application of methods such as modeling and simulation in science is conducive to addressing interrelationships and, to some extent, cumulative effects. A holistic perspective also is consistent with the trend, in some quarters of scientific thought, toward conceptual unity (Wilson, 1998). Still, a scientific EIA process is far from *comprehensive*. Scientific approaches provide only a partial basis for determining issues, for identifying options, for making interpretations, and for evaluating choices. These "value-full" activities require the infusion of nonscientific perspectives (Morgan, 1998). Scientific approaches tend to be especially weak in addressing qualitative, subjective concerns—for example, a wilderness experience, changes to a way of life, and community stress and stigma (Lemons and Brown, 1990). An analytical approach to science also is unable to deduce from the parts the behavior of complex systems (Caldwell, 1988; Miller, 1993).

A scientific EIA process is *systematic* when addressing those tasks for which it is best suited, most notably effects prediction and measurement. It tends to be less systematic in defining the problem, in identifying and evaluating options, and in coming to conclusions and recommendations. The quantitative orientation of scientific approaches can contribute to a bias toward options, and effects that can readily be measured. A preoccupation with precise prediction and measurement can result in insufficient attention being devoted to enhancing benefits and to preventing and minimizing adverse effects (Morgan, 1998).

A scientific EIA process can be environmentally *substantive* by building on existing natural and social science knowledge (Bowler, 1992). The stress on testing and refining explanations, within the scientific community, provides an opportunity to generalize beyond individual circumstances and personal intuition. Scientific EIA processes can contribute to the EIA learning curve. Some forms of scientific research (e.g., ecological science) are supportive of and consistent with environmental imperatives such as sustainability. Scientists often are adherents to and advocates of environmental values. Science, however, has a mixed record in advancing environmental values. An expectation of scientific and technical certainty is sometimes connected to a desire to dominate, exploit, or overly manage the environment (O'Riodan, 1995). Separation of the observer from the observed and the parts from the whole, as exemplified by analytical science, runs counter to a holistic environmental vision encompassing all organisms, including humans (Porritt, 2000). A preoccupation with applying scientific standards could also divert attention away from the environmental roles of EIA (i.e., environmental protection and management) (Morgan, 1998).

Properly presented scientific analysis can be *practical*. It can be focused and efficient. It can anticipate and respond to the concerns and priorities of government and of peer reviewers. It can provide succinct, clearly defined, and substantiated analyses. Independent scientific analysis is open to reinterpretation and is potentially credible to all parties (Lee et al., 1992). Scientific priorities do not always coincide with decision-making needs. Scientists may hesitate to be prescriptive or fail to appreciate that scientific concerns must be combined with other decision-making considerations. A preoccupation with precise predictions and verification can result in insufficient attention to both basic choices [e.g., whether a proposal should proceed, which option(s) are preferred, how to prevent and mitigate adverse effects] and to the broader imperative of protecting and enhancing the environment (Morgan, 1998). There could be situations where decisions need to be made but where there are too many uncertainties and confounding variables to predict or measure patterns of cause–effect relationships (Morgan, 1998). It could be unrealistic to expect that adequate time and resources will be made available to meet scientific standards.

A scientific EIA process has the potential to facilitate *democratic* local control. Independent scientists can provide valuable information and analysis. Community groups can be more influential in decision making when scientists are working with or for them. The decision-making effectiveness of community groups is further reinforced if they participate directly in data collection and analysis (Heinman, 1997). An emphasis on the primacy of scientific knowledge and methods, however, can shift power toward scientific and technical specialists and away from the local community. Scientists may not be aware of the extent to which their ideological biases influence their work (Dawkins, 1998; Miller, 1993). Scientists can be constrained (some would argue co-opted) by the political priorities of employers (Porritt, 2000). These political implications and influences can be hidden beneath an apolitical, value-neutral facade (Greer-Wooten, 1997).

Scientific research, and by extension a scientific EIA process, is *collaborative*. The respectful sharing and testing of knowledge, respect for colleagues, and reasoned argumentation are characteristics common to both scientific and democratic practice (Rothman and Sudarshan, 1998). Sound scientific research can also provide an information base for consensus building and conflict resolution initiatives. But participation beyond the scientific community is not inherent to scientific EIA practice. It is, at best, an add-on, crudely attached to a process focused on applied scientific research and methods (Morgan, 1998). As interpreted by scientists, scientific knowledge, tends to be viewed as the only legitimate decision-making basis (Porritt, 2000). The public is either excluded from or is peripheral to the process. Public involvement is further inhibited if scientific language and methods are not explained, if public knowledge, values and interpretations are discounted, and if scientists are arrogant in their actions and demeanor (Power and Adams, 1997; Rowson, 1997).

Procedural fairness can be facilitated in a scientific EIA process to the extent that the process is unbiased, explicit, and transparent. A scientific EIA process tends to clearly define *ethical* responsibilities and to substantiate assumptions and interpretations. It can also provide the analyses necessary to support actions to offset distributional inequities. Scientific EIA approaches do not tend to address directly fairness, equity, or social and environmental justice issues or perspectives. The value-free posture of analytical science runs counter to establishing an ethical foundation for the process or to explicitly addressing ethical issues, trade-offs, and dilemmas (Porritt, 2000). Contextual biases are unlikely to be revealed, and communications regarding ethical questions are likely to be distorted (Lemons and Brown, 1990).

Scientific practice is often creative and *adaptive*. Hypotheses, models, and theories are formulated, tested, and refined in open, iterative, interactive, and synergistic forums. Methods are applied, refined, and adapted. Adaptations respond to constructive advice from peer reviewers and from the scientific community. Adaptive management and applied research approaches, built around chaos and complexity theories are highly sensitive to contextual change and to emerging system-level properties. Analytical science has a tendency to be more rigid. Once an experimental design is formulated, it tends to be assumed that it will be applied in a linear fashion. Limited provision is made for creative alternations or for adaptations to respond to changing circumstances. It is further assumed that problems are understandable, testable, and manageable (Barrow, 1998). Certainty tends to be overestimated. Limitations and uncertainties tend to be underestimated. These expectations can be especially problematic when dealing with "wicked" problems (more than one correct problem formulation and groups of variables unique in time and space) and "messes" (complex and dynamic situations characterized by changing and interdependent problems) (Patterson and Williams, 1998). The limitations and complexities associated with EIA practice are probably even greater than those associated with applied scientific research. Hence there is an even greater need for flexibility and for proactive efforts to address the implications of uncertainty and complexity.

A scientific EIA process can facilitate *integration*. Bridges to the scientific community are built through literature reviews, peer review procedures, and the circulation and publication of findings and methodological adaptations. Links are established and maintained between the scientists on EIA study teams and scientific and technical reviewers within government agencies. Data from EIA analyses and from monitoring and follow-up procedures can extend from and feedback to broader environmental monitoring and management initiatives. Scientific EIA is conducive to broadening and extending the EIA knowledge base. Scientific EIA procedures can build on scientific theories and research that cut across disciplines. They can help forge links between EIA and other forms of environmental management. However, the integrative role of scientific EIA processes can be undermined by a tendency to discount or even reject nonscientific forms of knowledge and perspectives (Lemons and Brown, 1990). By maintaining the illusion of objectivity, scientific EIA processes tend to be especially weak in addressing value questions. Also, most scientific research tends to remain confined within disciplinary boundaries (Whitney, 1986). Scientific EIA processes tend not to generate the interdisciplinary and transdisciplinary perspectives and frameworks so essential to effective EIA practice.

3.7 SUMMING UP

In this chapter we test the premise that EIA processes, documents, and methods should be more scientifically rigorous. We present two practice-based stories where the role of science in the EIA process is an issue, and provide the conceptual underpinning for a scientific EIA process. We describe a scientific EIA process at it might be applied at the regulatory and applied levels, and then assesses the scientific EIA process against ideal EIA process characteristics. Table 3.6 is a checklist for formulating, applying, and assessing a scientific EIA process.

The two practice-based stories demonstrate that scientific knowledge and standards can be addressed directly through EIA requirements or indirectly by means of administrative discretion. The stories also illustrate that independent advice, consistent with a scientific perspective, can be a valuable addition to an EIA process. The effort to formulate a scientific EIA process begins with an overview of the major criticisms advanced by critics. The thrust of these arguments is that EIA should be treated as a form of applied research, consistent with prevailing standards and

Table 3.6   Checklist: Scientific EIA Process

The response to each question can be yes, partially, no, or uncertain. If yes, the adequacy of the process taken should be considered. If partially, the adequacy of the process and the implications of the areas not covered should be considered. If no, the implications of not addressing the issue raised by the question should be considered. If uncertain, the reasons why it is uncertain and any associated implications should be considered.

LESSONS FROM THE SCIENCE WARS DEBATES

1. Does the process achieve a reasonable balance between contextual adaptations and contributing to the EIA knowledge base?
2. Are the degrees, standards, and forms of rigor appropriate and reasonable?
3. Does the process minimize bias and make explicit and substantiate subjective judgments?
4. Does the process advance environmental values while guarding against negative tendencies that might undermine such efforts?
5. Does the process advance democratic values while guarding against negative tendencies that might undermine such efforts?
6. Does the EIA process establish a reasonable espoused standard and at the same time build on good applied research and EIA practice standards?
7. Is the process reasonable and realistic in its efforts to predict and control environmental change and in its efforts to anticipate and manage risks and uncertainties?
8. Does the process effectively undertake, balance, and blend analytical and integrative EIA activities?
9. Does the process effectively undertake, balance, and blend explanatory and prescriptive EIA activities?
10. Does the process facilitate integration while recognizing the limits to integration, the value of a plurality of theories, and the need for an open but critical perspective?

MANAGEMENT AT THE REGULATORY LEVEL

1. Do the regulatory requirements support such principles as:
   a. Interdisciplinary analysis?
   b. Assurance of scientific integrity?
   c. Description, application, and review of credible scientific methods?
   d. Use of objective and unbiased language?
   e. Use of explicit, substantiated, and conservative assumptions?
   f. Use of scientific criteria and methods for determining uncertainty and error?
   g. Use and explanation of scientific notations and scientific references?
2. Does the regulatory level:
   a. Sponsor and coordinate research to address EIA knowledge and methodological gaps?
   b. Help build and test a core body of good EIA practice?
   c. Extend science-based approaches beyond pre-approvals and link EIA to related environmental management instruments?
   d. Ensure that qualified social and natural scientists are involved in EIA document preparation?
   e. Ensure that sound and independent scientific advice is available as needed for screening, scoping, report review, and monitoring?
   f. Provide science-related methodological guidance for specific EIA activities, types of environments, and types of effects?

Table 3.6 (*Continued*)

g. Provide scientific research guidelines and limits?
h. Undertake initiatives to better integrate science and EIA practice?
i. Acknowledge the limits of science when applied to EIA practice?

MANAGEMENT AT THE APPLIED LEVEL

1. Does the process incorporate such startup activities as:
   a. Problem, need, and context determination?
   b. Objective identification?
   c. Constraint and assumption determination?
   d. Proposal characteristics?
   e. Methods identification?
   f. Hypothesis determination?
   g. Experimental design determination?

2. Do the analysis-related EIA activities:
   a. Use multiple and reliable environmental and social criteria and indicators?
   b. Select levels of detail appropriate to effects detection and prediction?
   c. Focus on sensitive and significant environmental components, functions, and processes?
   d. Collect data over a range of temporal and spatial scales (suitable for aggregation and disaggregation)?
   e. Address trends and patterns over time and space?
   f. Establish a context for interpretations?
   g. Collect data in a form suitable for subsequent testing?
   h. Refine hypotheses using baseline and comparable situation data to deduce predictions and implications?
   i. Facilitate predictive precision and causal hypotheses?
   j. Control confounding variables?

3. Do the synthesis-related EIA activities:
   a. Describe, justify, and apply models and theories to address interrelationship among environmental components and impacts and to test alternative hypotheses?
   b. Focus on critical environmental components, impacts, and interactions?
   c. Systematically address interconnections among individual models and theories?
   d. Adapt and refine models and theories to contextual characteristics?
   e. Incorporate qualitative analyses and alternative perspectives and knowledge where appropriate?

4. Do the hypothesis-testing EIA activities:
   a. Rigorously test alternative hypotheses both prior to and subsequent to approvals?
   b. Apply hypotheses testing to both predicted impacts and to the options and mitigation measure analysis?
   c. Explicitly identify gaps, limits, and uncertainties and their implications?

5. Do the interpretation EIA activities:
   a. Make appropriate use of statistical significance tests?
   b. Systematically explain and substantiate the basis for all interpretations?

6. Do the approval and post-approval EIA activities:
   a. Provide a sound scientific basis for decision making?
   b. Use a scientific experimental design to guide and structure monitoring and follow-up activities?

(*Continued*)

Table 3.6 (*Continued*)

---

   c. Share monitoring results with government review and stakeholder scientists and through peer review?

   d. Circulate and publish, where practical, findings and methodological insights, consistent with good science practice?

   e. Contribute to, where practical, enhanced EIA practice, an enhanced EIA knowledge base, and an enhanced environment?

7. Do the EIA input, output, and interaction EIA activities:

   a. Extend from natural science, social science, and interdisciplinary knowledge?

   b. Draw upon the experiences and insights of applied research scientists and EIA practitioners?

   c. Systematically use comparative, control, and pilot studies?

   d. Address knowledge gaps through targeted research?

   e. Employ scientific tools such as surveys, field investigations, modeling, computer simulations, and statistical analysis?

   f. Involve a team of natural and social scientists?

   g. Employ peer review to check the correctness of procedures and the plausibility of findings and conclusions?

   h. Provide opportunities for interactions among scientists?

   i. Provide links to the broader scientific community?

   j. Provide for agency, political, and public participation opportunities prior to decisions, especially those related to defining the process, to making interpretations, and to reaching conclusions and recommendations?

   k. Provide timely, succinct, explicit, and traceable documentation?

   l. Fully and fairly represent assumptions, methods, and findings?

   m. Subject documentation to independent evaluation?

   n. Circulate and publish, where practical, research findings and methodological innovations?

8. Does the process thoughtfully and critically:

   a. Draw upon holistic scientific approaches when creativity, flexibility, and trans-scientific perspectives and frameworks are needed?

   b. Draw upon management scientific approaches when it is necessary to place EIA-related science more firmly into the context of decision-making priorities, requirements, and limits?

   c. Draw upon complexity scientific approaches when uncertainty and the need for adaptability are especially important?

   d. Draw upon civic science when stakeholders should assume a more central role in the process?

   e. Employ a multiparadigm scientific approach where warranted?

PROCESS EFFECTIVENESS

1. Is the process rigorous without masking biases?

2. Is the process comprehensive, appreciating the matters that scientific EIA processes cannot address or address adequately?

3. Is the process systematic, recognizing that scientific EIA processes tend to treat some EIA activities more systematically than others?

4. Is the process environmentally sound, appreciating the negative tendencies that can reduce the environmental effectiveness of scientific EIA processes?

Table 3.6 (*Continued*)

5. Is the process practical, appreciating that some scientific process tendencies can be unrealistic?

6. Does the process facilitate local influence, recognizing that some scientific process tendencies can inhibit local influence?

7. Does the process facilitate stakeholder collaboration, recognizing that some scientific process tendencies can inhibit stakeholder involvement?

8. Is the process fair, recognizing that fairness does not tend to be a priority with scientific EIA processes?

9. Is the process flexible, recognizing that sometimes, especially analytical scientific processes can be rigid and can underestimate complexity and its implications?

10. Does the EIA process facilitate integration, recognizing that scientific EIA processes can inhibit integration with nonscientific knowledge and perspectives?

protocols of analytical science. This view is not shared uniformly. Some suggest that scientific standards are inappropriate. Others argue that the standards should be tempered.

Most commentators use analytical science as the touchstone in advancing their arguments. Consequently, analytical science is treated as the departure point. Key analytical science terms are defined. Examples of characteristics commonly ascribed to analytical science are listed. In attempting to establish a foundation for a scientific EIA process, it quickly became evident that the discussion surrounding the role of science in EIA is part of a much larger, protracted, and often heated series of debates. A highly selective and simplified version of these debates is presented. Ten sets of opposing positions are presented, together with middle-ground positions. The debates concern such matters as whether science (and by extension a scientific EIA process) should strive for absolute truth, whether it should be rigorous or relevant, whether it is objective or subjective, whether it is beneficial or detrimental to the environment, and so on. EIA process management implications are identified for each debate. The arguments are all potentially instructive for EIA practitioners. In most cases EIA process management will probably occupy a middle-ground position, but one tempered by a need to move closer to one position or the other, depending on the EIA activities involved and on contextual characteristics.

Regulatory scientific EIA process management to this point has consisted largely of identifying general science-related principles for application in EIA practice. Examples of these principles are cited. Other selectively applied science-related provisions and initiatives are noted. Four examples of studies initiated to better integrate science and EIA or environmental management practice are highlighted briefly. A detailed depiction is presented of a scientific EIA process. The EIA process is treated as an experiment, consistent with analytical science. The focal point of the startup activities is an initial set of hypotheses and an experimental design. The initial hypotheses are preliminary alternative explanations suitable for testing.

The experimental design is a research program for testing the hypotheses. The latter integrates such matters as problem definition, objectives, context, need, and methods.

The baseline analysis involves selecting and applying multiple and reliable environment criteria and indicators. The indicators preferably can be aggregated or disaggregated. They provide a dynamic picture of trends and patterns. The analysis focuses on sensitive and significant environmental components, functions, and processes. Predictions are deduced from the hypotheses. The predictions pertain to effects from the proposed action and from options and mitigation measures. Inter-relationships are addressed through conceptual and quantitative models and theories. The models and theories trace patterns of causal connections. A second round of model and theory building is usually necessary to address interconnections among models and theories. Hypotheses are tested prior to approvals using comparative, control, case, and pilot studies and after approvals using monitoring and follow-up studies. Gaps, limits, and uncertainties are explicitly identified together with implications. Interpretations, preferably supported by statistical analyses, are explicitly identified and substantiated. The scientific findings and interpretations are assumed to provide a sound decision-making basis. Post-approval activities involve both further hypothesis testing and contributions to the scientific and EIA knowledge base.

The EIA process extends from existing natural and social science knowledge. Ample use is made of scientific methods. Independent, skilled, and qualified scientists drive the process. The scientists strive to ensure a rigorous, open, and objective process, consistent with scientific principles and protocols. Peer reviewers assess the findings and procedures. Scientists interact and maintain contact with the broader scientific community. Stakeholders are involved prior to decisions and in reviewing documentary outputs. Documentation is consistent with good scientific practice. Research findings and methodological innovations are circulated and published wherever practical. The process is adapted, as needed, by drawing upon alternative scientific paradigms: holistic, management, complexity, and civic scientific approaches, for example.

The scientific EIA process is assessed against ideal EIA process characteristics. Not surprisingly, it is most effective in injecting a greater degree of rigor into EIA practice. It is weakest in facilitating fairness and in contributing to greater local democratic control and to more collaborative planning and decision making. It offers a mix of positive and negative tendencies for making EIA practice more comprehensive, systematic, substantive, practical, flexible, and integrative. In sum, a scientific EIA process has much too offer but is prone to several negative tendencies.

CHAPTER 4

HOW TO MAKE EIAs MORE RATIONAL

4.1 HIGHLIGHTS

In this chapter we address the question of how EIA processes, documents, and methods can become more rational. The suggestions to make EIA more rational are informed and tempered by the debate surrounding rationality both within and external to EIA.

- The analysis begins in Section 4.2 with two anecdotes. The stories describe applied experiences associated with efforts to make EIA practice more rational.
- The analysis in Section 4.3 then defines the problem, which is either (depending on one's perspective) an EIA process that is insufficiently rational, consistent, and systematic or an EIA process that is too technical, rational, and autocratic. The possibility is raised that both positions are overstated and that there might be some fertile middle ground.
- In Section 4.4 we provide a context for the rational EIA process. We define rationality and identify various rationality forms and describe the major characteristics of a typical rational planning process. Attributed strengths and limitations are highlighted. Various adaptations to and alternatives to the typical process are described briefly. How the debate has played out in EIA literature is summarized.

Environmental Impact Assessment: Practical Solutions to Recurrent Problems, By David P. Lawrence
ISBN 0-471-45722-1   Copyright © 2003 John Wiley & Sons, Inc.

- In Section 4.5 we build on Section 4.4. We detail how a rational EIA process could be implemented at the regulatory and applied levels. In Section 4.5.1 we address how rational EIA processes can be facilitated and structured at the regulatory level, and in Section 4.5.2 we demonstrate how a rational EIA process might be expressed at the applied level.

- In Section 4.6 we assess how well the rational EIA process presented in Section 4.5 satisfies ideal EIA process characteristics.

- In Section 4.7 we highlight the major insights and lessons derived from the analysis. A summary checklist is provided.

4.2 INSIGHTS FROM PRACTICE

4.2.1 Taking Rationality Off the Table

An EIA was prepared for the enlargement of a private hazardous waste landfill near Sarnia, Ontario, Canada. The EIA documents addressed options and impacts systematically and rationally. An EIA proposal described the purpose of the process and proposal, the EIA process, proposed study areas and impact zones, the proposal opportunity, proposed screening, comparative and site assessment criteria, proposed impact management policies and measures, links to related decisions, proposed public and agency consultation procedures, and identified issues and concerns.

A subsequent document explicitly, systematically, and consistently generated, screened, and compared alternatives to the proposal, alternative locations, alternative waste management methods, and engineering alternatives. Procedures for both deriving and evaluating the alternatives were presented. The alternatives represented all the reasonable choices that could meet the opportunity and were available to the proponent. The alternatives analyses provided detailed descriptions of the evaluation process, the rationale for and nature of the alternatives, and the criteria and criteria application procedures. Comparative criteria and indicators were described and ranked. Both qualitative and quantitative evaluation procedures were used. The analysis took into account mitigation potential. Each conclusion was fully substantiated. Responses were provided to each public and agency comment and suggestion.

A two-volume site assessment document included a scoping analysis, a baseline analysis, a facility characteristics description, an analysis of individual and cumulative impacts, an impact management strategy, and a detailing of public and agency consultation activities. The scoping analysis used a series of network diagrams to identify all potentially relevant impacts. The impact analyses involved the consistent and explicit application of a scaling system for characterizing the magnitude and importance of impacts, with and without mitigation. The analysis took into account changes from baseline and from current operations by project stage and for varying operating conditions. It addressed the frequency, spatial extent and duration of impacts, uncertainty of predictions, public and agency concerns, compliance with government requirements policies, criteria and guidelines, significance and sensitivity of receptors, and interconnections with other impacts. A clear rationale was provided for each study area, time horizon, method, criterion, and indicator employed. Conclusions were fully substantiated. A separate design and operations report was also produced.

The methods for conducting the alternatives and site assessment analyses did not become issues, largely because of the systematic and explicit nature of the analyses. Government reviewers were generally satisfied with the study methods and analyses. They focused instead on the prediction and management of a small number of impacts, largely related to the potential for groundwater contamination. Public concerns generally revolved around impact management and public consultation procedures. Both agency and public concerns were resolved to the satisfaction of most parties. The project was approved without the need for a public hearing. It is unlikely that the rational analysis of options and impacts, by itself, would have resulted in project approval. However, it did make it possible to narrow and focus the agenda for action and consultation to a manageable and resolvable set of issues. It also contributed to an acknowledgment by most parties that the EIA analysis and documentation were completed in a comprehensive and professional manner, which, in turn, contributing to the credibility of both the proponent and the consultant team.

DAVID P. LAWRENCE
Lawrence Environmental

4.2.2 Rationality Out of Context

A special authority was established to locate suitable landfill sites in three regions surrounding Metropolitan Toronto. This massive exercise involved three parallel but methodologically consistent processes. The technical siting process used involved two screening (using constraint mapping) steps, a site boundary identification step, and three progressively more detailed comparative evaluation steps. The comparative evaluation steps employed quantitative evaluation procedures, which relied heavily on expert judgments, as represented by impact ratings and criteria rankings and weightings as produced largely by a team of consultants. Criteria were grouped by discipline. The role of the public was largely to react to the analyses by the consultants and to make suggestions regarding criteria group and criteria rankings and weightings. The process generated a storm of controversy and eventually was abandoned.

There were numerous problems that contributed to the demise of the exercise. One of the most fundamental was a basic disconnect between the rational procedures employed by the proponent and consultant team and the reality or context as perceived and as experienced by the local residents. The consultants focused on the systematic application of what they considered to be rational evaluation procedures, supported by extensive analyses of secondary sources and field investigations, all within the parameters established by the terms of reference defined by the proponent. The local residents questioned the legitimacy of the assumptions built into the terms of reference (e.g., study boundaries, treatment of alternatives to land-fill, one landfill per region, public-sector facilities only, design assumptions). The analyses also were at odds with their knowledge of their communities. They recognized that once clearly unsuitable areas are excluded, siting a landfill is largely a case of finding suitable hydrogeological conditions and/or finding sites where impacts on local residents, farms, businesses, and natural environmental features within the site, around the site, and along access routes to the site are minimized. They were aware that large areas within or near industrial areas or highways, which would minimize community impacts, were not represented among the short-listed sites. The sites that were identified seemed minimally different from adjoining properties. All appeared to have significant site vicinity and access route impacts. The disciplinary distinctions seemed artificial and the abstract ranking and weighting of criteria groupings and criteria came across as a meaningless exercise. They received no comfort from the consultants' individual and collective impact ratings. The entire exercise seemed contrived, autocratic, and from their perspectives, irrational.

Although no guarantee of success, it is possible that the local residents would have felt more comfortable had a more serious effort been made to evaluate need and front-end options. Perhaps the siting process could have begun with an overview of regional conditions to identify and then compare siting opportunities (e.g., highly suitable areas, voluntary communities and areas). Such an approach would have given the public a clearer sense that the process was tailored to suit the conditions of the region rather than representing the force-fit application of a standardized method. Perhaps criteria could have been grouped by study area or receptors rather than by discipline. In this way the pattern of interrelated impacts on people and on other receptors would have been more immediately evident. Possibly using systematic qualitative procedures would have produced a more tangible representation of actual impacts and trade-offs. It could have helped if social equity, community, and natural system-level impacts, mitigation potential, interrelationships among impacts and uncertainties had been considered more systematically and explicitly. And most important, if local groups and residents had been true partners in the process, perhaps there would have been a genuine opportunity for a convergence of perceptions, experiences, and definitions of what is rational.

This story does not describe an EIA process where the public was irrational, as is too frequently assumed. It does demonstrate that perceptions of what is rational will vary and that to be truly rational, any methods must be adapted to and appropriate for the context. They must also effectively integrate the knowledge and perspectives of interested and affected parties.

DAVID P. LAWRENCE
Lawrence Environmental

4.3   DEFINING THE PROBLEM AND DECIDING ON A DIRECTION

The two stories presented in Section 4.2 illustrate that rationality can be either a positive or a negative force in an EIA process, depending on how it is applied, the rationality perspectives that it represents, and the fit between process and context. If EIA processes are rational, they clearly define the problem. They explicitly identify goals and objectives. Alternatives are identified and evaluated systematically against goals and objectives. The planning process is integrated, from early on, into organizational planning and decision making. The preferred alternative is implemented. The achievement of goals and unintended consequences is monitored and managed. Up-to-date and appropriate technical methods are integrated into the process systematically. Formal checks of document quality are instituted.

EIA effectiveness analyses provide the first clues as to whether EIA processes are rational. The environmental aspirations expressed in EIA legislative goals are not being realized to the extent hoped for (Clark, 1997). Effectiveness ratings for problem definitions, objectives formulation and preparing adequate terms of reference are low (Sadler, 1996; Spooner, 1998). The identification and evaluation of alternatives is a recurrent weakness in EIA guidelines and documents (Barker and Wood, 1999; ERM, 2000; Sadler, 1996; Spooner, 1998). Postapproval monitoring and management receive low performance ratings (Sadler, 1996; Spooner, 1998). Agencies integrate EIA too late in their internal planning processes and to an insufficient extent (Andrews, 1997; Clark, 1997; US CEQ, 1997a). Methods selection and application could be greatly improved, as could document quality (ERM, 2000; Sadler, 1996).

EIA processes, documents, and methods in practice fall well short of rationality ideals. Most EIA texts and much of EIA literature seeks to correct this deficiency (Canter, 1996; Gilpin, 1995; Morgan, 1998; Morris and Thérivel, 1995; Smith, 1993; Westman, 1985). EIA requirements, they suggest, should include objectives. Proposals and alternatives, they argue, should be systematically evaluated against objectives. Specific measures are described to facilitate more effective procedural guidance and control, especially regarding the treatment of alternatives (Ortolano, 1993). Explanations are provided concerning how environmental considerations can be integrated into project management and into organizational decision making. A diversity of largely technical and scientific methods are identified, described and compared. The methods apply to different EIA activities, impact types and environmental components. They also address interdisciplinary analysis and study team coordination. Quantitative and computer-based systems often receive particular attention (Julien, 1995). The net result is assumed to be a more rational, scientifically and technically sound, comprehensive, and objective decision-making basis (Culhane et al., 1987). Rational decisions and achievement of the environmental objectives of EIA requirements are assumed to flow naturally from the greater and more effective application of the recommended procedures and methods.

Some critics, in contrast, maintain that EIA practice either is or strives to be too rational. Rational processes are seen as politically naive in their failure to appreciate how decisions are made within organizations and how power is wielded. They are considered impractical because of their inability to recognize cognitive and decision-making limitations and to operate effectively across disciplines and professions. They are considered inflexible and poorly equipped to deal with uncertainty and conflict (Boothroyd and Rees, 1985). They are viewed as oblivious to the implications of operating in a multiactor, multi-interest sociopolitical environment (Greer-Wooten, 1997). They are described as not appreciating the ubiquitous role of values and interests. They are condemned for excluding extrarational forms of knowledge. They are labeled as technocratic and antidemocratic on the grounds that they marginalize the public's role in the decision-making process. These critics argue that the rational EIA process should be abandoned and replaced with more political, interpretive, intuitive, contextual, dynamic, collaborative, transdisciplinary, and value-full planning and decision-making models and procedures (Boothroyd and Rees, 1985; Craig, 1990; Greer-Wooten, 1997; Torgerson, 1981).

A third group of commentators focus on how rationality is defined and how it is applied. They suggest that ecological, practical, and communicative rationality forms are more directly relevant to EIA practice than are analytical rationality forms (Bartlett, 1997). They maintain that rationality tenets and assumptions should be relaxed and replaced with a more practical, political, social, flexible, and collaborative "reasoning" process (Barrow, 1997; Greer-Wooten, 1997; Kørnøv and Thissen, 2000). They argue for placing rationality in a broader context (i.e., vary rationality standards to match contextual conditions, constraints, and uncertainties) and for adjusting the level of detail to suit the scale and type of proposal (Dalal-Clayton and Sadler, 1998). They recommend selectively applying rationality. Rationality is considered especially appropriate when there is a high degree of certainty and control and a low degree of conflict (Kørnøv, 1998). They favor modifying and adapting rationality to accommodate knowledge and insights derived from experience, intuition, emotion, and imagination. They point to the need for decision making to combine the objective and analytical with the subjective and integrative (Kørnøv and Thissen, 2000). These commentators see value in rationality for EIA but only a form of rationality that is carefully defined, tightly circumscribed, selectively applied, and integrated both with other forms of knowledge and into pragmatic, sociopolitical planning, and decision-making processes (Culhane et al., 1987; Partidário, 1996).

As is evident from the above, there are at least three definitions of both the problem and the preferred direction. Applying valid rationality strengths, minimizing legitimate rationality deficiencies, and drawing on alternative rationality definitions and applications can help integrate these problem definitions.

4.4   SELECTING THE MOST APPROPRIATE ROUTE

4.4.1   Definitions and Distinctions

Rationality has been a central theme of Western thinking since the Renaissance (Alexander, 1986). Definitions of rationality encompass many elements, as illustrated in Figure 4.1. Ideally, a rational EIA process displays such *attributes* as purposeful, sensible, orderly, lucid, logical, coherent, transparent, explicit, replicable, consistent, reflective, reasoned, verifiable, and objective. It strives for accuracy, reliability, and accountability. It is technically sound and scientifically rigorous

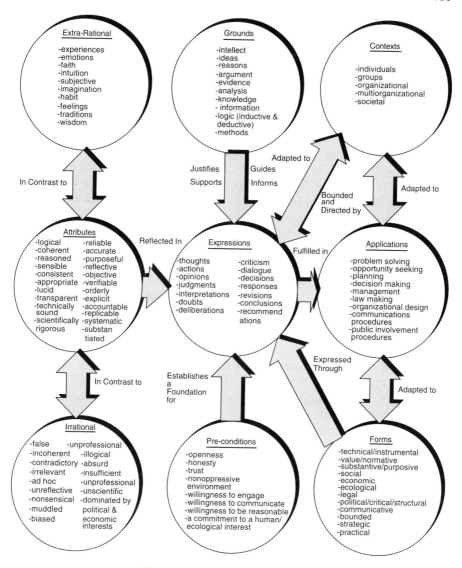

Figure 4.1 Rationality definition.

(see Chapter 3). It seeks to minimize such *irrational* factors as errors, falsehoods, contradictions, and incoherence. It tries to avoid nonsensical, irrelevant, ad hoc, illogical, unreflective, and inadequately supported arguments. Emotions, feelings, experiences, intuition, imagination, wisdom, habits, traditions, faith, and subjectivity are excluded because they are nonrational or *extrarational*.

Rational attributes are *expressed* in thoughts, actions, opinions, judgments, interpretations, criticisms, reflections, decisions, conclusions, and recommendations. They are evident in deliberations and dialogue. Rational expressions are *supported* by reasoned intellectual analysis, logic, and argumentation. They are *informed* and *guided* by ideas, information, knowledge, and evidence. They are *justified* by the sound and systematic application of technical and scientific methods, as conducted by qualified specialists. The expression of rationality in an EIA process occurs more fully when such *preconditions* as openness, honesty, and trust prevail. Also essential is a nonoppressive environment where interested and affected parties are willing to participate in the process, engage in reasoned dialogue, and commit to a broader human or ecological purpose.

Rational expressions are *fulfilled* in problem solving, in opportunity seeking, in planning, and in decision making. They also are *applied* in management, in law-making, in organizational design, and in communications and public involvement procedures. The expression and application of rationality is dependent on *context*. It varies for individuals, groups, and organizations. It operates differently at the multiorganizational and societal levels.

As indicated in Table 4.1, there are many rational *forms* relevant to EIA practice. Purpose, need, and objectives are often defined before the EIA commences, consistent with instrumental or technical rationality. EIA requirements and practices are structured around a process, in accordance with procedural rationality. Sometimes, EIA legislation and SEA practice identify and explore purpose and direction questions, as is the case with purposive (or value or normative) rationality. Occasionally,

Table 4.1   Potentially Relevant Forms of Rationality

Forms	Key Characteristics
Instrumental/ technical rationality	Search for the best possible means (how to do things) for given ends (what could be achieved)
	Stresses efficient, logical, and systematic goals achievement
	Based on causal explanations
	Emphasis on efficiency, measurement, and analysis
	Goals and objectives determined externally
Procedural rationality	Rationality of the process
	Procedures used to choose actions
	Acceptance or rejection of a claim base on procedures or rules followed
Purposive/value/ normative rationality	Rationality of ends
	Based on moral judgments
	Synthetic
	Evaluation and choice among goals
Functional rationality	Rationality inherent in the functioning of systems, societies, or organizations
	Clearly defined and calculable goals
Substantive rationality	Rationality of ends and means
	Rationality of the outcome of the process
	Applies to individual decisions or actions
Analytical rationality	Understanding by breaking things into parts and by studying differences and links
	Additive (sum of parts)

Table 4.1 (*Continued*)

Forms	Key Characteristics
Systems rationality	Understanding in terms of purpose and relevance
	Order flows from sense of whole
Social rationality	Seeks integration in social relations and social systems
	Makes social action possible and meaningful
	Assumes social formation prior to individual; identity from group; reason exercised for group
Economic rationality	Utilitarian
	Entails the maximum achievement of a plurality of goals
	Underlain by a principle of efficiency
	Assumes orderly measurement and aggregation
Political/critical/ structural rationality	Rationality of decision-making structures
	Preserves and improves decision structures
	Emphasis on practical capability for facing societal problems
	Requires an open, honest, informed debate
	Concerned with identifying and redressing structural inequities
Legal rationality	Reason inherent in clear, consistent, and detailed formal rules for preventing disputes and for providing solutions
Market rationality	Unconstrained pursuit of self-interest by individuals and organizations
Ecological rationality	Rationality of living systems
	Order of relationships among living systems and their environment
Communicative rationality	Organized dialogue to promote democracy and personal growth
	Concern with the quality of the communications
	Stress on mutual understanding and counteracting of communications barriers and distortions
Bounded/limited rationality	Search for satisfactory solution (good enough)
	Not all alternatives known or consequences considered; alternatives considered sequentially
	Not all preferences evoked
	Contingent on environmental conditions
Practical rationality	Starts with real, everyday life
	Pragmatic
Strategic rationality	Selective and contingent
	Adapted to local context and specific situation

Sources: Alexander (1986, 2000), Bartlett (1990), Braybrooke and Lindblom (1963), Etzioni (1967), Forester (1999), Friedmann (1987), Habermas (1993), Healey (1997), Kørnøv (1998), Sager (1994), Verma (1998).

EIA, in common with substantive rationality, considers both alternative ends and means. EIA both integrates individual effects (analytical rationality) and considers systems levels concerns and impacts (systems rationality). EIA addresses and adopts social (social rationality), economic (economic rationality), political (political rationality), and ecological (ecological rationality) perspectives. EIA is generally an action-forcing legal environmental management instrument (legal rationality). EIA considers market implications (market rationality). It involves a highly participative process (communicative rationality). EIA is constrained, focused, and pragmatic; consistent with bounded, practical, and strategic rationality forms, respectively.

4.4.2   Core Characteristics

The rational planning and decision-making process began (at least in the postwar period) with the suggestion that decision makers agree on goals, identify available alternatives for achieving goals, evaluate the consequences of alternatives, and select the alternative that comes closest to achieving the goals (Banfield, 1955; Simon, 1976). Implementation of the preferred alternative was assumed. This process was refined through the 1960s and 1970s, as illustrated in Figure 4.2. The revised process begins with a problem, need, or opportunity to be addressed. An appropriate constellation of values (the public interest) was determined (Davidoff and Reiner, 1962). General values were distilled into goals, principles, objectives, and criteria—progressively more precise measures of progress toward the public interest. Goals, objectives, and criteria were ranked (Boyce, 1971). Methods for assembling and analyzing data, operating within resources, responding to pertinent constraints and opportunities, determining present and predicting future conditions and deriving, screening, and comparing alternatives were formulated (Alexander, 1986; Friedmann, 1987). Uncertainties and variations in preferences were taken into account. Subsequently, more details were added concerning how the preferred alternative would be implemented. Allowance was made for interactions among process steps (scanning forward and feedback loops) (McLouglin, 1969). Provision also was made for public and agency involvement, often prior to decision making and occasionally as inputs to each stage in the process.

Numerous assumptions have been ascribed to the rational process, as detailed in Table 4.2. It tends, for example, to be assumed that problems are well defined, that the environment and available choices are predictable and controllable, that a unitary public interest can be defined and that decision makers are rational. It is expected that they will select and implement preferred alternatives based on the comprehensive and objective analyses of technical and scientific specialists. The rational process is seen as systematic, largely sequential, and optimizing (i.e., all alternatives considered and best alternative selected). These assumptions should be approached with caution. Part of the procedure for formulating a rational EIA process involves determining which ascribed assumptions are intrinsic to the rational process and which either apply only to specific rationality forms or could be relaxed, adjusted, or abandoned.

4.4.3   Attributed Strengths and Limitations

The rational process has been described as simple, explicit, logical, consistent, systematic, and adaptable (Caldwell, 1991; Sager, 1994). It helps to clarify future

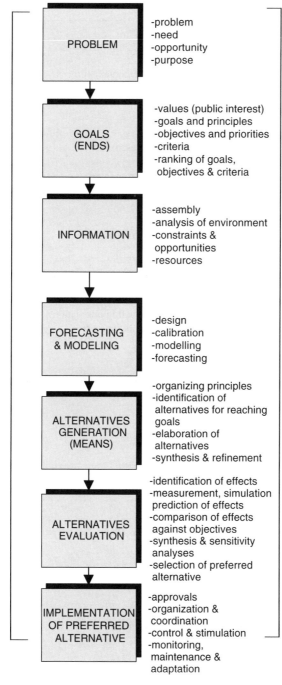

Figure 4.2   Rational process.

Table 4.2   Ascribed Rational Process Assumptions

General

Reason is applied systematically (central to problem)
The process is independent of the problem and of the context (process can be universally applied)
Adequate time, skill, and resources
Comprehensiveness
All actors in process rational
EIA specialist as technician (objective, apolitical, unbiased)
Sequential, analytical process
Pluralistic society (all competing interests have access to power)
Collective rationality is the aggregation of individual rationality (utilitarian)
Downward direction of control

Goals (Problems, Goals, Objectives)

Well-defined problem (susceptible to analysis and diagnosis)
People have preferences and act in accordance with them
Goals and objectives can be identified and articulated
There is a unitary public interest (value, goal, and objective consensus is possible)
Goals and objectives guide process (also basis for evaluating alternatives)
Specialists are value-neutral and can determine the public interest

Information

Supremacy of technical and scientific knowledge (as determined by independent specialists)
Complete and fully accessible baseline information
Manageable uncertainties

Forecasting and Modeling

Well-defined action space (all relevant variables)
Probability of occurrence can be predicted based on available data
Predictable and controllable environment
Stable society

Alternatives Generation (Alternatives, Plans, Strategies)

Well-defined alternatives
All reasonable alternatives available
All alternatives examined
Best alternative can be identified

Evaluation of Alternatives

Well-defined and well-known outcome space
All relevant consequences of each alternative can be determined
Preferences are transitive (goals and objectives can be ranked)
Alternatives can be assessed against goals, objectives, and criteria

Table 4.2 *(Continued)*

Predicted consequences and value preference differences can be amalgamated to select a preferred alternative (using formal evaluation methods)

Uncertainties in predictions and value preferences can be addressed (e.g., using sensitivity analyses)

Decisions based on evaluation of alternatives conducted by specialists

Implementation

Full approval is obtained

Environment is controllable and is controlled

Possible to monitor all key variables and to make appropriate adaptations

Interrelationships

Facts and values can be separated (objective from subjective)

Ends and means can be separated

Independence of probabilities and utilities (what is expected is unaffected by what is wanted)

Independence of analysis and evaluation

Separation of analysis/evaluation from implementation (technical from political)

Sources: Alexander (1986), Banfield (1955), Boyce (1971), Boyer (1983), Damasio (1994), Davidoff and Reiner (1962), Forester (1984), Friedmann (1987), Harper and Stein (1992), Healey (1997), Kørnøv and Thissen (2000), Mintzberg (1994), Sager (1994), Simon (1976), Smith (1993).

directions, establish priorities, and explore potential courses of action (Faludi, 1986). It provides a clear, coherent, comprehensive, unbiased, and defensible basis for decision making (Briassoulis, 1989; Caldwell, 1991; Healey, 1997). It systematically integrates scientific and technical knowledge. The role of specialists in the process is clearly defined and legitimized (Benveniste, 1989). Although the process objectives may not be fully realized, it is still considered beneficial to seek to be unbiased, comprehensive, consistent, and systematic (Briassoulis, 1989; Faludi, 1986).

The rational process has been attacked on a host of fronts. It is labeled as unrealistic, ineffective, incomplete, and inappropriate. It is considered *unrealistic* because human cognitive limits are not adequately considered (Webber, 1983). It fallaciously assumes that problems are well structured; goals and beliefs are clear and unambiguous; adequate, largely quantitative, environmental information is available; existing environmental conditions can be extrapolated into the future; all alternatives are available; all consequences can be determined; and the preferred alternative can and will be implemented (Forester, 1989; Healey, 1997; Mintzberg, 1994; Webber, 1983). It incorrectly assumes that all actors in the process are rational and that adequate resources are available to support a comprehensive analysis (Briassoulis, 1989; Forester, 1989). It fails to recognize the extent to which ends and means, reason and emotion, and analysis, evaluation, and implementation are necessarily interwoven (Damasio, 1994).

The rational process tends to be *ineffective* because it does not consider practical, commonsense considerations and solutions (Saul, 1992). It fails to focus the limited, available resources (Benveniste, 1989). The rational process concentrates on internal analyses, assuming that external environmental conditions are stable and can be controlled (Benveniste, 1989). As a result, it is not well adapted to contextual characteristics and does not respond promptly to changing circumstances (Mintzberg, 1994). It is especially ineffective on the political front. It assumes implementation. Consequently, it fails to consider and to address bureaucratic, political, and structural implementation obstacles and opportunities (Alexander, 1986).

The rational process is *incomplete*. It lacks social and environmental content or substance (Boyer, 1983). It seeks to attain goals (as a generic concept), but it is not driven, guided, and bounded by specific tangible social and environmental ethical principles and imperatives (Beauregard, 1987). It is conducive to systematic analysis but lacks a holistic perspective (Mintzberg, 1994). The image of people as rational decision makers is especially constraining. No provision is made for the contributions of extrarational insights, knowledge, experiences, wisdom, and methods (Alexander, 2000; Friedmann, 1987; Healey, 1997). The implications for process design and management of the subjective, social, and political nature of decision making are not addressed (Saul, 1992; Webber, 1983). The likelihood that perspectives and interests will clash is not considered (Boyer, 1983). The net result is a highly circumscribed and artificial view of people, how they reason, how they interact, and how, collectively, they reach and implement decisions.

The rational process is abstract. No effort is made to fit the process to the context. Consequently, process and context are often poorly matched. The rational process is especially *inappropriate* in situations characterized by high levels of complexity, uncertainty, and conflict (Briassoulis, 1989; Damasio, 1994; Healey, 1997). The rational process and the technical and scientific "experts" who support the process are presumed to be objective, unbiased, and value-free. In truth, both the experts and the process are prone to numerous, often hidden biases (Boyer, 1983; Mintzberg, 1994). "Objective" technical and scientific knowledge and methods are valued over subjective knowledge (Poulton, 1990). Analysis is favored over synthesis. Efficiency takes precedence over effectiveness. The process is more mechanistic than humanistic or ecocentric. Experts are the primary custodians of knowledge. Professional "mystifications" and rationalizations can inhibit public understanding and involvement (Forester, 1989; Saul, 1992). They can also lead to contempt by the specialists for the people (Saul, 1992). Often, "depoliticized," expert-driven processes become autocratic. They can mask political purposes. They tend to reinforce the existing distribution of power (Benveniste, 1989). Sometimes they compound existing or even create new inequities. The rational process, according to many of its critics, has a propensity to be highly undemocratic.

The foregoing ascribed strengths and limitations are not necessarily inherent to the rational process. They could simply be tendencies. Positive tendencies can be reinforced. Negative tendencies can be offset. Still, it is prudent to take these tendencies into account when designing and managing EIA processes with rational elements.

4.4.4 The Response

Several responses to the identified shortcomings seek to make the rational process more realistic and effective. Incrementalism advocates a bounded or limited rational process where satisfactory (rather than ideal) decisions are made in a sequential, informal, and interactive bargaining process in a highly constrained and uncertain environment (Braybrooke and Lindblom, 1963; Lindblom, 1965). A few alternatives are sequentially assessed based on the test of agreement. Mixed scanning envisions a two-tier planning process with incremental problem solving at the operational level and strategic-level policy making to address major changes and issues (Etzioni, 1967, 1986). Effective planning concentrates on building and applying practical political skills to facilitate implementation and to manage uncertainties (Benveniste, 1989). Strategic planning is selective and issue, action, and implementation oriented. It scans external and internal environmental conditions systematically to maximize opportunities and to minimize threats (Mintzberg, 1994). Contingency planning seeks to match procedural characteristics and environmental conditions (Alexander, 1986). Theory-in-action and reflection-in-action explore how practitioners pragmatically design, reflect on, reframe, and implement policies in practice (Schön, 1983; Schön and Rein, 1994). Strategic choice involves a collaborative, highly iterative problem-structuring process that manages uncertainties continuously (Friend and Hickling, 1997).

Other responses make the rational process more substantive and democratic. Advocacy planning, extending from the legal model (legal rationality), focuses on the needs of the poor in a pluralistic society (Davidoff, 1965). Social learning and related organizational development and societal guidance concepts offer more humanistic, organic, interactive, and adaptive planning and organizational models (Friedmann, 1987). Critical planning and related concepts, such as social justice, social mobilization, equity planning, progressive planning, and structural planning, seek to identify and redress social injustices and power inequities (Forester, 1989; Friedmann, 1987; Harper and Stein, 1992; Rawls, 2001). Substantive planning processes attempt to realize and operate within tangible humanistic, ecological, communitarian, and sustainability principles, limits, and imperatives (Beatley, 1995; Etzioni, 1995; Friedmann, 1987). Communicative and collaborative approaches integrate reasoned, ethical, and practical discourse and argumentation into interactive and value-full collaborative forums (Forester, 1999; Goldstein, 1984; Healey, 1997). They also minimize communication distortions; facilitate participation, consensus building, and conflict resolution; and justify moral norms (Habermas, 1993; Innes, 1995; Sager, 1994).

The debate surrounding the rational process has cycled through multiple iterations. No consensus has emerged nor is likely to, given the clash of perspectives and interests. Many of these perspectives are integrated into the rational EIA process presented in this chapter. Approaches that cannot be fully incorporated into a rational EIA process are integrated into other EIA processes presented in subsequent chapters.

4.4.5 EIA and Rationality

Some EIA literature, especially the sources that advocate the wider application of scientific and technical methods, are either oblivious to the debates surrounding rationality or come down firmly in the technical analytical camp (Canter, 1996; Gilpin, 1995; Morris and Thérivel, 1995). They tend to maintain that the process should be comprehensive, scientific, rational, and objective. They generally focus on the appropriate application of technical, often quantitative, methods by specialists (Julien, 1995).

Many EIA process characterizations (as described in Chapter 2) truncate, perhaps not consciously, the rational process by moving directly to criteria application to a proposed action and to reasonable alternatives. This tends to occur because the process is triggered only after a proposal is well defined. Greater attention could be devoted to problem structuring, to formulating goals and objectives (substantive rationality), to formulating alternative goals and objectives (purposive or value rationality), and to procedures for generating alternatives. SEA process depictions give more attention to front-end activities, such as problem definition, goal setting, and alternatives formulation.

EIA has partially avoided some rationality limitations. EIA, in common with social and ecological rationality, is driven and shaped by an environmental and social ethic (Bartlett, 1997). Process and substance are married, increasingly, in an effort to further both sustainability and social/ecological justice (Sadler, 1996). EIA operates within limits (scoping), is focused (reasonable alternatives, significant effects), appreciates the needs for synthesis (cumulative effects), adaptively manage risks and uncertainties and extends beyond decision making (monitoring and auditing) (Barrow, 1997; Glasson et al., 1999; Holling, 1978). EIA requirements and process descriptions demonstrate that it is impractical to identify all alternatives, to select the best alternative, and to assess all consequences (Bartlett, 1997; Culhane et al., 1987; Kørnøv and Thissen, 2000). They appreciate the potential contributions of the extrarational (e.g., traditional knowledge). They also recognize the value-full, social, collaborative, and political nature of the EIA process (Interorganizational Committee, 1994).

Many rationality debates are mirrored in EIA literature. There are lively discussions concerning whether SIA should be technical, political, or collaborative (Bartlett, 1997; Craig, 1990; Greer-Wooten, 1997; Lockie, 2001). Rational/technical and adaptive/ecological approaches are compared and contrasted (Boothroyd and Rees, 1985). There are debates regarding whether reason rather than rationality should guide the process (Torgerson, 1981). The validity of emotions and experiences as a decision-making basis is raised as an issue. There are discussions regarding whether EIA should be comprehensive and rigorous or is necessarily practical and constrained (Kørnøv, 1998). SEA characterizations range from a close parallel to the rational process to processes that share many of the characteristics of strategic planning, the strategic choice method, mixed scanning, and effective planning (Glasson et al., 1999; Kørnøv and Thissen, 2000; Partidário, 1996).

EIA literature and practice could benefit from a closer scrutiny of the rationality debates. EIA practitioners could assess whether the assumptions and limitations ascribed to rational processes apply to current and proposed EIA processes. The systematic integration of different rationality forms into the EIA process could help guard against some of the excesses of technical/analytical rationality. Practitioners and other process participants could reflect on how they individually and collectively apply reason to build theory in practice. The efforts to foster reasoned, practical ethical discourse could be especially appropriate for collaborative EIA processes. Practitioners could consider the role of the extrarational in planning and decision making. They could seek a better match between process and context by appreciating the contingent nature of rationality. SEA practitioners could learn from the experiences of strategic planning and the strategic choice method.

4.5 INSTITUTING A RATIONAL EIA PROCESS

4.5.1 Management at the Regulatory Level

The four jurisdictions (the United States, Canada, the European Union, and Australia) all include a broad purpose or policy basis for EIA requirements. They also require a consideration of alternatives, the environmental consequences of alternatives, and measures to prevent or minimize adverse effects. Each, to varying degrees and probably deliberately, falls short of the rational ideal. Various approaches are taken to bound alternatives. The United States refers to "reasonable," Canada to "technically and economically feasible," the EU to "main," "realistic," and "genuine," and Australia to "feasible" alternatives. Implied by such terms is a distinction between all alternatives and a narrower range of reasonable alternatives. There also is an apparent expectation that some possible alternatives will be screened out as impractical, possibly on nonenvironmental grounds (e.g., not technically or economically feasible). It is realistic to bound the alternatives analysis by screening out unreasonable alternatives. This does not imply that it is sufficient to summarily dismiss some potential alternatives as unrealistic, infeasible, or unreasonable. It could be helpful to require that alternatives be screened based only on clearly defined, consistently applied, and fully justified thresholds of acceptability, taking into account mitigation potential. Guidelines for screening analyses could assist in ensuring an adequate basis for differentiating between reasonable and unreasonable alternatives for different classes of proposal types.

The range of alternatives that must be addressed varies. The U.S. requirements are the broadest and least qualified. All reasonable alternatives must be examined, including the no-action alternative, the agency's preferred alternative, the environmentally preferred alternative, and reasonable alternatives not within the jurisdiction of the lead agency. Reference also is made to, where appropriate, alternative locations, alternative technologies, alternative means of transportation,

environmental release choices, cost-effective waste minimization and pollution prevention activities, economically beneficial landscape practices, obvious alternatives, and alternatives identified by the public. Europe only requires an analysis
of the main alternatives considered. However, guidelines give examples of many
types of alternatives. Under the Canadian requirements, alternative means must
be considered for large or complex projects. A consideration of functionally different ways of meeting the need and achieving the purpose (alternatives to) is recommended for large or complex projects. The Australian legislation only requires the
assessing of feasible alternatives, including the no-action alternative. The types of
alternatives available will naturally vary among proposal and location types. One
approach might be to identify the types of alternatives that must be considered (perhaps extending from the U.S. requirements) and the types of alternatives that should
normally be considered, unless an adequate rationale can be provided for their
exclusion. More precise requirements and recommendations could be included in
proposal type and proposal specific, and location type and location specific, requirements and guidelines.

Purpose and need must be addressed in the United States. There is no requirement to specify project goals or objectives. The purpose and need should not inappropriately narrow the range of reasonable alternatives. How each alternative will
achieve National Environmental Policy Act (NEPA) goals must be specified. The
environmentally preferred alternative is expected to promote NEPA's national
environmental policy. Europe does not require that need and purpose be addressed.
Alternatives are defined as ways in which the developer can feasibly meet the project's objectives. Guidelines refer to addressing the need for and objectives of the
project. Purpose is a requirement under the Canadian EIA legislation. The consideration of need is strongly encouraged. The preamble to the Canadian EIA legislation includes broad environmental objectives such as sustainable development and
environmental quality. Two recent panel decisions stipulate that the extent to which
sustainable development is promoted must be considered. The Australian legislation includes specific environmental, social, sustainability, and cultural objectives
and ecologically sustainable development (ESD) principles. Project-specific
guidelines refer to compliance with the objectives of the legislation and with
ESD principles.

A range of specific environmental goals and principles within EIA legislation (as
occurs in the Australian legislation) can provide a consistent litmus test for proposal
acceptability, provided that the loop is closed by stipulating that each alternative
must be assessed against the goals and comply with the principles. Such provisions
also could facilitate alternatives comparison. EIA requirements could go even
further and insist that the purpose of the proposal be identified, objectives be specified, and alternatives be assessed against the objectives. However, this shifts the
orientation away from the advancement of specific environmental goals and toward
the required application of the rational process. Such an orientation shift should be
approached with caution, given the many ascribed deficiencies associated with
rational planning processes.

The United States requires the rigorous exploration and objective evaluation of all reasonable alternatives. Mitigation must be considered. Reference is made to providing a clear basis for a choice, to the substantial treatment of alternatives considered in detail, to the use of an analytical rather than an encyclopedic approach, to the inclusion of relevant information (including the implications of data gaps), to scientific and professional integrity when selecting methods, and to avoiding post hoc rationalizations. The European EIA project directive requires that the main alternatives studied be outlined and that the main reasons for choosing the preferred alternative be described, taking into account environmental effects and mitigation. The guidelines provide examples of alternatives and highlight the potential roles of methods. The Canadian requirements have little to say about methodology beyond the scope of effects and alternatives that must and could be considered. The guidelines outline an analysis and evaluation process, which would identify alternatives (includes a screening step), determine environmental criteria, apply criteria, and select a preferred alternative, taking into account environmental effects and mitigation potential. The Australian legislation and regulations require the identification of alternatives, the assessment of impacts, the determination of mitigation measures, a comparative description, and a clear basis for the preferred alternative. References also are made to different impact types (e.g., relevant, unknown, irreversible, short- and long-term significance) and to preparing an impact management plan (mitigation, monitoring, and management). More details are provided in project-specific guidelines. These guidelines refer to comparing short-, medium-, and long-term advantages and disadvantages, providing reasons for choosing the preferred alternative and complying with the objectives of the legislation and with the ESD principles.

The inclusion of process and methodological requirements is a delicate balancing act. The U.S. requirements clearly seek to minimize bias and unsystematic procedures (i.e., irrationality). However, references to rigor, objectivity, and analysis are more appropriate when applied to impact identification and prediction than to subjective and integrative activities such as alternatives evaluation. Arguably, there is a role for experience, wisdom, emotions, and imagination (i.e., the extrarational), in combination with reason in evaluation. The rational procedural requirements and guidelines of the other jurisdictions could be more specific in identifying and applying rational evaluation process performance standards (e.g., systematic, explicit, logical, consistent, substantiated).

EIA requirements and guidelines, drawing on the rationality debates, could seek to foster the conditions necessary for reasoned, ethical, and practical procedures for formulating and evaluating alternatives. More attention could be devoted to problem and goal definition activities (value and substantive rationality), to matching process and context (contingency planning), and to facilitating and accommodating ecological, social, political, and communicative forms of rationality. EIA requirements could be reviewed to determine if they exhibit the ascribed rational process assumptions and could potentially be conducive (however unwittingly) to the ascribed rational process negative tendencies. Requirements should be

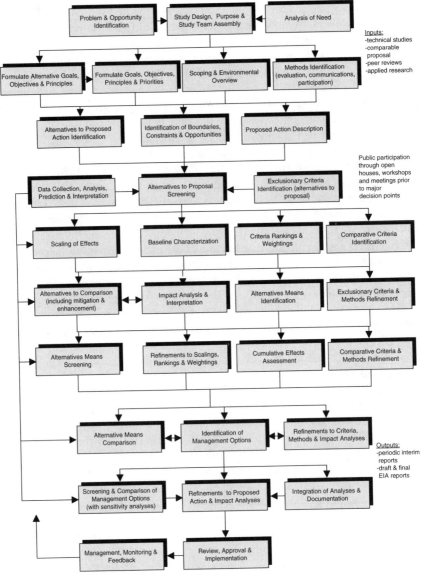

Figure 4.3 Example of a rational EIA process.

sufficiently flexible to retain and foster reasoning in EIA processes without precluding the potential contributions of variations of and alternatives to rationality.

4.5.2 Management at the Applied Level

Figure 4.3 shows an example of a rational EIA process. Figure 4.3 and the process description that follows integrates and builds on suggested rational process elements as advanced in EIA and related literature. EIA process managers and participants can pick and choose the relevant and appropriate elements.

Startup The process begins by characterizing the problem or opportunity and analyzing the need to solve the problem or take advantage of the opportunity. These analyses make it possible to identify the purpose for the process and the purpose for any proposed actions. A study design is then prepared to describe how the purpose is to be realized and the need met. Management and specialist teams are assembled to fulfill the study requirements. Broad goals, objectives, principles, and priorities are formulated to guide the process. They also provide a preliminary vision of conditions as they might exist if the problem is solved or the opportunity met. The goals flow directly from the problem or opportunity. The objectives refine the goals. The principles are broad performance standards. The priorities are system characteristics most directly and immediately relevant to problem resolution. The goals and objectives are specific and substantive. They address such concerns as sustainability, environmental quality, social and environmental justice, biodiversity, heritage, resource conservation, and energy efficiency.

A public scoping program, a form of practical rationality, focuses the process on key public and agency concerns and issues, major stakeholders, potentially significant impacts, shared interests, potential perspective, value and interest differences, and probable alternatives. Alternative goals, objectives, and principles are formulated, where necessary, to address major perspective, value, and interest differences (i.e., purposive or value rationality). A preliminary list of potentially applicable methods is compiled. These methods could support such activities as alternatives formulation and evaluation, data collection and analysis, impact prediction and interpretation, and public communications and participation. Both quantitative and qualitative methods could be identified. An environmental overview ensures that the goals, methods, and scoping activities are relevant and appropriate to the situation (i.e., the matching of process and context). The basic characteristics of the proposed action are identified. The proposed action is shaped to meet the need, fulfill the goals and objectives, and be consistent with the principles and priorities. Temporal, spatial, and jurisdictional boundaries for the EIA are established. Consistent with strategic rationality, the external and internal environments are scanned to determine constraints, opportunities, and limits.

Alternatives to the Proposed Action The major proposed action characteristics are described. Potential impact sources, stemming from the action, are identified. Aspects of the action, where choices exist, are highlighted. Potentially reasonable alternative ways of meeting the objectives and satisfying the principles are identified. If, for example, the need is transportation related, consideration might be given to such alternatives as no change, deferment, land-use planning changes, demand control procedures, growth management, and alternative modes. Clearly defined and fully substantiated exclusionary criteria are formulated to ensure a consistent basis for rejecting unacceptable alternatives. A screening procedure is formulated. Data are collected and complied to support the screening analysis. The exclusionary criteria are applied to the alternatives to the proposal.

Alternatives are rejected only if they clearly meet or exceed exclusionary criteria, taking into account mitigation potential and uncertainties.

The alternatives to the proposed action, remaining after the screening analysis, are compared. This analysis is undertaken at a broad level of detail consistent with the diverse nature of the alternatives. Supplementary data collection occurs to support the comparative analysis. Where practical, impacts are scaled. This ensures a consistent approach to impact magnitude. Comparative evaluation methods are formulated appropriate to the available data, the level of detail, and the nature of the alternatives. Comparative evaluation criteria are formulated. Care is taken to ensure that the criteria make it possible to address whether and the extent to which the alternatives contribute to the achievement of the goals and objectives and are consistent with the principles and priorities. Objectives and criteria are ranked and, where necessary to support the evaluation methods, weighted. Alternative rankings address value, interest, and objective variations. Both qualitative and quantitative evaluation methods are employed to combine the criteria rankings and the scaled impact data. Multiple sensitivity analyses address areas of uncertainty, the implications of mitigation and enhancement measures and variations in criteria rankings. Carefully reasoned arguments justify the alternative selected. The implications of any residual uncertainties are explored and explained.

A further data collection and analysis round provides the basis for identifying, analyzing, predicting, and interpreting impacts potentially associated with the alternative selected. Both technical and nontechnical forms of knowledge and experience are considered. Baseline environmental conditions are characterized. Probable future environmental conditions, assuming no proposed action, are predicted. Sensitive and significant social and ecological components, processes, interactions, and systems are identified. The characteristics of the action proposed are refined. Potential individual impacts are identified and then predicted. Criteria are applied to ensure a consistent approach to characterizing the magnitude and distribution (over time, over space, and among social groups) of potential impacts. Quantitative (e.g., quantitative models), semiquantitative and qualitative methods (e.g., conceptual models, scenarios) characterize baseline conditions and potential future impacts. Impact predictions are refined after incorporating mitigation and enhancement measures into proposed action characteristics. Key environmental interrelationships and patterns of direct and indirect impacts are determined. Uncertainties associated with both baseline conditions and predicted impacts are identified and their implications explored. Criteria are applied to ensure the consistent treatment of impact significance. The impact significance criteria take into account such matters as impact magnitude, impact distribution, public and agency priorities, mitigation potential, and levels and types of risks and uncertainties. A thorough rationale is prepared for all findings, interpretations, and conclusions.

Alternative Means of Carrying Out the Proposed Action The alternative means analysis employs the same steps and methods as the alternatives to analysis but at a greater level of detail. If the preferred alternative, emerging from the alternatives to analysis, for example, is a highway, alternative means could include route alternatives, alternative vertical and horizontal alignments, alternative intersection locations, and alternative intersection designs. Supplementary data collection and analysis are undertaken during this stage to support the cumulative effects analyses. Refinements are first made to the individual impact analysis to take into account changes to the proposed action characteristics. The cumulative effects analysis then addresses multiple additive and nonadditive effects on individual environmental components and on broader social, economic, and ecological systems. The effects result from the proposed action in conjunction with other historical, current, and likely future actions and activities. The cumulative effects analysis focuses on such concerns as temporal and spatial crowding and discontinuities, indirect, growth inducing and threshold effects, biomagnification, and feedback effects. As with the individual impact analysis, effects are quantified to the extent practical. Explicit criteria are applied to ensure consistent treatment of impact magnitude and significance. Areas of uncertainty and related implications are highlighted. Adjustments are made to proposed action characteristics to avoid and minimize potentially significant cumulative effects and uncertainties. Actions with other parties are coordinated. A clear rationale is provided for all findings, interpretations, and conclusions.

Approvals and Post Approvals   A management program is prepared once the proposed action characteristics are largely determined. The management program integrates and coordinates mitigation, enhancement, monitoring, feedback, contingency, and auditing procedures and methods. There are likely to be some options concerning management program elements. Management options are screened and compared using methods comparable to those associated with the alternatives analyses. Responsibilities and commitments are detailed. Some fine-tuning of the proposed action occurs to further enhance benefits and to minimize adverse effects.

The analyses are integrated, refined, and consolidated into draft and final EIA documents. Points of confusion are clarified. Concerns and objections are thoroughly considered and addressed. Further refinements are made to the proposed action based on inputs received during review and approval and (if the proposal is approved) as a result of monitoring and feedback. Obstacles to implementation are anticipated. Once and if approval occurs, a systematic effort is made to facilitate implementation. Methodological improvements are made for subsequent application based on the auditing of the EIA process.

Inputs, Outputs, and Interactions   The EIA process is supported by technical studies, reviews of comparable proposals and environments, peer reviews, and applied research. The public is involved in identifying concerns and preferences; in reviewing analyses and preliminary findings and conclusions; in identifying alternatives; in expressing opinions regarding criteria rankings, acceptable and preferred alternatives, significant impacts, and conclusions; and in responding to interim and draft documents. The public participates through open houses, workshops, and meetings prior to major decisions. The communications and consultation methods are jointly formulated and adapted with interested and affected parties. A proactive effort (e.g., participant funding, additional resources) is made to involve groups and organizations less likely or able to participate in the process. Care is taken to minimize communications and involvement distortions and inequities, consistent with communicative rationality principles. Close and frequent contact is maintained with regulatory review agencies.

Periodic interim reports are released as the process unfolds. Draft and final EIA summary and detailed reports are broadly circulated. All documents are designed to be lucid, unbiased, traceable, technically sound, scientifically rigorous, accurate, and consistent. Inputs received from agencies and the public are recorded and addressed. Changes made to documents as a result of inputs received are clearly specified. A clear rationale is provided for suggested changes not made.

4.6   ASSESSING PROCESS EFFECTIVENESS

Table 3.5 (in Chapter 3) presents criteria and indicators for assessing the positive and negative tendencies of the EIA processes described in Chapters 3 to 10. In this section we summarize examples of the positive and negative tendencies of the rational EIA process.

The rational EIA process, although not scientific, is generally compatible with the *rigorous* application of scientific standards, knowledge, methods, and procedures. It is conducive to participation by scientists and technical specialists. The process can ensure that assumptions, findings, interpretations, conclusions, and recommendations are explicit and substantiated.

Relevant physical, biological, social, cultural, and economic effects can be incorporated into a rational EIA process. It can address interrelationships and cumulative effects. It tends not to be fully *comprehensive* because it generally favors analysis over synthesis and is often biased toward effects that can readily be measured and scaled. Rational EIA processes tend not to be conducive to holistic perspectives. Rational EIA processes tend to place considerable emphasis on being *systematic*. A clear and traceable decision-making basis is provided. Options, effects, management measures, and methods are systematically generated, assessed, and applied. Rational EIA processes tend to devote less attention to problem analysis, to the assessment of alternative objectives, and to option identification.

Rational processes have often been criticized for emphasizing process over *substance*. Rational EIA processes are more substantive. They are guided by environmental values and deliberately seek to integrate social and environmental content into decision making. Nevertheless, they still have a tendency to focus on the process at the expense of demonstrating tangible contributions to environmental quality and sustainability. Rational EIA processes also tend to discount nontechnical, nonscientific environmental knowledge and perspectives.

Rational EIA processes often emphasize being comprehensive and systematic over being *practical*. The inclusion of scoping and the increased emphasis on good practice and regulatory compliance have resulted in more focused, efficient, effective, and realistic rational EIA processes. There remains, however, a continuing propensity to neglect practicality, political implications, and implementation in the drive to be thorough and systematic. Rational EIA processes generally provide the systematic justification that regulators prefer.

Stakeholders tend to have a peripheral position in rational EIA processes. *Local democratic influence* tends to be limited to identifying concerns and responding to the analyses of technical and scientific specialists. Rational EIA processes can be adjusted to facilitate a greater role for the public and to accommodate traditional knowledge. But the technical orientation of such processes tends to work against such adjustments. It is very difficult for a process that places such a high premium on specialist knowledge and methods to be conducive to delegating authority to stakeholders and local communities. Political implications and stakeholder acceptance tend to receive limited attention.

It is difficult for rational EIA processes to be highly *collaborative*. They can effectively communicate the bases for analyses, interpretations, and conclusions, which is helpful for collaboration. Rational EIA processes tend to favor technical and analytical knowledge and methods. Specialists are generally treated as the chief (if not the only) sources of knowledge and insight. Consequently, stakeholders tend to be pushed to one side in such processes. Rational EIA processes can provide data and analyses in support of consensus building and conflict resolution efforts. But the technical tone of such analyses can inhibit collaborative efforts.

Rational EIA processes can be designed to address distributional and other *ethical* concerns systematically. They can also incorporate procedural fairness measures and explicitly recognize rights and responsibilities. They are equitable in their search to minimize overall impacts and to provide an explicit and consistent decision-making basis. Rational EIA processes are not generally guided and shaped by ethical imperatives and standards. Equity and fairness tend instead to be secondary considerations when compared to the drive to be comprehensive and systematic.

The many variations of the rational EIA process suggest a degree of *adaptability*. Rational EIA processes are, however, prone to a "one size fits all" mentality that can inhibit matching the process to the context. They can incorporate risk concerns. They can address uncertainties with, for example, sensitivity analyses. They tend to assume or overestimate certainty. Rational EIA processes tend to be inflexible. They often do not adapt rapidly to changing circumstances. They tend to have particular difficulty with complex ill-defined problems and when operating in open and turbulent environments.

By seeking to exclude the extrarational, rational EIA processes can inhibit the *integration* of diverse values, forms of knowledge, perspectives, and ideals. Some strategic forms or adaptations to the rational EIA process address links to related decisions and can be linked to related environmental management forms and levels. More conventional, rational EIA processes are less conducive to such initiatives. Rational EIA processes can facilitate integration with proposal planning. Despite many efforts to bridge disciplines and professions, the analytical orientation of rational EIA processes tends to work against integrating and transcending individual disciplines, professions, and EIA types.

4.7 SUMMING UP

In this chapter we address the question of whether and how EIA processes, documents, and methods can become more rational. The two stories demonstrate that there is a potential role for rationality in the EIA process but that the role can be positive or negative, depending on the methods of application, the rationality perspectives encompassed by the process, and the match between process and context. Table 4.3 is a checklist for formulating, applying, and assessing a rational EIA process.

Table 4.3 Checklist: Rational EIA Process

The response to each question can be yes, partially, no, or uncertain. If yes, the adequacy of the process taken should be considered. If partially, the adequacy of the process and the implications of the areas not covered should be considered. If no, the implications of not addressing the issue raised by the question should be considered. If uncertain, the reasons why it is uncertain and any associated implications should be considered.

CONCEPTUAL DISTINCTIONS

1. Does the process exhibit rational attributes?
2. Does the process minimize irrationalities?
3. Does the process incorporate the extrarational in a manner complementary to reasoned analysis and synthesis?
4. Is the process justified, guided, supported, and informed by the reasoned application of ideas, reasons, arguments, evidence, analysis, knowledge, information, logic, and methods?
5. Are appropriate preconditions for the expression and application of reason in the process established?
6. Is the process adapted to varying contexts?
7. Does the process effectively incorporate and integrate different rationality forms?
8. Does the process systematically integrate core rational process characteristics?
9. Does the process share any of the ascribed rational process assumptions? If so, are the assumptions appropriate to the process, the proposal, and the context?
10. Does the process exhibit the ascribed strengths?
11. Does the process avoid or minimize the ascribed limitations?
12. Does the process incorporate appropriate elements of the response to the rational process?
13. Does the process build on the insights of EIA literature and practice in adapting and applying the rational EIA process?

MANAGEMENT AT THE REGULATORY LEVEL

1. Do the regulatory provisions and/or guidelines:
a. Establish a purpose and a policy basis for the EIA requirements?
b. Require the consideration of alternatives and the environmental consequences of alternatives?
c. Require the consideration of measures to prevent and minimize adverse effects associated with alternatives?

Table 4.3   (*Continued*)

---

d. Provide for screening of unreasonable alternatives based on clearly defined, consistently applied, and fully justified thresholds of acceptability?

e. Provide additional guidance for differentiating between reasonable and unreasonable alternatives for different classes of proposal types?

f. Provide for a broad range of potential alternatives?

g. Require the consideration of the no-action alternative, the environmentally preferred alternative, the alternative preferred by the proponent, and reasonable alternatives outside the jurisdiction of the proponent?

h. Require the consideration of both alternatives to the proposal and alternative means of carrying out the proposal?

i. Provide guidance for the types of alternatives often associated with various proposal and setting types?

j. Require the consideration of purpose and need?

k. Include specific environmental and sustainability goals and principles?

l. Require that the proposed action advance the goals and be consistent with the principles?

m. Require that reasonable alternatives be assessed against the goals and principles?

n. Provide for proposal objectives?

o. Provide for the assessment of alternatives against the proposal objectives?

p. Seek to minimize bias and nonsystematic procedures?

q. Apply rationality performance standards such as systematic, explicit, logical, consistent, and substantiated?

r. Foster the conditions necessary for the reasoned, ethical, and practical treatment of alternatives?

s. Provide for the exploration of problems and opportunities?

MANAGEMENT AT THE APPLIED LEVEL

1. Does the process incorporate such startup activities as:

    a. Problem and opportunity identification?

    b. Study design?

    c. Formulation of a purpose for the proposal and for the process?

    d. Analysis of need?

    e. Formulation of goals, objectives, principles, and priorities?

    f. Formulation of alternative goals, objectives, and principles?

    g. Scoping?

    h. Environmental overview?

    i. Identification of methods?

    j. Identification of boundaries, constraints, and opportunities?

2. Do the alternatives to the proposed action activities:

    a. Describe the proposed action?

    b. Systematically identify alternatives to the proposed action?

    c. Systematically collect, analyze, predict, and interpret the data necessary to support the analysis?

    d. Identify exclusionary criteria?

    e. Consistently apply the exclusionary criteria?

    f. Scale the effects?

    g. Characterize baseline conditions?

(Continued)

Table 4.3   (*Continued*)

h. Identify comparative criteria?

i. Rank and weight the criteria?

j. Systematically integrate the scaled data and criteria rankings/weightings to compare the alternatives?

k. Allow for mitigation potential in the comparative analysis?

l. Use sensitivity analyses to test uncertainties and variations in criteria and criteria rankings?

m. Provide a clear rationale for the preferred alternative?

3. Do the alternative means of carrying out the proposed action activities:

   a. Refine the proposed action?

   b. Systematically identify alternative means of carrying out the proposed action?

   c. Systematically collect, analyze, predict, and interpret the data necessary to support the analysis?

   d. Identify exclusionary criteria?

   e. Consistent apply the exclusionary criteria?

   f. Scale the effects?

   g. Characterize baseline conditions?

   h. Identify comparative criteria?

   i. Rank and weight the criteria?

   j. Systematically integrate the scaled data and criteria rankings/weightings to compare the alternatives?

   k. Allow for mitigation potential in the comparative analysis?

   l. Use sensitivity analyses to test uncertainties and variations in criteria and criteria rankings?

   m. Provide a clear rationale for the preferred alternative?

   n. Systematically analyze and interpret impacts?

   o. Systematically assess cumulative effects?

4. Do the approvals and post-approval activities:

   a. Identify management options?

   b. Systematically and consistently assess management options?

   c. Include the preparation of a management program?

   d. Provide a thorough, sound, and systematic basis for approvals?

   e. Include a monitoring program?

   f. Anticipate and address obstacles to implementation?

   g. Provide for methodological improvements through monitoring?

5. Do the input, output, and interaction activities:

   a. Support the process with technical studies, reviews of comparable proposals and environments, peer reviews, and applied research?

   b. Provide for public involvement in identifying concerns and priorities, in reviewing analyses and preliminary findings and conclusions, in identifying alternatives, and in expressing opinions regarding criteria rankings, acceptable and preferred alternatives, significant impacts, conclusions, and concerning interim, draft, and final reports?

   c. Proactively involve groups and organizations less likely or able to be involved in the process?

   d. Minimize communications and involvement distortions and inequities?

   e. Maintain close and frequent contact with regulatory review agencies?

Table 4.3 (*Continued*)

    f. Provide for the periodic release of interim reports?

    g. Provide lucid, unbiased, traceable, technically sound, scientifically rigorous, accurate, and consistent draft and final reports?

    h. Provide a clear rationale for all changes to documents?

    i. Respond to all agency and public suggestions and concerns regarding documentation?

PROCESS EFFECTIVENESS

  1. Is the process compatible with the rigorous application of scientific standards, knowledge, methods, and procedures?

  2. Does the process provide for the comprehensive assessment of effects and alternatives?

  3. Does the process systematically address defining the problem, identifying goals, generating and assessing alternatives, and predicting and managing effects?

  4. Is the process guided by and does it systematically integrate substantive environmental concerns?

  5. Does the process balance the desire to be comprehensive with the need to be practical?

  6. Does the process facilitate public involvement and accommodate traditional knowledge?

  7. Does the process effectively communicate the bases for analyses, interpretations, and conclusions?

  8. Does the process actively seek to facilitate collaboration among interested and affected parties?

  9. Does the process incorporate procedural and substantive fairness concerns?

10. Is the process adapted to context?

11. Does the process systematically address risks and uncertainty?

12. Does the process attempt to anticipate and to deal with changing circumstances?

13. Does the process attempt to integrate diverse values, forms of knowledge, perspectives, and ideals?

14. Does the process attempt to blend the rational with the extrarational?

15. Does the process attempt to link and transcend disciplines, professions, and EIA types?

16. Is the process linked to related environmental management forms?

    The problem can be posed in three ways: (1) EIA processes in practice are insufficiently rational; (2) EIA processes in practice are too rational; and (3) the ways in which rationality is defined and applied in EIA processes need to be modified. There are valid arguments in support of all three positions. The problems are addressed by exploring the potential to apply rationality strengths, minimize rationality deficiencies, and draw on alternative rationality definitions and applications.

    Rationality attributes (e.g., logical, consistent, systematic) are identified and contrasted with irrational and extrarational decision-making factors. Rationality attributes are expressed in many ways. They are fulfilled in numerous applications. They function best when supported by grounds (such as evidence, analysis, and methods) and when key preconditions (such as openness and trust) are satisfied. There are many rationality forms relevant to EIA process management. Rationality expressions and applications vary depending on context. As commonly described, the rational process involves defining a problem, identifying goals, collecting and analyzing information, forecasting and modeling future conditions, generating and evaluating alternatives, and implementing the preferred alternative. Public and agency involvement tends to take place prior to major decisions in the process. A great many assumptions have been ascribed to rationality. It tends to be presumed that the process can and should be comprehensive, systematic, and sequential, the analysis objective, the parties rational, and the environment predictable and controllable. These assumptions are not necessarily inherent to a rational EIA process. Ascribed rationality assumptions should be considered carefully because they are often implicit in EIA processes.

Many strengths and limitations are attributed to the rational process. It is considered simple, explicit, logical, consistent, and systematic. It provides a clear sense of direction and a sound decision-making basis. The rational process has been attacked on many fronts. It is considered unrealistic (regarding human limitations, decision-making constraints, and decision-making procedures). It is seen as ineffective (lacking focus, not adapted to context and not politically astute). It is viewed as incomplete (lacking in content, incomplete image of people, excludes the contribution of the extrarational). It is characterized as inappropriate (not suited to uncertainty or conflict, technically and scientifically biased, autocratic tendencies). These strengths and limitations are tendencies that can be offset or reinforced. Several responses to the shortcomings identified focus the process in an effort to make it more realistic and effective. Others selectively adapt and combine social, political, legal, ecological, and communicative rationality forms to make the process more substantive and democratic. The debates surrounding rationality have cycled through multiple iterations. Some EIA literature ignores or discounts the debate. Many EIA process characterizations truncate the rational process by moving directly to applying criteria to a proposed action and reasonable alternatives. EIA has avoided some rationality limitations partly because of EIA process characteristics (e.g., environmental ethic, scoping, cumulative effects assessment) and partly through deliberate efforts to offset negative tendencies. Many rationality debates are mirrored in EIA literature. EIA literature and practice could derive additional benefits from a closer examination of the rationality debates.

EIA requirements in the four jurisdictions are rational to the extent that they include a policy basis and require the consideration of alternatives, the environmental consequences of alternatives, and measures to prevent or minimize adverse effects. Exclusionary decision rules could be better defined. Each jurisdiction bounds rationality with provisions to screen out unreasonable alternatives. The range of types of alternatives that must be considered varies. The U.S. requirements are the most inclusive. More precise requirements and recommendations could be included for classes of proposals and for individual proposals. The four jurisdictions, to varying degrees, provide specific legislative environmental objectives and principles as benchmarks for evaluating proposals. Such requirements can contribute to more substantive EIA processes. Requiring that individual proposals include objectives would make EIA processes more systematic but could reinforce negative rational process tendencies. The jurisdictions vary in the extent to which procedural and methodological requirements mirror rational process requirements. The U.S. requirements appear to go too far in seeking to exclude the extrarational. The requirements in the other jurisdictions are very general. All the jurisdictions could devote more attention to problem and goal definition activities, to contextual adaptations, to incorporating different rationality forms, and to deriving lessons from the rational process debate.

The example rational EIA process characterizes the problem or opportunity, analyzes need, determines the process and proposal purpose, and assembles a study team. Goals, objectives, principles, and priorities are determined. Alternative goals, objectives, and principles are formulated where warranted. Methods are identified. An environmental overview is conducted. A scoping program is formulated and applied. Boundaries for the process are identified. Internal and external constraints and opportunities are identified. The proposed action is described. Potentially reasonable ways of meeting the objectives and satisfying the principles are assessed. Unreasonable alternatives are excluded by applying exclusionary criteria. Reasonable alternatives are compared using evaluation methods, which combine scaled effects with criteria rankings and weightings. Uncertainties and variations in preferences are addressed through sensitivity analyses. Mitigation potential is integrated into the analysis. The analysis is supported by methods refinements and by data collection, analysis, prediction, and interpretation. The alternative means of carrying out the action proposed are also assessed. The same basic steps are followed but at a greater level of detail. Baseline conditions are characterized and individual and cumulative impacts, stemming from the proposed action, are identified, predicted, and interpreted in parallel with the alternatives to and alternatives means analyses.

An impact management program refines and facilitates implementation of the proposed action. Options associated with the management program are generated and assessed in a manner comparable to the alternatives analyses. A clear and consistent decision-making basis is established, taking into account agency and public comments and suggestions. If approved, monitoring and auditing programs are undertaken. The monitoring program minimizes adverse impacts and maintains or enhances benefits. The auditing program facilitates methodological refinements. The process is supported by technical studies, reviews of comparable proposals and environments, peer reviews, and applied research. The public identifies concerns and suggestions and responds to analyses and documents. Communications and involvement distortions and inequities are minimized. Periodic interim reports are released. The draft and final report are distributed broadly. The documents provide a clear, unbiased, systematic, and accurate decision-making basis.

Rational EIA processes tend to be consistent, systematic, and supportive of the rigorous application of scientific standards, knowledge, and methods. They often are not conducive to holistic perspectives. They can integrate substantive environmental concerns but sometimes emphasize process over substance and often discount nontechnical, irrational environmental knowledge and forms. They sometimes sacrifice practicality in the drive to be comprehensive. The emphasis on technical and scientific knowledge and methods can inhibit local democratic control and can undermine the potential for collaborative efforts to build consensus and resolve conflicts. Rational EIA processes can incorporate procedural and substantive fairness measures. But equity and fairness tend to be secondary concerns. Rational EIA processes can assume many forms but can be inflexible in practice. They often are not well adapted for addressing complexity and uncertainty. The emphasis on analysis (as compared with synthesis) and the exclusion of the extra-rational can inhibit integrating diverse forms of knowledge, perspectives, and ideals. It can also undermine efforts to span and transcend EIA types and professions and to establish and maintain links to other environmental management forms.

CHAPTER 5

HOW TO MAKE EIAs MORE SUBSTANTIVE

5.1 HIGHLIGHTS

This chapter is concerned with designing and managing regulatory and applied EIA processes to better integrate environmental perspectives, values, and knowledge.

- The analysis begins in Section 5.2 with two applied anecdotes that describe applied experiences associated with efforts to make EIA practice more substantive.
- The analysis in Section 5.3 then defines the general problem, which is a shortfall between EIA environmental aspirations and achievements. The more specific problem is the role that the EIA process assumes in widening or narrowing that gap. The direction is an EIA process more conducive to integrating environmental perspectives, values, and knowledge and to furthering environmental objectives.
- In Section 5.4 we provide an overview of a range of ecological, social, and sustainability concepts. Attributes pertinent to EIA process management are highlighted. Methods that could facilitate the integration process are also described briefly.
- In Section 5.5 we extend from the concepts presented in Section 5.4. We detail how an environmentally substantive EIA process could be implemented at the regulatory and applied levels. In Section 5.5.1 we address how more

Environmental Impact Assessment: Practical Solutions to Recurrent Problems, By David P. Lawrence
ISBN 0-471-45722-1   Copyright © 2003 John Wiley & Sons, Inc.

environmentally substantive EIA processes could be facilitated and structured at the regulatory level, and in Section 5.5.2 we demonstrate how an environmentally substantive EIA process might be expressed at the applied level.

- In Section 5.6 we assess how well the environmentally substantive EIA process presented in Section 5.5 satisfies ideal EIA process characteristics.

- In Section 5.7 we highlight the major insights and lessons derived from the analysis. A summary checklist is provided.

5.2 INSIGHTS FROM PRACTICE

5.2.1 Public Participation and Influence as a Means to Make EIA More Substantive

An environmental impact assessment (EIA) was completed successfully in March 2002, in accordance with the Austrian EIA act, for a waste management facility in Frohnleiten, a municipality of the Austrian province of Styria. The facility proposed will employ mechanical and biological waste treatment methods. During 2001, several engaged citizens of the municipality founded an active group to facilitate their involvement in the public participation process for the EIA. The group had very intensive communications with both the developer and the EIA authority both before and during the EIA process. It also reviewed the environmental report, published in October 2001 thoroughly and critically. The know-how of the group was displayed in its highly successful and intensive communications, coordination, information, and advisory activities.

The group was able to strongly influence the EIA decision-making process and to contribute to essential, substantive environmental changes. The group helped bring about a reduction of the total input for the facility from 97,500 tons to 76,276 tons. It also contributed to restrictions in mechanical process measures, to limits on the time periods for transport from and to the facility (to minimize adverse transport effects), and to improvements in the chemical air cleaning process of the facility. Also, after fruitful negotiations sessions, the group was able to convince the developer to publish yearly reports on the facility and to organize regular information days. In addition, the agreement provided for mandatory regular monitoring of the facility emissions, both during test operations (planned for July 2003) and during full operations (planned to begin in 2004).

This case study demonstrates that an active and engaged public can both improve the EIA process and contribute to substantive environmental improvements. It points to an important link between process and substance. A more conventional, more technical EIA process is unlikely to result in substantive environmental improvements beyond those required by regulators. A more democratic EIA process, where the public strongly influences decision making, can often contribute to an enhanced level of environmental quality. The combination of an EIA process characterized by collaborative or democratic decision making and substantive environmental improvements could, in turn, be conducive to a greater level of public involvement in and acceptance of both the process and its outcomes. It could also enhance the public credibility of both the proponent and the government. This case study describes a process where the public largely took the initiative. Such a proactive and organized public cannot be taken for granted. Ideally, the EIA process should foster heightened public influence and control. As this case study illustrates, such an approach could have both procedural and substantive benefits.

RALF ASCHEMANN
Austrian Institute for the Development of Environmental Assessment

5.2.2   A Role for the Social Sciences in Making the Process More Substantive

During the 1990s, the coastal retirement and holiday community of Mandurah experienced rapid population growth as it became a popular bedroom community to the Perth metropolitan area in Western Australia. The rapid urbanization of the area resulted in strong land competition within the region and a growing demand for upgraded infrastructure. In 1996, the State Government Cabinet established the Mandurah Ocean Marina Taskforce to progress concept plans for the development of a major new marina complex on a 30-hectare site near the town center as a means to boost tourism and local employment. Despite the site being one of the most strategically important land parcels in Mandurah, it had for years remained largely undeveloped, due to the inability to resolve conflicts with the community, in particular with the residents of the Peninsula Caravan Park, who would be displaced by the proposal. In an attempt to resolve the conflict, the regional development commission hired a consultant to conduct a social impact assessment (SIA) of the impact of the marina proposal on the permanent residents of the caravan park. The 40-year-old 4-hectare caravan park included 32 permanent households and 129 semipermanent households at the time of the study. The average age of the permanent residents was 64.8 years, with 82.4 percent over the age of 60. The majority of residences were modest holiday caravans (i.e., trailers), some up to 30 years old.

The design of the SIA drew heavily on the psychology and sociology literature. It employed concepts such as *person–environment fit*, sense of community and place, personal control, social well-being and quality of life, and the effects of involuntary relocation on the aged. In-depth interviews were conducted with each of the households as well as with a range of local and state government agencies. Semipermanent park residents were surveyed via a mail questionnaire. Despite the perception of outsiders that the caravan park provided a poor quality of life to residents, consistent with the research literature, the park residents themselves were experiencing a high level of satisfaction. They demonstrated a strong sense of community and attachment to place. They had forged strong friendships and social support networks within both the park and the surrounding community.

The permanent residents expressed powerlessness in their ability to influence government decision making because of their age and their belief that outsiders consider caravan park residents to be second-class citizens.

The SIA practitioner and regional development commission explored a range of housing alternatives and compensation options with the permanent residents but few attractive options existed. The aged residents were fearful of losing their autonomy and security if they had to go into public housing or a retirement home. Other caravan parks in the region did not offer the locational advantages of the Peninsula Caravan Park or the potential to accommodate all of the permanent residents at a single location. Although the SIA did not lead to easy or completely satisfactory outcomes for the permanent residents, it did result in more compassionate handling of their situation. It allowed a process that to the extent possible, attempted to give residents a sense of control over their destinies. At a minimum the SIA gave the aged residents a legitimized voice in putting forward their claims of impact; claims which to that point had been largely dismissed by decision makers. Construction of the marina began in 1999, with management of the marina handed over to the city of Mandurah in 2001.

The SIA also highlighted several issues for land-use planners in the state. Due to unsupportable stereotypes of caravan park residents, planning policies have resulted in new caravan parks being situated on the fringes of urban areas, isolated from community services, while established caravan parks in proximity to social amenities are being replaced by "higher-value" land uses. The SIA brought to light the negative social impacts of such land-use planning practices and the suitability of well-located caravan parks as an accommodation option for some retirees. More broadly, this story demonstrates that appropriate use of social scientific methods by qualified SIA professionals can integrate substantive insights and constructive suggestions into the process. Such insights and suggestions are less likely to be provided when social concerns are treated as a subset of environmental factors, a subset not requiring the involvement of SIA specialists.

JO ANNE BECKWITH
Department of Resource Development
Michigan State University

5.3 DEFINING THE PROBLEM AND DECIDING ON A DIRECTION

The preceding stories demonstrate that to be effective, EIA processes must be substantive. As one story demonstrates, substance can be integrated into the EIA process through the application, by natural and social scientists, of natural and social scientific knowledge, concepts, theories, frameworks, and methods. As the other story illustrates, a more substantive EIA process can also result from enhanced public participation and influence, which broadens and reorients the process to better accommodate a wider range of substantive knowledge and perspectives. The public also can be a source of substantive knowledge and insights (e.g., community and traditional knowledge). This is much more likely to occur with a collaborative-democratic EIA process. The stories, although indicative of the importance of integrating process and substance within the EIA process, provide only an initial sense of how that task might be accomplished.

The relationships between the process and substance in EIA practice, as illustrated in Figure 5.1, can be approached from many perspectives. The position

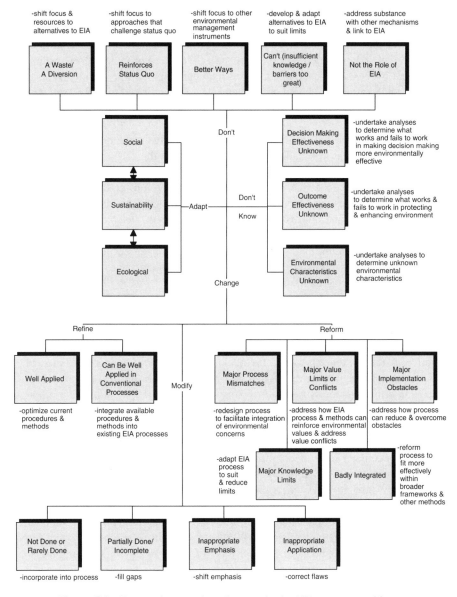

Figure 5.1 Perspectives on the substance in the EIA process problem.

can and has been taken that EIA is a negative force in the EIA movement. Some label it a deceitful procedure that legitimizes, without excluding or altering, environmentally unsound projects and practices. EIA is considered an empty and wasteful paper-processing exercise. Resources devoted to EIA should, it is argued, be redirected to redefining agency missions and to tangible initiatives that directly advance the cause of environmental quality and sustainability. Alternative environmental management approaches (such as the ecosystem approach and adaptive environmental management) are sometimes suggested as tools that could replace EIA. The point is occasionally made that EIA presumes a degree of predictability and control that is so inconsistent with the knowledge base and the institutional structure that it is doomed to failure as an effective environmental management instrument. At best, it is concluded, EIA is a procedural instrument with no substantive content. Substance is added only when and if EIA is linked to and placed within the context of substantive environmental management frameworks and tools (Dennis, 1997).

The argument that EIA is invariably either a negative force or that it serves no substantive purpose is not made now as often it was in the early days of EIA practice, at least not by those knowledgeable of EIA practice. Nevertheless, the arguments are useful because EIA can and too often does reinforce the status quo and waste resources. It also can be of limited value if knowledge and control limitations are not recognized and addressed, and if the relationship between EIA and other environmental management instruments is not considered. The substantive role of EIA is shared with numerous other environmental management forms. Thus, making the EIA process more substantive requires complementary roles among instruments and within broader frameworks.

It is difficult to reach firm conclusions concerning the EIA process–substance problem because of uncertainties surrounding both the knowledge base for EIA and the consequences from EIA. EIA has been advanced as an effective mechanism for integrating social and environmental values and perspectives into institutional practice. Although there is ample favorable (and unfavorable) anecdotal evidence, it is an overstatement to conclude that EIA has been either highly effective or ineffective in bringing about such a transformation (Bartlett, 1994). It also is unclear if the results without EIA would have been substantially different. A similar aura of uncertainty surrounds whether EIA appreciably reduces adverse environmental impacts and whether it greatly enhances natural and social environmental conditions and benefits. These uncertainties stem in part from a spotty, albeit improving record of monitoring environmental impacts and mitigation effectiveness (Clark, 1997). They also result from a mixed record in furthering EIA environmental aspirations. EIA effectiveness ratings concerning decision-making benefits, reduced social and ecological impacts, and contributions to sustainability all leave considerable room for improvement (Sadler, 1996). It is difficult to reach firm conclusions regarding EIA benefits because of the myriad of uncertainties surrounding the analysis, prediction, and management of ecological, social, and administrative systems (Dearden and Mitchell, 1998). These uncertainties are compounded when social and ecological systems are linked to assess cumulative and sustainability effects.

EIA is often credited with contributing to a greater sensitivity to and account-ability for environmental consequences by agencies (Andrews, 1997; Caldwell, 1997). It also is viewed as helping keep environmental issues before the public (Moore, 1992). Although there is a sense of overall progress, there is considerable variability in how well environmental concerns are integrated into decision making and regarding the extent to which environmental quality is enhanced (Dennis, 1997). Most EIA texts assume that disciplinary and interdisciplinary knowledge and methods can readily be incorporated into conventional EIA process activities and stages (Canter, 1996; Morgan, 1998; Petts, 1999). The point is commonly made that *tiering* (fitting project-level EIA within a SEA framework) greatly expedites the process (Morris and Thérivel, 1995; Vanclay and Bronstein, 1995).

Many commentators suggest that the marriage of process and substance will necessitate both procedural and substantive adjustments. Some argue that EIAs commonly lack an ecosystem perspective and fail to explore how social impacts are constructed socially (Beanlands and Duinker, 1983; Brooke, 1998; Burdge, 2002; Greer-Wooten, 1997). They suggest the need for a greater effort to include ecological principles, to fully address biodiversity impacts, to more effectively integrate social concerns and knowledge, to apply nonpositivist social science approaches and perspectives, to consider the benefits of conservation, and to devote more attention to global warming, cumulative effects, and other sustainability-related concerns (Andrews, 1997; Burdge, 2002; Clark, 1997; Kaufman, 1997; Lockie, 2001; Moore, 1992; Treweek, 1999). They emphasize the need to imple-ment EIA policy goals, to monitor social and ecological effects, and to ensure bureaucratic accountability (Bronfman, 1991; Dearden and Mitchell, 1998, Kaufman, 1997; Treweek, 1999). EIA processes and practices should, they stress, be more interdisciplinary, place-based, and adaptive (Dearden and Mitchell, 1998; US CEQ, 1997a). They highlight such methodological shortcomings as inadequate study designs, inappropriate temporal and spatial boundaries, inadequate data, inappropriate statistical techniques, critical impacts not predicted, and an insuffi-cient decision-making basis (Alberti and Parker, 1991; Beanlands and Duinker, 1983; Freudenburg, 1986). Many methodological limitations are traced to a poor understanding of social and natural systems and of available social and ecological scientific concepts, models, and methods (Beanlands and Duinker, 1983; Burdge, 2002; Craig, 1990; Treweek and Hankard, 1998). These shortcomings imply an iterative EIA process–substance relationship, with cycles of adjustments in an ongoing effort to match procedural and substantive characteristics.

Some commentators maintain that it is premature to reorient EIA requirements and practices to meet substantive environmental ends. They argue that fundamental value shifts are a prerequisite to an EIA process driven by ecological, social, and sustainability ethical principles (Euston, 1997; Kaufman, 1997). EIA can play a supporting role in advancing such values and in addressing basic value conflicts, but not through conventional procedures. Much ecological and social knowledge, upon which EIA depends, is fraught with uncertainties, especially when cause–effects relationships must be discerned and future environmental effects must be predicted (Deaden and Mitchell, 1998). Additional complexity is added when cumulative ecological and social system effects must be determined and conclusions reached regarding sustainability implications. The substantive effectiveness of EIA is further severely inhibited by "balkanized" government environmental and resource responsibilities (Weiner, 1997). Effective partnerships must be established and maintained among agencies, with nongovernment organizations, and with the public (Dennis, 1997; IEMTF, 1995). These substantive value, knowledge, implementation, and institutional obstacles and challenges necessitate, they argue, fundamental reforms to the EIA process.

The problem is not the same for ecological, social, and sustainability concerns, although EIA effectiveness reviews suggest that there is substantial room for improvement in all three areas (Denq and Altenhofel, 1997; Freudenburg, 1986; Sadler, 1996). Effectiveness ratings tend to be lower for treating social concerns and lower still for treating sustainability concerns. The nature of the concerns naturally varies. An EIA process conducive to addressing ecological concerns will not necessarily be appropriate for incorporating social or sustainability concerns. Adaptations will be necessary for each type of concern as well as for interactions among concerns.

There is no simple answer to the question, Which of these perspectives is valid or most valid? A greater understanding of environmental conditions and of decision making and outcome effectiveness is required. Pending such knowledge, it must be assumed that all problem perspectives and solutions, depending on the circumstances and to varying degrees, are valid. Sometimes EIA is more trouble than it is worth. Sometimes the substance–process relationship needs refinement, sometimes modification, and sometimes reform. Sometimes the issues and solutions vary depending on whether ecological, social, or sustainability concerns are being addressed. The direction then is to explore how EIA process management can better address this constellation of interrelated problems and solutions.

5.4   SELECTING THE MOST APPROPRIATE ROUTE

This overview of substantive environmental concepts establishes the foundation for enhanced EIA process management. The analysis is necessarily abbreviated and selective. Only ecological, social, and sustainability concepts are considered. EIA effectiveness ratings for treating such concerns tend to be low. The choice of concepts is admittedly arbitrary. Only concept characteristics directly relevant to EIA process management are identified. Key concept attributes are first described. Then EIA process management implications are explored.

5.4.1 Ecological Concepts

Ecology is a branch of the biological sciences concerned with the relationships between organisms and their environments, including relationships with other organisms. Ecological impact assessment is a formal process of identifying, quantifying, and evaluating the potential impacts of defined actions on ecosystems (Treweek, 1999). Table 5.1 identifies and describes briefly several ecological

Table 5.1   Examples of Potentially Relevant Ecological Concepts

Applied ecology (also ecosystem sciences)	Requires an ecological perspective, adherence to basic ecological concepts, and an appropriate interdisciplinary conceptual framework
	Ecological systems: self-controlled within constraints, evolving and complex; ecosystems part of a larger sociobiophysical system (human cultures and environments part of system)
	Importance of temporal and spatial boundaries which reflect ecological processes
	Seeks to better understand self-organizing structures and processes (management challenge to protect self-organizing capacities—ecosystem integrity)
	Recognizes highly dynamic systems, extreme variation and major predictive and management constraints; strong interest in scale, patterns, rhythms, and thresholds in biophysical systems
	Focuses on key variables, key processes, and ecosystem tolerance; importance of habitat and biological diversity
	Focuses on questions relevant to decision-making choices
Ecological impact assessment	Tends to fit within major EIA stages; scoping, focusing, and ecological monitoring critical
	Focuses on interactions between ecological stressors and receptors; a systems perspective
	Concerned with state of the environment (e.g., biodiversity loss); addresses impacts on ecosystems and their components (valued ecosystem components)
	Use of applied ecological concepts, principles, and methods (e.g., surveys, taxonomic classification, GIS, modeling, statistical analysis, ecological evaluation, monitoring)
	Takes into account barriers, limits, and uncertainties (institutional, knowledge, methodological, natural variation)
	Closely linked to socioeconomic impacts, risk assessment, pollution control, and land-use and resource planning
Environmental indicators	Methods and measures to monitor environmental status (trends and conditions)
	Important for framing problems and for determining solutions
	Physical, chemical, and biological variables used to construct environmental change indicators; incorporated into environmental statistics (state of environment); a decision-making input
	Example indicator categories: response indicators (overall biological conditions), exposure or habitat indicators (ecosystem exposure to pollutants and habitat degradation), and stressor indicators (human and natural processes that change exposure and habitat conditions)
	Can help monitor environmental problems; responses also depend on social and political considerations
Biodiversity	The array of interacting, genetically distinct populations, and species in a region, the communities they comprise and the variety of ecosystems of which they are functioning parts; relationships and interactions are critical

(Continued)

Table 5.1　(*Continued*)

	Components: regional ecosystem diversity, local ecosystem diversity, and genetic diversity
	Factors contributing to biodiversity decline: physical alteration, pollution, overharvesting, introduction of exotic species, natural processes, and global climate change
	Biodiversity principles (e.g., big picture or landscape perspective, protect communities and ecosystems, minimize fragmentation, promote native species, protect unique or sensitive environments, maintain or mimic natural ecosystems processes, structural diversity and genetic diversity, restore ecosystems, communities, and species)
	Intentionally multispecies emphasis; seeks to protect broader habitats and ecosystems that support biodiversity
	Stresses need to think in terms of comprehensive multiscale ecological networks and to adopt nested hierarchical conservation strategies
Ecosystem approach (also ecosystem management)	Place-driven environmental protection strategy; use of natural boundaries and ecological indicators
	Whole system and broad regional and temporal perspectives (multiple scales and time horizons); appreciates the dynamic nature of ecosystems
	Based on ecosystem integrity and sustainability principles, and values; incorporates such concepts as carrying capacity, resilience, self-organization, community diversity and stability, and the precautionary principle
	Seeks to restore and sustain health, productivity, and biological diversity of ecosystems and quality of life (humans part of environment)
	Stresses need for dynamic, transdisciplinary, visionary (explicit ecosystem goals), proactive, adaptive, and participatory planning process
	Recognizes importance of institutional arrangements (especially coordination and communications) and of integration with social and economic goals and context
Environmental planning and management (also resource planning and management)	Largely rational planning process adapted to integrate environmental knowledge and methods; methods have roots in ecological and social sciences
	Encourages inclusion of ecological perspectives
	Stresses need to reach consensus on environmental issues
	Multi- to interdisciplinary; stress needs for comprehensive approach
	Usually advisory and participatory
	Increasing recognition of need to address environmental justice and equity issues

Table 5.1   (*Continued*)

Integrated environmental and resource management and assessment	Advocates need for a more effective, integrated, and coordinated approach
	Holistic-, regional-, and ecosystem-based perspective; stresses preservation of natural systems integrity
	Interconnective, goal-oriented, and strategic; involves both human and natural resources in ecosystem
	Greater attention to social, political, economic, and institutional factors operating in ecosystem (including opportunities and barriers stemming from institutional arrangements) and to links to sustainability
	Supported by integrated management systems (e.g., database management, GIS, expert systems)
	Recognizes importance of stakeholder collaboration and of conflict management
	Recognizes importance of context and of links to urban and regional planning
Adaptive environmental assessment and management	Iterative decision-making process; mimics the dynamic, cyclic, and surprise-ridden state of nature; decisions and assumptions revisited; long-term research, monitoring, and management critical; seeks more resilient policy
	Generally involves a series of workshop facilitated by core groups of experts; focuses on building and testing (usually, computer) models as tools for generating and testing options; ongoing data acquisition
	Combines scientific information with a forum for interested and affected parties; a minimum regrets planning tool
	Emphasizes interdisciplinary communications and collaboration; integrates societal and ecosystem goals and values
	Carries EIA into ongoing management; highlights importance of monitoring and of adaptive management in face of uncertainty and complexity; an open and continuous leaning processes—learning by doing
Traditional knowledge	Way of knowing and thinking about relationships of living beings (including humans) with one another and with the environment (a way of life)
	Cumulative body of knowledge and beliefs, handed down through generations by cultural transmission
	Relies on observation and knowledge of indigenous peoples
	Holistic: a form of environmental knowledge that integrates social, ethical, cultural, technical, scientific, historic, ecological, and spiritual; emphasis on interrelationships; avoids scientific reduction
	Includes interrelationships among physical, biological, and human; humans as participants in environment rather than only as observers
	Fluid and flexible; importance of understanding how operates in indigenous contexts; often misunderstood and misapplied

Sources: Alberti and Parker (1991), Armitage (1995), Bagri et al. (1998), Barrow (1997), Beanlands and Duinker (1983), Beatley (2000), Berkes (1993), CEAA (1996), Coleman (1996), Dearden and Mitchell (1998), Hegmann and Yarranton (1995), Hollick (1993), Holling (1978), Hooper et al. (1999), IEMTF (1995), Kozlowski (1990), Lou and Rykiel (1992), Margerum (1997), Sallenave (1994), Slocombe (1993), Smith (1993), Treweek (1995, 1999), US CEQ (1993), Wackernagel and Rees (1996), Wieringa and Morton (1996), Wiles et al. (1999).

concepts potentially relevant to EIA process management. The recurrent themes exhibited in the concepts imply an EIA process distinctly different from the conventional process assumed in most EIA texts. The concepts all begin from an ecological systems perspective. They see planning and management as shaped by ecological visions, goals, and principles. Disciplinary boundaries are spanned and transcended. Less emphasis is placed on comprehensive disciplinary analyses and single function institutions. More stress is placed on selective, transdisciplinary synthesis and place-based coalitions of agencies and stakeholders. Temporal and spatial boundaries are extended to match natural patterns and rhythms. Multiple spatial and temporal horizons and boundaries are employed. Natural systems are seen as dynamic, self-organizing, complex, evolving, and uncertain. Planning processes, to match such characteristics, are viewed as necessarily open, adaptive, creative, collaborative, iterative, selective, and action-oriented.

Project-induced stresses can result in ecological thresholds being exceeded, notwithstanding resiliency. Severe prediction and control limits are noted. Thresholds are difficult to discern and often change. Major implementation barriers and obstacles are identified. The value of scientific and rational knowledge and methods is recognized. But the need to integrate extrarational perspectives, values, and interests also is acknowledged. The distinction between natural and human (e.g., social, political, and economic) environments is seen as forced and inhibiting. Distinctions among environmental management instruments, of which EIA is only one, also are seen as artificial. Pre-approval analysis is no longer the preoccupation. Instead, continuous management approaches are advocated that extend through implementation and that rely heavily on monitoring and adaptive management.

The scientific and rational EIA processes (described in Chapters 3 and 4) display few of these characteristics. They can assume a valuable supportive role. However, they appear poorly suited to integrating ecological substance and the EIA process. The assumption that process and substance are independent cannot be supported. A substantive EIA process will be conducive to integrating ecological perspectives, knowledge, and methods only if it is designed and managed with a sound appreciation of the procedural implications of substantive characteristics. Judging from the characteristics of the concepts presented in Table 5.1, it will be almost impossible to imbue an ecological perspective into a comprehensive, rigid, closed, top-down, and lineal EIA process that assumes implementation and a high degree of predictability and control.

It does not follow that the themes and concepts presented in Table 5.1 can be transferred directly into a substantive EIA process. Concepts, such as the ecosystem approach and adaptive management, have a mixed track record. Many concepts and principles are in dispute. Application in practice has sometimes proven problematic. The concepts are general, with a limited empirical foundation (Slocombe, 1993). In common with scientific and rational EIA processes, they tend to be expert-centered. They are often hampered by data limitations (US CEQ, 1997a). There are significant differences between ecological and social systems (Slocombe, 1993; US CEQ, 1997a). Procedures for overcoming institutional and implementation barriers, for managing uncertainty and for facilitating adaptation, creativity and collaboration are far from fully developed (Hooper et al., 1999). The concepts encompass such a wide range of aspirations (several of which are pursued in other chapters) that they run the risk of becoming either overly general or overly complex. While appreciating and addressing these potential limitations and obstacles, these concepts still can help make EIA processes more substantive.

5.4.2 Social Concepts

Social, as applied in impact assessment, is a term or concept that is broad and therefore difficult to define. It also is defined and applied in different ways (Burdge, 2002). SIA can refer to the distinguishing characteristics of people, communities, and society (e.g., demographic, cultural, institutions, customs, traditions, political systems). It can involve multiple levels of human aggregation (e.g., families, groups, organizations, communities, society). It can include perceptions, attitudes, norms, values, aspirations, and beliefs. It can pertain to patterns of association, interactions, and interdependencies. It can refer to health and social well-being, quality of life and living environment, economic and material well-being, cultural values and integrity, personal and property rights, and gender relations. Depending on the definition, it can apply theories, concepts, and methods from such disciplines as economics, anthropology, political science, psychology, history, philosophy, and archaeology and such professions as land-use planning, social planning, landscape analysis, health planning, risk management, resource management, public involvement, and environmental management. It is highly interactive with the physical (both natural and built), with resources, and with the ecological. It is both a field of study (e.g., social sciences) and a field of application (e.g., social impact assessment). It is an evolving field that embraces a diversity of distinct, partially overlapping and partially conflicting concepts, models, theories, perspectives, and frameworks. There are numerous frameworks available for structuring social criteria.

EIA texts generally treat social impacts as a distinct discipline, but for the most part, as a subset of EIA. Sometimes social and economic impacts are considered separately. Sometimes they are combined under the umbrella of socioeconomic impacts. Occasionally, health impacts and cultural, historical, and archaeological impacts are addressed separately. The track record of integrating SIA into agency decision making and into the assessment process has at best been mixed (Bronfman, 1991). Too often, SIA has assumed a marginal decision-making role (Burdge, 2002; Lockie, 2001).

The major stages in the SIA process largely parallel those of the EIA process. Sometimes SIA processes begin with public involvement. They often include a separate stage for predicting public responses to impacts (Finsterbusch, 1995; Interorganizational Committee, 1994). Social impacts are not the same as ecological impacts (Barrow, 1997). People react in anticipation of and adapt to change. Human reactions vary greatly among individuals and groups and over time. Social phenomena are difficult to predict (Finsterbusch, 1995). Social units are not fixed structures. Social phenomena involve adaptive interactions. SIA involves both social change processes (intervening variables that may lead to impacts) and social impacts (intended and unintended consequences on the human environment from planned interventions) (Vanclay, 2002). The EIA process can influence how people anticipate and adapt to change.

Table 5.2 identifies and briefly describes several social concepts potentially relevant to EIA process management. In considering these concepts the dangers of preconceptions and implicit assumptions about the conduct of the process, about the choice and application of methods, about the perspectives of potentially interested and affected parties and about potential social impacts, are immediately apparent. The analysis and interpretation of social impacts should be approached with caution. Assumptions should be carefully scrutinized. Ongoing adaptations will be required. The world and proposed actions should be seen through the eyes of and as experienced by potentially affected parties. If social impacts are, in part, socially constructed, this suggests a socially constructed EIA process (i.e., collaboratively designed and managed with interested and affected parties). Perspectives, norms, perceptions, beliefs, and values will change over time and can vary greatly depending on individual and group characteristics and depending on the level of social aggregation. This suggests iteratively exploring social impacts from multiple perspectives and at multiple levels. The magnitude and nature of social impacts are partly dependent on the process. The process is both an end (e.g., to provide a sound decision-making basis) and a means (e.g., a way of facilitating community empowerment, of avoiding and ameliorating adverse social impacts, of generating and enhancing social benefits). This implies a process sensitive to public perceptions and perspectives, that actively seeks to manage positive and negative impacts from the outset (i.e., impact management as an continuous function), and that facilitates the achievement of community objectives.

Meaning and value are socially determined and are adjusted through social interactions. Dialogue is central to social interactions. Distortions in dialogue can exacerbate social impacts. The EIA process is a form of social interaction. Dialogue is a central attribute of the process. This suggests designing and managing the process to facilitate dialogue, to contribute to co-learning, and to minimize communications distortions (Lockie, 2001). It also points to the need to understand how the EIA process, as a form of social interaction, fits within and potentially affects existing social interaction patterns. Social interactions and impacts are both political and ethical. This suggests an EIA process consistent with procedural and ethical principles and standards (see Chapter 9) and conducive to attaining political objectives (see Chapter 7). A reorientation of the SIA process, consistent with the view that SIA is primarily a mechanism for facilitating constructive social and political interaction and change, could result in less emphasis on impact prediction and more stress on co-learning and impact management (Lockie, 2001).

A taxonomic EIA approach (which assumes minimal interactions among impact categories) is highly inappropriate, given the dynamic nature of social interactions and impacts. What is required instead is an EIA process built around conceptual models, frameworks, and stories that explore and trace through patterns of interaction and available choices, from multiple perspectives (Vanclay, 2002).

Table 5.2 Examples of Potentially Relevant Social Concepts

Technical SIA	Relies heavily on natural and social sciences
	Seeks to maximize net social welfare (utilitarianism)
	Employs an adapted rational planning process
	Sees SIA as a technical component of the planning and decision-making process
	Uses a reductionist and objective research mode
Political SIA (also social conflict)	Conflict over resources and interests central to social life
	Interest-based approach; decision-making value-laden and political
	Social life is diverse; social order is based on manipulation and control by dominant forces
	Tends to be issue-oriented; emphasizes openness, rectifying inequities, and empowerment
	Seeks to understand basis for conflicts and how conflicts escalate
	Seeks to manage or contain conflict sufficient to identify mutually acceptable actions
	Stresses need to strengthen the local institutional base (i.e., capacity building, community development)
	May be a realistic approach when positions are polarized and world views conflict
Positivistic social science	Modeled after natural sciences; hypotheses tested by carefully analyzing the "numbers"
	Researcher as detached, neutral, and objective
	Stresses the value of an experimental, objective research approach, which seeks to explain cause–effect relationships logically
	Although flawed in its assumption of objectivity, is still instructive in terms of systematic and explicit research procedures; helpful in detecting methodological bias but contains own, often implicit, assumptions
	Social sciences can be difficult to apply in SIA because of inconsistencies in units of analysis, theoretical models, and language; social scientific traditions tend to be critical and discursive rather than predictive and explanatory
Functional, ecological, and systems theory	Assumes shared norms and values in society
	Assumes a stable, cohesive, consensus-based, and orderly social system; based on reciprocity, cooperation, and recognition of authority
	Assumes that system units are functionally related; change seen as an outside disturbance to an otherwise harmonious system; change accommodated by subtle shifts in system parts
	Reflected in most rational and participatory EIA processes
Interpretative social science	Adopts a practical approach; not value-free; common sense a vital information source
	Seeks to understand how people manage their everyday lives and construct meaning in natural settings
	Recognizes that people experience social reality in different ways
	Sees the unique features of specific contexts as essential to an understanding of social meaning
	May be helpful in addressing community-level impacts

(Continued)

Table 5.2 (*Continued*)

Critical social science	Sees social science as critical and action-oriented; a political, moral activity
	Research conducted to critique and transform social relations
	Focuses on identifying and rectifying distortions and inequities
	Argues that social reality has multiple layers (illusions, myths, distortions, false consciousness)
	Potentially useful for addressing community empowerment issues
Exchange theory	Assumes human behavior reflects peoples' attempts to maximize rewards (utility) for involvement
	Expects people will only become involved (and will continue to be involved) if they will benefit (or are rewarded) for their involvement; interaction seen as an exchange of rewards
	Requires a careful analysis of rewards (e.g., monetary, prestige, power, appreciation) that a setting is offering and how the rewards and patterns of exchange will be affected by a given change
	Sometimes basis for expectation that level of community acceptance will increase with the level of local benefits and compensation offered; of dubious validity and can be ethically problematic
	Has been incorporated into some siting approaches and explains some behavior
Symbolic meaning	Focuses on the inferred meaning attached to actions rather than to actions themselves
	People learn meanings and symbols in social interactions; can also alter meanings through introspection (their own interpretations of situations) and through interactions
	Conflicts may be exacerbated by definitions of situations
	Definitions of the situation by groups and individuals highly relevant to EIA practice
Social learning	An approach for linking social concerns and public participation
	Includes both cognitive enhancement (e.g., learning about problem, learning about the values and interests of others) and moral development (e.g., developing a sense of self-respect and respect for others, developing moral reasoning skills)
	Potentially useful procedure for integrating social and moral considerations into participatory planning approaches
Phenomenological sociology	Focuses on describing and studying one's own and others, experiences without preconceptions
	Importance of avoiding preconceptions about external causes and consequences
	Opposed to objectivism, positivism, the acceptance of unobservable matters, and unsupported speculative thinking
	Seeks to analyze and describe everyday life; assumes that people create the world rather than being formed by social forces
	Although highly theoretical underscores the need to begin with as few preconceptions as practical and to start from public and other stakeholder experiences and perspectives

Sources: Burdge and Vanclay (1995), Craib (1984), Craig (1990), Halstead et al. (1984), Lee (2000), Manring et al. (1990), Newman (1997), Ritzer (1996), Webler et al. (1995).

The social sciences can make an important contribution to designing and applying an SIA/EIA process. But it is not a simple case of directly applying social science methods and models. There are multiple overlapping and conflicting social scientific models and methods available. There are numerous interpretations of the appropriate purposes for and conduct of applied social research. Moving from the theoretical and the explanatory to the prescriptive and the practical can be very difficult. Contextual adjustments are essential. It is especially important to identify and appreciate the implications of knowledge, resource, and control constraints and obstacles.

The often-marginal role of social considerations in EIA processes suggests the need for SIA practitioners to proactively advocate and extend the role of SIA within and among organizations (Bronfman, 1991). The impartial analyst role is insufficient. It also implies the need for clearer definitions, enhanced methods, more follow-up research, a concerted effort to enlarge the SIA knowledge base, and a reconsideration of the nature of the SIA process (Burdge, 2002; Lockie, 2001).

In common with the ecological concepts, the social concepts can shape and be influenced by the EIA process. Process and substance are intertwined. It is not possible to identify a precise set of performance standards for a substantive EIA process. However, general procedural characteristics, more and less conducive to combining substance and process in the EIA process can be identified.

5.4.3 Sustainability Concepts

The roots of sustainability or sustainable development, as a concept, have been traced well back into the nineteenth century and beyond. The definition most commonly used as a point of departure is that of the World Commission on Environment and Development (WCED): "development that meets the needs of the present without compromising the ability of future generations to meet their own needs" (WCED, 1987, p. 8). There is a continuing debate surrounding the definition of *sustainable development* or *sustainability* (*sustainability* is used here for the sake of brevity). The debates have concerned whether, for example, the definition should be broadened or adapted to address intrageneration inequities, spatial inequities, human aspirations, other species needs, public participation in decision making, ecological limits, relationships among sustainability forms, and sustainability instruments. Common to most definitions is a desire to maintain, over an indefinite future, necessary and desired attributes of the sociopolitical system and of the natural environment (Deakin et al., 2002; Robinson et al., 1990).

Some key interrelationships among sustainability elements are highlighted in Figure 5.2. The definition is refined through sustainability forms and ethical perspectives. Sustainability forms concern overlapping and interdependent value systems (e.g., ecological, social, economic) (Sadler, 1996). There is much debate concerning where the greatest emphasis should be placed and regarding how best to address interactions and interdependencies. Underlying these debates are a multiplicity of institutional, ideological, and academic sustainability perspectives and world views (Mebratu, 1998). These perspective differences are reflected in

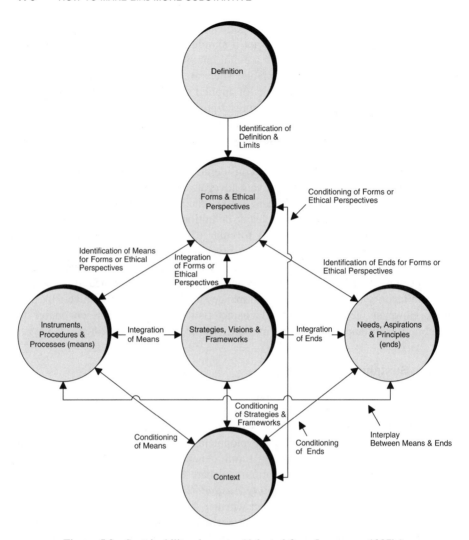

Figure 5.2   Sustainability elements. (Adapted from Lawrence, 1997b.)

varying positions regarding such matters as the treatment of growth, the role of government and the market, and the importance of ecological limits and social justice (Constanza, 2000). The net result is multiple versions of sustainability (Robinson, et al., 1990). Notwithstanding such differences, there are broad principles and imperatives common to many sustainability perspectives.

Sustainability forms and perspectives provide a basis for identifying sustainability ends (needs, aspirations, and principles) and sustainability means (instruments, procedures, and processes). Sustainability ends are both procedural (e.g., openness, fairness, participation) and substantive (e.g., ecosystem integrity, protection of biological diversity, enhanced quality of life, satisfaction of basic human needs, social justice). Sustainability ends can assume different forms (e.g., goals, principles) and can operate at varying levels of detail (e.g., broad goals, specific objectives, or priorities). Sustainability means are the mechanisms by which stakeholders work separately and together to move toward sustainability ends (e.g., green planning, legal instruments, indicators, financial incentives and penalties, institutional reforms, direct citizen mobilization, applied research, consumption and lifestyle choices, forums for joint planning and cooperation). Sustainability forms, ends, and means are brought together in sustainability strategies, visions, and frameworks. It is through such integrative mechanisms that core sustainability principles, themes, limits, decision rules, approaches, and methods are identified and applied (Brooke, 1998; Devuyst, 1999; Gibson, 2001). Part of application involves adapting ends, means and strategies for different situations (Shearman, 1990).

The general interrelationships between sustainability and EIA are illustrated in Figure 5.3. Sustainability and EIA can be integrated at three levels: the conceptual level (theory and research), the regulatory level (sustainability related EIA requirements), and the applied level (integrating sustainability concerns into EIA practice). Both EIA and sustainability are concerned with maintaining and enhancing ecological, economic, and social environments. They reform, manage, and apply science and technology, institutional arrangements, and human environmental interventions. Both address interrelationships within and between environments and human activities. Sustainability can and should provide a means for redefining EIA. EIA can and should be an instrument for facilitating sustainability.

EIA texts tend to assume that sustainability is an input to and an output from the conventional EIA process. The essential features of the process remain unchanged. The relationships between sustainability and the EIA process, however, are more complex, as illustrated in Figure 5.4. (*Note:* The numbers in Figure 5.4 are explained below.) Sustainability offers the potential to extend and, in many senses, complete the EIA process. Project-level EIA and SEA, for example, identify, predict, and manage direct, indirect and cumulative effects (1). Cumulative effects tend to be addressed incrementally (e.g., project effects in combination with the effects of related activities). Effects are projected into the future, usually assuming current trends persist. Sustainability starts with holistic images (often both desirable and undesirable) of the future (7). It begins from the whole and moves to the parts. It adopts a long-term perspective. It does not assume that there is a single potential future or that trends define the future. It both extends from the present and traces connections back from the future.

Conventional EIA and SEA approaches usually address significance in terms of the importance of individual environmental components, effects, or interactions. Relationships between significance determinations and context are often poorly defined. Sustainability visions and limits provide a context and a touchstone for significance interpretations and impact management actions (2) (Sadler, 1996). Effects can be assessed for their contribution to sustainability visions, goals, targets, and principles (3) and, where practical, regarding whether they jeopardize environmental or social carrying capacities or thresholds (4) (Noorbakhsh and Ranjan, 1999; Sadler, 1996). Proposed actions or alternatives can be a catalyst for sustainability.

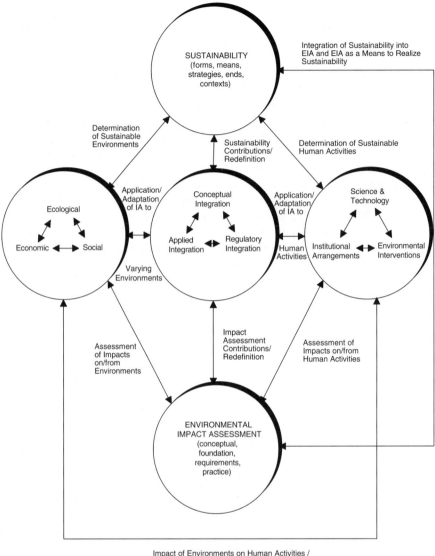

Figure 5.3 Integrating EIA and sustainability. (From Lawrence, 1997b.)

Sustainability assessment could be considered the highest "rung" in the assessment ladder (6). It provides a context for SEA and links local actions to global concerns (Berke, 2002). Broad strategies that seek to integrate individual SEAs could be subjected to a sustainability assessment. Sustainability analyses could incorporate global and transboundary effects and priorities into integrative strategies, into

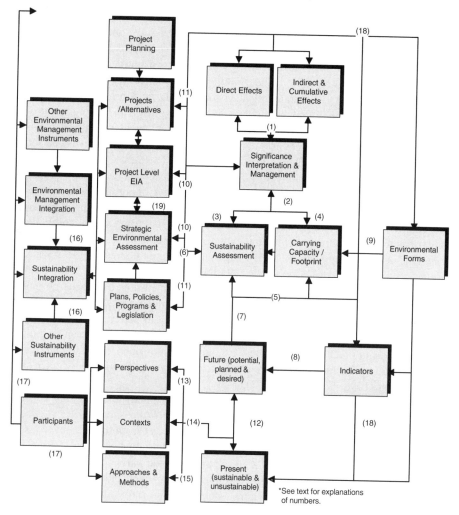

Figure 5.4   Role of sustainability in the EIA process.

lower-order SEAs and into project-level EIAs. The database for EIA and SEA often has gaps and inconsistencies, especially as scales are broadened (to, for example, address cumulative effects) and as time horizons are extended. Sustainability indicators, which combine and supplement environmental, social, and economic indicators, can address these gaps (8). Sustainability indicators also help determine whether ecological and social thresholds are being approached and whether progress is being made toward sustainability targets (Jepson, 2001).

Conventional EIA and SEA commonly treat social, economic, and ecological effects separately. Critical links among such effects are usually considered when addressing cumulative effects. Sustainability recognizes that social, economic, and ecological systems are highly interdependent (9). It uses holistic visions and integrative frameworks to address interdependencies from the outset. Sustainability decision rules often focus on interdependencies (e.g., economic growth within ecological carrying capacity). Sustainability can help focus project-level EIAs and SEAs (i.e., contributing to or undermining of sustainability) (10). EIAs and SEAs can be guided by sustainability visions, goals, and principles. They can broaden spatial and temporal boundaries to address global and intergenerational impacts. They can characterize baseline conditions in terms of sustainable and unsustainable activities and environments. The generation of alternatives can focus on choices likely to be conducive to sustainability (11). Rather than being viewed only as environmental intrusions (to be ameliorated to acceptable levels), proposed actions can be treated instead as potential sustainability opportunities or catalysts. Unsustainable alternatives can be screened from consideration (e.g., threat to carrying capacity). The remaining alternatives can be evaluated using sustainability decision rules (e.g., maintenance of natural capital, waste generation within assimilative capacity, renewable resources within regeneration rate, nonrenewable resources equal to substitution rate) (Goodland, 1993; Noorbakhsh and Ranjan, 1999; Sadler, 1996). Mitigation, compensation, and local benefits measures can focus on maintaining and enhancing sustainability (e.g., in-kind compensation for natural capital loss). Before-and-after comparisons can be undertaken to determine sustainability-related changes (Thérivel and Minas, 2002). The overall EIA process can be guided by sustainability procedural principles (e.g., keep options open, precautionary principle, a fair, accessible, efficient, and effective process) (Beatley, 1995; Gibson, 2001; Sadler, 1996; Slocombe, 1993).

Sustainability recognizes that there are multiple perspectives concerning how the world is, how the world is likely to be, and how the world should be (13). These perspective differences result in many different pathways from the present to a sustainable or unsustainable future (12). Sustainability initiatives seek to identify the "overlapping consensus" among interested and affected parties that will provide a basis of action (17) (Rawls, 2001). A variety of approaches and methods can both define a sustainable future (e.g., visions, scenarios) and assess the contributions of individual proposals (e.g., apportionment techniques, sustainability indices, footprint analysis) (14) (George, 1997; Wackernagel and Rees, 1996). Adjustments also are made for different settings and situations (15). Multiple perspectives and methods are available for characterizing present and potential future conditions.

Project-level EIA and SEA operate largely independently from other sustainability instruments. Connections are made to other instruments but usually only after the process is well advanced, often during the review and approval stage. Sustainability initiatives recognize that many mutually supportive instruments are required (16). Efforts are made to ensure complementary visions, actions, and monitoring systems (18). The EIA process could be modified to more effectively address the advantages and constraints associated with integrating sustainability concerns into both EIA and SEA (19). These efforts could be broadened to embed, from the outset, SEA and EIA within the full network of sustainability instruments (16, 19).

Many issues, obstacles and dilemmas remain concerning how best to integrate sustainability and the EIA process. The theoretical base for sustainability is not well developed. Many questions are still being raised regarding how best to determine what is sustainable, over what area, and for how long a period (Briassoulis, 1999; Shearman, 1990). There are many debates concerning who is to decide what is and is not sustainable (Robinson et al., 1990). Apportionment procedures, how to consider uncertainties, and the treatment of compromises and trade-offs are difficult issues requiring further attention (Gibson, 2001). The fragmentation of disciplines, sectors, and institutions continues to hinder integration efforts. Some argue that sustainability is either not possible or is a smokescreen for "business as usual." Others suggest that more fundamental changes in values and behavior are necessary before any discernible progress toward sustainability can be made. On the bright side, the range of sustainability initiatives is enormous and the record of tangible improvements from these initiatives is considerable. Sufficient experience in undertaking sustainability assessments or appraisals has already been acquired, so that effectiveness factors (e.g., broad local involvement, early in the process, adequate resources) have been identified (Thérivel and Minas, 2002). An optimistic interpretation would be that EIA practice can build on the successes while appreciating and addressing the constraints. It is an overstatement to suggest that EIA has made more that a minor contribution to sustainability to this point. EIA process reforms, along the lines described above, could increase that contribution.

5.4.4 Methods

Formulating and applying a substantive EIA process requires numerous methods sensitive to ecological, social, and sustainability characteristics and objectives. Examples of potentially relevant methods are described briefly in Table 5.3. The

Table 5.3   Examples of Potentially Relevant Substantive Methods

Network analysis and systems diagrams	Network diagrams used in causal analysis; traces in two dimensions links between actions and environments
	Integrates impact causes and consequences by identifying primary, secondary, and tertiary interrelationships
	Systems diagrams: conceptual models of environmental systems (e.g., energy flows among environmental components)
Modeling	Mathematical equations and computer simulations of an environmental system or system component
	Integrates standard natural and social sciences modeling approaches with increased computer use and capabilities
	Simplified system representation and explicit system behavior assumptions
	Can describe, explain, and/or predict characteristics
Projection and forecasting	Projection: hypothetical assumptions entered into a mechanistic quantitative procedure; tendency to assume that projections are the most probable or the most desirable future; often neither

(Continued)

Table 5.3   (*Continued*)

	Forecast: a best guess about the future; adds judgment regarding future behavior and other assumptions; tendency to be dry and technocratic (e.g., based on economic, demographic, and environmental data predictions, limited number of variables, statistical emphasis, trend extrapolation, narrow range of probabilities, near future); lacking in vision
	Conveys impression that value neutral and apolitical; often neither; often weak on social and prone to conservative bias
	Can be helpful for testing alternative assumptions and model parameters; based on present choices and behaviors
Backcasting	Futures method in which a desirable future endpoint is identified so that it is possible to identify the decisions and actions necessary to achieve the desired endpoint; makes human behavioral assumptions explicit
	Requires iterations to resolve inconsistencies and to mitigate adverse economic, social, and environmental impacts
Visioning	Seeks to create a shared vision of a desired future; must be shared to be responsible
	Collaborative process for melding images of an ideal future state; aspirations of a group; optimistic picture rather than a fantasy; has generally emphasized goals and process over means
	Can be supported by visualization techniques; basic techniques: drawings, maps, photographs, physical models; more computerized/contemporary methods—electronic pen and paper, GISs (geographic information systems), CAM (computer-aided mapping), MIMS (mapping information management systems), image-editing programs, motion picture, video, three-dimensional digital models, virtual reality, simulators
	Focuses on what really want (clear goals); has to be flexible and evolving; can map out future possibilities
	Often based on worldviews (e.g., technologically optimistic vs. technologically pessimistic, big centralized government vs. small, localized government, high consumption vs. low consumption); assumptions, characteristics, and uncertainties can be tested against sustainability criteria; important that feasible and politically realistic
Scenario writing	Stories about events that would affect decisions; stories invented about plausible futures and their implications; describes in narrative form the unfolding of events, reactions of key actors, and consequences, including measurable costs and benefits as well as intangible changes; can be combined with gaming
	Distinctions: state (future at year *x*) vs. process (sequence of events leading up to); end state– vs. beginning state–driven; for planning vs. for prediction; process planning end state (idealization), process prediction end state (prophecy), process planning beginning state (developmental), and process prediction beginning state (simulation)
	Can address such matters as economic activity, demographic factors, social values, and governance styles; often distinguished on ideological basis (e.g., conservative, reformist, radical)
	Examples of forms: expert planning (use of consultants), morphological approaches (based on driving forces) and cross-impact approaches (explores interrelationships and dependencies between potential events and conditions)

Table 5.3 (*Continued*)

	Value in responding intelligently to plausible future; heart—the rehearsal of scenario implications; can accelerate response rates to eventualities
Story telling	Meant to convince people to adopt storyteller's preferred course of action
	Illustrates whole problem: actors and plausible chains of events; coherently integrates images of past, present, and future, supported by evidence and encompasses opposing points of view
	Ideally consistent, testable, morally acceptable, actionable, and beautiful (graceful, subtle, elegant)
	Helpful in explaining the significance and implications of present and future events
Ecological footprint (also carrying capacity and ecological thresholds)	Footprint analysis: land or water area required to support a defined human population and material standard indefinitely; a biophysical measure for natural capital and a method to monitor human use of ecological capital
	Carrying capacity—ecological: threshold of stress below which populations and ecosystem functions can be sustained; in social context generally measured by level of services
	Assumes that natural thresholds exist; can be derived from expert opinions, government authorities, or surveys; can also use mathematical equations to estimate critical levels for environmental parameters of most concern; projects assessed in relation to carrying capacity; includes monitoring of unused capacity
	Useful for assessing cumulative effects; possible basis for determining sustainability
	Typically, a high degree of uncertainty about whether thresholds breached; difficult to measure directly; may be multiple thresholds; often, necessary regional data absent
Life-cycle analysis	An analytical environmental management tool that considers the environmental impact of a product, process, or human activity over its entire life cycle (cradle to grave); implementation has rarely been complete
	Based on a scientific understanding of process inputs (e.g., retrieval and consumption of raw materials) and outputs (e.g., the fate of all pollutants and residuals and environmental effects)
	Permits evaluation of the environmental consequences of alternative processes and design concepts
	Quantifies inputs and outputs, characterizes environmental and health effects, and evaluates improvement opportunities
Rapid rural and participatory rural appraisal	RRA aims to extract information from outside community; focuses on selected variables using triangulation process (secondary sources, reconnaissance surveys, and field data gathering)
	PPA: enables local people to conduct own analyses and take action; reflects recognition of legitimacy and practical value of traditional environmental responses
	Both address time constraints—the rapid assembly of diverse information

Sources: Al-Kodmany (2002), Bardwell (1991), Canter (1996), Cole (2001), Constanza (2000), Dearden and Mitchell (1998), Gilpin (1995), Hegmann and Yarranton (1995), Hirschorn (1980), Ison and Miller (2000), Jepson (2001), Morgan (1998), Myers and Kituse (2001), Noorbakhsh and Ranjan (1999), Patton and Sawicki (1993), Sadler (1996), Shoemaker (1994), Skea (1999), Smith (1993), US CEQ (1997b), Van Der Vorst et al. (1999), Wachs (2001), Wackernagel and Rees (1996), Walker and Johnston (1999).

ecological, social, and sustainability concepts, described in previous subsections demonstrate the need to address interrelationships systematically. EIA practice makes considerable use of network analysis, systems diagrams, and modeling to address interconnections and interdependencies. The systematic consideration of interrelationships finds its fullest expression in cumulative effects assessment (CEA). As outlined in Table 5.4, all four jurisdictions provide guidance for selecting and applying CEA methods. Methods for addressing interrelationships among disciplines are not as fully developed as those for considering interrelationships within disciplines.

Table 5.4   Examples of CEA Methods Guidance

UNITED STATES

CEQ handbook on cumulative effects assessment (CEA) methods (US CEQ, 1997b)

Sees analyzing cumulative effects as enhancing the traditional components of EIA; focuses on scoping (what counts), determining baseline and thresholds of environmental change (environmental description), and tracing through cause–effect relationships (consequences)

Identifies general principles, including the sustainability of resources, ecosystems, and human communities; also stresses the need to assess accumulative effects or synergistic effects and capacity of each resource, ecosystem, and human community affected to accommodate additional effects

Identifies and provides examples and types of cumulative environmental effects

Describes processes for scoping cumulative effects, describing the affected environment, and for determining the environmental consequences of cumulative effects

Presents overview of characteristics, strengths, and weaknesses of major categories of methods (i.e., questions, interviews and panels, checklists; matrices, network and systems diagrams, modeling, trends analysis, overlay mapping and GIS, carrying-capacity analysis, ecosystem analysis, economic impact analysis, social impact analysis); includes examples

CANADA

Federal reference guide and practitioners' guide (CEAA, 1994; CEAWG and AXYS, 1999)

Reference guide presents overview of concept, describes regulatory requirements, addresses general issues (advice and consultation, documentation, uncertainty, level of effort), and presents a CEE framework (treatment of CEE during scoping, analysis, mitigation, significance determination, follow-up); appendix describes procedures for identifying future projects); references provided for methods

Practitioners' guide elaborates on concepts and framework; includes methods for each component, describes procedures for preparing and completing a CEA, addresses adaptations (small actions, regional planning, and land-use studies), includes case studies and provides overview of cumulative effects history in Canada

Methods described in practitioners' guide include procedures for establishing temporal and spatial boundaries, options for selecting future actions, interaction matrices, impact models, spatial analysis using GIS, indicators, numerical models, and significance determination procedures

Alberta guide provides an overview of CEA (definitions, concepts, other jurisdictional approaches), addresses relationship to regulatory requirements and describes CEA toolbox (disciplines, concepts, frameworks, techniques, and technical aids) (Hegmann and · Yarranton, 1995)

Table 5.4 *(Continued)*

Techniques addressed in Alberta guide include information organizers, analysis methods, teamwork methods, socioeconomic impact assessments, comprehensive economic approaches, comprehensive socioeconomic models, risk assessment, and CEA techniques; over 50 techniques described

EUROPEAN UNION

Guidelines describe practical approaches and methods for addressing indirect and cumulative effects (Walker and Johnston, 1999)

Describes regulatory requirements; defines, provides rationale, and discusses integration of indirect impacts, cumulative impacts, and impact interactions; part of rationale the promotion of sustainable development

Describes and identifies advantages and disadvantages of expert opinion, consultations and questionnaires, checklists, spatial analysis, network and systems analysis, matrices, carrying capacity, and threshold analysis and modeling methods; presents examples of each type of method; describes how to apply tools and methods, and provides case studies

Describes treatment of CEE in scoping, project characteristics, the receiving environment (boundaries, data collection), and impact assessment (scoping, magnitude and significance, mitigation, monitoring, problems and uncertainties, reporting)

Identifies potential role of tools and methods for each EIA activity

Case study analyses in appendices include background to the project, overview of methodology, types of cumulative and indirect impacts, and how indirect and cumulative impacts identified and evaluated

AUSTRALIA

Assesses potential role of cumulative impact assessment and strategic assessment in EIA; a background study for the review of the Commonwealth EIA process (Court et al., 1994)

Provides rationale for EIA to address cumulative impacts and strategic assessment; describes study approach and methodology, analyzes responses to consultations and review of current practice, addresses legal aspects, explores options for future application of CIA and SEA in EIA, presents case studies, and offers recommendations

Provides a systematic rationale for addressing cumulative impacts and SEA [general need, links to ecologically sustainable development (ESD), scientific needs, institutional needs] in commonwealth EIA process

Brief references to CIA methods (e.g., ad hoc, checklists, matrices, networks, mapping and overlays, modeling, weighting/evaluative methods, adaptive procedures, biogeographic theory)

Overview of overseas theory and practice highlights conceptual issues, definitions, methodological approaches, assessment and predictive tools in CIA, interactions of SEA and ESD, methodological approaches to SEA, and institutional matters

Includes two case studies with descriptions of methods applied for addressing CEEs

EIA is about decision making for the future. Substantive EIA processes must consider long-term implications and explore pathways toward and back from sustainable futures. EIA generally relies on projection and forecasting techniques when anticipating future conditions. Although helpful, such techniques provide only a partial picture of a potential future. They also are weak on social concerns, are lacking in vision, often underestimate uncertainties, and are prone to quantitative and conservative biases. Visioning, scenario writing, and storytelling are better able to integrate qualitative, social, ecological, and political considerations. They can also provide multiple images of a desired future and of varying routes to that future. Backcasting helps work back through decisions and actions from a desired future to the present.

The EIA process should establish proximity to thresholds, assess progress toward ecological, social, and sustainability ends, and compare alternative courses of action. Ecological footprint analysis, carrying capacity analysis, and environmental indicators (see Table 5.1) can help assess status and choices. The process must adapt to and manage uncertainties and data gaps. Rapid rural appraisal, scenario writing, and adaptive environmental assessment (see also Chapter 10) are well suited to addressing uncertainties. A high level of community participation is essential for making EIA processes more substantive. Visioning, storytelling, participatory rural appraisal, the ecosystem approach, and social learning can all help in involving interested and potentially affected parties.

5.5   INSTITUTING A SUBSTANTIVE EIA PROCESS

5.5.1   Management at the Regulatory Level

All four jurisdictions (the United States, Canada, the European Union, and Australia) explicitly combine procedural and substantive requirements, albeit in different ways and with varying degrees of success. In the United States, the goals of the National Environmental Policy Act (NEPA) are consistent with sustainability. NEPA applies to a broad range of federal actions. Explicit reference is made to environmental and resources impacts over the short and long terms and to addressing irreversible and irretrievable resource commitments. Social and economic effects are considered only if they are related to physical or natural environmental changes. Specific substantive environmental requirements are addressed largely through a host of regulations, related legislation, executive orders, and directives. Both general and specific guidance is provided. The requirements and guidelines concern such matters as protecting human health and valued and sensitive natural and built environmental features, species, habitats and resources, energy conservation, environmental quality, pollution, and waste prevention. Ecological and sustainability concerns considered include biodiversity, cumulative effects, ecological processes and habitat protection, ozone depletion, ecological risk assessment, life-cycle impact assessment, biobased products and bioenergy, transboundary effects, and effects and activities in the antarctic. Social impacts are partially addressed with requirements and guidelines related to heritage resources, noise control, and environmental justice. An interagency committee has provided some substantive SIA guidance (Interorganizational Committee, 1994). It is unclear the extent to which the latter report is applied in regulatory practice. Applied research and coordinative initiatives have considered such matters as regional, interagency, and intergovernmental ecological management, environmental data and trends, integrating EIA with environmental regulations and permitting, and the application of the ecosystem approach and of adaptive management and monitoring.

Substantive environmental concerns are integrated into U.S. EIA procedures partially through significance interpretations. Requirements and guidelines also stress the need to incorporate substantive environmental concerns into existing NEPA levels of analysis (e.g., national, regional, site-specific) and analysis components (e.g., scoping, public participation, affected environment, impact analysis, alternatives, record of decision, monitoring). The U.S. approach assumes that cross-cutting substantive environmental requirements can be married with largely procedural EIA requirements (except in the sense of general goals and requirements), without redefining the EIA process. The U.S. approach provides considerable substantive guidance. However, the pieces do not necessarily fit together into a coherent whole. Unified visions of sustainable futures are largely lacking. Data gaps are inevitable. Requirements can be interpreted too narrowly, as when cultural resources are considered to be equivalent to rather than distinct from historical resources (King, 2002).

The dispersal of responsibility across multiple agencies, geographic areas, and levels of government can exacerbate existing coordination and consistency problems. Place-based planning can be problematic. These difficulties are recognized and efforts are being made to address them. The Council on Environmental Quality's effectiveness review of NEPA (US CEQ, 1997a) stresses the value of place-based ecosystem approaches and of adaptive management. Similar arguments are made in a report on the ecosystem approach prepared by an interagency task force (IEMTF, 1995). Both documents describe knowledge and implementation obstacles and issues. They also suggest approaches for addressing these concerns and provide numerous positive case examples. It is unclear how far this suggested reorientation has permeated into U.S. regulatory EIA practice.

The Canadian federal approach to combining substance and process is both similar to and different from the U.S. approach. The purposes stated for the *Canadian Environmental Assessment Act* refer to promoting sustainable development. However, the definition of the environment is narrow (only indirect socioeconomic effects are included). The requirements in the main body of the act address sustainability concerns only partially. Cumulative effects must be addressed. Comprehensive studies are required to consider renewable resource capacity. The act includes transboundary and international effects provisions. There are a few environmental triggers for the act (e.g., national parks and protected areas, activities in wildlife areas, effects on migratory birds, arctic waters pollution, fisheries impacts). The timing and application of these triggers has been uneven.

Guidelines address some substantive environmental concerns, including cumulative effects, biodiversity, physical, and cultural heritage resources and health effects. Guidelines refer to some broader strategies (e.g., Canadian Biodiversity Strategy). Selective references are made to links between EIA and the precautionary

principle, thresholds, traditional knowledge, sustainable development, risk management, health, SIA, and environmental management. A 1983 study (Beanlands
and Duinker, 1983) sponsored by the Canadian Environmental Assessment Agency
(CEAA) provides detailed guidance concerning the integration of ecology into EIA.
This study is sometimes referred to in generic and project-specific guidelines. Some
substance guidance also has been provided, through studies sponsored by the Canadian Environmental Assessment Research Council in the 1980s, concerning such
topics as SIA, CEA, mathematical models, risk management, EIA and economics,
and native knowledge. The guidelines generally assume that substantive environmental requirements can be integrated into conventional EIA activities.

The Canadian EIA requirements and guidelines (at the project level) address
substantive environmental concerns selectively and at varying levels of detail. A
coherent and clearly defined role for EIA requirements, within the context of a
broader strategy for achieving specific substantive ecological, social, and sustainability objectives, has yet to be defined. Some elements of such a strategy are emerging. The federal government has a nonlegislated EIA process for all federal policy
and program initiatives submitted for cabinet review. This SEA requirement has a
broad definition of the environment (ecological, social, economic), seeks to
optimize positive environmental effects, and is intended to implement agency
sustainable development strategies. Each federal agency is required to prepare a
sustainable development strategy. The federal government has a green plan. The
Commissioner of the Environment and Sustainable Development performs an oversight function. Substantive CEAA research priorities include climate change, biodiversity, transboundary pollution, regional environmental effects, and human
impacts. A government-wide assurance program is being instituted. Guidelines
for the treatment of traditional knowledge are being prepared.

Government agency SEAs and sustainability development strategies have varied
greatly in application and completeness. Coordination and implementation of substantive, place-based environmental initiatives across agencies and governments
remains problematic, despite some notable successes. Progress toward sustainability is, at best, halting, or more realistically, unclear. Also unclear is the role of EIA
requirements within such efforts. Yet to be addressed are the EIA process implications of these varying efforts to combine EIA procedural and substantive requirements. Two recent panel decisions stipulate that evidence be provided that the
precautionary principle is respected and that the undertaking makes a positive contribution to sustainability (Gibson, 2000). A recent court decision in British Columbia (BC) concerning the application of BC's EIA legislation contained similar
requirements. A discussion paper containing proposed changes to Manitoba's
Environment Act suggests expanding the effects assessment to include (among
other things) project sustainability (Manitoba Conservation, 2001). These types
of decisions and proposals could accelerate efforts to ensure that EIA requirements
contribute explicitly to tangible environmental enhancement and, more specifically,
to the broader goal of sustainability.

The European approach (at the European Economic Community level) to the
integration of EIA process and substance is similar in many respects to those of

the United States and Canada. The Project Directive (85/337/EEC and 97/11/EEC) includes a largely physical definition of direct and indirect effects. Mandatory triggers all refer to projects. A partial bridge between process and substance is established through selection criteria for determining whether the directive should be applied. The selection criteria refer, for example, to land use and abundance, the quality and regenerative capacity of natural resources, wetlands, coastal zones, mountain and forest areas, protected areas, community environmental quality standards, densely populated areas, and landscapes of historical, cultural, and archaeological significance. The Project Directive also cross-references the Directive for Conservation of Wild Birds (79/409/EEC) and the Habitats Directive (92/43/EEC). The Project Directive stresses the importance of promoting biodiversity, consistent with the UN Convention of Biological Diversity, 1992, and related guidelines (UNEP, 2002). The screening, scoping, and EIS review guidelines (June 2001) provide more details regarding sensitive and significant locations, types of ecological and social impacts, and potentially relevant methods.

The European approach, at the project level, specifies overall process requirements, selectively and flexibly introduces environment types, cross-references more specific substantive requirements and sponsors methodological research. This broad-brush and adaptable approach is perhaps inevitable because of the diversity of regulatory approaches across the European Union and the need to maintain national autonomy. However, given the general nature of the requirements, it may be optimistic to expect that the Project Directive will lead to more than limited improvements in combining environmental substance with European EIA requirements. Some additional measure of consistency could be facilitated by higher-level requirements and initiatives, by detailed guidelines, and by further methodological research (Teller and Bond, 2002; UNEP, 2002).

The purpose of the Plans and Program Directive [COM (96) 511 + COM(99)73] refers to environmental quality, human health, and the prudent use of natural resources, consistent with the precautionary principle. Sustainable development is a community objective. Progress toward sustainable development is a directive objective. The directive applies to plans and programs that set the framework for future development consents (e.g., energy, waste, industry, telecommunications, tourism, certain transport infrastructure plans and programs, town and country planning and land use). The directive does not require that social and economic effects be considered. It does recommend that "the value and vulnerability of the area likely to be affected due to special natural characteristics or cultural heritage" be considered. Only negative environmental effects must be addressed. The SEA directive's narrow definition of the environment and effects contrasts with sustainability assessment approaches. Sustainability assessments tend to integrate ecological, social, and economic concerns and are objectives driven (Smith and Sheate, 2001; Thérivel and Minas, 2002). This suggests the need to reconcile the two approaches, possibly by integrating the SEA directive requirements into sustainability assessment (Smith and Sheate, 2001) and through impact assessments of major Commission initiatives, both linked to the European Sustainable Development Strategy (CEC, 2002).

The European Commission has sponsored considerable guidance and research regarding SEA methods and practices, cumulative effects, coastal zone management, and sustainable resource management. A major recently completed study addresses SEA and the integration of the environment into strategic decision making (ICCL et al., 2001). A multidisciplinary network (BEQUEST: building environmental quality evaluation for sustainability) of scientific and professional communities also is being sponsored. BEQUEST has sought to develop a framework for the analysis of sustainable development and to compile a directory of methods for assessing the sustainability of urban development (Deakin et al., 2002).

SEAs could provide an environmental context and direction for project-level EIAs. Substantive guidance also is provided through a variety of requirements (e.g., impact assessments for legislative proposals, conventions on long-range air pollution, transboundary impacts, and integrated pollution prevention and control) and initiatives (e.g., sustainable development strategy, environmental information, observation network) (CEC, 2002). A variety of models (e.g., constitutional/legislation, government led strategies, ad hoc institutional procedures) and tools (e.g., SEA, strategic environmental analysis, E-test of legislation, sustainability assessment or appraisal, integrated environmental assessment, economic tools, green accounting, environmental management systems, objectives, targets and indicators, environmental monitoring and reporting, public participation, education and awareness, matrices and appraisal tables) are applied in various European Union countries to integrate environmental concerns into public and private decision making (ICCL et al., 2001). EIA quality and effectiveness analyses also have been applied extensively to assess EIA requirements, guidelines, procedures, and documents.

It is too early to tell if all these elements will be mutually supportive. The optimistic interpretation is that this loose network of requirements, guidelines, and initiatives will gradually coalesce into a broad strategy or into a mutually consistent but loosely affiliated series of strategies. This strategy or strategies would, in turn, lead to a more central role for environmental considerations in decision making and to tangible environmental improvements in the direction of sustainability. EIA could assume a pivotal role in this effort. The EIA process would be progressively reformed to be more fully conducive to realizing substantive environmental ends. The pessimistic interpretation is that the present "patchwork quilt" will remain largely unchanged. Major gaps and inconsistencies will continue. The EIA and SEA requirements will be too general to make an appreciable difference. Progress (or not) toward sustainable futures will be unclear. The implications for EIA process management will remain largely unexplored. Only time will tell which interpretation is more accurate.

The commonwealth of Australia has come at the issue of how best to combine process and substance in EIA in a different way from the other three jurisdictions. It starts from environmental substance and builds EIA requirements around substantive environmental imperatives. The objectives of the Environmental Protection and Biodiversity Conservation Act 1999 and Regulations 2000 (EPBCA) include environmental protection (especially matters of national environmental significance), promotion of ecologically sustainable development through the conservation and ecologically sustainable use of natural resources and the promotion of biodiversity conservation. The EPBCA protects native species (especially threatened and migratory species), establishes a whale sanctuary, protects ecosystems (reserves, ecological communities, off-reserve conservation measures), and identifies processes that threaten all biodiversity levels. It identifies ecologically sustainable development principles, including the integration of short- and long-term economic, environmental, social, and equity considerations, the precautionary principle, intergenerational equity, the conservation of biological diversity and ecological integrity, and improved valuation, pricing, and incentive mechanisms. The EPBCA applies to Australians outside the exclusive economic zone and to the land, water, seabed, and air over Australia, the external territory, the exclusive economic zone, and the continental shelf. It also applies to vessels and aircraft.

The EPCBA explicitly identifies matters of national environmental significance (e.g., World Heritage properties, Ramsar wetlands, listed threatened species and communities, listed migratory species, nuclear actions, the marine environment). EIA requirements (subject to referral) are triggered by actions that affect matters of national environmental significance and proposals involving the commonwealth (commonwealth land and action), which will or are likely to have significant environmental impacts. Requirements for environmental approvals, economic and social matters, principles of ecologically sustainable development, and people's environmental history must be taken into account. The EPBCA cross-references Australia's obligations under various UN conventions pertaining to world heritage sites, wetlands, threatened species, endangered communities, and migratory species. The EPBCA objectives and principles are applied to projects through administrative and project specific guidelines. The EPBCA includes discretionary provisions for undertaking strategic assessments. The EPBCA also includes detailed conservation of biodiversity provisions and an extensive range of enforcement provisions. The minister is required to consider the precautionary principle when making a decision.

The regulations to the EPBCA include criteria and scaling systems for matters of national environmental significance, document content requirements, activity limitations, and management principles. The EPBCA and regulations are supported by administrative guidelines for determining significance (specific criteria for matters of national environmental significance) and guidelines for various activities in selective sensitive and significant environments. A discussion paper and draft regulations have been prepared for a greenhouse trigger. Complementary environmental activities include guidelines and criteria for determining the need for and level of EIA in Australia, various national environmental strategies (e.g., sustainable development, biodiversity, greenhouse gas emissions, forests, oceans), and an intergovernmental agreement on the environment.

The Australian commonwealth approach integrates process and substance by building EIA requirements around matters of national environmental significance and specific ecological and sustainability objectives and principles. These requirements are refined through specific criteria and scaling systems. EIA requirements also are merged with other requirements to protect biodiversity and to protect and enhance sensitive and significant species and habitats. The integration of process and substance is further reinforced by explicit cross-references to international conventions, by the broad territorial application of the requirements, and by national environmental strategies and agreements. The linking of process to substance is less clear for environmentally significant actions by the commonwealth or on commonwealth land, which are not matters of national environmental significance. Although the EPBCA refers to social, economic, and equity matters, the requirements and guidance provided are very general. The discretionary application of the SEA provisions could limit the potential for SEAs to provide a substantive environmental bridge between the EPBCA and national environmental strategies. The net result is a strong union of process and substance but within a narrowly defined area.

Figure 5.5 highlights several ways in which EIA substance and process can be integrated, based on the experiences in the four jurisdictions. Procedural EIA requirements can be linked to related substantive environmental requirements and placed within the context of sustainability strategies and plans. Substantive requirements can be built directly into EIA legislation and regulations. EIA requirements can be formally merged with other environmental requirements and/or informally linked to strategic and regional planning and management efforts. Substantive requirements can be comprehensive (e.g., ecological, social, economic, sustainability). They can focus on specific priorities (e.g., ecological sustainability, environmental justice). Procedural and substantive requirements can be addressed through tiering (e.g., regulatory/sustainability assessment, SEA, project-level EIA). An effective blending of the procedural and the substantive will incorporate elements of each of these approaches. The four jurisdictions include aspects of each approach, albeit in different ways and to varying degrees. Distinguishing between effective and less effective combinations would require systematic effectiveness analyses, tempered by jurisdiction-specific adjustments. Still, at least on the surface, directly integrating substantive requirements into EIA legislation, regulations, and guidelines, coupled with procedural guidance conducive to realizing substantive objectives and requirements, would seem the most direct route to more substantively effective EIA practice.

5.5.2 Management at the Applied Level

Figure 5.6 illustrates an example of a substantive EIA process. The figure and the process description that follows draw on the concepts and methods presented in Section 5.3. EIA process managers and participants can pick and choose the relevant and appropriate elements.

Startup EIA, and by extension the EIA process, is an instrument for realizing tangible ecological, social, and sustainability objectives. The process begins with an overview of pertinent environmental and sustainability plans, strategies, programs, and other public, private, and multistakeholder initiatives. These initiatives could influence the current and future environmental and community conditions in the geographic areas potentially affected by proposed actions. They provide a

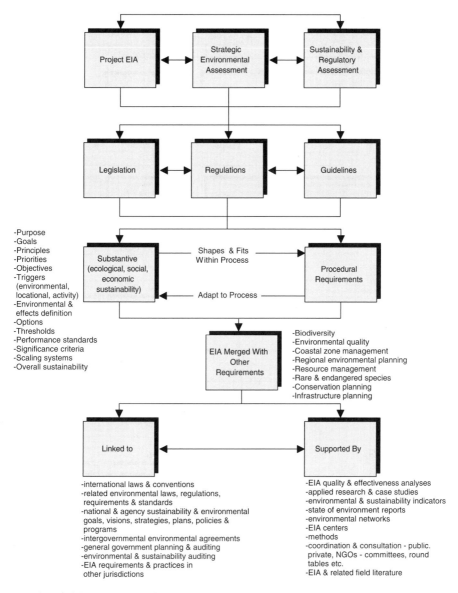

Figure 5.5 Examples of regulatory approaches to integrating EIA and substance.

context. They could be undermined, complemented, or unaffected by any proposed actions. An analysis of need is undertaken. Need is addressed both in the conventional sense (e.g., a market opportunity, a public service need) and in the sense of identified sustainability problems (i.e., shortfalls between sustainability objectives and expected future conditions). Parties potentially interested and affected by

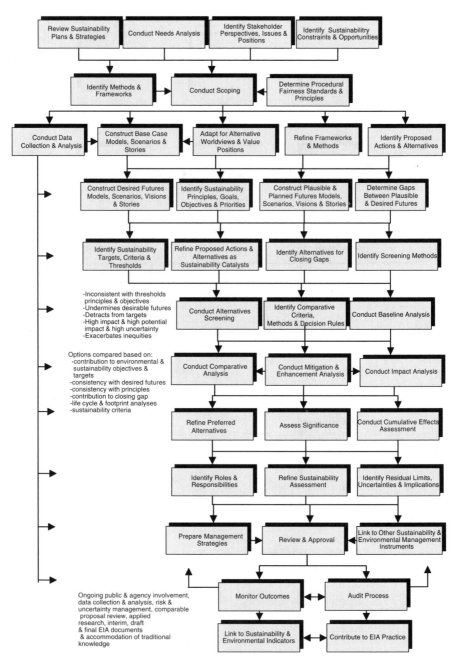

Figure 5.6 Example of a substantive EIA process.

potential actions are identified. The perspectives, concerns, and positions of these parties are identified. Varying world views regarding current environmental conditions and desired future conditions are determined. Known sustainability constraints and opportunities are identified.

These background analyses feed into study design and scoping. Study design and scoping focus the process on major issues, needs, and participants, within the context of relevant sustainability initiatives, constraints, opportunities, and perspectives. Frameworks for guiding the process and types of methods likely to be applied in the EIA process are identified. Procedural fairness standards and principles are jointly determined to facilitate the full and equal participation of interested and potentially affected parties.

Sustainability Analysis The sustainability analysis establishes the foundation for the sustainability assessment. The sustainability analysis employs multiple qualitative and quantitative methods to characterize the present (sustainable and unsustainable activities and environments) and the future (plausible, planned, and desired)—all from multiple perspectives. Conventional EIA processes tend to view proposed actions (and their alternatives) as intrusions on social, economic, and ecological systems—the negative effects of which are to be minimized. In this case proposed actions and alternatives are envisioned as potential catalysts for a sustainable future. Methods and frameworks for undertaking the sustainability analysis are formulated and then progressively refined and adapted. A preliminary set of proposed actions and options are identified, refined, and adapted in an ongoing effort to meet the identified need while facilitating sustainability.

A series of base case (i.e., past and present conditions without proposed actions or their alternatives) scenarios, models, and stories are first constructed. Conceptual and quantitative network diagrams and models explore spatial and temporal patterns of interrelationships among historical and current environmental conditions and activities. Alternative assumptions and inputs address uncertainties and explore varying interpretations. Scenarios and stories describe and explain how present conditions evolved from the past. These integrative tools are jointly formulated and applied with interested and affected parties. The analysis draws on scientific and traditional data and knowledge. Different base-case characterizations reflect alternative worldviews and value positions. Frameworks are constructed to explore overlaps and interconnections among disciplinary models and systems. Group consensus building and conflict resolution techniques identify common ground and residual areas of dispute among the characterizations. The analysis is undertaken at multiple levels (e.g., regional, community, local). It focuses on sensitive and significant environmental components and processes that could potentially be affected by proposed actions and alternatives. Activity patterns that appear sustainable and unsustainable, positive sustainability initiatives (that have and are making a difference), and sustainability constraints and opportunities (building on the startup analysis) are identified. Multiple iterations are required to explore gaps, uncertainties, links, and varying interpretations. The outcome from these efforts is a small number of partially overlapping and partially interconnected base-case characterizations.

The core values and underlying assumptions of each are identified. Varying interpretations and perspectives are noted. Where practical, ecological and social carrying capacities surpassed or in jeopardy are indicated. Major uncertainties and potential implications are highlighted.

The analysis next addresses future conditions, again in multiple ways and from multiple perspectives. Conventional forecasts identify trends in key conditions pertinent to sustainability, appreciating that trends are often a poor predictor of the future. Key attributes of the planned future, as envisioned in public policy and spatial planning documents, are highlighted. Plausible future conditions (pertinent to both sustainability and to proposed actions and their alternatives) are addressed using models, visions, scenarios, and stories. These plausible futures address how conditions (e.g., demographic, ecological, social, economic, political, institutional, technology) might evolve taking into account baseline conditions, forecasts (including those prepared by others), proposed activities in the area, sustainability instruments in operation and the planned future. Visions, models, scenarios, network diagrams, backcasting and stories also characterize desired, sustainable futures. The desired futures include both end states (for various time horizons) and sequences of events (from the present to the future and back from the future to the present). The plausible and desired futures encompass multiple worldviews and value positions, alternative assumptions and interpretations and varying perspectives regarding the nature and implications of uncertainties. Group processes are again used to search for common ground, to build consensus and to resolve disputes. A complete consensus does not emerge. Residual differences are highlighted for subsequent application in sensitivity analyses and in contingency measures. The analysis focuses on identifying and characterizing discrepancies among plausible planned futures and sustainable futures. Instances where plausible futures suggest that social or ecological thresholds could be surpassed are highlighted. The gaps and thresholds provide the basis for identifying sustainability principles, goals, objectives, and priorities. Major remaining uncertainties and potential implications are noted. The analysis is placed within the context of broader government and multistakeholder sustainability initiatives. It also builds on other efforts to characterize plausible and desired future conditions.

Sustainability Assessment The proposed action and the alternatives to the proposed action are reconsidered, taking into account the gaps (including thresholds) between the plausible and the desired futures. Ways in which the gaps could be closed are assessed. The proposed action and the alternatives are modified, to the extent practical, to assume the role of a catalyst for narrowing the gaps. The possibility of generating additional alternatives, which might better address the sustainability shortfalls, also is explored.

The sustainability principles, goals, objectives, and priorities are refined into more specific targets, criteria, and thresholds. Methods and exclusionary criteria for screening out unacceptable alternatives are formulated. Alternative screening criteria are formulated to reflect varying perspectives where a consensus cannot be reached among interested and affected parties. Decision rules are formulated for addressing situations where the results from applying alternative screening criteria conflict. The screening criteria are applied to the proposed action and to the alternatives. Alternatives that are, for example, inconsistent with sustainability thresholds, principles, and objectives undermine desirable futures, detract from targets, result in major impacts, and exacerbate inequities for socially disadvantaged groups are excluded from further consideration. A precautionary approach is adopted. Thus marginal alternatives, characterized by major potential impacts and high levels of uncertainty, also are excluded.

Criteria, methods, and decision rules for comparing the remaining alternatives are formulated. The alternatives are compared for their contributions to environmental and sustainability objectives, for their consistency with desired futures and with sustainability principles and criteria and for their potential contribution to closing the gaps between plausible and desired futures. The comparison of alternatives is supported by techniques such as life-cycle and footprint analysis. Uncertainties and alternative interpretations are addressed with sensitivity analyses. A precautionary approach is applied. The implications of applying mitigation and enhancement measures are explored. A preferred alternative is selected, supported by a clear rationale.

Baseline conditions, pertinent to the preferred alternative, are characterized. Individual and cumulative impacts associated with the preferred alternative are identified, predicted, and interpreted (in terms of significance). Mitigation and enhancement measures are introduced to prevent and offset adverse impacts and to enhance benefits. These activities largely mirror those associated with conventional EIA processes. However, unlike conventional processes, the analysis focuses on using and refining the proposed action as a sustainability instrument or catalyst. The impact analysis also builds toward an overall assessment of contribution to sustainability (e.g., contribution to or amelioration of global and transboundary problems, maintenance of ecological integrity, maintenance of natural capital, waste generation within assimilative capacity, reduced energy and material use, maintenance of environmental quality, biodiversity maintenance, amelioration of intergenerational and intragenerational inequities, pollution prevention, avoidance of risk to carrying capacity). Implementation, monitoring, and auditing roles and responsibilities are specified. Residual limits (e.g., knowledge, resource, institutional) and uncertainties, together with associated implications, are detailed from a precautionary perspective (e.g., minimum regrets). Interested and affected parties are fully involved in all facets of the sustainability assessment.

Approvals and Post Approvals   The process culminates in final conclusions regarding the acceptability, on sustainability grounds, of the proposed action. The proposed action includes a management strategy and links to other EIA tiers and to other environmental management and sustainability instruments, networks, and strategies. Measures to ensure follow-up coordination, communications, cooperation, public involvement, and coalition/capacity building are instituted. Consideration is given to how best to overcome implementation barriers. The implementation measures could include institutional reform.

If the proposed action is acceptable, with or without approval conditions, environmental changes, impacts, and mitigation/enhancement are monitored. Monitoring results provide the basis for ongoing environmental and sustainability management. The environmental change monitoring results are incorporated into broader environmental and sustainability indicator systems. The EIA process is audited. The lessons learned from the auditing analysis are broadly circulated as a contribution to EIA practice.

Ongoing Activities The EIA process is highly iterative, dynamic, and collaborative. It provides for both continuous (e.g., advisory committees) and periodic (e.g., workshops, forums, conferences, open houses) public and agency involvement opportunities. The workshops, conferences, and forums support generating, refining, and integrating visions, scenarios, models, and stories. They also contribute to generating and applying goals, objectives, criteria, principles, and decision rules. Surveys, interviews, meetings, and focus groups help identify perspectives, concerns, and initiatives. Periodic opportunities (e.g., open houses, meetings) are provided for broader public and government official (elected and nonelected) involvement. Provision is made for public and agency involvement during the post-approvals stage. Participant assistance is provided to ensure the effective involvement of all interested and affected parties. Specialists participating in the process are part of and maintain contact with broader environmental management and sustainability networks.

Data and analysis, including traditional knowledge and reviews of comparable situations, are incorporated into each activity. Particular consideration is given to related sustainability experiences. Technical (e.g., model building, scenario generation) and procedural (e.g., consensus building, conflict resolution) advice and applied research are provided whenever needed. Impact and uncertainty management are continuous functions in the process. Numerous interim documents are generated as the process unfolds. The results of the process are consolidated into draft and final EIA documents, which provide a complete decision-making basis. Documentation extends into post approvals with the production and circulation of monitoring and auditing reports. The process is knit together by the central purpose and theme of facilitating substantive, sustainable environmental improvement.

5.6 ASSESSING PROCESS EFFECTIVENESS

Table 3.5 (in Chapter 3) presents criteria and indicators for assessing the positive and negative tendencies of the EIA processes described in Chapters 3 to 10. In this section we summarize examples of the positive and negative tendencies of the substantive EIA process example.

A substantive EIA process is consistent with a holistic scientific approach. Model building can *rigorously* apply scientific methods. The substantive EIA process can integrate scientific analyses and knowledge. But the focus tends to be on trans-scientific issues less amenable to scientific analysis. Thus scientific analyses tend to be broad-brush and selective. Assumptions, findings, methods, interpretations, conclusions, and recommendations can be explicitly presented. The qualitative, evolving, and extrarational nature of techniques such as visioning, scenario writing, and storytelling, as well as the procedures for drawing them together, are less rigorous and can inhibit a precise tracing of how outcomes are generated. The process focuses on facilitating tangible environmental improvements rather than on contributing to the scientific knowledge base.

A substantive EIA process is well suited to treating physical, biological, social, cultural, and economic effects *comprehensively*. This open and broad process can readily accommodate different forms of knowledge and experiences. It adopts a holistic perspective. It begins with a broad definition of the problem. It strives to place actions and consequences within a broader context. It explicitly addresses interrelationships and cumulative effects.

A substantive EIA process is *systematic*. It is directed toward well-defined substantive goals. It employs multiple methods. It systematically formulates and evaluates alternatives, identifies and predicts effects, and interprets and manages positive and negative impacts. Some inconsistencies inevitably arise in the effort to integrate numerous and very different methods and perspectives. The complex and conflicting images of the present and of the future can result in some confusion and can inhibit traceability.

This process is, by definition, *substantive*. It is guided by and designed to integrate multiple values, ideals, perspectives, and forms of knowledge. It is highly conducive to advancing environmental quality and sustainability objectives. The breadth of the analysis necessarily leads to some compromises in level of detail. Objectives and basic principles are well defined. Sometimes the procedures for applying and implementing the objectives and principles are too broad.

A substantive EIA process is generally *practical*. It focuses on tangible environmental improvements. It is well suited to multistakeholder environmental management situations. Efficient and effective resource use can be difficult with such a large and complex approach. Exceptional management skills are required. There is considerable experience with applying each method. There is much less experience with linking the methods together into a coherent whole within an EIA process. Time will be needed to agree on the methods and to determine how they are to be applied and integrated. The process proceeds at its own pace, especially the efforts to integrate methods and to identify common ground among diverse perspectives. Roles and responsibilities will overlap and may be difficult to define with precision. The process can focus on major issues and trade-offs if well managed. A gulf can emerge between abstract images and reality. The process tends to be stronger on long-term perspectives and fundamental change than on short-term perspectives and incremental change. The process and process outputs may not be familiar to regulators, a problem that can be ameliorated if regulators are fully involved in the process.

A substantive EIA process can be *democratic*. The process is consistent with the views advocated by many stakeholder groups. A substantive EIA process can accommodate many perspectives and forms of knowledge, including traditional knowledge. Some delegation of authority to stakeholders can occur by applying techniques such as participatory rural appraisal. Techniques such as scenario writing, story telling, and visioning also are conducive to community influence. This open, collaborative approach seeks to both integrate stakeholder perspectives and advance ecological, social, and sustainability ends. Substantive EIA processes have a tendency to be politically naïve and to underestimate political and institutional barriers to implementation.

A substantive EIA process can be *collaborative*. It is consistent with many public perspectives. The public is directly involved as partners and as sources of knowledge and experience. Stories, scenarios, and qualitative concepts and models are formulated jointly. Quantitative models tend to be generated by specialists and can be more difficult for the general public to comprehend. Multiple methods and perspectives on the future are sometimes confusing and difficult to follow. This constructive integrative approach is generally conducive to public involvement. It also lays a foundation for consensus building and conflict resolution. It devotes less attention to the mechanisms by which consensus is to be built and conflicts resolved. It tends to be less effective when positions are highly polarized or when expectations regarding what can be realistically achieved through the process are raised to unrealistic levels.

Substantive EIA processes often include measures to facilitate procedural fairness. They explicitly address distributional fairness concerns, including intergenerational inequities and the needs of the disadvantaged. *Ethical* concerns are likely to be a priority with many interested and affected parties. Under the umbrella of sustainability the process seeks constructively and creatively to identify "win–win" solutions. Such solutions are not always possible. Although a consideration, ethical matters are rarely the priority with substantive EIA processes.

A substantive EIA process is generally *adaptive*. The process is open and employs many techniques for testing alternative inputs. It adaptively addresses multiple interrelationships. As the process acquires momentum, because of its size, duration, and complexity, it can be difficult to revisit earlier steps. The process fosters creative problem solving and opportunity seeking. Complex and "wicked" problems may elude characterization. It is an open process that systematically addresses interactions with context. It considers risks and uncertainties. Risks and uncertainties are not usually addressed in depth.

A substantive EIA process is highly conducive to *integrating* diverse values, forms of knowledge, perspectives, and ideals. It explores relationships with related decisions. It seeks constructive links to related environmental management and sustainability instruments. It actively strives to bridge and transcend disciplinary, EIA type, and professional boundaries. It can integrate proposal planning and the EIA process. The apportionment issue (e.g., the significance of a marginal increase in greenhouse gas emissions) can be problematic when determining when major proposal modifications are required or whether the proposal is unacceptable.

5.7 SUMMING UP

This chapter is concerned with managing regulatory and applied EIA processes to better integrate environmental perspectives, values, and knowledge. The two stories describe two ways in which the EIA process can become more substantive: (1) through the application, by natural and social scientists, of natural and social scientific knowledge, concepts, theories, frameworks, and methods; and (2) through the efforts of the public to broaden and reconfigure the EIA process to more effectively define and achieve substantive objectives and to better accommodate substantive community and traditional knowledge and perspectives.

The general problem addressed is a shortfall between EIA environmental aspirations and achievements. The more specific problem is the role that the EIA process assumes in widening or narrowing that gap. Numerous ecological, social, and sustainability concepts and methods are presented. These concepts and methods provide the basis for characterizing how a substantive EIA process might be applied at the regulatory and applied levels. The substantive EIA process is then assessed against ideal EIA process characteristics. Table 5.5 is a checklist for formulating, applying, and assessing a substantive EIA process.

Table 5.5 Checklist: A Substantive EIA Process

The response to each question can be yes, partially, no, or uncertain. If yes, the adequacy of the process taken should be considered. If partially, the adequacy of the process and the implications of the areas not covered should be considered. If no, the implications of not addressing the issue raised by the question should be considered. If uncertain, the reasons why it is uncertain and any associated implications should be considered.

ECOLOGICAL CONCEPTUAL DISTINCTIONS

1. Does the process assess ecosystem impacts systematically and comprehensively?
2. Does the process reflect an ecological systems perspective?
3. Does the process recognize the limits to ecological knowledge?
4. Is the process conducive to addressing ecological concerns and impacts?
5. Are biodiversity impacts addressed?
6. Is the potential for exceeding ecological thresholds considered?
7. Is the process conducive to applying an ecosystem approach?
8. Is the process conducive to integrated environmental and resource management?
9. Is the process conducive to applying adaptive environmental assessment and management methods?
10. Is the process conducive to achieving ecological objectives?

SOCIAL CONCEPTUAL DISTINCTIONS

1. Does the process assess social impacts systematically and comprehensively?
2. Does the process predict public responses to impacts?
3. Is an effort made to see the proposal and potential impacts through the eyes of and as experienced by potentially affected parties?
4. Is the EIA process socially constructed?
5. Does the process extend from and adapt diverse social concepts?
6. Are social impacts approached from multiple perspectives and at multiple levels?
7. Is the role of the process in shaping the nature and magnitude of social impacts recognized?

(*Continued*)

Table 5.5   (*Continued*)

8. Does the process accommodate multiple and changing perspectives, norms, perceptions, beliefs, and values?

9. Does the process recognize the limits of social knowledge?

10. Is the central role of dialogue and social interaction recognized?

11. Is the process conducive to integrating social knowledge and experiences?

12. Is the process conducive to achieving social objectives?

SUSTAINABILITY CONCEPTUAL DISTINCTIONS

1. Is the process directed toward and conducive to enhanced sustainability?

2. Does the process address social, ecological, and economic sustainability forms as well as interdependencies among sustainability forms?

3. Does the process incorporate sustainability ends, means, strategies, visions, and frameworks?

4. Is the process adapted to contextual characteristics?

5. Does the process distinguish between sustainable and unsustainable environments and human activities?

6. Is the process linked to other sustainability instruments?

7. Does the process distinguish between sustainable and unsustainable future conditions?

8. Is the process supported by, and does it provide, outputs to sustainability indicators?

9. Are sustainability considerations integrated into each EIA process activity?

10. Does the process recognize and incorporate multiple perspectives on sustainability?

11. Does the process recognize constraints and limits to sustainability?

12. Is the process conducive to achieving sustainability objectives?

SUBSTANTIVE METHODS

1. Does the process employ methods sensitive to ecological, social, and sustainability characteristics and objectives?

2. Are appropriate methods used to address interconnections, interdependencies, and cumulative effects?

3. Are appropriate methods used to predict plausible futures?

4. Are appropriate methods used to formulate desired futures?

5. Are appropriate methods used to determine pathways to and from desired futures?

6. Are appropriate methods used to determine proximity to thresholds?

7. Are appropriate methods used to assess progress toward ecological, social, and sustainability ends?

8. Are appropriate methods used to adapt to and manage uncertainties?

9. Are appropriate methods used to facilitate the involvement of interested and affected parties?

10. Are the methods applied appropriately?

11. Are the limits of methods acknowledged, and are the implications of those limits considered?

MANAGEMENT AT THE REGULATORY LEVEL

1. Do the EIA regulatory provisions and/or guidelines:

Table 5.5 (*Continued*)

a. Directly incorporate substantive ecological, social, economic, and sustainability provisions into:
  (1) Project-level EIA requirements?
  (2) Strategic environmental assessment requirements?
  (3) Sustainability and regulatory assessment requirements?
b. Require that proposed actions be assessed against specific and substantive goals, principles, priorities, and performance standards?
c. Require that proposed actions be assessed in terms of overall sustainability?
d. Include environmental, locational and activity triggers?
e. Broadly define environment and effects (ecological, social, economic, interrelationships)?
f. Adjust procedural requirements to be conducive to integrating substantive environmental concerns?

2. Are the EIA regulatory provisions and/or guidelines:
  a. Merged, where appropriate, with other substantive environmental requirements?
  b. Linked to, as appropriate, with:
    (1) International laws and conventions?
    (2) Related environmental laws, regulations, requirements, and standards?
    (3) National and agency sustainability and environmental goals, visions, strategies, plans, policies, and programs?
    (4) Intergovernmental environmental agreements?
    (5) General government planning and auditing?
    (6) Environmental and sustainability auditing?
    (7) EIA requirements and practices in other jurisdictions?
  c. Supported, as appropriate, with:
    (1) EIA quality and effectiveness analyses?
    (2) Applied research, case studies, and EIA and related field literature?
    (3) Environmental and sustainability indicators and state-of-the-environment reporting?
    (4) Environmental networks, EIA centers, and methods?
    (5) Coordination and consultation with interested and affected parties?

MANAGEMENT AT THE APPLIED LEVEL

1. Does the process incorporate such startup activities as:
  a. Overview of sustainability plans and strategies?
  b. Broadly based analysis of need?
  c. Identification of stakeholder perspectives, issues, and positions?
  d. Identification of sustainability constraints and opportunities?
  e. Scoping and identification of methods and frameworks?
  f. Determination of procedural fairness standards and principles?

2. Does the process incorporate such sustainability analysis activities as:
  a. Construct base-case models, scenarios, and stories?
  b. Adapt the models, scenarios, and stories to account for varying world views and value positions?
  c. Refine frameworks and methods?
  d. Identify proposed actions and alternatives?
  e. Construct plausible and planned future models, scenarios, visions, and stories?
  f. Construct desired futures models, scenarios, visions, and stories?

(*Continued*)

Table 5.5   (*Continued*)

---

    g. Determine gaps between plausible and desired futures?

3. Does the process incorporate such sustainability assessment activities as:
    a. Identify sustainability principles, goals, objectives, and priorities?
    b. Identify sustainability targets, criteria, and thresholds?
    c. Refine proposed actions and alternatives as sustainability catalysts?
    d. Identify alternatives for closing gaps?
    e. Identify and apply sustainability-based screening methods and criteria?
    f. Identify and apply sustainability-based comparative evaluation methods and criteria?
    g. Integrate sustainability considerations into baseline, impact, mitigation, enhancement, cumulative effects, and significance interpretation activities?
    h. Integrate the analyses into an overall sustainability assessment?
    i. Identify roles and responsibilities?
    j. Identify residual limits, uncertainties, and implications?

4. Does the process include such approval and post-approval activities as:
    a. Preparation of management strategies?
    b. Links to other sustainability and environmental management instruments?
    c. Monitoring the process outcomes?
    d. Auditing the process?
    e. Links to sustainability and environmental indicators?
    f. Contributions to EIA practice?

5. Does the process provide for such ongoing activities as:
    a. Comparable proposal reviews?
    b. Data collection and analysis?
    c. Applied research?
    d. Public and agency involvement?
    e. Accommodation of traditional knowledge?
    f. Environmental and uncertainty management?
    g. Interim, draft, and final report preparation?

PROCESS EFFECTIVENESS

1. Does the process effectively integrate scientific knowledge, standards, procedures, and methods?

2. Does the process assess the environment comprehensively from a holistic perspective?

3. Is the process guided by substantive environmental goals?

4. Is the process conducive to advancing environmental quality and sustainability objectives?

5. Does the process focus on tangible environmental improvements?

6. Does the process facilitate local influence?

7. Is the process collaborative?

8. Does the process facilitate procedural and substantive fairness?

9. Is the process adaptive and conducive to managing risks and uncertainties?

10. Does the process integrate diverse values, forms of knowledge, perspectives, and ideals?

11. Does the process explore links with related decisions?

12. Does the process bridge and transcend disciplinary, EIA type, and professional boundaries?

The relationship between EIA process and substance has been approached from several perspectives. Some say that EIA is an unnecessary diversion of resources. They argue that it reinforces the status quo or that it cannot be applied properly or that there are better ways to bring about environmental improvements. Others argue that the substantive benefits of the EIA process are unknown because of knowledge gaps concerning decision-making effectiveness, outcome effectiveness, and environmental characteristics. Still others submit that the EIA process can be conducive to environmental advancement but that refinements or modifications or major reforms are needed. Many acknowledge that issues and solutions vary depending on whether ecological, social, or sustainability concerns are being considered. This chapter explores how EIA process management can better address this constellation of interrelated problems and solutions.

An overview of ecological, social, and sustainability concepts and methods provides the basis for the substantive EIA process. The ecological concepts explored include applied ecology, ecological impact assessment, environmental indicators, biodiversity, the ecosystem approach, environmental planning and management, integrated environmental and resource management and assessment, adaptive environmental assessment, and management and traditional knowledge. Examples of recurrent themes displayed by these concepts include a number of needs: for an ecological systems perspective, to adopt a place-based approach, to employ sound ecological knowledge, to transcend disciplinary boundaries, to recognize ecological stresses and limits, to acknowledge knowledge and control limits, to manage impacts continuously and adaptively (both pre and post approval), and for the process to be open, adaptive, creative, collaborative, iterative, selective, and action-oriented.

The social concepts considered include technical SIA, political SIA, positivistic social science, functional, ecological and systems theory, interpretative social science, critical social science, exchange theory, symbolic meaning, social learning, and phenomenological sociology. The overview analysis recognizes that people react in anticipation of and adapt to change and that social phenomena are very difficult to predict and influence in predictable ways. It demonstrates that there are multiple potentially applicable but partially overlapping and conflicting social models, theories, perspectives, and frameworks. It points to the often-peripheral position of SIA in decision making. It describes the gulf between social sciences and applied fields such as SIA. It underscores the importance of being cautious regarding preconceptions and to design and adapt the process to fit the context. It emphasizes the need to see the world through the eyes of potentially affected parties. It acknowledges the value of a socially constructed EIA process. It stresses the importance of exploring social impacts at multiple levels and from multiple perspectives. It demonstrates that the SIA/EIA process can be beneficial or can exacerbate negative impacts. It illustrates how meaning and value are socially determined. It shows the central role of dialogue and social interactions in the process. It stresses that frameworks and methods should address social interactions and choices systematically, from multiple perspectives.

The overview of sustainability concepts describes the varying perceptions of the nature and purpose of sustainability. It illustrates how sustainability is refined through sustainability forms and ethical perspectives, directed by needs, aspirations, and principles, applied through instruments, procedures, and processes, integrated by strategies, visions, and frameworks and adapted to contexts. It demonstrates that EIA and sustainability are applied to varying environments and activities and can be integrated at conceptual, regulatory, and applied levels. It describes how sustainability extends and completes EIA. Sustainability adds to the EIA process a sounder basis for interpreting significance, determining environmental limits, integrating measures of environmental change, interpreting present conditions, determining plausible, planned, and desirable conditions, integrating diverse perspectives and methods, adapting to context, and linking to other environmental management and sustainability instruments. Most important, sustainability helps makes the EIA process more effective in advancing substantive EIA aspirations.

Several methods, potentially conducive to a more substantive EIA process, are described briefly. The methods described include network analysis, systems diagrams, modeling, projection, forecasting, backcasting, visioning, scenario writing, storytelling, ecological footprint analysis, life-cycle analysis, rapid rural appraisal, and participatory rural appraisal. The major characteristics of each method are outlined. Potential roles for each method are indicated. Collectively, the methods effectively address interrelationships, interpret past and present conditions, identify ecological and social limits, portray plausible and desirable future conditions, determine how the gaps between plausible and desirable future conditions can be narrowed, manage uncertainties, and facilitate involving interested and potentially affected parties.

Substantive EIA requirements and guidelines in the four jurisdictions are described briefly. Each jurisdiction integrates process and substance in different ways, although there are many parallels. There are many positive and negative features and examples associated with how substantive environmental concerns are addressed in each jurisdiction. The United States, European Union, and Canada generally combine largely procedural EIA requirements with links to more substantive requirements. This indirect treatment of substantive environmental concerns can contribute to coordination difficulties and to inconsistencies. The Australian approach is more direct but more tightly circumscribed. None of the jurisdictions fully address social concerns. Only limited consideration is given to how the EIA process can be more conducive to realizing environmental objectives. The appropriate mix of approaches will vary by jurisdiction. Additional effectiveness analyses are required. It seems advantageous for environmental substance to be integrated into EIA requirements directly at the project, strategic, and regulatory levels. The selective merging of EIA with other substantive environmental requirements can sometimes be beneficial. Further consideration should also be given to EIA process adaptations that enhance the effectiveness of links to other environmental requirements and the potential for substantive environmental enhancements.

An example substantive EIA process is described. A context is established. Sustainability plans and strategies, sustainability constraints and opportunities, the need for action, and stakeholder perspectives, issues, and positions are reviewed. The process is scoped. Potentially appropriate methods and frameworks are identified and refined. Procedural fairness standards and principles are determined.

Proposed actions and alternatives are identified. Base-case models, scenarios, and stories are constructed and adapted to encompass alternative world views and value positions. Models, scenarios, visions, and stories are used to construct plausible and desired futures. Gaps between plausible and desired futures are determined. The gaps provide the basis for identifying sustainability principles, goals, objectives, and priorities. More specific sustainability targets, criteria, and thresholds are then formulated. Alternatives for closing the gaps are identified. Proposed actions and alternatives are refined and treated as potential sustainability catalysts.

The alternatives are screened and compared using sustainability thresholds, criteria, and decision rules. A sustainability assessment of the preferred alternative is undertaken, extending from such conventional EIA activities as baseline analysis, impact analysis, cumulative effects assessment, mitigation and enhancement analysis, and significance interpretations. Appropriate roles and responsibilities are determined. Residual limits, uncertainties, and implications are identified. Overall impact management strategies are prepared. Links to other sustainability and environmental management instruments are specified. These analyses provide the basis for proposal review and approval or disapproval.

Outcomes from the process are monitored and linked to sustainability and environmental indicators. The EIA process is audited. The auditing results are circulated widely to help improve EIA practice. The EIA process is supported by such ongoing activities as public and agency involvement, comparable proposal review, data collection and analysis, applied research, the accommodation of traditional knowledge, and the preparation of interim, draft, and final documents.

A substantive EIA process is consistent with a holistic scientific approach. It can integrate scientific methods and knowledge but tends to focus on applying transscientific methods, which facilitate tangible environmental improvements. A substantive EIA process encompasses many effects and experiences. It is especially effective in addressing interconnections and in placing the process within a broader context. It is directed toward well-defined substantive goals and uses multiple methods. Inconsistencies sometimes arise in the complex task of drawing together diverse methods and perspectives. A substantive EIA process can help achieve environmental objectives and can integrate multiple values, forms of knowledge, perspectives, and ideals. It has a clear sense of direction. But it can be difficult to manage and can be hampered by complexity, a broad level of detail, political naivety, and insufficient attention to short-term requirements, institutional barriers, and regulatory perspectives. It generally facilitates local influence and is usually collaborative.

This complex process can be difficult to follow. The process can break down in highly polarized positions or if expectations regarding what can practically be achieved through the process are raised to unrealistic levels. A substantive EIA process can facilitate procedural fairness and can address distributional fairness, risk, and uncertainty concerns. Such concerns tend to be addressed at a broad level of detail. The process is open and generally flexible. As with any complex process, adaptability diminishes as the process acquires momentum. A substantive EIA process can effectively establish links to related environmental management and sustainability instruments and can effectively bridge and transcend disciplinary, EIA type, and professional barriers. Determining the appropriate contributions of individual proposals to environmental and sustainability targets can be problematic.

CHAPTER 6

HOW TO MAKE EIAs MORE PRACTICAL

6.1 HIGHLIGHTS

This chapter portrays a streamlined, efficient, and effective EIA process—a practical process based on realistic expectations and competent practice.

- The analysis begins in Section 6.2 with three applied anecdotes. The stories describe applied experiences associated with efforts to make EIA practice more practical.

- The analysis in Section 6.3 then defines the problem, the tendency for EIA processes to be unfocused, disconnected from reality, weak on implementation, of variable quality, and slow to learn from experience and practice. The direction is toward ways of making the EIA process more focused, relevant, feasible, competent, and effective.

- In Section 6.4 we introduce a diversity of concepts bearing on how the EIA process can become more focused (on what matters), realistic (in terms of how management and decision making takes place), feasible (in terms of decision making and implementation follow-through), competent (in process execution), and effective (in facilitating EIA process management learning).

- In Section 6.5 we draw together the insights and lessons presented in Section 6.4. We describe the properties of a practical EIA process at both

Environmental Impact Assessment: Practical Solutions to Recurrent Problems, By David P. Lawrence
ISBN 0-471-45722-1   Copyright © 2003 John Wiley & Sons, Inc.

the regulatory and applied levels. In Section 6.5.1 we explore how EIA requirements could be more practical and in Section 6.5.2 illustrate how a practical EIA process could be expressed at the applied level.

- In Section 6.6 we assess how well the practical EIA process presented in Section 6.5 satisfies ideal EIA process characteristics.

- In Section 6.7 we highlight the major insights and lessons derived from the analysis. A summary checklist is provided.

6.2 INSIGHTS FROM PRACTICE

6.2.1 Combining SIA and Public Consultation to Provide Practical Solutions

The Uluru Kata Tjuta National Park in Central Australia is a World Heritage–listed area. A management plan must be prepared every five years. A social and cultural impact study was commissioned and a social impact assessment (SIA) consultant was engaged to identify the impacts of tourism and national park management on the local community of indigenous traditional owners. The outcomes from the cultural and social impact study were used to provide management strategies to maintain cultural heritage values.

The consultant used participative SIA methods combined with a standard SIA evaluation framework. She undertook one-on-one consultations with key elders, women, and youth, and conducted several workshops for traditional owners and for other stakeholders such as park staff and tourism operators. Numerous concerns and aspirations were identified. Practical recommendations were formulated to mitigate concerns and maximize aspirations. The indigenous traditional owners were concerned about maintaining ritual and ceremonial activities, protecting and being able to visit sacred sites, and providing tourism management and employment opportunities for youth to counter substance abuse problems. Tourism operators, in contrast, were concerned primarily about increasing visitor access and the range of activities available to tourists.

SIA and community consultation methods, appropriate to the situation and applied carefully, laid the groundwork for practical strategies to address the concerns and aspirations of traditional owners and stakeholder groups. These strategies were incorporated into the five-year management plan. Many measures were implemented immediately. Prompt and successful implementation of mutually acceptable strategies demonstrates that a properly designed and executed cultural and SIA process can facilitate logical and desirable initiatives. All that was required was a suitable forum to bring them to the surface.

A monitoring and evaluation framework also was developed to monitor social and cultural impacts. The framework will make it possible to measure the "state of the culture." It applies indicators considered critical by traditional owners for maintaining cultural values. The indicators address such concerns as health outcomes, housing and services provision, employment and training outcomes, and the maintenance of language and ceremonial activity, especially for younger traditional owners.

This case study demonstrates that SIA and public consultation methods can be practical and effective but must focus on the priorities of interested and affected parties, be appropriate to the local context, and be realistic about what is possible and practical. All key stakeholders must be recognized and empowered. In this case the indigenous traditional owners (a group usually disempowered) met with other stakeholders on an equal footing, supported by a land-rights and legislative framework that recognized their right to negotiate. Substantial and meaningful strategies acceptable to all parties resulted. The interests and perspectives of all parties need to be recognized and endorsed. Expertise and skill in methods application are essential. A collaborative planning and decision-making process appears especially conducive to identifying mutually acceptable and advantageous solutions, even where surface-level conflicts appear to preclude such solutions. A monitoring and evaluation framework enhances the potential to implement and refine agreed-upon solutions both successfully and effectively. It also provides a means to adjust strategies, again in a mutually beneficial manner, to address emerging and changing circumstances.

ANNIE HOLDEN
ImpaxSIA Consulting, Brisbane, Australia

6.2.2 Collaboration in Focusing the Process Without Sacrificing Substance

Over the last five years, oil sands development projects in the Ft. McMurray region of Alberta, Canada, have necessitated the preparation of about a dozen EIAs. Each project generally entailed surface mining or in situ production methods, usually involving a "green field" site. Developments had to meet both provincial and federal (Canadian Environmental Assessment Act) EIA requirements. As proponents worked their way through the EIA process, it was commonly necessary to compile an extensive array of regional baseline information and to conduct numerous monitoring plans and projects. The terms of reference (TORs) for each project often involved 30 pages of requirements. As a result, the EIAs often cost over $2 million and took years to work their way through the full regulatory process. The EIA standard for projects of this type seemed engrained in large documents, with a heavy reliance on monitoring and adaptive management to fully address predicted effects.

Shell Canada's Peace River Complex is a heavy oil in situ production facility that has existed at the site for more than 30 years. Over the years, different subsurface production techniques have been used to improve bitumen recovery from the oil sands. The Complex is over 300 kilometers from the "boom" area of Ft. McMurray. Consequently, examination of the environmental effects of the complex have been seen as quite distinct from those associated with the regional effects of an intensely developed area. Recently, Shell Canada Limited sought regulatory approval for new development that would increase production from the Peace River Complex. Alberta Environment advised that an EIA would be required. Shell has, through its discussions with regulatory agencies, sought their cooperation in developing a TOR approach that differed markedly from the typical TOR that was becoming "generic" for the other oil sands developments in the province. The cooperation of government staff was critical to designing a TOR customized to the specific nature of the project (i.e., an existing facility albeit expanding in size and production rates). It was also evident that the Peace River facility was isolated from the few other industrial activities in the area. The public in the area potentially affected by the facility, particularly the town of Peace River and the First Nations people in the area, participated in the TOR preparation by detailing their concerns and interests.

The result of this effort was a TOR that focused on those development aspects most likely to affect that environment, on the most significant and sensitive environmental components, and on the most potentially significant and uncertain effects. The analysis also reflected the concerns and priorities of all parties interested in and potentially affected by the proposed development. Far fewer resources were devoted to gathering extensive and largely irrelevant (to the project and its effects) baseline data. The Peace River EIA process also largely relied on a strategic approach because project details were not known at the time that the EIA was prepared. The proponent committed to address site-specific development aspects during detailed engineering design. The focused TOR, in combination with a strategic-level EIA approach, resulted in a much more focused EIA. The cooperation of regulatory agencies in the effort was critical. Ultimately, the effort of all parties resulted in appreciable cost and time savings without compromising the integrity of the EIA process or the environmental outcomes from the process.

This story demonstrates the importance of not succumbing to the temptation to duplicate the same set of requirements for each successive project, regardless of need and regardless of whether the requirements are appropriate to the situation. It suggests that agency requirements are sometimes more related to a desire to construct a regional database than to a genuine commitment to identify the information requirements needed to provide a sound decision-making basis. The story also makes it clear that focusing can only be effective with the cooperation and active involvement of all interested and potentially affected parties. It underscores the message that costly, time-consuming, and unfocused data collection can detract from, rather than contribute to, the realization of environmental objectives, especially when potentially significant concerns are lost or masked in the wealth of data. There are, of course, dangers associated with focusing and strategic-level decision making that are not fully offset by the cooperative involvement of all parties. Careful attention needs to be paid to lessons from comparable projects in comparable environments. Particular attention needs to be given to tailoring requirements to the proposed project and to the potentially affected environment. A concerted effort should be made to identify and fill critical data gaps, to manage uncertainties, to integrate stakeholder concerns and knowledge, and to focus continually on impact significance, from multiple perspectives. A systematic impact management strategy, including monitoring, contingency measures, and continuing stakeholder involvement, takes on even greater importance. Overall, however, as long as the proper safeguards are in place and the process is open and collaborative, a focused and strategic EIA approach can have both economic and environmental benefits.

ROGER CREASEY
Shell Canada Ltd.

6.2.3 Streamlining Involves More Than the Direct Application of Available Tools

The Canadian Environmental Assessment Act (CEAA) establishes a federal environmental assessment process to consider the environmental effects of proposed projects where there is a federal decision or responsibility, whether as a proponent, land administrator, source of funding, or regulator. Most projects and activities that trigger an environmental assessment under CEAA are assessed through a self-directed screening. Recognizing the large number of screenings, many of which concern routine projects with a well-understood range of predictable and mitigable effects, CEAA provides for a *class screening* mechanism. A class screening process is a planning tool that theoretically simplifies and streamlines the environmental assessment, regulatory review, and approval processes for routine or repetitive projects or activities. A class screening process is expected to improve the predictability, efficiency, and certainty of the environmental assessment process under CEAA for some types of projects.

Environmental specialists and advisors at the Canadian Environmental Assessment Agency continue to promote class screening as a tool to streamline the environmental assessment process for both industry players and the federal regulatory authorities. Experiences in Ontario indicate that class screening is a well-used tool at the provincial level to streamline the environmental assessment and approvals process for routine activities. For the past five years, Ontario has had a class screening process in place for transportation facility projects and for municipal road, water, and wastewater projects. Continued use of the class screening tool suggests that it is effective in the provincial context. Few federal agencies, however, have taken advantage of the class screening provision under CEAA. Only four class screening processes have been instituted to date—Parks Canada, the Canadian Food Inspection Agency, Fisheries and Oceans (British Columbia and Yukon), and the Prairie Farm Rehabilitation Agency—but over 15 more are in development or review.

There are currently some challenges to the development and use of class screening at the federal level, and presumably, these challenges constitute a barrier to advancing class screening under CEAA. For example, one federal authority commented that although a class screening process could conceivably improve the clarity and certainty of the environmental assessment process for industry by defining information requirements and the steps in the process, the efficiency of the process would probably not improve because the federal regulators are still required to review and make a screening decision for each project or activity proposed. In addition, development of a class screening process entails some capital expenditure on the part of federal authorities. Either a consultant must be hired to develop the process, or an employee(s) must be dedicated to the task on a full-time basis. Either way, human and financial resources must be allocated for the task. Previous federal authorities have taken about two to three years to develop and declare a class screening process. Thus the benefits of a class screening process, upon implementation, are typically not apparent until a few years after the initial commitment and expenditure. In this time of government cutbacks it is often difficult for federal authorities to rationalize the expenditure to develop a class screening process when there are limited short-term benefits.

Although there are no empirical data to indicate that class screenings under CEAA have increased the efficiency of the environmental assessment process, practice suggests that the trend toward developing and implementing this process is positive. At a minimum, the development of class screening processes clarifies the information required for proponents to submit to the federal authority to complete a screening under CEAA, and clearly defines the responsibilities for all parties involved in the environmental assessment process. Pending amendments to CEAA provide additional clarity and guidance, thereby encouraging the use of class screenings. Presumably, with the new amendments to CEAA, class screenings will be easier to formulate and implement, allowing federal authorities to use limited resources more efficiently.

This story suggests that streamlining EIA requirements and processes involve more than simply applying available tools, such as class assessments. There needs to be an understanding of where and the extent to which current inefficiencies occur (which would be addressed or not addressed by the tool), how current operating procedures are likely to change, the associated startup and long-term costs and benefits, the likelihood of institutional resistance and inertia (and the means to overcome such challenges), the effectiveness lessons from applying the tool to date and in other settings (appreciating contextual differences), and the positive and negative implications for decision making and implementation. Also, with any streamlining tool it is essential to understand what the implications might be for maintaining and enhancing the environment and for open and participative decision making. In the case of screening, for example, are adequate safeguards built in to "bump up" a proposal when the environment is especially sensitive or the project especially controversial? Is public access to and involvement in the EIA process inhibited or maintained? Are the ascribed benefits likely to be realized, and to what extent? Collectively, these questions and uncertainties underscore the need to carefully monitor and adapt the institution of streamlining procedures in an open and interactive process. It also suggests the need to carefully scrutinize historical experiences and experiences in other jurisdictions.

LESLEY MATTHEWS
Jacques Whitford Environment Limited

6.3 DEFINING THE PROBLEM AND DECIDING ON A DIRECTION

The three stories address practicality in different ways. The first story describes a collaborative EIA process structured and managed to produce and implement practical solutions to real problems shared by an array of stakeholders. The second story describes a scoping-related experience that successfully overcame an engrained approach to EIA requirements without sacrificing substantive objectives or stakeholder interests. The third story describes the practical benefits and obstacles associated with implementing class assessment procedures—a tool for streamlining EIA regulatory requirements. The stories provide, at best, a partial and very preliminary sense of the potential role of practicality in the EIA process.

Practicality has many dimensions, as illustrated in Figure 6.1. Each dimension encompasses numerous elements relevant to practicality in the EIA process. The EIA process, as expressed in EIA theory and practice, can be much more practical. Despite the widespread advocacy of scoping (especially in the United States), too often EIA documents remain excessively descriptive, lengthy, and unfocused (Barrow, 1997; Ensminger and McLean, 1993). EIA processes continue to take too much time and to consume too many resources (Sadler, 1996; Wolfe, 1987). Alternatively, they operate within such severe time and budget restrictions that the potential for good practice is seriously inhibited (Clark, 1997; Offringa, 1997). Scoping, although demonstrably beneficial, is sometimes a difficult concept to apply in practice (Clark, 1997; Morgan, 1998; Sadler, 1996; Wood et al., 1996). Rather than being focused and streamlined, shorter EIA reports tend to be of poorer quality (Wood et al., 1996). Practitioners continue to struggle with identifying priorities and impacts of real concern (Sadler, 1996). Effectiveness ratings for study design activities such as problem definition, objectives determination, and terms of reference formulation leave considerable room for improvement (Sadler, 1996). Ill-defined or excessive requirements, often in combination with gaps, overlaps, and coordination problems across government levels and agencies, are still problems in many jurisdictions (Anderson, 2001; Ensminger and McLean, 1993; Offringa, 1997). In short, EIA documents, processes, and institutional arrangements could be more efficient and *focused*.

EIA is plagued by such *reality*-related problems as (1) a gulf between how policy making and project planning take place and how EIA processes and practices assume that they take place, (2) EIA theory not well grounded in or derived from EIA practices and experiences, (3) EIA practices poorly suited to the contexts in which they are applied, and (4) EIA processes and practices that fail to appreciate the contributions and implications of varying stakeholder values and perspectives. These problems suggest that EIA theory and practice need a reality check.

The rational assumptions (see Chapter 4) embedded in most EIA process characterizations are rarely realistic. Human, institutional, and political characteristics, constraints, and behavioral patterns need to be better understood, especially regarding how decisions are made and implemented (Weiner, 1997). More consideration could be given to policy and decision-making models that mimic planning and decision making as they are (such as bounded rationality), in contrast to how rationality

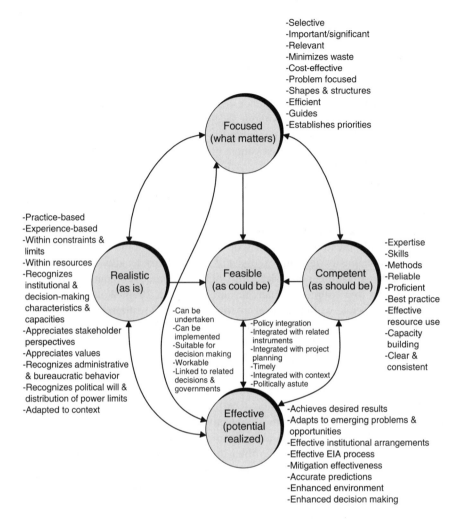

Figure 6.1   Examples of dimensions of practicality.

advocates would like them to be (Nilsson and Dalkmann, 2001; Nitz and Brown, 2001). Decision-making and implementation constraints are not insurmountable—but first they must be understood. The naïve expectation that rational and/or scientific EIA documents and processes lead inevitably to environmentally sound decision making and implementation is questionable at best and at worst can reduce the relevance of EIA outputs to major project and policy decisions (Nitz and Brown, 2001). The late (in the decision-making process) initiation of EIA requirements and the large number of major decisions not subjected to EIA requirements remain recurrent problems in EIA practice (US CEQ, 1997a). More systematic consideration needs to be given to how EIA systems affect policy making and project planning (Bartlett, 1989). EIA practitioners need to be better informed about the nature of policy-making and project planning processes (Nitz and Brown, 2001).

There is an urgent need to learn from experience and good practice (Glasson et al., 1999; Sadler, 1996). Lessons and insights, derived from good practice, need to be better integrated into EIA regulatory requirements, guidelines, and practices (ERM, 2000; Spooner, 1998). EIA regulators and theorists should strive for an enhanced understanding of stakeholder (e.g., bureaucrats, politicians, proponents, practitioners, nongovernment organizations, members of the public), perspectives, interests, and needs (Rowson, 1997). The EIA process (or more exactly, multiple EIA processes) needs to be designed to match contextual characteristics more closely (Greer-Wooten, 1997; Nilsson and Dalkmann, 2001; Nitz and Brown, 2001).

A practical EIA process must also be prescriptive. The shift to the prescriptive does not mean an abandonment of the real. A practical EIA process remains realistic but also seeks out *feasible* actions that can be undertaken and implemented in varying contexts. EIA practice needs to balance practicality and prescription. Too often, EIA practice (1) neglects the needs of decision makers, (2) fails to facilitate implementation, and (3) is not integrated effectively with other environmental management instruments, project planning, and public policy making.

Providing relevant and sound environmental information and advice to decision makers, although necessary and improving, is far from the whole picture. The needs, values, and perspectives of decision makers and other stakeholders are often neither identified nor addressed (Spooner, 1998; Wood, 1995). Post-approval management and follow-up remain more the exception than the rule (Sadler, 1996). Strategies are required to ameliorate implementation obstacles such as overlapping mandates, delays, late triggers, unclear, incomplete, or contradictory requirements and guidelines, coordination and consistency difficulties, and the propensity of agencies to treat EIA requirements as a rigid paperwork exercise (Anderson, 2001; Clark, 1997; Sadler, 1996; US CEQ, 1997a; Weiner, 1997). The potential for EIA as a strategic decision-making tool is not fully realized (Clark, 1997). EIA requirements are still rarely applied to major government decisions or to nongovernmental actions, with potentially significant environmental consequences (Andrews, 1997). EIA documents, when prepared, are commonly treated as decision implementation rather than decision-making documents (Ensminger and McLean, 1993). Too little attention is devoted to establishing complementary links between EIA and SEA and project planning, policymaking, and other environmental management and sustainability instruments (Nitz and Brown, 2001). By neglecting these practical decision-making and implementation considerations, the EIA process often falls short of its potential.

Sound execution of the EIA process requires *competence*. Much advice regarding methods and procedures is offered in EIA texts and literature. But EIA practice, as reflected in requirements, guidelines, and documents, too frequently fails to meet even minimum good practice performance standards (Glasson et al., 1999; Sadler, 1996; Spooner, 1998). The extreme variability in the quality of EIA documents, from project to project and from region to region, is difficult to reconcile with the image of a maturing field of theory and practice. This shortfall could simply be the result of a failure to apply available knowledge and insight. Perhaps the variability in quality could be largely explained by differences in the experience and expertise of practitioners (Barker and Wood, 1999).

Alternatively, EIA literature could be missing the mark in meeting practitioner needs. Perhaps the guidance provided is too superficial, too scattered across numerous sources, too difficult to access, and too difficult to understand or apply (ERM, 2000). There may be insufficient time or money to apply state-of-the-art practice. Possibly, the management skills and expertise required to tie the pieces together are insufficiently developed or inconsistently applied (Glasson et al., 1999). More capacity building could be required before EIA practitioners achieve the necessary proficiency levels (Offringa, 1997). Good-practice standards could be too general, contradictory, or unrealistic. Methods may need to become more cost-effective and more conducive to applying new technologies (Offringa, 1997). The skills and expertise required of practitioners could be more complex than the methods purveyors realize (Webster, 1997). Simply assuming that the necessary knowledge is available and that the problem will resolve itself as more experience is acquired is a dubious strategy given the continuing quality disparities after some 30 years of EIA practice. A more prudent strategy is to assume that EIA competence deficiencies require an array of responses.

EIA practicality "problems" are little more than impressions and the "solutions" offered no more than speculations if EIA *effectiveness* is not addressed systematically. Considerable progress has been made in formulating and applying EIA quality and effectiveness criteria and performance standards to documents, procedures, methods, and institutional arrangements (ERM, 2000; Sadler, 1996; Wood et al., 1996). These efforts, although laudable, barely "scratch the surface" in terms of what is required to close the loop from experience to learning (Glasson et al., 1999; Wood, 1995). The monitoring and auditing of actual environmental impacts (as compared with effects predicted) is still more the exception than the rule (Clark, 1997; Culhane, 1993; Morgan, 1998; Sadler, 1996). The effectiveness of mitigation measures is rarely determined (Clark, 1997). A much greater effort could be made to assess and compare EIA methods, process designs, management strategies, and institutional arrangements, in varying contexts (Glasson et al., 1999).

The contribution of EIA to more environmentally sound decision making and to a more sustainable environment is more often an assumption than a demonstrated outcome (Andrews, 1997; Welles, 1997). The magnitude and nature of the contribution and which strategies and tactics effectively operate within constraints and overcome implementation obstacles are even less clear. In the absence of demonstrated contributions, it is difficult to argue for the continued allocation of resources to EIA and for a well-defined role for EIA within environmental management and sustainability strategies. What is required is a substantiated case that EIA achieves desired results and adapts to emerging problems and opportunities.

The problem then is five clusters of interrelated problems, all bearing on the issue of practicality in the EIA process. The direction is concepts and approaches for making the EIA process more focused, realistic, feasible, competent, and effective. These approaches establish a foundation for practical EIA regulatory requirements and practical EIA processes.

6.4 SELECTING THE MOST APPROPRIATE ROUTE

6.4.1 Focused

The EIA process, in its fullest expression, is, by definition, impossible. More environmental components and interactions, alternatives, and direct and indirect effects can always be suggested. The level of detail can always be increased. More parties can be involved. More participation can occur. More research can be undertaken of uncertainties. In short, there is no *stopping rule*. As a result, the EIA process can never be comprehensive. The real issue is how to focus the EIA process to balance environmental objectives, available resources, and decision-making requirements. Focusing or scoping (used interchangeably here) has been integral to U.S. federal EIA requirements for more than 20 years. Scoping is featured less prominently in the EIA systems of other jurisdictions. Notwithstanding the extensive experience with scoping (at least in the United States), there remains considerable room for improvement in EIA scoping requirements and practices. Figure 6.2 provides an overview of the major elements associated with focusing regulatory and applied EIA processes.

Scoping focuses EIA institutional arrangements, the EIA process, and EIA documents (Morgan, 1998). It determines what will and will not be examined (Wolfe, 1987). It establishes appropriate levels of detail for various analyses (Wolfe, 1987). It directs and structures the EIA process, EIA operational procedures, and institutional reforms. Scoping establishes priorities (Eccleston, 1999a). The benefits ascribed to scoping are considerable. When it works effectively, scoping reduces the duration of EIA planning and review processes, abbreviates EIA documents, ensures efficient resource use, and identifies key issues, priorities, and problems early enough in the process to take appropriate action (Glasson et al., 1999; Sadler, 1996; US EPA, 1998b; Wolfe, 1987). It is conducive to early stakeholder involvement and can reduce the likelihood and severity of conflict among stakeholders (ERM, 2000; Sadler, 1996; US EPA, 1998b). Scoping focuses the process on potentially significant issues and impacts (Bond and Stewart, 2002). It reduces the likelihood that resources will be wasted on insignificant concerns (ERM, 2000; US EPA, 1998b). Scoping contributes to higher-quality EIA documents and to more environmentally sound decisions (ERM, 2000; Sadler, 1996).

Narrowing the scope of the EIA process, based on overview analyses and preliminary stakeholder discussions, however, can lead to the premature rejection of alternatives and to unanticipated effects (Erickson, 1994). The level of detail may not be sufficient to justify a clear distinction between significant and insignificant issues and impacts. Once an EIA process is scoped, the resulting study designs could be treated as blueprints, to be followed regardless of changing circumstances. Sometimes, a scoping process is unduly influenced by vested interests. These

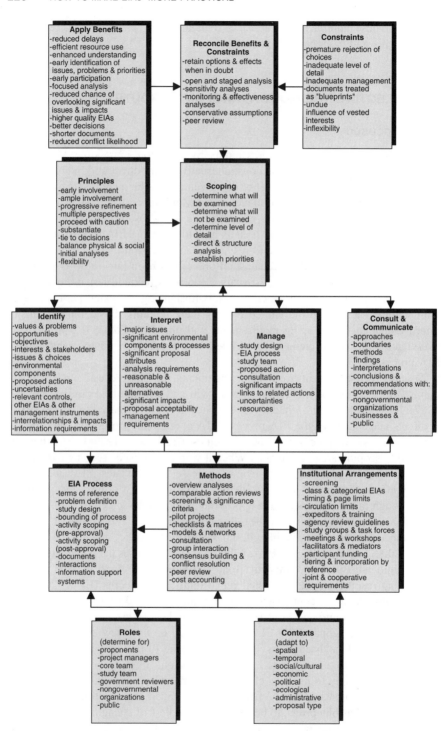

Figure 6.2 Focusing the EIA process.

constraints can be ameliorated if alternatives and potential impacts are retained for further consideration if there is doubt regarding their suitability and significance. An open and staged process, with a high level of stakeholder participation, in combination with conservative assumptions, sensitivity analyses, peer review, and a careful scrutiny of the results of monitoring and effectiveness analyses can reduce the likelihood of inadequately supported decisions (Morgan, 1998). Scoping works best when there is early and ample stakeholder involvement (US EPA, 1998b; Morgan, 1998). Multiple perspectives should be brought to bear on scoping interpretations and decisions. Scoping need not be only a stage near the outset of the EIA process. Instead, it can be a scanning–focusing phase preceding more detailed analyses and prior to each EIA process decision (Brown, 1998; Kennedy and Ross, 1992). Caution is essential given the broad-brush nature of the analyses. Sometimes, this means scanning ahead. Sometimes, previous decisions need to be reconsidered (Brown, 1998). Flexibility to adjust to changing circumstances is critical. Adequate consideration should be given to both biophysical and socioeconomic concerns (Erickson, 1994; Morgan, 1998).

Scoping can be applied at the outset to define the problem, establish the terms of reference, design the overall EIA process, and set the study boundaries (Barrow, 1997; Sadler, 1996; Wood, 2000). It can also focus and structure each EIA process activity and document leading up to (e.g., alternatives formulation and evaluation, baseline analyses, impact identification and prediction, public and agency consultation, document preparation), and subsequent to (e.g., monitoring, mitigation, auditing) proposal acceptance or rejection (Kennedy and Ross, 1992).

Scoping helps reform EIA institutional arrangements. Screening distinguishes between actions subject and not subject to EIA requirements. It also determines applicable approval streams. EIA screening and scoping requirements are streamlined and focused by class or categorical EIA requirements and by significance thresholds and criteria. Documents can be simplified by page limits, incorporation by reference, report format requirements, and page limits (Kreske, 1996). Review and approval can be expedited by timing and circulation limits, by merged and cooperative interagency and intergovernmental requirements, and by agency review guidelines. The latter directly link EIA requirements to agency mandates, policies, programs, and priorities (Kreske, 1996). Meetings, workshops, study groups, task forces, expeditors, facilitators, mediators, and participant funding can constructively bring together interested and affected parties, both within and external to the government review process.

6.4.2 Realistic

A practical EIA process is necessarily grounded in practice and experience. It is realistic. Figure 6.3 illustrates examples of distinctions potentially relevant to making the EIA process more realistic. Table 6.1 presents a summary overview of the characteristics of several potentially relevant realism concepts. Collectively, these concepts suggest that knowledge is often subjective, pluralistic, experience-based, and socially constructed. Practice is concrete, action-oriented, critical, and

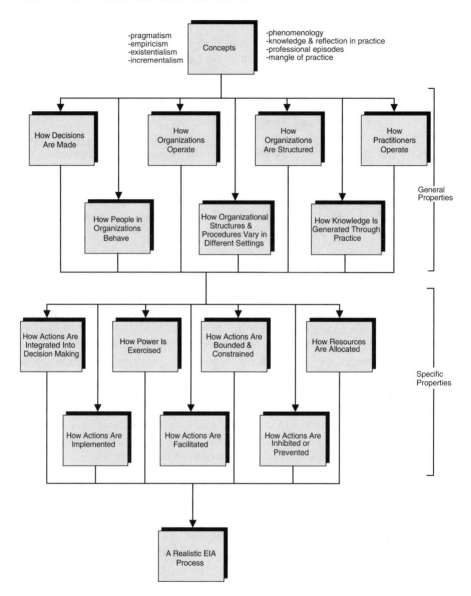

Figure 6.3 Realism in the EIA process.

experimental. Planning, policy making, and decision making are frequently constrained, decentralized, incremental, collaborative, communicative, political, and pluralistic. Society and environment (the context that circumscribes policy and decision making) are commonly fragmented, uncertain, complex, ambiguous, and unpredictable. Distinctions between theory and practice, ends and means, facts and values, and objectivity and subjectivity are artificial. These portrayals of reality

Table 6.1   Examples of Potentially Relevant Realism Concepts

Pragmatism	Philosophy of everyday life; antifoundational; importance of dissent and irreverence
	Plurality of shifting truths grounded in concrete experiences and language
	Actions structured by subjective interpretations of the world; interpretations evaluated in terms of their practical implications; intent is to solve human problems
	Concepts are socially constructed; truth not understandable outside the social and psychological processes and community that makes truth possible; justification from prior experience
	Concepts, terms, and assumptions tentative and provisional—always open to further interpretation and criticism—fallibilism; prediction possible but limited
	Focus on concreteness, action, adequacy, facts, and power; turns from abstractions, verbal solutions, a priori reasons, closed systems, origins, fixed principles, and absolutes
	Pluralistic: a plurality of traditions, perspectives, and philosophical orientations
	No definitive formulation to problems and no clear solutions; all knowledge contingent
	Learning by doing (learning and doing indivisible); learning a collaborative experience
	Problems solved by common sense and experimentation; guided by changing experience
Empiricism	Reliance on experience and observation alone
	Founded on belief that all knowledge originates in experience or in the practice of relying on observation and experiment
	Limited consideration of system or theory
	Focuses on collecting facts and observation
	Emphasis on information derived from human senses
	Exemplified in studies of practice
Existentialism	Point of departure experience rather than generalized concepts; all concepts derived from human perceiving, pattern forming, symbolizing, comparing, and conceptualizing
	Emphasis on immediate experiences and individuals as autonomous moral agents
	Terms described not defined; consistency only possible through repetition of experience; not possible to state assumptions and conclusions only reached based on implications
	Different people reach different conclusions based on same information
	Propositions have multiple meanings; communications failures expected
	Reality only partially conveyed by symbols
	Existence contingent (not independent of situation); only here and now meaningful and present experiences—complex, unique, correlated, uninterpretable, and uncommunicable

(Continued)

Table 6.1 (*Continued*)

Phenomenology	All knowledge is subjective
	Analyzes and identifies basic features of subjective knowledge to understand individual and to make life more significant
	Belief that people should be studied free from any preconceived theories and suppositions about how they act
	Search for understanding of nature of act rather than explanation
	Belief that for people world exists only as a mental construction; created in acts of intentionality
Incrementalism (also bounded rationality)	Margin-dependent choices; successive limited comparisons; a process of gradual change (muddling through)
	Restricted number of values, alternatives, and consequences; available means and solutions
	Objectives adjusted to policies (means and ends overlap and reciprocal); no coherent set of goals
	Analysis and evaluation: serial, remedial, socially fragmented, and unpredictable
	Assumes ambiguous and poorly defined problems, incomplete information (baseline conditions, values, alternatives, consequences); thinking inseparable from context and experience
	Decision making fragmented and largely reactive to external circumstances; not value-free; decision makers avoid uncertainty and adverse consequences
	Appreciates human (especially expert) knowledge and control limits; political, social, and economic environments complex, uncertain and unstable; planning incomplete, partial, collective, and episodic
	Atomistic society and decentralized decision-making structures and procedures; policy making a negotiation and bargaining process involving a plurality of competing interests and values
	Test of a good policy: agreement; driven by political circumstances; focus on political negotiations and coalitions
	No to limited reliance on theory; adapted to limited cognitive capacities; influenced by free competition model of economics
	Rationality bounded by cognitive limits, social differentiation, pluralistic conflict, and structural distortion
Knowledge and reflection in practice	Each person develops his or her own way of framing (taken-for-granted assumptions) role (e.g., writer, organizer, advocate); may choose from profession's repertoire or fashion own interpersonal theory of action
	Knowing in practice (common sense) directs and limits reflection in practice (thinking about what we are doing); self-reinforcing system in which role frames, action strategies, relevant facts, and interpersonal theories are bound together
	Behavior understood in terms of problems set for self
	Roles evolve in conversation with situation; practice as exploratory experiment (probing, playful)
	Policies sometimes reframed in action; often as a result of reflecting on frame conflicts

Table 6.1 (*Continued*)

	Double visioning: awareness of own perspective and that of others
	Rhetorical frames: underlies persuasive use of stories and arguments; action frames inform policy practice
	Study of strategies to resolve frame conflicts (e.g., resistance, appealing to consensus, mapping one frame over another)
	Policy design inevitably social, political, pragmatic, and communicative
Professional episodes	Schematic framework for analyzing professional practice episodes
	Theories are socially constructed; knowledge derived from action
	Episodes analogous to dramas; practitioners construct performances with constituent others
	Distinctions: institutional professional espoused theory vs. practitioner espoused theory, practitioner theory-in-use vs. practitioner espoused theory
	Communications (talk is action) at core of professional episode
	Practitioner strives for one or more of enhanced self-esteem, mastery of professional domain, cognitive and value consistency, self-actualization, or significant impact on world of contemporaries
	Practitioner in performing concrete professional tasks is the ultimate theorist for each episode
	Importance of concrete situation, language and communications, ambiguity, and evaluation
Mangle of practice	Human and nonhuman material agents (e.g., tools) intertwined and coevolve
	Simultaneously objective, relative, and historical
	Dialectic of resistance and accommodation
	Favors antidisciplinary synthesis and multidisciplinary eclecticism
	Multiple rather than monolithic conceptualizations, models, and approximation techniques
	Data and theory not necessarily connected; approximations toward the truth
	Practice aims to make associations (translations, alignments) between diverse elements

Sources: Blanco (1994), Bolan (1980), Braybrooke and Lindblom (1963), Etzioni (1967), Forester (1989), Friedmann (1987), Hainer (1968), Lindblom (1965), Menard (1997), Nilsson and Dalkmann (2001), Pickering (1995), Schön (1983), Schön and Rein (1994), Simon (1976), Verma (1998).

challenge the value and validity of preconceived theories and suppositions, abstractions, fixed principles, absolutes, and symbolizing.

These realism concepts are largely a reaction against the rational assumptions (see Chapter 4) inherent in most policy, planning, and EIA theories. Although they overstate decision-making constraints, they are a closer approximation of the environment within which most EIA practitioners operate than the antiseptic versions of the EIA process presented in most EIA texts. These realism concepts, however, provide only impressions rather than a firm foundation for a practical EIA process. What is required is a more detailed characterization of the reality of EIA practice. EIA literature, together with the literature of related fields such as planning, offer a sense of many aspects of EIA practice. A more complete picture would draw heavily upon related fields, such as political science, and would integrate relevant distinctions and concepts from decision making, public policy, organizational, and administration theory.

A realistic EIA process would recognize how decisions are made, how organizations are structured, how people behave in organizations, and how organizational structures and procedures vary depending on contextual characteristics. How EIA practitioners operate (effectively and ineffectively) and how knowledge is generated through practice would be understood. The mechanisms by which EIA-related actions are integrated within decision making would be appreciated. Factors that promote and impede the integration of EIA and organizational planning and decision making would be evident (Keysar and Steinemann, 2002). How power is exercised, how resources are allocated, and how actions are bounded and constrained would be acknowledged. The various ways in which actions are implemented, facilitated, inhibited, or prevented would be understood.

A realistic foundation has been partially constructed within EIA and even more so in related fields such as planning and public policy. The relevant analyses are widely scattered. Cognitive limits and the inherent knowledge uncertainties associated with decision making have been considered (Beanlands and Duinker, 1983; Nilsson and Dalkmann, 2001). The constraints and opportunities posed by institutional arrangements, the exercise of political power, bureaucratic behavioral patterns, and the implications of ecological, economic, social, and cultural conditions are sometimes noted.

Institutional arrangements concern the government structures and procedures pertaining directly (e.g., legislation, regulations, policies, guidelines, staff, and budgets) and indirectly (e.g., related policies, programs and activities, departmental and agency jurisdictions and responsibilities, interactions among agencies and with other government levels, controls, resources, coordination and information transmission mechanisms, antagonisms, procedures for mediating conflicts and for representing interests, general efficiency, accountability, and flexibility) to EIA (Nilsson and Dalkmann, 2001; Pressman and Wildavsky, 1973; Rickson et al., 1990; Shoemaker, 1994; Smith, 1993).

EIA practitioners often see themselves as objective, independent, and apolitical advisors. But the EIA process is inherently *political*. Politics is a major determinant of if and how EIA requirements are applied. The lack of political will is a major impediment to achieving environmental objectives (Caldwell, 1997). Realistic EIA practice appreciates how political power is used and misused (Smith, 1993). If political power is highly dispersed, evasion and the dilution of reforms is the usual result (Pressman and Wildavsky, 1973). Highly concentrated political power can be authoritarian, coercive, and narrowly focused. Inequities in the distribution of prestige, power, and equity can make it difficult to realize social objectives (Rickson et al., 1990a). A centralized and hierarchical style of governance can prevent or severely inhibit the negotiation and consensus building needed to facilitate public understanding and possibly support (Conçalves, 2002). Stakeholder roles, formal and informal procedures for forming alliances, intervention rules and political structures, and decision processes need to be considered (Smith, 1993). Control (e.g., procedural, judicial, evaluative, development aid agency, professional, and direct public and agency) mechanisms need to be taken into account (Ortolano, 1993).

Bureaucratic structures, procedures, and patterns of behavior strongly influence EIA effectiveness. Bureaucracies exhibit such characteristics as fixed official jurisdictional areas, rationalistic division of labor, official duties, hierarchical structure, management by written rules, and expert management (Hummel, 1977). The role of bureaucracies can be negative or positive. When negative, there is a gap between what is important to the bureaucracy (e.g., precision, stability, formal rationality, formalistic impersonality) and what is important in society (e.g., justice, freedom, poverty, illness) (Hummel, 1977). A failure to institutionalize new forms of public participation into EIA requirements and practices, for example, can result in a gap between public aspirations and expectations and available participation forms (Conçalves, 2002). Such gaps often reinforce public distrust and contribute to community opposition.

Negative bureaucratic behaviors that can impede environmental initiatives include, for example, rigidity, classification (oversimplifying the world), interagency and intergovernmental antagonisms, favoring routine and prescribed rules over policy, displacing ends with means, preferential treatment of some client groups over others, "empire building," overcommitment, and ill-defined criteria (Pressman and Wildavsky, 1973; Rickson et al., 1990; Sorenson and Auster, 1989).

Often, government participants will agree with substantive environmental ends but will still oppose or fail to facilitate a proposed action. Pressman and Wildavsky (1973) identify several reasons for this behavior, including (1) direct incompatibility with other commitments, (2) no direct incompatibility but a preference for other programs, (3) simultaneous commitments to other projects, (4) dependence on others who lack a sense of urgency in the program, (5) differences of opinion on leadership and proper organization, (6) legal and procedural differences, and (7) agreement coupled with lack of power. Bureaucracies do not always exhibit such tendencies. Frequently, government officials assume a positive and proactive role in advancing environmental objectives. Still, it is prudent to be aware of general bureaucratic tendencies (positive and negative) and of the specific constraints and opportunities posed by the structures, procedures, and behavioral patterns of each government department and agency involved in the EIA process.

A realistic EIA process considers interrelationships between process and ecological, social, economic, cultural, and political *contexts*. The dangers associated with uniformly applying standardized definitions of good EIA practice and appropriate institutional arrangements are acknowledged increasingly. Particular attention has been devoted to the adaptations required to meet the needs of developing countries (Barrow, 1997; Lee, 2000; Rickson et al., 1990b; Smith, 1993). Comparable adaptations have been suggested for transitional economics, for northern environments, and for numerous other setting types. It is also necessary to make adjustments to suit the unique circumstances associated with a proposed action in a particular setting. The EIA process is not simply designed to fit contextual realities. EIA is an instrument for change. How contextual characteristics are likely to change, both positively and negatively, in response to changing EIA requirements and practices should be considered.

Institutional, political, bureaucratic, and contextual constraints and opportunities are strongly influenced by the quality and effectiveness of EIA practice. As documented in the "defining the problem and deciding on a direction" sections of Chapters 2 to 10, there remains a considerable shortfall between EIA aspirations and achievements. Good EIA practice can alleviate EIA practice deficiencies. Part of good EIA practice includes accounting for, and ameliorating where practical, institutional, political, bureaucratic, and contextual constraints. It includes taking advantage of opportunities. The full incorporation of realism into the EIA process requires integrative frameworks, concepts, and distinctions presented in a user-friendly format. It also necessitates a thorough canvassing of sources bearing on distinctions such as those presented in Figure 6.3. Most important, it requires empirical studies of EIA practice—studies that systematically draw out positive and negative experience-based lessons and insights, with potential for broader application.

6.4.3 Feasible

A feasible EIA process such as the one illustrated in Figure 6.4 is workable. It can be undertaken. It provides a decision-making basis. It can be implemented. It can be managed. It is appropriate to the context. It is built on a realistic foundation (i.e., it is experience and practice based). Consistent with realism, it is social, political, subjective, uncertain, and constrained. It is guided by strategies, informed by concepts, and aided by tactics—all of which are realistic and practical. It overlaps and is merged with decision making, implementation, management, and context. It contributes and adapts to EIA reforms. Appropriate links are made to related tools and methods. It is integrated with related policies, programs, and plans. It is blended with organizational operations. It is linked to other decision-making levels, related environmental management instruments, and the relevant actions of other governments, the private sector, and nongovernmental organizations. It crosses disciplinary and professional boundaries. It is integrated, where practical, within synthesis (e.g., sustainability) frameworks.

Table 6.2 describes briefly a cross section of relevant feasibility strategies, concepts, and tactics. The EIA process depicted in these sources is selective, cyclical, open, fluid, decentralized, and evolving. It is incremental but progressive. It learns through experimentation, reflection, and dialogue. It scans ahead and reconsiders past decisions. It continuously explores uncertainties, interconnections, complexities, and conflicts. It is reasonable rather than rational. It operates at multiple levels of detail. Preferred choices are not compared systematically and comprehensively. Rather, they are tested for agreement and feasibility. Then they are refined, adapted, and embellished to better meet agreed-upon needs, consistent with stakeholder

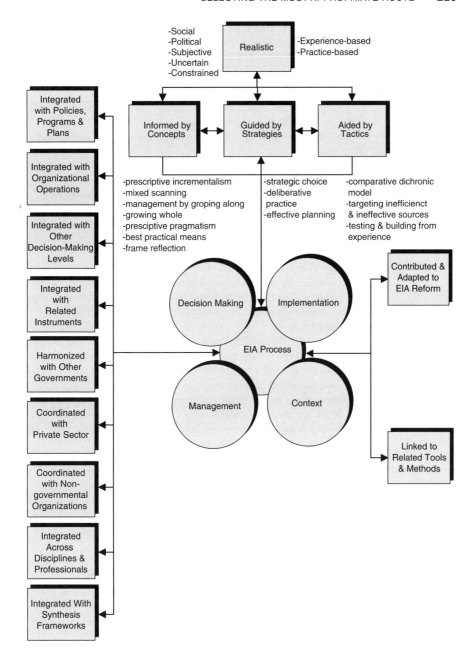

-Social
-Political
-Subjective
-Uncertain
-Constrained

Realistic

-Experience-based
-Practice-based

Informed by
Concepts

Guided by
Strategies

Aided by
Tactics

Integrated
with Policies,
Programs &
Plans

Integrated with
Organizational
Operations

Integrated with
Other
Decision-Making
Levels

Integrated
with
Related
Instruments

Harmonized
with Other
Governments

Coordinated
with
Private Sector

Coordinated
with Non-
governmental
Organizations

Integrated
Across
Disciplines &
Professionals

Integrated With
Synthesis
Frameworks

-prescriptive incrementalism
-mixed scanning
-management by groping along
-growing whole
-presciptive pragmatism
-best practical means
-frame reflection

-strategic choice
-deliberative
 practice
-effective planning

-comparative dichronic
 model
-targeting inefficienct
 & ineffective sources
-testing & building from
 experience

Decision Making

Implementation

EIA Process

Management

Context

Contributed &
Adapted to
EIA Reform

Linked to
Related Tools
& Methods

Figure 6.4   Feasibility in the EIA process.

Table 6.2   Examples of Potentially Relevant Feasibility Strategies, Concepts, and Tactics

	Strategies
Strategic choice	Strategic choice: choosing in a strategic way
	Complementary aspects of any planning approach: technology, organization, process, and product
	Four decision-making modes: shaping, designing, comparing, and choosing
	Explores uncertainties about the working environment, guiding values, and related decisions
	Distinguishes among decision areas, links, schemes, and options
	Sequence of structured workshops (can be supplemented by software)"
	Detailed guidance and practical advice provided (based on extensive experience from a variety of planning and development decisions)
	Oriented toward interactive participation; a learning process
	Issue oriented, cyclical, selective, and subjective
	Systematically addresses lateral connections; addresses web of relationships between technical and sociopolitical streams
	Addresses skill requirements and practicalities for each decision-making mode
Deliberative practice	Process must be simultaneously interpretive, practical, political, and ethical (need to integrate theory, practice, pragmatism, and ethics)
	Ethics not as standards to follow but as pragmatic action (the allocation and recognition of values)
	Rationality is an interactive and argumentative process of marshaling evidence and giving reasons
	Consensus building created on existing political stages (but also addresses power imbalances)
	Need to improvise in complex and novel situations
	Necessary to empathize with other parties and remain politically neutral at the same time
	Critical listening, reflection-in-action, and constructive argumentation all interact
	Importance of practical storytelling (letting stories supplement our limited rationality)
	Challenge not to avoid, transcend, or displace conflict but to deal with practical differences in and through conflictual settings
	Fluid process (issues formulated and reformulated)
	Streams of choices, problems, solutions meet in unpredictable ways to shape ongoing, complex, and "messy" organizational outcomes
	Argues for activist mediation (concern with process, efficiency, stability, and well-informed character of outcome)
Effective planning	Planning as organizational learning
	Importance of networking, learning from errors, experimentation, research, and pilot projects
	Links strategic thought to implementation; a political and management activity

Table 6.2   (*Continued*)

	Need to build coalitions and networks to support proposal (building on shared values, importance of negotiations)
	Technical and political considerations married
	Stresses need to pay attention to logistics, reduce derailment potential, build trust, demystify, make contextual adjustments, and democratize
	Process depends on transactions; continually evolving and engaged
	Experience from the field the point of departure
	Planning intertwined with management; argues for greater use of incentives and risk taking
	Stresses value of making good on promises, correcting errors, removing mask of expertise, and delegating authority

Concepts

Prescriptive incrementalism	Various adaptations to incrementalism to make more prescriptive
	Dialogical incrementalism (a dialogical process aimed at mutual understanding and agreement)
	Purposive incrementalism (directed toward a purpose or vision and learning based)
Mixed scanning	Comprehensive broad-angle analysis
	Focusing on areas revealed by broad-angle analysis for more detailed scrutiny
	Fundamental decisions set context for incremental decisions that lead to new fundamental decisions
	Can be at several levels of detail and coverage
Management by groping along	Experiment: determine what works and does not work
	Progressively moves toward objectives; objectives well defined but means not; successes create new capabilities and motivate (strategy of small wins—facilitates learning and adaptation)
	Test ideas before different audiences and gauge results; try different permutations and combinations
Growing whole	Goodness of fit between proposed action and context; sensitivity to stress and misfit
	View of proposed action as a part of a growing whole (ecological, built, social, economic, cultural)
	Assess how well preserves and enhances wholeness at many levels and in many ways
	Move iteratively between ends and means, analysis and synthesis, rational and irrational, constraints and ideas; adds refinements, adjustments, and embellishments—all within a complex, ambiguous, organic, and evolving process
	Involves multiple process designers; communications critical
Prescriptive pragmatism	Philosophy of action; encompasses both doing good (moral and political) and being right (coherent and accurate technical analysis)
	Includes human experience, practical activity, and democratic experience; importance of achieving and maintaining trust

(*Continued*)

Table 6.2   (*Continued*)

	Practical endeavor that links satisfying human needs with application experience
Best practical means	Common approach to pollution control requirements
	Practical taken to mean "reasonably practical" having regard to the state of technology, local circumstances, and financial implications
Frame reflection	Scrutinize day-to-day tasks of practitioners; lessons from best practice and practice failure (practice wisdom, craft knowledge, experiential knowledge)
	Transmit and exchange practical knowledge; policy evolves dialectically; policy discourse
	Critically examine underlying assumptions, ideas, and beliefs; act from one perspective but be aware of others
	Frame criteria: true, beautiful, just, coherent, utility, or fruitfulness

Tactics

Comparative diachronic model	Series of snapshots over time as development progresses; attempt to fill in what happens in between
	Use of comparative and control studies to provide basis for impact study
	Use of impact study (supported by comparative and control studies) as decision-making basis
	Use of control study and post-approval impact analyses to manage impacts
Targeting inefficiency and ineffectiveness sources	Counter negative bureaucratic tendencies by opening up systems, going outside the bureaucracies, and with feedback and accountability loops
	Undertake implementation analysis (consider implementation feasibility at early stages, anticipate implementation, backward mapping)
Testing and building from experience	Use of pilot studies
	Staged approvals
	Identification and documentation of best practices (experimental knowledge, craft knowledge, true statements)
	Use of empirical studies to provide insightful knowledge and accounts of practical constraints

Sources: Alexander et al. (1987), Behn (1988), Benveniste (1989), Burdge (1994), Etzioni (1967), Forester (1999), Friend and Hickling (1997), Gilpin (1995), Hoch (1984), Hummel (1977), Kørnøv (1998), Patton and Sawicki (1993), Sager (1994), Schön and Rein (1994), Sorenson and Auster (1989).

perspectives. The process draws heavily upon knowledge derived from experience and practice. It operates within resource and other constraints. It transcends such false dichotomies as ends and means, technical and political, and objective and subjective. It crosses disciplinary and professional boundaries freely in the search for practical solutions to real problems. The process unifies planning, management, decision making, and implementation. It is connected, as needed, to related decisions, methods, and instruments. It is carefully matched to context—a context that is uncertain, complex, ambiguous, and subject to rapid and erratic change.

Unification of the EIA process and *decision making* means that the process is built around decisions and the information, analysis, and interpretative needs of all parties involved in decision making. EIA and organizational planning are ideally merged and concurrent (Keysar and Steinemann, 2002). At a minimum the EIA process should strongly influence agency planning and decision making. Decision-making needs are anticipated, refined in consultation with stakeholders, and adapted as positions evolve and as new concerns emerge. EIA documents cross-reference all requirements and comments. Reviewers and other interested and affected parties can, from the EIA documents, readily determine how and where their concerns and requirements are considered. Reasons are provided for concerns and suggestions not addressed. The treatment of each requirement and suggestion is discussed with each party before documents are finalized.

A feasible EIA process anticipates *implementation* requirements from the outset. Requirements are refined jointly with all parties likely to influence implementation directly or indirectly. Implementation commitments are clearly specified prior to approval. Close contact is maintained with agencies and departments likely to be involved in approvals and likely to impose conditions of approval. Conditions of approval are integrated into environmental management plans and strategies. Implementation includes such technical tasks as preparing monitoring reports, quality assurance, and assessing mitigation effectiveness. It includes building coalitions of support, identifying and offsetting implementation obstacles, ensuring adequate resources to facilitate effective implementation, and making an effective, merit-based case, adapted to the needs and perspectives of each party associated with implementation (Wandesforde-Smith, 1989). Sometimes institutional capacity building is necessary prior to implementation.

The EIA process is not designed and then adapted to the *context*. Process design begins only after the nature and potential implications of contextual characteristics are considered. The views of all interested and affected parties are actively solicited to ensure that the perspectives of each are reflected in the process (Rowson, 1997). The process is carefully designed and managed to account for EIA regulatory requirements and institutional, ecological, social, cultural, political, and economic conditions, constraints, and opportunities (Barrow, 1997; Lee, 2000). Alternative EIA process design types (e.g., rational, adaptive, conflict management) are appropriate for different contextual categories (e.g., varying levels of certainty and social conflict) (Nilsson and Dalkmann, 2001). The role of the EIA process as an instrument for changing the context is considered. Changing contextual characteristics are monitored up to and through implementation. The process is adjusted and refined as the context evolves. Multiple scenarios and sensitivity analyses ensure that the EIA process and process outcomes are sufficiently robust to respond rapidly to changing conditions.

A feasible EIA process treats impact *management* (project management is addressed in Section 6.4.4) as an ongoing function rather than as a stage at or near the end of the process. From the outset, consideration is given to how to avoid or minimize adverse effects, how to enhance benefits, how to offset inequities, and how to manage uncertainties. Mitigation is integrated into the alternatives analyses and into the proposal characteristics. Proponent and proposal-related impact management, compensation, and monitoring policies and strategies are formulated near the beginning of the process. They are refined jointly with stakeholders. Baseline analyses set up the environmental monitoring. Comparable action reviews, comparable environmental analyses, and pilot projects establish a foundation for impact analysis and management extending through the action life cycle.

Individual impact management measures (e.g., mitigation, compensation, local benefits, monitoring, contingency measures, financial security, funding, environmental liability) are consolidated within an impact management strategy. The strategy specifies management objectives and principles, variables, spatial and temporal boundaries, resources, responsibilities, testing protocols, methods, contingency measures, reporting requirements and stakeholder involvement, and conflict resolution procedures (Canter, 1996; Glasson et al., 1999; UNEP, 1997). It is linked to proponent policies, programs, and environmental management systems, to government requirements (e.g., compliance monitoring), and to environmental monitoring systems (Canter, 1996). Commitments to communities are formalized, where warranted, in impact management agreements. The impact management strategy is refined and adapted prior to approvals and throughout the implementation period. Impact management outcomes are documented in a form suitable for controlling impacts, for assessing mitigation effectiveness, and for validating and refining methods (Canter, 1996). Results are shared with stakeholders.

EIA is an evaluation tool. Evaluation methods also are used within the EIA process. *Additional relevant methods* sometimes are applied outside or partially overlap with the EIA process. Examples include feasibility studies, needs assessments, life-cycle analyses, risk assessments, technology assessments, futures research, total quality management procedures, economic and social cost–benefit analyses, policy and program evaluations, environmental management systems, and conflict management procedures (Gilpin, 1995; Mayda, 1996; Ridgeway, 1999; UNEP, 1997). A realistic EIA process addresses links to methods used outside the EIA process. Cross-referencing can minimize duplication. Inconsistencies are identified. Integration potential may be considered, at least to the point that a clear and consistent basis is provided for decision making and implementation.

A feasible EIA process has a reciprocal relationship with *EIA reform*. The EIA system does not stand still while the EIA process for a single proposal unfolds—especially for a process that takes years to complete. The rules of the game change. New requirements are instituted. Additional guidelines are issued. Perspectives and positions change. Sometimes a proposal, when caught in midreview, is "grandfathered." More frequently, the changes are subtle, particularly in terms of evolving agency perspectives and positions. A feasible EIA process strikes a balance between consistent review positions over time and adaptations to changing circumstances. This generally means constructive discussions, some reconsideration of previous decisions, and some refinements to analyses and documentation. The potential for major changes can be greatly ameliorated if close contact is maintained with review agencies and if the EIA process is undertaken in accordance with good practice standards in addition to meeting regulatory requirements. There is usually a lag between EIA requirements and good practice. Sometimes an EIA process (especially for a large, complex proposal, one involving new technologies or where there are major environmental uncertainties) cannot be reviewed or managed adequately without EIA system changes (Lee, 2000). Auditing the experiences associated with individual EIA proposal reviews contributes to EIA system reforms.

EIA requirements and the EIA process are interwoven with the actions of others. A feasible EIA process is necessarily boundary spanning. Project-level EIA requirements and procedures tend to be more effective when defined within the context of national, regional, and local sustainability, environmental, resource management, social and economic policies, strategies, programs, and plans (i.e., tiering) (Gilpin, 1995; Lee, 2000; Sadler, 1996). The auditing of project-level experiences can contribute to policy-, program-, and planning-level reforms. EIA works best when EIA roles (e.g., EIA preparation, EIA review) are a natural extension of agency objectives, policies, and operating procedures. Ideally, agencies consistently apply explicit environmental and resource quality performance criteria and standards. The concurrent application of environmental approvals and permitting requirements can expedite the EIA process (Sadler, 1996). The EIA process is further facilitated if the proponent has an environmental management system (EMS) in place (Barrow, 1997). An EIA is often the impetus for instituting an EMS (Glasson et al., 1999).

Government EIA responsibilities are often subdivided between head office and regional offices. There tends to be a greater concentration of specialists at the head office. Occasionally, both head office and regional office specialists comment on an EIA, not always consistently. Such divisions of responsibility need to be closely scrutinized. The process is more complex when there are multiple government levels (see Section 6.4.5). Intergovernmental agreements, informal coordination, area-wide planning and management, and procedures to ensure a single process, a single EIA document, and consistent timing requirements have all contributed to improved coordination and a clearer division of responsibility. The increased application of regional sustainability strategies and of regional environmental and resource management tools (e.g., the ecosystem approach, integrated environmental management, adaptive management) have further facilitated joint planning and management among government levels (Margerum, 1997). Some coordination difficulties will always remain. A feasible EIA process ensures that such coordination difficulties do not unnecessarily bog down EIA preparation and review. Interconnections among the disciplines and professions involved in EIA preparation and review are identified and explored (UNEP, 1997).

6.4.4 Competence

Competence is a key aspect of practicality. When things go wrong with an EIA process, the tendency is to blame unforeseeable circumstances. More often than the participants care to admit, the problems that arise are foreseeable and are resolvable through competent EIA practice. Competence is much more than the appropriate application of the methods and models presented in most EIA texts. The knowledge of specialists extends well beyond the overviews of disciplinary analyses presented in such references. If properly selected and coordinated, the specialists do not tend to be the problem. More frequently, problems arise with the ways in which individual analyses are guided, integrated, and applied. Sometimes, roles and responsibilities are poorly defined. Table 6.3 lists examples of roles and responsibilities for various participants in the EIA process. Good practices are not always applied for activities that transcend individual specialties. Table 6.4 provides good practice examples for study team management, study team participation, database management,

Table 6.3 Competency—Examples Roles and Responsibilities

---

*Project Management (project manager, project coordinator(s),
technical writer, editor, administrator)*

Formulate (with proponent and study team) overall approach, study design, EIA process (activities, events, inputs, outputs), and general methods [identification, prediction, evaluation, interpretation, cumulative effects assessment (CEA), participation, mitigation, compensation, monitoring, management]; clear rationale for each

Establish management structure and determine appropriate level of detail for each activity

Assemble study team and related resources; establish team roles, norms, and environment for joint action; arrange, with proponent, contracts

Control and manage team organization, activities, budgets, timing, and schedule; set work standards

Monitor and update project plan continually; keep project and progress (task completion, budget completion, schedule) records; communicate progress

Establish priorities, objectives, and milestones, solve problems, manage conflicts, negotiate trade-offs, and remove roadblocks

Manage core team and support staff; identify stakeholders

Determine report formats, hardware and software requirements, mapping scales and database management and GIS requirements; establish tracking and sign-off procedures

Identify, with team, analysis gaps and research, and training requirements

Coordinate analysis (proposal, environment, proposal–environment interactions), synthesis (data, criteria, significance, CEA, conclusions, recommendations), and interactions (internal—technical and management, external—agencies, elected representatives, groups, and individuals)

Guide and challenge study team—scope, level of detail, database, assumptions, interpretations, judgments, conclusions, recommendations

Document and present (with input and review by study team)—study design, study team organization, general frameworks and methods, general conclusions and recommendations, overall interim documents, draft and final overall EIA documents, and summary documents

Guide: documentation and presentation by individual study team members; review and edit each input for consistency, quality, and substantiation

Coordinate documentation consistent with study schedule and decision-making requirements

Organize, with proponent and with public consultation specialists, public involvement program

Table 6.3 (*Continued*)

Participate in interactions with management, agencies, elected representatives, and public; act as spokesperson for team
Ensure overall efficiency, relevance, and adequacy
Identify uncertainties and risks and develop a management strategy to address; decide, with proponent how to address unforeseen circumstances; prepare change orders

SPECIALISTS (DESIGN AND ENGINEERING, DISCIPLINARY, PROFESSIONAL, METHODOLOGY, PUBLIC CONSULTATION, MEDIATION, CONFLICT MANAGEMENT, LEGAL)

Formulate own methods and assumptions
Manage internal organization, time, budget and tasks
Undertake data collection, analysis, and interpretation
Document and present methods, analysis, role in synthesis, conclusions, and recommendations
Participate in formulation of overall approach, synthesis, and interactions
Specialist advisors address discrete problems, methodology, applied research, comparable proposals, and environments
Peer reviews of interim and draft documents (for proponent, reviewers, or for other participants)

PROPONENT (CO-PROPONENTS, LEAD AND SECONDARY AGENCIES, SPONSOR AGENCIES)

Overall schedule and budgets
Corporate policies, programs and operations
Characteristics of existing operations
Priorities and requirements
Proposal characteristics
Terms of reference
Commitments
Higher-level agency interactions
Participation in public involvement process

AGENCIES AND GOVERNMENTS (EIA AGENCIES, SPECIALIST REVIEW AGENCIES, OTHER GOVERNMENT LEVELS, INDIGENOUS PEOPLES' GOVERNMENTS)

EIA requirements and guidelines interpretations
Technical requirements and guidelines interpretations
Data provision, analysis, and interpretation
Technical expertise
Policy, program, and priority interpretation
Experience with comparable proposals and environments
Output review

PUBLIC (NONGOVERNMENTAL ORGANIZATIONS, DIRECTLY AFFECTED GROUPS AND INDIVIDUALS, INDIRECTLY AFFECTED AND INTERESTED GROUPS AND INDIVIDUALS)

Provision of data
Participation in scoping
Data review and interpretation
Participation in determining criteria importance, alternatives preference, impact management measures, conclusions, and recommendations

(*Continued*)

Table 6.3   (*Continued*)

INTERACTIONS

Within study team

Between project managers and proponent

Between project management, study team members, and specialist advisors

Between project management, and study team members, and agency representatives

Between project managers, proponents and study team members, and elected representatives and public

Mechanisms: management committees, steering committees, advisory committees, task force, workshops built around frameworks, or models and meetings

Sources: Erickson (1994), Glasson et al. (1999), Greenall (1985), Harrop and Nixon (1999), Holling (1978), Kreske (1996), UNEP (1997).

the application of geographic information systems (GISs), report writing and documentation, financial control and budgeting, and the preparation of work programs and schedules.

Competence-related problems continue to occur in the EIA process, notwithstanding the ample, readily available advice and guidance, which should minimize

Table 6.4   Competency: Good Practice Examples

STUDY TEAM MANAGEMENT

Involve team in designing and scoping; clearly define objectives, approach, and anticipated inputs and outputs

Ensure that roles and responsibilities well defined; set priorities and maintain momentum

Provide guidelines for text and table formats to ensure consistent inputs

Ensure a coordinated approach to external contacts

Provide for reciprocity of influence (manager and specialists); facilitate dialogue and integration; recognize different mindsets of specialist types

Emphasize early drafts and initial outputs; scan ahead and "test water"; early opportunity for internal and external review

Sketch out alternative approaches for dealing with problems and conflicts

Test and challenge basis for all interpretations and conclusions; be aware of own limitations and those of others; take corrective actions

Focus on and manage a collaborative, constructive, and creative response to all problems and disputes

Ensure that project manager is not the bottleneck; employ core team on larger projects

Allow for regular meetings and workshops at key decisions

Use subgroups to address problems and to address interconnections among specialists

Often management functions shared between internal (proponent) and external (consultants, secondments, term contracts); tends to be more effective if ongoing proponent involvement

Keep a record of findings, events, directives, changes in direction, comments, concerns, agreements, and decisions

Leadership skills: analysis, integration, management, communications, presentation, negotiation, problem solving, general knowledge of each specialty, ability to ensure quality of work, detailed EIA knowledge and experience, ability to delegate

Leadership style (e.g., command and control, empowerment, learning) must match situation

Table 6.4   (*Continued*)

Leadership qualities: action and results oriented, self-confident, self-starter, visionary, enthusiastic, energetic, reliable, mature, even-tempered, adaptive, politically astute, tolerant of uncertainty, sense of humor, and patience

STUDY TEAM PARTICIPATION

Study team selection: availability, expertise, proposal-type experience, EIA experience, local environmental knowledge, study team experience, personality and attitude, receptivity to viewpoints of others, work traits, range of interests (broader better), writing and communications skills, listening skills, adaptability, ability to interact with public and politicians, ability and experience with hearings, oriented to work to schedule, willingness to travel and make site visits, professional credibility, adaptability; often prudent to make process competitive

Study team style: interdisciplinary (coordination at higher level) and transdisciplinary (coordination at all levels) rather than disciplinary (specialization in isolation), multi-disciplinary (no cooperation), or cross-disciplinary (rigid polarization)

Often core team, each member of which spans a few disciplines; a useful middle ground between project management and full team of specialists when large project

Prompt and ongoing attention to small group problems (e.g., leadership—authoritarian or leadership struggles, blocks in group development, poor decision making, interpersonal conflicts, communications difficulties, goal ambiguity)

Importance of clear purpose, expectations, and accountability; clear terms of reference for each team member

Participate and contribute to overall team activities (e.g., team discussions, agency consultations, alternatives analysis, significance interpretations, public involvement, synthesis, and summary document preparation and review, presentations at events, links to related disciplines)

Undertake specialist analyses in accordance with good practice standards of field, guidance from project management and expectations of regulators

Adhere to scope of work, budgets, and timing requirements; address implications of limitations and uncertainties

Respond promptly and fully to all questions and concerns raised about analyses from study team, regulators, public, and peer reviewers

Work with related specialties in addressing interconnections across disciplines and in formulating and applying integrative frameworks (e.g., modeling, CEA, impact management strategy)

DATABASE MANAGEMENT

Tie information to decision-making requirements

Database management involves determining what data are to be collected, when, by whom, at what level of detail, and how to be collected, compiled, analyzed, interpreted, integrated, applied, supplemented, refined, presented, and monitored

Data collected throughout process; dependent on requirements of activities and decisions

Data management: continuous, evolving, and dependent on context

Ensure that all data complete, accurate, and properly referenced

Should reflect priorities, should be guided by data management strategy, and should identify and explore systematically implications of errors (correct), gaps (fill when necessary), inconsistencies (resolve), and uncertainties (allow for)

Interpret data reliability (sources, methods for collecting, methods for compiling)

(Continued)

Table 6.4 (*Continued*)

Involve stakeholders in data collection, analysis, and interpretation; make effective use of local and traditional knowledge

Important that data are retrievable, cross-checked, and updated; important that dated and referenced

Consider environmental data available in computerized information and retrieval systems (e.g., government agency information systems, environmental databases, and electronic bulletin boards)

GEOGRAPHIC INFORMATION SYSTEM (GIS) APPLICATION

Can store, retrieve, analyze, and display spatial data

Importance of availability and quality of spatial data

Useful for mapping, overlays, baseline analyses, modeling, monitoring (regular updating), visual displays, video imaging, testing of alternatives, route and site selection, CEA (incremental impacts, biodiversity), and public consultation

Takes time to set up; high training and technical requirements; data often not available in digital format; potential data and user-related errors; weak analytical capabilities

Can be combined with global positioning systems, imagery from satellites and aircraft and Internet

Assumes importance of environmental impacts dependent on spatial distribution of impacts

Pitfalls: not taking into account purpose of map, zooming in to improve accuracy, neglecting map projections and coordinate systems, failing to document and evaluate map sources, not including necessary map elements, presenting too much information, inappropriate typefaces, misrepresenting qualitative and quantitative data, mapping absolute values, and neglecting data collection effects

REPORT WRITING AND DOCUMENTATION

Design to suit audience

Build around preliminary and then detailed outline

Focus on what is important; space devoted to topic should be consistent with importance for decision making

Ensure that audience can readily determine how major issues were addressed

Ensure that regulators can readily determine that all requirements satisfied

Use simple and familiar language; be succinct and clear; minimize generalities

Ensure well-structured and visually attractive presentation (ample use of visual displays)

Check for technical errors and mistakes; ensure that factually accurate; avoid plagiarism and bias (use neutral language)

Ensure a consistent writing and presentation style; review and edit for consistency

Identify limitations and uncertainties; identify implications and strategies for addressing

Allow for planning, organizational, and editing mistakes

Use consistent referencing and numbering system, indentations, titles, headings, and margins

Avoid clichés and jargon; avoid defensive language

List acronyms and sources; define technical terms

Provide reasons for data, methods, assumptions, findings, interpretations, conclusions, and recommendations

Provide summaries and use appendices, cross-references, and tiering to streamline text

FINANCIAL CONTROL AND BUDGETING

Match staff to available budgets over time

Table 6.4 (*Continued*)

Track project expenditures regularly
Recognize that expenditures build to a peak
Conduct a postmortem of budgeting experience
Tie each expenditure to decision-making priorities
Team leader to monitor resource use
Apply, as needed, graphs, charts, and computerized techniques to track expenditures and to compare task completion with budget expended

WORK PROGRAMS AND SCHEDULING

Work program addresses goals, issues, and problems
Includes activities, tasks, events, inputs, and outputs; purpose, sequence, duration, personal, hours, disbursements, and budgets for each
Need to address interactions among activities
Allow sufficient time for unforeseen circumstances (float time)
Maintain flexibility; provide additional time for agency and public involvement activities; dangerous to cut short
Use of graphs, charts, and computerized techniques to chart actual progress against scheduled progress
Critical paths methods can help determine overall structure; often helpful to provide a range of time estimates (worst, most likely, quickest)
Allow sufficient time for internal and external review, editing, word processing, and consideration of interconnections

Sources: Antunes *et al.* (2001), Barrow (1997), Bendix (1984), Buckley (1998), Canter (1996), Greenall (1985), Harrop and Nixon (1999), Hodgson and White (2001), Jantsch (1971), João (1998), Kent and Klosterman (2000), Kreske (1996), Thérivel et al. (1992), UNEP (1997), US EPA (1998b), Verma (1995), Webster (1997).

such problems. Perhaps this shortfall between knowledge and execution can be partially explained by a failure to focus on recurrent, avoidable, competence-related problems. Twenty examples of such problems follow.

1. *Project managers as "bottlenecks."* On a large project, a project manager can be overwhelmed if she or he attempts to take on all the project management responsibilities. A core team approach is more appropriate for a large project. The same problem can occur on even an intermediate-sized project if the project manager micro-manages every aspect of the EIA process. A good team and effective delegation are essential. Effective delegation means strategic management, not the absence of control and guidance.

2. *Project managers as "autocrats."* In an EIA process, project management is not simply giving orders. Close and ongoing communications and consultation should be maintained with proponents, with other study team members, and with stakeholders. The project manager should provide a clear rationale for all instructions. Often, others have useful advice to offer. The project manager should

be a good listener and should actively seek constructive advice and criticism. Openmindedness, flexibility, and an even temperament are all part of leadership.

3. *Project managers as "doormats."* Project managers need to have a clear vision of where the EIA process is to go and how objectives are to be achieved. The EIA process cannot be allowed simply to drift. The project manager has to have sufficient self-confidence, experience, and general knowledge to challenge specialists when inputs are unsubstantiated, incomplete, inconsistent with requirements, misdirected, badly written, poorly structured, or of dubious quality. She or he also has to ensure adherence to budget, scope, format, and timing requirements. The project manager should exercise such responsibilities firmly and calmly.

4. *Team members who aren't team players.* Sometimes, specialist team members see their role as no more than undertaking and documenting their analyses. They see team interactions, compliance with document format requirements, and other general project activities as unnecessary distractions to be avoided where possible or, if necessary, reluctantly tolerated. EIA is a highly interdisciplinary, often transdisciplinary activity. This necessitates the full participation of specialists in such joint EIA activities as scoping, alternatives analysis, significance interpretation, cumulative effects assessment (CEA), agency and public involvement, impact management, and document preparation and review. A unified and consistent documentation approach also is essential.

5. *Not up to the task.* Sometimes, specialists involved in an EIA process do not have sufficient relevant expertise and experience in their field, in EIA, in applied knowledge situations, concerning the local environment, regarding the proposal type, or in working on a team. This type of problem can generally be minimized with careful team selection and effective project management. The competency problem is more problematic at the project management level. Having extensive project management experience, as is often the case with engineers and designers, is not the same as having extensive EIA project management experience and expertise. Similarly, lawyers may have a working familiarity with EIA laws and regulations but be lacking in EIA and project management experience and expertise. Also, lawyers occasionally treat EIAs as advocacy documents. Sometimes lawyers, engineers, designers, and specialists in other fields are competent EIA project managers. However, an in-depth understanding of EIA as a field of theory and practice coupled with extensive EIA project management experience are essential.

6. *A failure to focus.* As detailed in Section 6.4.1, a practical EIA process is necessarily focused. Without focus important concerns receive too little attention, and unimportant concerns receive too much attention. The net result is a protracted and costly EIA process and EIA documents of dubious quality. Unfocused documents tend to be highly descriptive and very lengthy. Decision makers and stakeholders have difficulty determining if and how their concerns and priorities are addressed. EIA is a decision-making tool. As such, the EIA process should

concentrate on providing a sound basis for making and implementing environmentally sound decisions.

7. *Gaps and blind spots.* EIA practice is sometimes subject to tunnel vision. Occasionally, as detailed in Chapters 2 and 4, the analysis of alternatives is too narrow, too superficial, and too abbreviated in the rush to concentrate on predicting and managing the effects of the action proposed. Social and cultural effects tend to receive insufficient attention. More attention still needs to be devoted to indirect, cumulative, and sustainability effects, although as described in Chapter 5, the situation is improving. EIA practice sometimes concentrates exclusively on meeting EIA regulatory requirements. As detailed in Chapters 7 and 8, the appropriate treatment of stakeholder concerns and perspectives is frequently just as important in determining whether an EIA will be approved and implemented effectively.

8. *A failure to integrate.* EIA documents, which represent little more than a compilation of specialist inputs, are of limited decision-making value. Competent EIA processes and documents trace through the interactions among disciplinary inputs. They systematically undertake such integrative activities as alternatives assessment, model building, assessing cumulative effects, and formulating impact management strategies. Integration also entails creatively accommodating multiple study team, proponent, regulator, and stakeholder perspectives and interests.

9. *A failure to substantiate.* Sometimes, EIA documents are full of unsupported assertions, claims, interpretations, and conclusions. Occasionally, specialists are under the mistaken impression that their professional judgment provides a sufficient basis for an interpretation or conclusion. It does not. Interpretations and conclusions should always be supported by evidence and explicit reasons. In this way, judgments can be tested and evaluated independently.

10. *Artificial timelines and false economies.* A focused and well-structured EIA process can be executed expeditiously and economically. Occasionally, there are hard deadlines, emergency situations, and severe resource constraints, which necessitate implementation of an abbreviated, selective, broad-level, and streamlined EIA process. But there are limits. Sometimes, artificial time and budget constraints are imposed either at the outset of a process or when a process is taking longer than expected. These constraints can result in superficial, error-prone, and inadequate analyses and truncated agency and public consultation procedures. The most common outcome from artificial limits is a much more time-consuming, controversial, and costly review and approval process and a much greater likelihood of process failure.

11. *Quantify everything.* The desire for precise, verifiable predictions and consistent comparisons is laudable. However, the database must be capable of supporting such efforts. Forcing the quantification of qualitative data can distort the analysis of impacts and inhibit the reasoned comparison of alternatives. The inappropriate application of quantitative methods can imply a greater level of precision and control than can be supported and can make it more difficult for decision makers and stakeholders to understand or participate in the EIA process.

12. *A failure to quantify.* Appreciating the limits of quantification does not mean abandoning all efforts to quantify. It can be extremely exasperating to read an EIA document full of vague generalities and ambiguous statements. Quantified predictions should be provided wherever practical, with due allowance for uncertainties. In this way, predicted impacts can be monitored, the accuracy of predictions determined, and the suitability of predictive methods evaluated. Precision in specifying mitigation measures is necessary for the measures to be implemented and for mitigation effectiveness to be determined.

13. *Bias and advocacy.* The standard of EIA success should not be approval. Instead, it should be an environmentally sound decision-making basis and an enhanced environment. EIA professionals cannot be objective or value free. However, consistent with professional codes of practice, they can work toward EIA objectives in a manner consistent with good practice standards. It is essential to the credibility of the EIA process and documents for the professional integrity of the study team to be maintained. EIA documents should be checked scrupulously to ensure that there is no bias.

14. *A failure to adjust.* Except on the simplest EIA projects, a "carved-in-stone" approach to EIA process management is rarely effective. Modifications occur in proposal characteristics, environmental conditions, available alternatives, and stakeholder positions. Unanticipated events occur. The rules of the game evolve. An EIA process also must evolve and adjust in response to changing circumstances. A gulf between what is needed of a process and what it can provide will emerge and progressively widen with an inflexible EIA process, usually to the point that a major crisis occurs. The outcome from the crisis will tend to be either the termination of the process or major, costly, and time-consuming modifications. Such crises can be avoided or greatly ameliorated with an adaptive EIA process (see Chapter 10).

15. *A failure to anticipate.* EIA practitioners sometimes complain when things go wrong that they were blindsided by unanticipated events and changing circumstances. Sometimes the complaints are valid. Often, however, there are ample early warning signs. These early warning signs can frequently be detected by scanning ahead, by frequent consultations with other parties, through pilot projects, with systematic assessments of comparable situations and by pretesting interpretations, options, and conclusions. A flexible EIA process also makes it easier to anticipate and respond rapidly to change.

16. *A failure to communicate.* An EIA process can be greatly hampered by poorly structured, badly presented, and awkwardly written EIA documents, even if those documents are technically sound. Competent EIA documents and presentations should be clear, succinct, and tailored to the audience. Effective communications channels into the EIA process from regulators and from other interested and affected parties are also essential.

17. *Participation without involvement.* A sure sign of a questionable EIA process is the tendency to count the number of meetings, attendees, and submissions (i.e., inputs) without detailing the changes to the process and

documents resulting from stakeholder comments and suggestions (i.e., outputs). Involvement also is inhibited if participation consists largely of presentations (i.e., one-way communications). Events conducive to two-way communications (e.g., workshops and open houses) and continuous involvement procedures (e.g., advisory committees) are less likely to result in an EIA process characterized by participation without involvement (see Chapter 8).

18. *A lack of perspective.* Environmental specialists, proponents, regulators, nongovernmental organizations, and indigenous people will often interpret the significance and acceptability of impacts and proposed actions very differently. The EIA process and documents should reflect and accommodate this multiplicity of perspectives. There are many ways of looking at the world and how it should be. It is especially important that judgmental activities such as scoping, significance interpretation, evaluation of alternatives, proposal acceptability and the determination of appropriate mitigation, compensation, and monitoring be interpreted from the perspective of each interested and affected party in the process. Consultation programs should also be tailored to a variety of needs and perspectives.

19. *One size does not fit all.* An EIA process that operates effectively in one setting can be entirely inappropriate in another. Context matters. The EIA process should be designed to suit the characteristics of the proposal and the setting. Further adjustments to suit unique project and environmental characteristics are also essential. The goal should be an EIA process that (1) fits the context (e.g., ecological, social, political, institutional, economic) and (2) selectively and positively influences the context (i.e., EIA as an instrument for environmental enhancement and sustainability).

20. *Neglect of follow-through.* A well-designed and executed process and sound EIA documents are necessary. They are not sufficient. Adequate attention must be devoted to follow-through issues, procedures, and requirements. Such concerns need to be addressed both prior to and throughout implementation.

These competence-related pitfalls are largely avoidable. They are not always obvious. Care must be taken to minimize the likelihood and severity of their occurrence.

6.4.5 Effectiveness

The final aspect of practicality is effectiveness. Feasibility addresses the workability of the EIA process (i.e., can it be undertaken, and can it be implemented?). Effectiveness considers how well it was undertaken. Competence deals with adequate practice levels. Effectiveness "raises the bar." As illustrated in Figure 6.5, it addresses the quality of the inputs (e.g., institutional arrangements, processes, methods, participant performance, documents) and the effectiveness of the direct and indirect outputs (e.g., goals achievement, environmental changes, methodological performance, management performance, contribution to practice).

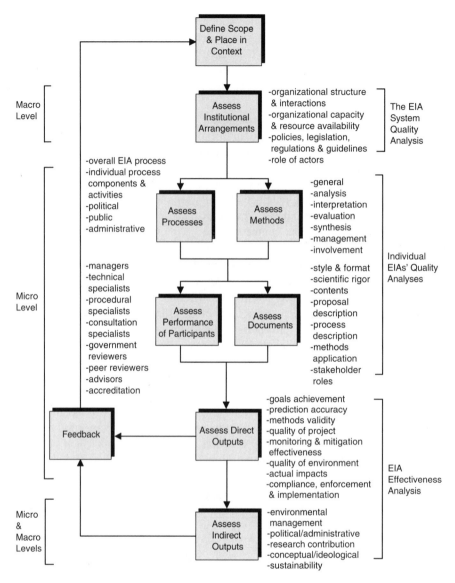

Figure 6.5 EIA quality and effectiveness analyses. (Adapted from Lawrence, 1997a.)

Reviews of *institutional arrangements* evaluate the adequacy of EIA and strategic environmental assessment (SEA) policies, laws, regulations, and guidelines (Halstead et al., 1984). Such reviews consider such matters as application to significant undertakings, environmental and effects definitions, scoping provisions, requirements to address alternatives and cumulative effects, public consultation requirements, transparent decision making, provisions for follow-up, enforcement and auditing, appeal and dispute settlement provisions, and methodological guidance (Gibson, 1993; Sadler, 1996; Spooner, 1998). The suitability of organizational structures and procedures to undertake EIA-related responsibilities can be assessed (Kreske, 1996). The capability and capacity of organizational systems to conduct good practice EIA regulation can be evaluated based on such considerations as EIA and environmental staff qualifications, workload, and the human, financial, and other resources devoted to EIA administration and enforcement.

The quality of *individual EIA processes* can be assessed overall and for individual EIA activities and components (Lee, 2000). The analysis of the overall EIA process can address consistency with good practice and appropriateness to context. The extent to which the EIA process supports transparent and accountable decision making can be evaluated. The appropriateness and effectiveness of the political, public, and government agency involvement procedures can be considered. The choice and manner of application of all methods (e.g., data collection, compilation and analysis, prediction, interpretation, CEA, management, involvement) can be assessed (Ortolano, 1993). The qualifications, roles, and role performance of process participants (e.g., managers, technical and procedural specialists, government reviewers, peer reviewers, advisors) can be analyzed. EIA documents can be evaluated for style, format, content, and the treatment of individual EIA activities, methods, and events (Barker and Wood, 1999; Wood et al., 1996). How well the documents focus on major concerns, comply with regulatory requirements, reflect stakeholder perspectives, and integrate public and agency concerns and contributions can be evaluated.

Direct and indirect outputs from EIA processes can be assessed. Output analyses interpret results, both intended (relative to expectations) and unintended (positive and negative). Direct output analyses provide the basis for follow-up actions and practice refinements. Indirect output analyses are the means by which EIA processes make substantive contributions to enhanced EIA practice. Direct output effectiveness analyses address whether EIA purposes, goals, and objectives have been achieved, the accuracy of environmental change and impact predictions, and the validity of methods (Barrow, 1997; Culhane, 1993; Tomlinson and Atkinson, 1987; UNEP, 1997). They can determine project modifications and quality, the suitability of monitoring measures, the effectiveness of mitigation and compensation measures, the quality of impact management, and the extent to which commitments are implemented, requirements are complied with, and adequate enforcement occurs (Glasson et al., 1999; Harrop and Nixon, 1999; UNEP, 1997; Wende, 2002). EIA effectiveness reviews also can isolate factors that result in or impede effectiveness gains (Wende, 2002).

Indirect output analyses address the role of institutional arrangements and individual EIA processes in furthering environmental management, environmental administration, and decision making, the EIA knowledge base, and societal goals such as sustainability (Barrow, 1997; Glasson et al., 1999). Evaluations are undertaken of the contribution by EIA to environmental objectives as compared to the costs and negative impacts incurred and relative to the achievements of other environmental management instruments. Such analyses can facilitate institutional arrangement reforms.

Numerous *methods* can be applied in effectiveness reviews (e.g., ad hoc procedures, checklists, applying principles, criteria or performance standards, the use of scaling levels) (Sadler, 1996; US EPA, 1998b). Effectiveness reviews can be undertaken by individual experts (internal or external, accredited or not accredited), panels of experts, public reviews, independent commissions, official inquiries, public reviews, or through legal proceedings such as court actions (Barrow, 1997; Tomlinson and Atkinson, 1987; UNEP, 1997). Various approaches can be adopted for conducting effectiveness reviews. A scientific-analytic, a management-efficiency, an interactive-interpretative, or an adaptive-evolving approach could be appropriate, depending on such considerations as project complexity, data availability, degree of uncertainty, degree of controversy, and the rate and predictability of changing conditions (Culhane, 1993; Lee, 2000; Serafin et al., 1992). The interpretative nature of EIA quality and effectiveness analyses underscores the importance of stakeholder involvement and perspectives (Spooner, 1998; UNEP, 1997; US EPA, 1998b).

Ideally, an effectiveness review should include (1) a screening step (to reject an unacceptable action or document), (2) a performance analysis step (to evaluate actions or documents considered adequate but not necessarily consistent with good practice standards), (3) supplementary analyses (to overcome deficiencies), (4) clarifications (to resolve misunderstandings), (5) the documentation of findings at each decision (to ensure decision-making transparency), (6) provisions for agency and public involvement at each decision (to ensure full public and agency involvement in each step in the process), (7) monitoring or auditing analyses (to assess outputs), (8) an approval step (to provide a decision-making basis and to determine conditions), and (9) a modifications step (to adapt implementation to changing conditions).

6.5   INSTITUTING A PRACTICAL EIA PROCESS

6.5.1   Management at the Regulatory Level

A practical EIA regulatory system should (1) minimize EIA duplication and overlap among government levels, (2) ensure that EIA roles among government departments and agencies are well coordinated, (3) focus on what is important and minimize unnecessary costs and delays, and (4) ensure a minimum level of EIA competence and contribute to an enhanced level of EIA practice. The EIA systems in the four jurisdictions (the United States, Canada, the European Union, and Australia) all seek to achieve these objectives, albeit in different ways.

EIA Harmonization Approaches   The need for and effectiveness of EIA harmonization measures depends on the division of environmental responsibilities among governments, the nature of EIA requirements at each level, the relationships between levels, and which measures are acceptable to all parties. The choice and application of harmonization measures should be approached with caution. There appears to be a fairly clear division of environmental responsibilities between the federal and state governments in the United States. Many states have modeled their EIA requirements after those of the federal government (Kreske, 1996). Federal U.S. EIA requirements provide for eliminating duplication with other governments (e.g., joint processes, documents, and public hearings), for addressing conflicts between federal and state or local planning, and for including state and local governments as joint or cooperating agencies. State and local agencies can participate in interagency agreements. There are numerous partnerships, shared databases, and joint place-based environmental planning and management initiatives involving federal, state, local, and tribal governments. A NEPA (National Environmental Policy Act) task force, recently established by the Council on Environmental Quality, is addressing, among implementation practices and procedures, intergovernmental collaboration, including joint-lead processes. The U.S. approach is appropriate, assuming that there is limited duplication or overlap, EIA systems are largely complementary and there is an ongoing need for intergovernmental cooperation.

Harmonization is potentially more problematic in Canada and Australia. Environmental responsibilities are shared between the upper (federal)- and lower (provincial, state, territorial)-level governments. The EIA systems in the two countries vary greatly at the lower level. There also are major differences between levels in EIA systems. Both countries have proceeded toward harmonization with an array of formal multilateral and bilateral agreements, transboundary regulatory provisions, and informal administrative procedures. There also is considerable senior-level (e.g., councils and committees of environmental ministers and of senior administrators) and project-specific cooperation.

In Canada there is little direct duplication between levels but some overlap problems. Project-specific arrangements have generally been reached for a single set of documents, joint public meetings, and a single hearing. The federal government generally believes that duplication and overlap problems are resolved or resolvable. Some provinces believe that significant problems still sometimes occur. They suggest that federal EIA "law list" triggers (late in process or a minor concern opening up a major process), conflicting process schedules, and overlapping and conflicting comments, perspectives, and requirements sometimes remain problems. Each of the three northern territories is moving toward delegation to a single EIA system administered jointly by the federal government, the territorial government, and indigenous governments. Several provinces and indigenous communities have indicated a preference for some form of accreditation, equivalency, delegation, or lead government system.

The issue in the European Union (EU) is EIA harmonization among countries and sometimes also among government levels within a single country. Harmonization at the EU level involves the countries jointly defining the project and strategic-level common ground among EIA systems. The areas of agreement provide the basis for directives to which all countries must conform. The directives define a minimum level of good practice, address EIA-related interactions among governments (extending from a transboundary EIA convention), and cross-reference related EU conventions, agreements, and requirements. The European Commission refines and facilitates the implementation of the directives with an extensive array of guidelines, applied research, effectiveness analyses, and data sharing. Harmonization efforts occur within the context of numerous EU environmental policies, treaties, laws, regulations, conventions, directives, and action programs. The European approach requires a delicate balancing act between establishing an adequate strategic-level performance standard and leaving sufficient discretion to member states with very different institutional, economic, social, ecological, and cultural conditions.

The Australian EIA legislation provides for accrediting or delegating EIA approvals based on bilateral agreements. There is a clear division of environmental responsibility in Australia. Matters of national environmental significance are explicitly identified. States or territories can refer proposals for commonwealth EIA review. In common with Canada, commonwealth actions, commonwealth funding, and activities on commonwealth lands trigger EIA requirements. But unlike Canada, other senior-level laws do not trigger the EIA legislation. The Australian approach seems less prone to harmonization problems.

EIA Coordination Approaches   EIA coordination across agencies and departments within the same government is a concern with all four jurisdictions. Coordination is particularly important where individual agencies and departments have a high degree of autonomy, which seems to be the situation in all four cases. Coordination is facilitated in many ways. The precise mix of measures varies among jurisdictions. EIA requirements and guidelines can specify EIA preparation, review and interaction roles, identify interagency consultation and circulation provisions, cross-reference related laws and permitting requirements, stipulate agency reporting requirements, and provide for interagency dispute resolution. The simultaneous triggering of EIA and other permitting requirements is addressed by such mechanisms as coordinated timing, merged requirements, joint documents, substitution agreements, tiering, cross referencing, interagency offices, interagency agreements, and circulation requirements. Coordinating agencies and EIA process managers help guide and bring the parties together. Guidelines to address coordination concerns, training sessions, and joint scoping and problem-oriented workshops and meetings contribute to enhanced understanding and help break down communications barriers. Agencies often work together on task forces and committees.

Coordination can be built around a common purpose (e.g., area-based environmental planning and management, the determination of significance, the management of cumulative effects, the maintenance of biodiversity) or defined within the context of upper-level policies, plans, and strategies. Common databases, joint applied research, and shared or common follow-up and auditing procedures help bring agencies together. Informal interagency cooperation and collaboration is common in all four jurisdictions. The coordination measures, which have evolved in each jurisdiction, are not necessarily optimal or appropriate. Additional effectiveness analyses would help determine which measures and combinations of measures work best under which circumstances.

EIA Expediting and Focusing Approaches All four jurisdictions have sought to focus on proposals likely to induce significant impacts and on significant impacts resulting from proposed actions. They also have strived to expedite the preparation and review of documents and the implementation of requirements. Screening procedures focus the system on significant proposals. Different types of assessment, class and categorical assessments, and separate requirements for minor projects address variations in degrees of significance among proposed actions. Scoping and significance requirements, procedures, and guidelines enable EIA systems to zero-in on significant impacts. Time limits, suggested timelines, document circulation limits, quick-test EIA triggers, and accelerated project planning reduce the likelihood of delays or extended timelines. Checklists, electronic registries, and clearly defined standards and performance criteria accelerate the review process. Page limits, combined documents, tiering, cross-referencing and document content, style, summary, and appendix requirements reduce unnecessary paperwork. Studies of EIA costs and competitiveness and valuation, pricing, and incentive mechanisms address the overall efficiency of the EIA system.

The U.S. EIA system seems to have gone farthest in streamlining the EIA system. It has formalized scoping. Detailed scoping requirements and guidelines are provided. It offers extensive guidance for reducing paperwork and for accelerating EIA document review. Timing is a concern in Canada. Timing-related reforms include screening requirement modifications, the enhanced use of class assessments, the identification of an EIA coordinator, a more formal approach to scoping, minor project regulations, and a procedures for reducing the likelihood of late triggers. An EIA competitiveness study also was undertaken. Some Canadian provinces (British Columbia, Newfoundland) include time limits for each stage in the EIA review and approval process.

The European streamlining approach largely revolves around identifying minimum requirements. General references are made to appropriate time frames, cost-effective solutions, and combining documents. Detailed screening, scoping, and document review guidance is provided. An effectiveness review of the project directive advocates screening reforms and a more formalized scoping approach. The U.K. Environment Agency has produced scoping guidance that addresses issues common to all development types and to four categories of site operations. It also considers specific impacts and mitigation measures for 72 different project types (Bond and Stewart, 2002).

The Australian EIA system focuses on matters of national environmental significance. There are timing requirements for each decision in the EIA process for each type of EIA document. There also are class assessment provisions, specific document content requirements, and general references to reasonably practical and feasible alternatives. It is unclear how well these measures focus and streamline the EIA systems in the four jurisdictions. Also unclear is the extent of adverse consequences stemming from the measures. Effectiveness reviews in each jurisdiction do provide anecdotal evidence of efficiency gains.

EIA Competence and Effectiveness Approaches   All four jurisdictions seek to enhance EIA competence and effectiveness. The United States, Canada, and Australia have each recently completed a major effectiveness review of the federal EIA system (Anderson, 2001; Auditor-General, 2002–2003; US CEQ, 1997a). In the United States, individual agencies (e.g., Department of Energy) have established NEPA quality improvement teams and have arranged for independent reviews of NEPA process management practices (Caldwell et al., 1998). The U.S. Council on Environmental Quality (CEQ) formed on May 20, 2002 a NEPA task force to review NEPA implementation practices and procedures in the areas of technology and information management, interagency, and intergovernmental collaboration, including joint lead processes, programmatic analysis and subsequent tiering documents, adaptive management, and a range of other implementation issues. The NEPA task force is composed of representatives from several federal agencies.

Effectiveness reviews have been undertaken of the project directive in Europe and of EIA guideline documents. Both the project and SEA directives provide for periodic effectiveness reviews (after five years and then seven years thereafter). The Australian EIA legislation provides for a report after five years concerning whether additional matters of national environmental significance should be added. It also requires a state-of-the-environment report every five years. An independent review of the act's operations and of the extent to which its objectives have been met is to be undertaken after 10 years. Annual reports on the implementation of EIA requirements are prepared in all four jurisdictions. The first annual report concerning the Australian EIA legislation, for example, points to benefits in such areas as increased certainty in the division of responsibilities, greater intergovernmental cooperation because of the accreditation procedures, effectiveness gains because of the strict timing requirements, and a focusing of commonwealth decision making on matters of national environmental significance (Environment Australia, 2001).

The four jurisdictions have many mechanisms in place for assessing the effectiveness of individual aspects of the EIA system. The U.S. Department of Energy issues a quarterly report on EIA lessons learned. An interdepartmental compliance monitoring framework and a multiyear interdepartmental program for determining the EIA contribution to better project planning are being developed in Canada. EU member states are required to communicate how they are ensuring the quality of EIAs and to monitor the significant effects of project, plan, and program implementation. Several effectiveness studies have been sponsored in Europe of the quality of EIA documents, of the effectiveness of consultation arrangements, of the integration of EIA and project authorization, of the effectiveness of EIA requirements, guidelines, and methods, and of the costs and benefits of EIA legislation. The Australian EIA legislation includes environmental audit provisions and requires that the environmental record of the person proposing the action be considered.

Numerous requirements, guidelines, and applied research initiatives address competence in the four jurisdictions. All four establish minimum EIA requirements and provide good practice documentation guidance and guidance for various EIA process activities, proposal types, impacts, and methods. U.S. requirements and guidelines pertain to such matters as agency roles, document evaluation, procurement procedures, conflict of interest provisions, and the assessment of bio-diversity, environmental justice, and cumulative effects. Competence levels are enhanced in the United States through coordinated research, EIA training, shared databases, and place-based environmental management initiatives. In Canada over 20 departments and agencies are to participate in a government-wide EIA quality assurance program. EIA training, sponsored research, and agency sustain-ability strategies further facilitate enhanced EIA practice. The European Union has sponsored many applied research studies and good practice guides, especially regarding strategic environmental assessment. EIA centers and institutes, scattered across Europe, play a key role in supporting enhanced practice. The Australian and New Zealand Environmental Conservation Council prepared detailed guidelines and criteria for determining the need for and level of EIA in Australia.

The diversity of EIA competence and effectiveness initiatives applied in the four jurisdictions underscores the potential benefits of knowledge sharing across juris-dictions. As noted in Figure 6.5, the effectiveness of coordination, harmonization, and streamlining measures also could be evaluated.

6.5.2 Management at the Applied Level

Figure 6.6 illustrates a practical EIA process. The figure and the description that follows depict a focused, realistic, feasible, competent, and effective EIA process. EIA process managers can pick and choose the relevant and appropriate elements.

Startup   Planning, decision making, and implementation are assumed to be inte-grated, constrained, decentralized, incremental, and partisan. It is recognized that many, often conflicting, parties and interests will need to be involved in the process. The context within which the EIA process operates is expected to be uncertain and unstable.

The EIA process is focused (through scoping) on what is relevant and important to regulators and other stakeholders. The initial scoping is supported by an over-view of regulatory requirements and priorities, a scanning of key environmental and proposal characteristics, and the identification of primary stakeholder issues and concerns. The problem to be solved or need to be met is clearly identified. Prio-rities, boundaries, roles and responsibilities, major choices, sensitive and significant environmental components, major anticipated impacts, key proposal characteristics, primary stakeholders, and critical issues (from the perspective of each stakeholder group) are determined. Scoping provides the basis for the EIA process approach. The approach identifies major activities, events, inputs, and outputs. A study team, appropriate for addressing the identified issues, is assembled. An initial study design is prepared. The study design determines study organization, tasks, roles and responsibilities, budgets, and schedule. The approach, the study design, and politi-cal, agency, and stakeholder participation involvement approaches are formulated jointly with interested and affected parties.

The approach and study design are defined at a broad level of detail. They are expected to evolve and change through the process. There are major uncertainties

Support Analyses
-feasibility studies
-empirical studies
-comparable action reviews
-implementation studies
-experiments
-baseline studies
-pilot project
-applied research

Ongoing & Recurrent Activities
-focusing/scoping
-grounding & reflection in practice & experience
-project, process & study team management
-financial control & scheduling
-data base management (collection, analysis, interpretation, application & management), including use of GIS
-consideration of implementation
-integration with related decisions, instruments, actions, frameworks & methods
-integration across disciplines & professions
-refinements, coordination, expediting & adjustments
-tests to ensure competence
-quality & effectiveness analyses
-scanning ahead & feedback
-documentation
-integration with EIA & instiutional arrangements reforms
-ensuring that adequate standards of competence are maintained

Communications & Participation
-communications & consultation with interested & affected parties
-political, agency & stakeholder involvement in a series of:
-workshops
-conferences
-meetings
-advisory committee meetings & simulation exercises
-possible use of alternative dispute resolution mechanisms

Figure 6.6 Example of a practical EIA process.

regarding both ends and means. The "operating room" within which the process unfolds is highly constrained. Unforeseen and unforeseeable circumstances are anticipated to emerge through the process that will require approach and study design modifications and refinements. Ample float time is provided. Contingency funds are set aside for changing conditions.

Planning, Decision Making, and Implementation   A practical EIA process is nonlinear. It is iterative, cyclical, and incremental. It involves multiple stakeholders debating, discussing, negotiating, reviewing, analyzing, comparing, and bargaining about choices and constraints. The process is built around a series of decisions. It cycles back and forth among process elements. It is characterized by continuous learning. It provides for multiple interactions, for scanning ahead, and for feedback. It merges and transcends such conventional dichotomies as ends and means, objective and subjective, technical and political, analysis and synthesis, planning and implementation, and process and context. It crosses disciplinary and professional boundaries.

The process reflects bureaucratic and political requirements, preferences, and priorities. It operates within boundaries, acknowledges constraints and seeks out opportunities. It is focused, experimental, and action-oriented. Roles are negotiated. Ends are a general direction rather than precise objectives. Means are reasonable, available, and practical choices. The process is built on a solid foundation of experience and practice-based knowledge, methods, insights, skills, and wisdom. Stakeholder perspectives, concerns, and preferences are integrated into the process. The process is designed to suit and refined to better match the context. Risks and uncertainties are freely acknowledged. The process proceeds cautiously and incrementally. Short-term time horizons largely predominate.

Choices that depart appreciably from current conditions, are highly uncertain, are potentially contrary to regulatory requirements, are controversial, are likely to be difficult or costly to implement, and are unlikely to be accepted by key stakeholders are quickly screened out. The key tests of a good option are regulatory compliance, stakeholder acceptance, ease of implementation, acceptable costs, and cost-effectiveness. Option comparison involves a reasoned exploration, from multiple perspectives, of implications and consequences rather than the formal application of evaluation methods. Once agreement is reached preferred choices are adapted, refined, and tested.

Outcomes from the process are formalized and documented in draft and final EIA documents, consistent with regulatory requirements and agency expectations. Uncertainties regarding impact magnitude, impact significance, and mitigation effectiveness are incorporated into impact management strategies and tactics. Strategies are formulated, refined, and applied to facilitate approvals and implementation. Review and implementation tend to be incremental (e.g., phased approval), adaptive (e.g., continued focus on managing uncertainties), and conditional (e.g., ample provisional for monitoring and contingency measures). Surprises are expected to emerge during post approvals. Surprises require both anticipation (to the extent practical) and prompt remedial action. The EIA process extends through implementation. Direct and indirect outputs are assessed through effectiveness analyses. Knowledge, practice, and experience are considered tentative, contingent, partial, ambiguous, and uncertain.

Support Analyses   Selective baseline analyses are undertaken periodically. These analyses focus on the requirements and expectations of regulators and of other stakeholders. Applied experience and practice-based knowledge are emphasized. Comparable projects, empirical studies, and implementation studies are reviewed to determine how power is exercised, how decisions are made, and how implementation obstacles and opportunities could be addressed. Experiments and pilot projects are used to "test the water." Feasibility studies are undertaken to ensure that choices are cost-effectiveness and capable of implementation. Applied research is undertaken, where essential, to fill data gaps, which might impede decision making and implementation. Residual uncertainties are highlighted. Decision making and implementation implications of uncertainties are explored.

Ongoing and Recurrent Activities Focusing occurs with each cycle in the process. Documents, events, and interactions also are scoped. Analyses, interpretations, and conclusions draw heavily upon experience and reflection of practice. Project, process and study team management, and financial and schedule control are maintained throughout the process. The database management system is continually updated and refined. Political feasibility and implementation requirements and implications are addressed, both prior and subsequent to approvals.

The EIA process is integrated with decision making. It is linked to related decisions, actions, and environmental management instruments. Tiering and cross-referencing reduce paperwork, simplify review, and place the process and documents within a policy and strategic planning context. Related decision-making methods, such as cost–benefit analysis, feasibility studies, risk assessment, quality assurance, and technology assessment, are summarized and referenced, as appropriate. Critical links across disciplines and professions are identified. Individual analyses are integrated into methodological frameworks (e.g., sustainability assessment, integrated impact assessment) where needed to address cumulative effects and to ascertain progress toward broader environmental objectives.

Succinct and readily understandable interim, working, background, applied research, and consultation papers are prepared. They provide a clear decision-making basis, record decision-making process, and establish the basis for draft and final EIA documents. Further refinements are introduced based on EIA quality analyses of procedures, methods, documents, and participant performance and EIA effectiveness analyses of interim outputs and comparable projects. EIA documents incorporate stakeholder perspectives, demonstrate regulatory compliance, substantiate all assumptions, interpretations, and conclusions, and respond to the comments and suggestions of process participants. Interactions among process participants are coordinated. EIA review, approval, and implementation are expedited. A minimum level of practice competence is maintained. Good practice is actively encouraged and facilitated. The EIA process evolves in conjunction with EIA and institutional reform.

Communications and Participation A practical EIA process is open and interactive. Communications and participation with interested and affected parties are recurrent activities. Interactions are especially intensive leading up to and immediately following major decisions. Major perspectives are reflected in the analysis. All parties have an opportunity to review and respond to interim and draft EIA documents. The concerns and priorities of all parties who could assume a significant role in approvals and implementation, in overcoming obstacles, and in building coalitions of support are solicited and documented.

Consultation methods conducive to identifying and accommodating differences (e.g., workshops, conferences, advisory committees) are applied as appropriate. Alternative dispute resolution methods, such as mediators and facilitators, are used when perspective and interest differences threaten the process. Close contact is maintained with regulators to minimize uncertainties regarding regulatory compliance. Consultation activities are both formal and informal.

6.6 ASSESSING PROCESS EFFECTIVENESS

Table 3.5 (in Chapter 3) presents criteria and indicators for assessing the positive and negative tendencies of the EIA processes described in Chapters 3 to 10. In this section we summarize examples of the positive and negative tendencies of practical EIA processes.

A practical EIA process emphasizes relevance over *rigor*. Scientific standards, protocols, knowledge, and methods are applied where needed to meet regulatory requirements. In common with science, knowledge is derived from or tested in practice. Unlike science, practicality is largely atheoretical. Scientists, in a practical EIA process, are only one among several interests at the table. A practical EIA process can ensure a minimum level of scientific competence. Competence is defined as soundness for decision making rather than the satisfaction of applied scientific performance standards. EIA quality and effectiveness analyses partially parallel scientific protocols. An EIA process built around stakeholder negotiations is sometimes not conducive to explicit and substantiated assumptions, findings, interpretations, and conclusions. Enhancing the scientific knowledge base is generally a low priority with practical EIA processes.

A practical EIA process *comprehensively* seeks compliance with regulatory requirements. It focuses on major issues, differences, and participants. Long-term and major changes receive less attention. Problems and opportunities are usually narrowly defined. A holistic perspective is rarely applied. Multiple perspectives are integrated into the process. Stakeholder involvement often results in a broad definition of the environment, of reasonable choices, and of potentially significant impacts. The satisfactory (to key stakeholders) treatment of short-term changes can establish rapport and trust conducive to longer term and more substantial changes. A practical EIA process is highly sensitive to interrelationships. It bridges and transcends artificial distinctions and barriers.

A practical EIA process *systematically* addresses decision making and implementation in a variety of ways and from multiple perspectives. The routes by which agreements are reached can be difficult to trace. The process focuses on a manageable number of problems and opportunities. Objectives emerge from rather than being determined at the inception of the process. Proven methods that produce results expeditiously are applied. State-of-the-art methods tend not to be a priority. The process is disjointed and partial. Stakeholders can increase the likelihood that a broad range of impacts will be addressed. Ongoing management, approvals, and implementation can be conducive to systematic impact management.

Practical EIA processes tend not to be environmentally *substantive*. Environmental values and ideals are a means (to facilitate approvals and implementation) rather than an end. The test of a good option is stakeholder agreement, not environmental enhancement. Environmental substance is introduced from external sources (e.g., through regulatory requirements or from stakeholders) rather than being a product of the process. Short-term and selective perspectives are generally at odds with long-term holistic visions of a sustainable future. But short-term successes and agreements can lay the groundwork for larger-scale environmental advancements. A sustainable future also can be built incrementally. An open process can be conducive to integrating environmental knowledge and perspectives. There is a danger that perspectives, not perceived as relevant to short-term decision-making needs, will lose out in negotiations.

A *practical* EIA process efficiently and effectively meets decision-making and implementation needs and requirements. It operates within and prudently allocates available resources. Proven methods, consistent with good practice, are used. Roles and responsibilities are well defined. Key issues and trade-offs are stressed. Knowledge and decision-making constraints are acknowledged. How bureaucracies operate and how decisions are made (e.g., collective, political, partial, incremental) are understood. Expectations and performance standards are realistic. Practical EIA processes can be less effective for major proposals with long-term consequences. They do not always produce the level of analysis and justification required for project approvals and implementation. Sometimes, they underestimate what is possible and necessary.

Practical EIA processes do not actively seek to maintain or enhance *democratic local influence*. But they appreciate that stakeholder understanding, involvement, and support (or at least acquiescence) are often essential prerequisites to approval and implementation. They are issue-oriented. They assume that decision making is decentralized, collective, interest-based, and political. The delegation of authority to stakeholders and local communities is occasionally a means but rarely an end. Increased local influence is sometimes an accidental by-product of practical EIA processes. By not seeking to redress power inequities, practical EIA processes can reinforce the existing distribution of power. The accommodation of traditional knowledge is consistent with the orientation toward experience-based knowledge.

Practical EIA processes are often *collaborative*. They recognize the need to integrate the concerns and perspectives of all parties with a stake in proposal approval and implementation. Rapport and trust, established through stakeholder discussions and negotiations, can help launch more ambitious joint planning endeavors. Consensus building and conflict resolution are facilitated only to the extent necessary to ensure expeditious approvals and implementation. Collaboration can be limited to satisfying regulatory requirements and to recognizing political priorities when stakeholders are poorly organized, lacking in resources or highly divided.

Fairness is an issue with practical EIA processes only when raised by major stakeholders. Procedural inequities are offset when they are obstacles to stakeholder agreements. Measures to reduce distributional inequities (e.g., a local benefits program) are introduced to ameliorate opposition to the action proposed. This hit-or-miss approach to fairness falls well short of a process guided and shaped by *ethical* standards and imperatives. Practical EIA processes define responsibilities but rarely recognize rights. Ethical issues, implications, trade-offs, and dilemmas are seldom addressed explicitly. Multiple perspectives are integrated. Social and environmental fairness, equity, and justice concerns are not emphasized.

Practical EIA processes *adaptively* acknowledge, anticipate, and respond to changing circumstances. They are designed to suit the context. The process evolves in parallel with changing external conditions. Identifying and exploring critical problems and opportunities are emphasized. Problems, opportunities, risks, and uncertainties are considered when pivotal to approvals and implementation. Practical EIA processes have a short-term and incremental orientation. This limits their adaptability to major changes with long-term implications.

Practical EIA processes are *integrated* with project planning, decision making, management, and implementation. Links to and from related decisions and other environmental management forms and levels are considered explicitly. Artificial barriers among disciplines, professions, and EIA types are transcended. The narrow focus on decision making and implementation can inhibit integration within synthesis frameworks.

6.7   SUMMING UP

In this chapter we portray a streamlined, efficient, and effective EIA process, a practical process based on realistic expectations and competent practice. The three stories illustrate practicality in the EIA process in different ways. The first story demonstrates that practicality can be an outcome from a collaborative planning process structured and managed to provide practical solutions to real stakeholder problems. The second story involves the application of a practicality-related method (scoping) as a tool to counteract engrained perspectives and behaviors and in a manner that results in a more efficient EIA process without sacrificing substantive environmental objectives or stakeholder interests. The third story illustrates that applying a regulatory streamlining procedure, class assessments, although providing practical benefits, can also have negative implications, can encounter serious obstacles, and is effective only if cautiously and thoughtfully adapted and refined to suit the situation. All three stories underscore the importance of matching process to context. The stories provide only a partial and preliminary impression of how an EIA process can become more practical.

The problem is the tendency for EIA processes to be unfocused, disconnected from reality, weak on implementation, of variable quality, and slow to learn from experience and practice. Several concepts are introduced to make the EIA process more focused, relevant, feasible, competent, and effective. The concepts provide the basis for practical EIA requirements and practical EIA processes. A practical EIA process is assessed against ideal EIA process characteristics. Table 6.5 is a checklist for formulating, applying, and assessing a practical EIA process.

Table 6.5   Checklist: A Practical EIA Process

The response to each question can be yes, partially, no, or uncertain. If yes, the adequacy of the process taken should be considered. If partially, the adequacy of the process and the implications of the areas not covered should be considered. If no, the implications of not addressing the issue raised by the question should be considered. If uncertain, the reasons why it is uncertain and any associated implications should be considered.

FOCUSED

1. Are scoping principles clearly defined and applied consistently?
2. Does scoping determine, with a clear rationale, what will and will not be examined, the appropriate level of level, how the analysis is to be structured and directed, and what will be the priorities?
3. Is scoping a broad and collaborative identification, interpretation, management, consultation, and communications effort?
4. Does scoping focus and structure the EIA process and documents?
5. Does scoping help reform and apply EIA institutional arrangements?
6. Are participant roles in scoping clearly defined?
7. Is scoping designed to match the context?

REALISTIC

1. Does the EIA process take in account how decisions are made, how organizations operate, how organizations are structured, and how practitioners operate?
2. Does the EIA process consider how people in organizations behave, how organizational structures and procedures vary in different settings, and how knowledge is generated through practice?
3. Does the EIA process take into account how actions are integrated into decision making, how power is exercised, how actions are bounded and constrained, and how resources are allocated?
4. Does the EIA process consider how actions are implemented, facilitated, inhibited, and prevented?

FEASIBLE

1. Is the EIA process informed by concepts, guided by strategies, and aided by tactics?
2. Is the EIA process integrated with decision making, implementation, management, and context?
3. Does the EIA process contribute to and is it adapted to EIA reform?
4. Is the EIA process linked to related tools and methods?
5. Is the EIA process integrated with policies, programs, and plans, with organizational operations, with other decision-making levels, and with related instruments?
6. Is the EIA process harmonized with other governments, coordinated with the private sector, and coordinated with nongovernmental organizations?

Table 6.5   (*Continued*)

7. Is the EIA process integrated across disciplines and professions and integrated within synthesis frameworks?

COMPETENCE

1. Does the EIA process ensure competency in the performance of roles and responsibilities by project managers, specialists, the proponent, agencies, governments, and the public?
2. Does the EIA process apply good practices in study team management, study team participation, database management, the use of geographic information systems, report writing and documentation, financial control and budgeting, and work program formulation and scheduling?
3. Does the EIA process actively seek to avoid and minimize recurrent, avoidable, competence-related problems?

EFFECTIVENESS

1. Is the quality of EIA institutional arrangements considered and assessed?
2. Is the quality of EIA processes, methods, documents, and participant performances considered and assessed?
3. Does the EIA process provide for the effectiveness assessment of direct and indirect process outputs?

MANAGEMENT AT THE REGULATORY LEVEL

1. Do the EIA requirements and guidelines minimize EIA duplication and overlap among government levels?
2. Do the EIA requirements and guidelines ensure that EIA roles among government departments and agencies are well coordinated?
3. Do the EIA requirements and guidelines focus the EIA system on what is important and minimize unnecessary costs and delays?
4. Do the EIA requirements ensure a minimum level of EIA competence and contribute to an enhanced level of EIA practice?

MANAGEMENT AT THE APPLIED LEVEL

1. Does the process incorporate such startup activities as:
   a. Initial scoping taking into account regulatory requirements and stakeholder issues?
   b. Formulation of an agency and political participation approach?
   c. Formulation of a stakeholder participation approach?
   d. Formulation of an overall approach and study design?
2. Is an iterative and incremental process undertaken involving debate, discussion, negotiation, review, reflection, analysis, synthesis, and comparison and bargaining, centered on decisions?
3. Does the process incorporate such planning, decision-making, and implementation activities as:
   a. Consideration of political and bureaucratic requirements, preferences, and priorities?
   b. Identification of reasonable and practical means?
   c. Determination of constraints, opportunities, limits, and boundaries?
   d. Determination of stakeholders' perspectives, concerns, and preferences?
   e. Identification of ends (needs, values, objectives, and criteria)?

(*Continued*)

Table 6.5   (*Continued*)

---

    f. Determination of roles, responsibilities, and available resources?

    g. Adaptations to context (ecological, social, cultural, political, and institutional)?

    h. Consideration of risks, uncertainties and probabilities?

    i. Analysis, interpretation and synthesis of information?

    j. Integration of experiences, wisdom, and good practice?

    k. Integration of knowledge, skills, and methods?

    l. Impact and uncertainty analysis, interpretation, and management?

    m. Determination of adequate and satisfactory options?

    n. Determination and refinement of proposed actions?

    o. Incremental reviews and approvals?

    p. Formulation of impact management and implementation strategies?

    q. Preparation of draft and final EIA documents?

    r. Systematic consideration of implementation-related issues?

    s. Preparation of direct and indirect output effectiveness analyses?

4. Is the EIA process supported by:

    a. Feasibility studies, empirical studies, comparable project reviews, and implementation studies?

    b. Baseline analyses?

    c. Experiments, pilot projects, and applied research?

5. Does the EIA process provide for such ongoing and recurrent activities as:

    a. Focusing and scoping?

    b. Grounding and reflection in practice and experience?

    c. Project, process, and study team management?

    d. Financial control and scheduling?

    e. Database management?

    f. Consideration of implementation?

    g. Integration with related decisions, instruments, actions, frameworks, and methods?

    h. Integration across disciplines and professions?

    i. Refinements, coordination, expediting, and adjustments?

    j. Tests to ensure competence?

    k. Quality and effectiveness analyses?

    l. Scanning ahead and feedback?

    m. Documentation?

    n. Integration with EIA and institutional arrangements reforms?

    o. Ensuring that adequate standards of competence are maintained?

6. Does the EIA process provide for communications and consultations with interested and affected parties?

7. Does the EIA process provide for recurrent and ongoing political, agency, and stakeholder involvement?

8. Does the EIA process provide for the use of alternative dispute resolution mechanisms, as appropriate?

PROCESS EFFECTIVENESS

1. Does the process balance rigor and relevance while ensuring an adequate level of applied scientific performance?

2. Does the process comprehensively address regulatory requirements, together with all other considerations pertinent to decision making and implementation?

Table 6.5 (*Continued*)

3. Does the process systematically address decision-making and implementation-related issues?

4. Does the process integrate substantive environmental concerns, consistent with regulatory requirements and stakeholder priorities?

5. Does the process efficiently and effectively meet decision-making and implementation needs and requirements?

6. Does the process maintain or enhance local influence, to the extent practical?

7. Does the process undertake planning, decision making, and implementation collaboratively with stakeholders?

8. Is the process procedurally and substantively fair, to the extent practical?

9. Is the process adaptive to changing circumstances and suited to the context?

10. Is the process integrated with project planning, decision making, management, and implementation?

11. Is the process linked, where practical, to related decisions, instruments, methods, and decision-making levels?

Too often, EIA documents are unfocused and excessively descriptive. EIA processes frequently take too much (or too little) time and consume too many (or too few) resources. Project planning and EIA processes and theory and practice are still widely separated. EIA processes are sometimes poorly adapted to context. They do not always adequately integrate stakeholder values and perspectives. EIA processes could make better use of experience and good practice. They neglect the needs of decision makers. They can fail to facilitate implementation. Sometimes they are poorly integrated with other environmental management instruments and with public policy making. There is too much variability in EIA competence levels. They do not adequately maintain and enhance EIA quality and effectiveness.

A practical EIA process *focuses* on what is relevant and important. Focusing or scoping can be applied to EIA institutional arrangements, to EIA documents, and to EIA process activities, inputs, and outputs. Scoping is based on clearly defined and consistently applied principles. It identifies the possibilities, decides what is important, shapes and structures the process, and involves interested and affected parties. Numerous scoping methods are available. Scoping roles are clearly defined. The conduct of scoping varies depending on context.

A practical EIA process is grounded in practice and experience. It is *realistic*. A realistic EIA process understands how decisions are made, how organizations operate, how organizations are structured, how people behave in organizations and how organizational structures, and procedures vary in different settings. It accounts for how practitioners operate and how knowledge is generated through practice. It is aware of how actions are integrated into decision making, how power is exercised, and how resources are allocated. It considers how actions are implemented, facilitated, inhibited, and prevented.

A practical EIA process is *feasible*. It is workable. It can be undertaken and implemented. It is informed by concepts, guided by strategies, and aided by tactics. It merges the EIA process with decision making, implementation, management, and context. It is linked to related tools and methods. It contributes to and is adapted to EIA reforms. It is integrated with policies, plans, and programs, with organizational operations, with other decision-making levels, and with related instruments. It is harmonized with other governmental requirements and is coordinated with private-sector and nongovernment organizational activities. It addresses interconnections among disciplines and professions. It is embedded within synthesis frameworks.

Competence is essential in a practical EIA process. It pertains to the qualifications of participants and to the conduct of analyses. It includes the choice and execution of roles and responsibilities by project managers, specialists, proponents, governments, and the public, both individually and collectively. It includes such joint EIA process activities as study team management, study team participation, database management, the application of geographic information systems, report writing, documentation, financial control, budgeting, and the preparation of work programs and schedules. It actively avoids and minimizes recurrent, avoidable, competence-related problems.

A practical EIA process is *effective*. Effectiveness addresses how well the process worked. It concerns the quality of inputs (e.g., institutional arrangements, processes, methods, participant performance, documents) and the effectiveness of direct and indirect outputs (e.g., goals achievement, environmental changes, methodological performance, management performance, contribution to practice). EIA quality and effectiveness analyses raise the level of EIA practice. They distinguish between acceptability and performance levels. They provide an opportunity to correct deficiencies. They involve interested and affected parties. They provide a clear rationale for all interpretations and conclusions. They document results in a form suitable for enhanced practice.

A practical EIA regulatory system minimizes EIA duplication and overlap among government levels, ensures that the EIA roles of government departments and agencies are well coordinated, focuses on what is important, minimizes unnecessary costs and delays, ensures a minimum level of EIA competence, and contributes to enhanced EIA practice. The four jurisdictions apply multiple methods to achieve these objectives. The diversity of approaches points to the potential benefits of knowledge sharing. Uncertainties regarding performance suggest the need for additional effectiveness analyses.

A practical EIA process is focused, realistic, feasible, competent, and effective. It focuses from the outset on regulatory requirements and stakeholder issues. It is based on a well-designed but flexible overall approach and study design. The approach includes procedures for involving governments, politicians, the public, and other stakeholders. The process is built around decisions. It is iterative and incremental. It involves debate, discussion, negotiation, review, reflection, analysis, synthesis, comparison, and bargaining among interested and affected parties. The process integrates ends, means, constraints, analyses, knowledge, methods, skills, experience, uncertainties, requirements, perspectives, and concerns. It operates within available resources, clearly defines roles and responsibilities, explores consequences and implications, provides a rationale for interpretations and conclusions, and is adapted to the context. It applies impact and uncertainty analyses, interpretation, and management to identify adequate and satisfactory options and to select and refine proposed actions. It prepares draft and final EIA documents. It formulates impact management and implementation strategies. It is merged with approvals and implementation. It evaluates the effectiveness of direct and indirect outputs.

A practical EIA process is supported by baseline analyses, by applied experience, and by practice-based knowledge. It applies experiments, pilot projects, feasibility analyses, and targeted research to refine and test interpretations and conclusions. It identifies residual uncertainties and their implications. Many activities in the process are recurrent or continuous (e.g., focusing, management, considering implementation, competence tests, integration, quality and effectiveness analyses, the incorporation of knowledge and experience, documentation). Close contact is maintained with interested and affected parties. Communications and participation focus on issues and perspectives bearing directly on decision making and implementation.

A practical EIA process emphasizes relevance over rigor. Scientific standards are maintained when identified in regulatory requirements and in good practice standards. Regulatory, decision-making, and implementation concerns are addressed comprehensively. The process is usually selective and less systematic in considering other matters. Environmental substance is integrated but usually as a means and in response to external pressures. Practical EIA processes are generally efficient and effective in meeting decision-making and implementation needs and requirements. They are less effective in addressing long-term and major changes and choices. They can facilitate greater local influence, stakeholder collaboration, and procedural and substantive fairness. Stakeholders generally raise such concerns. Such concerns are not intrinsic to the process. Practical EIA processes are generally open and flexible, except in adapting to major changes with long-term implications. The process is integrated with project planning, decision making, management, and implementation. The narrow focus on decision making and implementation can inhibit integration within synthesis frameworks.

CHAPTER 7

HOW TO MAKE EIAs MORE DEMOCRATIC

7.1 HIGHLIGHTS

In this chapter we describe a process where the people take the lead in EIA process management. The decision-making role of the people and communities most directly affected by the proposed action is enhanced. Groups, segments of society, and perspectives commonly excluded from or underrepresented in EIA processes are featured more prominently.

- The analysis begins in Section 8.2 with three applied anecdotes. The stories describe applied experiences associated with efforts to make EIA practice more democratic.
- The analysis in Section 8.3, then defines the problem. From the perspective of many public participants in the EIA process, the problem is that the public's decision-making role is too often nonexistent to negligible. Frequently, members of the public are, or believe themselves to be, powerless in decisions that greatly influence their lives. The direction is enhancing the decision-making role of the public, especially those most directly affected and most vulnerable to change.
- In Section 7.4 we provide an overview of democratic concepts and methods. We explore how (1) the decision-making role of the public can be enhanced, (2) the decision-making role of people affected by proposed actions can be enlarged; and (3) power imbalances in the EIA process can be rectified.

Environmental Impact Assessment: Practical Solutions to Recurrent Problems, By David P. Lawrence
ISBN 0-471-45722-1 Copyright © 2003 John Wiley & Sons, Inc.

- In Section 7.5 we apply the insights, distinctions, and lessons identified in Section 7.4. We describes the properties of a democratic EIA process at both the regulatory and applied levels. In Section 7.5.1 we explore how EIA requirements and guidelines could be more democratic and in Section 7.5.2 demonstrate how a democratic EIA process could be expressed at the applied level.

- In Section 7.6 we assess how well the democratic EIA process presented in Section 7.5 satisfies ideal EIA process characteristics.

- In Section 7.7 we highlight the major insights and lessons derived from the analysis. A summary checklist is provided.

7.2 INSIGHTS FROM PRACTICE

7.2.1 When the Public Takes Control of Its Own Fate

In Australia, social impact assessments (SIAs) frequently fail to be substantive. Often, it is because scoping and funding of SIAs are predetermined. Once the money runs out and the basic boxes have been ticked, the SIA is considered to be complete, regardless of whether the SIA has identified the need for more work to develop solutions to the sometimes complex and sensitive matters identified. Furthermore, engineering companies, which generally manage the EIS process, appear to both disregard and fear key stakeholder groups when they are community groups (as distinct from government or other industry stakeholders). So they fail to incorporate them fully into the SIA process, on the assumption that they will be disruptive rather than constructive.

This case study from central Queensland provides an example of how empowering community groups to participate effectively in the SIA process can produce effective and substantive responses to the problems identified by the SIA in a way that engineering firms and their consultants seldom do. The SIA undertaken for stage 2 of the Stuart Oil Shale Project (mine and processing plant) identified numerous potential negative impacts on the local fruit-growing industry. The SIA consultative process identified the possible impacts but did not attempt to offer possible solutions. The possible impacts identified included loss of critical mass, inability to raise capital due to negative impacts on land values, and inability to attract new growers into the region.

The development of solutions occurred only when the local fruit growers' association obtained government funding to undertake a strategic planning exercise. The consultant, who had prepared the SIA, also facilitated formulation of the strategic plan. The strategic planning process was extensive. It resulted in the exhaustive exploration of a full range of possible responses that the fruit growers could make in reaction to the potential impacts associated with the proposed mine and processing plant. Prior to this point, the growers had been accused of using the oil shale project as an excuse for their own problems—problems which, it was said, largely resulted from drought and changed market conditions. In the end the fruit growers were able to mount a convincing argument that the management options available to them had been entirely foreclosed by the project. A buyout of their properties was therefore the only solution to the problems they faced. A buyout package, funded by government and industry, was eventually developed.

By initiating the strategic planning process, the growers took informal control of the SIA process. They became the "experts." They were able to fully investigate multiple possible responses to the threats posed to their local industry. The possible responses they explored included diversification, amalgamation, improved marketing, implementation of an irrigation scheme, further training, and community development. The options they considered were well beyond what would have been considered within the SIA by the engineering company and the proponent. The growers found a way to broaden and redefine the SIA process. However, to have the necessary control over their own fate, they had to go outside the SIA process, employ a parallel planning process, and then reshape the SIA process based on the results of the parallel process. Ideally, it should be possible to design and apply a democratic EIA process where interested and affected publics have a high degree of influence and control over the decisions that affect them. However, as this story demonstrates, when the EIA/SIA process is too narrowly defined and provides for limited to no public influence, the public can be very creative and persistent in their efforts to redefine the process from the outside in, using whatever external means are available.

ANNIE HOLDEN
ImpaxSIA Consulting, Brisbane, Australia

7.2.2 Curing Affected Communities and Publics

An EIA was conducted in 1998 and 1999 on a large hog processing facility and its accompanying wastewater treatment plant in Brandon, Manitoba. The processing facility, owned and operated by Maple Leaf Pork, was ultimately licensed to slaughter up to 54,000 hogs per week under a single daily work shift scenario. The wastewater plant, owned and operated by the city of Brandon, treats the effluent from the slaughterhouse using covered anaerobic lagoons, aeration, and ultraviolet disinfection. Discharge from the plant goes into the Assiniboine River, which has a slow flow, meandering form, and gentle gradient. During the EIA, project supporters focused primarily on the potential economic benefits (such as the creation of up to 2100 jobs). Critics expressed concerns over potential adverse effects on the community (such as increased demands on social service systems) and on the natural environment (such as increased levels of phosphate, ammonia, and other nutrients in the Assiniboine).

The initial decision to locate the processing facility in Brandon, made prior to the start of the EIA, was highly political and involved direct negotiations among the president of Maple Leaf Pork, the premier of Manitoba, and the mayor of Brandon. A key factor in the negotiations was the provision of substantial provincial and municipal subsidies. The subsidies were part of a provincial effort to expand the hog industry in Manitoba, a strategic response to a perceived need for agricultural diversification.

The project was contentious from the outset. An early focus of concern was the lack of transparency in the decision to locate the processing facility in Brandon. This became the subject of an unsuccessful lawsuit launched by the Westman Community Action Coalition, a local community organization. For project critics and other community members, the political decision to locate the facility in Brandon was an indication that the ultimate outcome of the EIA decision-making process was a foregone conclusion. This belief was fuelled by how the EIA process was applied. First, the provincial environment ministry divided and phased the assessment. This involved treating the development as two distinct projects (i.e., the processing plant and the wastewater treatment facility). Each project was, in turn, divided into stages (e.g., preconstruction, construction, and operation), with separate assessments for each stage. As well, the ministry decided not to hold public hearings, despite considerable public demand and substantial uncertainty over potential impacts. Public involvement was limited to information meetings organized by the project proponents, operation of a public registry, and passive consultation (e.g., submission of letters and petitions).

The EIA was at best legitimating. It did not reflect community-based, democratic decision making. Using Arnstein's classic ladder of citizen involvement, participation perhaps reached the middle rungs (tokenism), characterized by one-way flows of information from managers to citizens and consultation in which citizens are given a voice but are not necessarily heeded. In fact, it closely approximated the bottom rungs of the ladder (nonparticipation), characterized by public relations exercises designed to educate or "cure" citizens of their ignorance, and thereby gain public support. In many respects, this was a classic case of decide–announce–defend, with little or no participation in decision making beyond the managerial, political, and business elite. In the end, because of its pro-development political context and narrow application of provincial law, this EIA failed to advance important social sustainability objectives: namely, participation, empowerment, and equity. This story also demonstrates that an EIA process, characterized by no public influence and minimal public involvement, is inherently unfair, is likely to exacerbate conflict, and will fail to benefit from community perspectives and knowledge. In particular, it underscores the need for early public involvement in key decisions pertaining to such matters as major facility characteristics, location, project-specific regulatory requirements, and the design of the EIA process.

ALAN DIDUCK
Environmental Studies Program, University of Winnipeg

BRUCE MITCHELL
Department of Geography, University of Waterloo

7.2.3 Local Control Gone Awry

This case is an example of a number of projects in Ontario that have had similar experiences in the last several years with projects dealing with waste management facilities: either expansions or new site development. The new legislation in Ontario passed as of January 1, 1997, which was intended partly to ensure broad public consultation, consists of two separate phases. Phase 1 requires the proponent to develop a terms of reference (TOR) (i.e., the workplan for the EA work with public consultation). The Terms of Reference (TOR) is submitted for approval by the Ministry of Environment, which must be satisfied that the TOR deals with all relevant concerns and there was acceptable public consultation in development of the TOR. Once the ministry reviews the workplan and hears public comment, the ministry decides to approve the workplan as proposed, to approve it with some modifications, or to reject it and send the proponent back and begin again. Upon approval of a TOR, the proponent can then begin the actual environmental assessment (EA) study work. Embarking on the EA work requires putting in place another public consultation program dealing specifically with the environmental assessment.

In both phases of this EA process, the public consultation program is usually developed to reflect the basic principles which ensure that consultation is clear and involves the stakeholders and all interested parties. The consultation principles usually applied are: (1) the process would be clear, open, and inclusive; (2) stakeholders concerns should be identified early in the process and addressed in the environmental assessment work; (3) the proponent will respond to all issues and concerns as part of the EA work; and (4) there should be multiple consultation opportunities utilizing a number of techniques at key decision-making points in the project.

The public consultation program of these two phases usually involves open houses, workshops, public meetings, a liaison or advisory committee, newsletters, and in some circumstances, community information centers. The process of developing a TOR has taken anywhere from 12 to 24 months. After a TOR is approved by the ministry, the environmental assessment work is undertaken, and again, this process can run anywhere from 18 months to three or four years.

The openness and multiple opportunities for public consultation provide helpful input to the proponent by identifying issues and concerns to address during the environmental assessment. But there is a downside. Because of the openness and the usually high visibility that such projects normally receive, the public consultation process becomes a vehicle for special interests (organized or otherwise) that see opportunities to either promote causes or to raise individual profiles for other potential activities (i.e., seeking local office). In addition, given that it is an open process without checks and balances to ensure accuracy of comments, people are able to make outrageous claims without any substantiation. Claims that are stark and/or shocking tend to receive substantial media coverage. The problem with such open consultation processes is that the proponent, or proponent's consultants, are not in a position to do more than respond in a balanced and measured way. But since the proponent is seeking to bring in a facility, which is usually not wanted in a community, proponents' responses are not given much weight. In contrast, many people in the community often accept the extreme and negative comments uncritically. This results in the public consultation process being taken over by special-interest groups, or in some cases only three or four people, who purport to speak for the community and community interests.

This often extends the consultation program, making it extremely difficult for proponents, regulators, and municipal council decision makers to deal objectively with the environmental assessment analysis and the predictions from the EA. Considerable pressure is put on municipal politicians to align themselves with opponents, even when the politicians realize that a small group or a few people are spreading disinformation about the proposed facilities. The end result is prolonging the EA process, often considerably. A proponent will often agree to additional meetings for the further discussions that will always be requested by project opponents.

The only venue where the proponent can obtain a reasonable and objective hearing is through the quasijudicial part of the regulatory process. If a public hearing is required by the Ministry of Environment to determine whether or not a project should be approved, this establishes a judicial context within which the EA is tested. It is in this stage within the EA approvals process where claims or charges made must be substantiated. Also, cross-examination is permitted. It is often at this level where the disinformation is stripped away and the focus finally is put on the environmental impacts of the proposed project. Recent experience shows that EA processes are being extended for months and years, resulting unnecessarily in additional costs. Further, the disinformation causes undue stress for some in the affected communities and a blurring of the lines among facts, hypotheses, assumptions, and fiction.

This case example shows the need for regulators to be more involved in the early stages and to respond to disinformation. There is also a need for open and honest communication and consultation in dealing with complex environmental proposals. More broadly, it demonstrates that local influence and control is not necessarily conducive to sound decision making and to environmental protection and enhancement, especially if decision making is based on disinformation and is controlled by a few. The enhancement of local influence and control, to be effective, should be coupled with concerted efforts to prevent and offset disinformation, to broaden the decision-making base, and to facilitate the advancement of environmental values and objectives.

The case example suggests further, especially given the recurrent nature of the problems that have arisen, that alternative public consultation/public influence models may have to be considered. Particular attention, given the inequities in the distribution of impacts and benefits, may need to be given to procedural and outcome equity. Both the lessons from the processes under the current model and more positive examples in other venues may need to be considered. Fine-tuning the existing model or railing against how circumstances tend to unfold is unlikely to prevent the negative pattern from continuing to repeat itself.

PETER HOMENUCK
IER Planning, Research and Management Services

7.3 DEFINING THE PROBLEM AND DECIDING ON A DIRECTION

The three stories, presented in the preceding section, approach the question of democracy in contrasting ways. The first story describes an SIA where a group, affected by a proposed project, increased its decision-making influence by going outside a narrowly defined SIA process to obtain knowledge and insights that it then used to broaden and reform the SIA process. The second story describes an EIA process that collapsed, largely because of a failure to permit even a moderate level of public participation and influence, particularly in early decision-making phases. The third story demonstrates that greater levels of local influence and control are not always conducive to more effective decision making and environmental protection and enhancement, especially when the process is distorted by disinformation and taken over by a few individuals or groups. Efforts to enhance local control should be coupled with and tempered by measures to offset disinformation, broaden the decision-making base, and advance environmental values and objectives.

The stories illustrate that a failure to provide for an adequate level of public decision-making influence and control is likely to result in either process failure or the public taking actions to ensure a greater level of decision-making influence, either within or external to the process. The stories are suggestive of both the advantages of greater public influence and control and the dangers associated with processes that fail to facilitate informed (by accurate data and sound analysis and interpretation) effective and representative public influence and control. The stories, however, provide only an initial and partial sense of how democratic values should and should not be approached in EIA process design and management.

The value of public participation in the EIA process is widely acknowledged. Public participation leads to better designed projects, an enhanced understanding of environmental conditions, more accurate predictions, a more focused EIA process (on stakeholder issues and reasonable alternatives) and more effective impact management (Barrow, 1997; Bisset, 1996; Hughes, 1998; Lee, 2000). Public participation clears up misunderstandings and resolves conflicts (Barrow, 1997; Lee, 2000). The public is more likely to accept and support projects when there is a high degree of public involvement in the EIA process (Barrow, 1997). Approval is expedited. Proponents are more likely to achieve their objectives (Lee, 2000). Decisions are viewed as more legitimate, both by regulators and by the public (Barrow, 1997; Hughes, 1998). The public is better informed about the project. Public issues and concerns are more likely to be addressed (Hughes, 1998). Public participation skills are enhanced (Barrow, 1997). Decision making becomes more open, transparent and less susceptible to lobbying by vested interests (Hyman et al., 1988). Everybody benefits.

This apparent consensus is more apparent than real. Participation can be defined in many ways. Perspectives vary greatly regarding the types and degrees of participation, which are necessary and appropriate. Suspicions concerning motivations abound. A seemingly shared interest in participation often masks deep-ceded value and interest differences. Absolute positions, not conducive to consensus building or conflict resolution, are commonplace. The positions and perspectives of other parties are often misunderstood and overstated. The major parties frequently differ dramatically in their assessments of the effectiveness of the participation measures applied in EIA practice.

Proponents have a tendency to be wary of public participation. They often see it as costly and time consuming (Kreske, 1996). They commonly are concerned that it will raise their profile, impede their relationships with authorities, and exacerbate rather than resolve conflicts (Bisset, 1996; Morgan, 1998). Many engage in public participation when it seems necessary for regulatory compliance and when it facilitates project approval and implementation. They are often adamant that they cannot share decision-making authority (UNEP, 1997). They shy away from conflict (Lee, 2000). They tend to be more comfortable with closed processes or at least processes that are not opened up until major decisions have been made (Glasson et al., 1999). They generally give greater weight to technical analyses (Morgan, 1998). They sometimes characterize public inputs as subjective and inconclusive (Morgan, 1998). Some are quick to label concerned groups and individuals as ill informed, unrepresentative, and selfish [i.e., "not in my backyard" (NIMBY)] adversaries (Canter, 1996). Such proponents see the role of public participation as educating the public, neutralizing opponents, and legitimating the process (Gerrard, 1995). They are frequently skeptical of collaboration. They tend to favor lobbying (of both regulators and politicians) and coercive measures (such as expropriation) for obtaining approvals (Gerrard, 1995).

Regulators often encourage participation, albeit within tightly circumscribed limits. Government agencies in many jurisdictions have a tradition of administrative discretion, secrecy, and limited public involvement (Glasson et al., 1999). Political interest groups often lobby government agencies and sometimes have considerable influence (Hyman et al., 1988). Public decision making is becoming more open and transparent, in part, because of EIA requirements and procedures (US CEQ, 1997a). But many regulators remain reluctant to open up policy and decision-making procedures more fully. Government agencies are commonly segmented and mission-oriented. EIA responsibilities are often secondary functions. Government officials, with EIA review responsibilities, rarely have the time to address matters not immediately relevant to administrating regulatory requirements. Personal advancement is much more closely tied to meeting the desires and requirements of senior government officials and politicians than to satisfying the concerns of interested and affected members of the public. Sometimes societal goals are displaced by internal organizational goals, such as organizational survival or expansion (Hyman et al., 1988). Government agencies often display a propensity toward centralized control, resulting in an inherent tension with the decentralizing tendencies of public involvement (Canter, 1996). Regulators tend to be comfortable with technical knowledge and less comfortable with conflict and confrontation (Canter, 1996). Regulators, in common with proponents, often preclude any possibility of shared or delegated decision making (Kreske, 1996).

EIA practitioners are often placed in the unenviable position of attempting to meet regulatory requirements, operate within time and budget constraints, and not offend proponents (March, 1998). At the same time, they attempt to adhere to good practice standards, including those related to the role of public participation in the EIA process. In the ongoing effort to juggle these competing demands, they have a tendency to control the process, at least to the extent necessary to perform their responsibilities adequately (Maynes, 1989). They may be reluctant to apply unproven methods or to engage in open-ended processes where outcomes are difficult to predict or manage (Solomon et al., 1997). They have a tendency to view public participation as the proficient application of the appropriate suite of methods. They are generally in favor of public participation. They often support the use of mediators, facilitators, and similar procedural specialists to address conflicts as they emerge (Bisset, 1996). They tend to view local control as unrealistic, except in a very limited range of circumstances (e.g., local community infrastructure) (Morgan, 1998). As technical specialists, they usually favor technical over nontechnical knowledge (Canter, 1996).

The *public* generally acknowledges that public participation in the EIA process can be beneficial. But they often are highly skeptical about whether and how often those benefits are realized in practice. They seldom see major differences among proponents, regulators, and practitioners. All are frequently seen as arrogant, patronizing, manipulative, and unworthy of trust or support (Arnstein, 1969; Gerrard, 1995; Parenteau, 1988). Proponents often are seen as relentlessly pursuing the implementation of predetermined projects, regardless of environmental and local community consequences. Regulators are viewed as consumed with administrative matters and highly susceptible to lobbying by vested interests (Maynes, 1989). Practitioners are seen as expeditors, who work on behalf of proponents, to meet minimal regulatory requirements. None of these parties is expected to consider public concerns seriously.

The public often sees itself as powerless or, at best, fulfilling a marginal role in the EIA process (US CEQ, 1997a). They note that they are commonly treated as adversaries (US CEQ, 1997a). Decisions have generally been made before they are involved. They see communications as one-way (US CEQ, 1997a). They often see their role in the process as tokenism. Frequently, they believe that they have no control over either the process or its outcomes. Occasionally, they hope that the EIA process can be a tool for defending themselves against aggression and a means of redistributing power (Torgerson, 1980). But power is so concentrated in existing structures that the changes wrought through EIA are usually seen as minimal (Torgerson, 1980). Sometimes they insist that they are being manipulated or coopted. Often, they expect that by the end of the process, power will be even more centralized in distant authorities. The EIA process, they commonly assert, does not prevent approval of environmentally unsound projects (Mittelstaedt et al., 1997). Often, the outcome, they point out, is the unfair distribution of facilities and services. Frequently, they conclude, it is the most vulnerable who bear the greatest burden and receive the least benefits (Gerrard, 1995).

Commentators on EIA practice echo many of the concerns raised by the public. They generally agree that EIA practice is increasingly open and collaborative. They acknowledge the many benefits that have accrued from increased participation by interested and affected parties in the EIA process. But they also suggest that participation typically occurs too late in the process, is too infrequent, and is too narrowly defined (Morgan, 1998). They suggest that EIA processes, all too frequently, are shaped and sometimes corrupted by narrow political interests (Rickson et al., 1990; Smith, 1993). EIA practitioners are characterized as, at worst, compromised and at best naive regarding EIA politics (Craig, 1990). Social, cultural, and procedural issues, they argue, receive too little attention. Community power implications, they maintain, are neglected (Thompson and Williams, 1992). Technical and quantitative concerns and methods are, they assert, overemphasized (Solomon et al., 1997). Excessive reliance, they point out, is placed on a narrow range of public consultation techniques (Solomon et al., 1997). Public involvement guidance is, they conclude, highly variable, too general, and often unclear (Hughes, 1998). As reflected in EIA requirements and guidelines, EIA regulators are admonished for not offering sufficient practical public involvement guidance and for not consistently and proactively supporting public participation (Hughes, 1998).

EIA practice commentators tend to be especially critical of the persistent use of coercive and reactionary (i.e., decide, announce, defend) involvement and implementation methods (Armour, 1990b; Halstead et al., 1984; Rabe, 1994). They point to the repeated failure of such approaches (Solomon et al., 1997). They describe NIMBY as a natural, reasonable, valid, and often constructive reaction to (1) a major threat to individuals and communities; (2) autocratic decision making; (3) a fundamental imbalance in the distribution of costs and benefits; (4) a legacy of poor communications, biased analysis, and inept management; (5) a failure to adequately address social impact, uncertainty, dread, stigma, and perceived risk concerns; and (6) belated, partial, and unduly restricted efforts to involve interested and affected parties (Mazmanion and Morell, 1994).

The commentators note that on a broader political front, the trend is toward more transparent, democratic, decentralized, and accountable decision making; greater access to information; more explicit and understandable documentation; earlier, more continuous and more collaborative public involvement; and more direct involvement of nongovernmental organizations (NGOs) (Bisset, 1996; Glasson et al., 1999; Lee, 2000; Wood, 1995). EIA, they argue, should be at the forefront rather than lagging behind such trends (IAIA and IEA, undated; Interorganizational Committee, 1994). EIA practice commentators have devoted particular attention to the positive and negative experiences associated with siting locally unwanted land uses. Although the issues are complex and the experiences mixed, they point out that there have been several notable "successes" when siting processes have been built around a blending of local control, social equity, and shared management (Gerrard, 1995; Rabe, 1994; Seley, 1983). They contrast these experiences with the more uniformly negative experiences associated with autocratic, technical, and coercive siting approaches (Halstead et al., 1984; Solomon et al., 1997).

The conclusions of EIA commentators are reinforced by the results of EIA effectiveness reviews. These reviews consistently conclude that public involvement occurs too late and too infrequently in the EIA process (Sadler, 1996; US CEQ, 1997a). Institutional provisions for public participation are highly variable. They are rated as excellent to good less than half the time (Sadler, 1996). More creative outreach is needed (US CEQ, 1997a). The performance of public involvement is bad as often as it is good (Sadler, 1996). Public involvement has only a significant influence on decision making about one-fourth of the time (Sadler, 1996). Clearly, the treatment of public involvement in the EIA process, from a practitioners' perspective, is falling well short of its potential.

How best to determine the direction in light of these differing perspectives on the problem? One approach is to decide that there is no problem (i.e., the EIA process is already sufficiently democratic). Many proponents would probably favor this approach on the grounds that they are already "bending over backward" to facilitate public consultation. They might also argue that the appropriate form of public involvement is highly dependent on the situation. Flexibility is therefore essential. As illustrated by the EIA effectiveness reviews, practitioners and regulators would suggest that there remains considerable room for improvement. The public and EIA commentators would probably maintain that the *status quo* is more than inadequate—it is unacceptable. Overall, the case for no change in current practices seems highly dubious.

A second approach is to decide that all that is required is polishing and refining current practice. There should be more and earlier opportunities for public involvement. Requirements and guidelines should be clarified and should offer more practical advice. A wider range of involvement methods should be applied. Regulators should support public participation proactively. Public involvement methods should be used more effectively. Most regulators and practitioners would probably favor this approach. The public and commentators would probably see this approach as necessary but far from sufficient. They would argue that such changes would not instill trust or alter the peripheral decision-making role of the public. The public would continue to have limited influence over the process or its outcomes. It would not be able to defend its interests adequately or to ensure that its concerns are addressed properly. Others would decide its fate in processes where democratic decision making is more illusionary than real. This approach is not likely to narrow the gulf appreciably between the public and other participants in the EIA process.

Assuming that the first two approaches are, respectively, unacceptable and insufficient, only two possibilities remain. One is to identify the problem as an undemocratic EIA process. The only appropriate response is to delegate or share decision-making authority with the public. Delegating or sharing decision-making authority with the public is the heart of the EIA process presented in this chapter. The second possibility is to accept that refining current public involvement practices is not enough but also to expect that there will be few situations where decision-making authority can be delegated to or shared with the public. The appropriate response is therefore a collaborative EIA process. The public is an active and ongoing participant in a collaborative EIA process. But final decision-making authority continues to reside with proponents and regulators. A collaborative EIA process is presented in Chapter 8.

7.4 SELECTING THE MOST APPROPRIATE ROUTE

7.4.1 Definitions and Distinctions

Democracy is rule by the people, either directly and/or indirectly through representatives elected periodically. It is both a system of government and a political theory. It is a collective, social and political (i.e., concerned with the exercise of power) endeavor (Nagel, 1987). It presumes or at least aspires toward such values and ethical principles as freedom of speech and assembly, equality of opportunity, minority rights, and majority rule. Democratic theory presumes that all citizens are or can participate equally in decision making and can be equally influential in the political system. Where this is not the case, democracy has a responsibility to correct power imbalances. Direct democracy is the ideal but is not always practical. Representative democracy is a compromise where there is a large population base and specialized roles. This does not mean that direct democratic principles, such as involving those affected by decisions, continuity of involvement, consensus building, discussion, action, and community, need to be abandoned (Nagel, 1987; Pateman, 1970). Instead, EIA practitioners should actively seek to ensure that direct democratic principles are expressed and, wherever practical, fulfilled. The onus should be on those seeking to circumscribe the application of direct democratic principles to demonstrate why such limitations are essential to good and effective governance.

Figure 7.1 is a highly simplified illustration of key democratic participants and interactions. The ultimate source of authority, for courts, politicians, and government officials, is the people. The courts, politicians, and government officials should all be responsive and responsible to the people. Ideally, they should share and delegate authority back to the people. Practical opportunities for direct democracy should be considered. The public should be fully informed of and involved in all decisions likely to affect them.

These ideals are commonly transgressed in practice. Politicians and government officials are often reluctant to inform, much less involve, share, or delegate decision-making authority back to the people. Frequently, tension develops between democracy and bureaucracy (the world of government officials) and between politicians (who believe that they are in charge once elected) and the public (who wish a continuing say in matters that affect them between elections). Increasingly, democracy is now seen as a continuous and dynamic process where even if governments retain the final authority, close public scrutiny is essential (Gilpin, 1995). Arguably, the trend is toward (or should be) a more participatory form of democracy. In a participatory democracy particular attention is devoted to the relationships between individuals and authority structures (Pateman, 1970). Public involvement is a right (Lee, 2000). The public fully understands the problems being addressed and the means proposed for addressing the problems (Canter, 1996). Public involvement

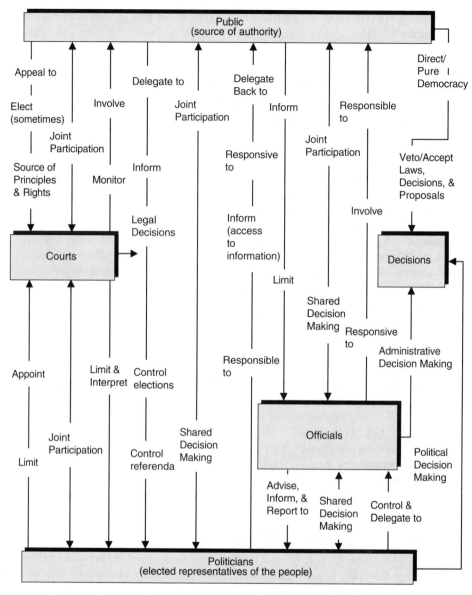

Figure 7.1   Examples of democratic participants and interactions.

occurs earlier. It is more frequent. It is sometimes continuous. The public has a direct and acknowledged role in decision making (Lee and George, 2000). The public has influence and, where practical, control over the decisions that affect them (Nagel, 1987).

EIA practice should be viewed as a microcosm of participatory democracy. There should be frequent, preferably continuous, public involvement provisions, full access to information, the right of appeal to an independent third party, the full involvement of interested and affected parties, and an explicit decision-making role for the public. Arguably, EIA practice has a responsibility to expand and extend democratic participatory tenets. Protection of the environment is a critical democratic responsibility. Proposals, subject to EIA requirements, invariably directly affect the environment and peoples' day-to-day lives. Thus there is a democratic responsibility to ensure that people have a major say in the decisions that affect them. Impacts tend to occur at a site-specific and community-scale level. A community scale is often conducive to applying direct democratic principles and methods. Power and influence is seldom equally distributed among EIA process participants. Therefore, power inequities must be offset. Environmental costs and benefits are rarely equally distributed. A proactive effort is required to ameliorate environmental inequities.

The EIA process should be an expression of participatory democracy. As illustrated in Figure 7.2, it should be built on a foundation of participatory democratic

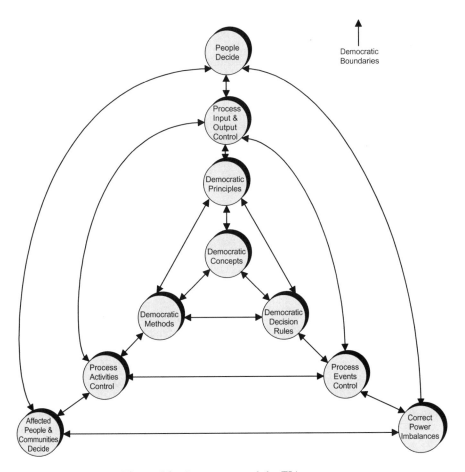

Figure 7.2   Democracy and the EIA process.

concepts. It should integrate democratic principles, methods, and decision rules. The public should have a major say over process activities, inputs, outputs, and events. The people should shape and guide the process. Particular attention should be devoted to the decision-making role of affected people and communities. Steps should be taken to prevent and ameliorate power imbalances. Every aspect of the process should be conducted within the boundaries of participatory democratic limits and requirements.

7.4.2 The People Decide

Letting the people decide is a simple concept. Implementing the concept is more complex. Do the people decide directly or indirectly through their representatives? Which people decide? Are the people's representatives members of community associations and other nongovernmental organizations, or are they elected representatives? What are the roles of different levels of elected representatives? Where do nonelected government officials fit in? If the goal is to retain decision-making control as close to local people as practical in an EIA process, as illustrated in Figure 7.3, members of the public most directly affected by the proposed action take the lead with the active participation of locally elected politicians. Alternatively, a team of ordinary citizens could take the lead. If the plan or agreement reached by the team receives broad public support, it is likely that political support will follow (Todd, 2002). As one moves down the ladder, decision-making control becomes progressively less direct and increasingly less subject to the control and influence of the local public and their representatives. The argument will be made that there are broader constituency interests represented by senior-level politicians and specialized areas of expertise possessed only by senior government officials. However, if the ideal is "rule by the people," it must be demonstrated why it is necessary to move down the ladder. Senior governments are also obliged to contribute to local capacity building (to facilitate local control), to ensure that the actions of regulatory officials are subject to local scrutiny and appeal, and to decentralize government operations wherever practical (to maximize local contact).

Letting the people decide also entails determining how decisions are to be made. Should they be made directly through referenda (e.g., local veto or acceptance)? Who sits on committees? Should decisions be made based on majority rule, or is a consensus preferable? Should decision-making authority be delegated to an independent third party (e.g., an arbitrator, an environmental court, a public inquiry) (Halstead et al., 1984; Westman, 1985)? Should an advocate, a public defender, or peer reviewers assist the public (Kasperson et al., 1984)? Should alternative dispute resolution (e.g., mediator, facilitator) aid decision making? Should the local people decide by themselves, or should decisions be made jointly with other parties?

Decisions must also be made regarding which EIA process choices will be made by or shared with the public. Figure 7.4 identifies examples of potential choices. The potential choices encompass the major decisions leading up to project

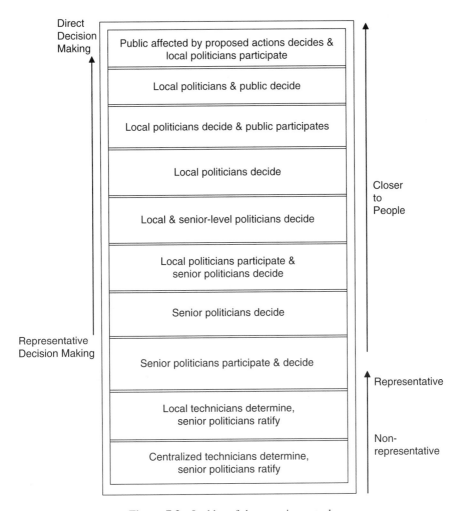

Figure 7.3 Ladder of democratic control.

approval, project approval or disapproval (including the determination of conditions), and major post-approval choices (operations, closure, and post closure). It is necessary to determine for each decision when a community has decided (e.g., majority rule, more than majority rule, majority rule for more than one geographic subarea, consensus among representatives). It is also necessary to decide if the range of choices available to the public should be bounded. The public, for example, could be permitted to choose among a set of alternatives all considered to be environmentally, technically, and financially acceptable. Alternatively, choices by the public could be subject to acceptability confirmation. Sometimes public choices are limited to certain project types (e.g., local community infrastructure) or to predefined geographic areas. The goal should be a consensus among the major parties concerning the public control "rules and boundaries."

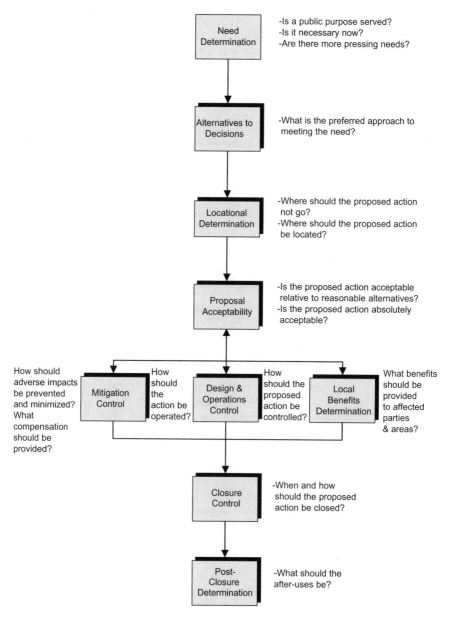

Figure 7.4   Examples of democratic decisions in the EIA process.

The distinction between local and shared public control has been addressed in natural resource management. Community-based natural resource management occurs when the community is allocated ownership or authority (a form of delegated decision making) for natural resource management. Local populations arguably have a greater interest in sustainable resource use than do distant govern-ment officials. Co-management runs the spectrum from almost complete self-man-agement to almost total state management. It has been widely applied in forest, fisheries, and wildlife management, especially with indigenous peoples. The parties make trade-offs among themselves and may adopt a legal agreement, which shares the legal authority for resource management (Harvey and Usher, 1996). Provision is made for funding, training, and staffing to support the partnership (Mitterstaedt et al., 1997). Co-management tends to work best when there is community consen-sus, a credible lead agency, a clear mission for the partnership, meaningful delega-tion, meaningful tenure, revenue autonomy, meaningful inclusion of interests, and reliance on existing structures (Harvey and Usher, 1996). Lessons and insights from both community-based resource management and resource co-management could greatly assist the design and management of democratic EIA processes (Todd, 2002).

Democratic EIA processes are sometimes applied. Citizens are centrally located in the process (Armour, 1990b). The public is a partner with the proponent and with regulators (Lee, 2000). EIA practitioners act as facilitators and collaborators (Armour, 1990b). The parties jointly solve problems, control the process, conduct analyses, and reach decisions (preferably, a consensus) (Lee, 2000; UNEP, 1997; Westman, 1985). The collaboration often extends through design and implementa-tion (Lee, 2000). Continuous (e.g., committees) and interactive (e.g., workshops, conferences) techniques are applied in an ongoing effort to reach and retain consen-sus and to maintain contact with and support from broader constituencies.

Democratic EIA processes have been most fully developed and tested in the voluntary siting of locally unwanted land uses (LULUs), especially nuclear and hazardous waste management facilities. Voluntary siting processes tend to begin with an extended period of community dialogue to discuss the problem and the gen-eral approaches available for solving the problem (Rabe, 1994). The siting process begins only after there is a broad consensus concerning the need, the facilities required to meet the need, and the essential attributes of the siting process. In a voluntary siting process, communities can express an interest in hosting a facility (SPTF, undated). They can also opt out of the process at any time (Rabe, 1994). An initial analysis is generally undertaken to determine if there are potentially envir-onmentally suitable locations within or near the communities that express an inter-est. A referendum is held in each environmentally acceptable community that has expressed an interest, to determine the level of community support (Gerrard, 1995). The determination of a preferred location(s) takes into account levels of community support, environmental suitability, and geographic fairness.

The siting process is highly collaborative and adaptive. The communities, the proponents, and the regulators work together in a partnership (Mazmanion and Morell, 1994). The community has a right to select among technically acceptable design and operational options (SPTF, undated). The process tends to be open, highly participative (e.g., community advisory committees, workshops, multiple meetings), and consensus-based. Risks and impacts are avoided and reduced to well below acceptable levels (Rabe, 1994). Often, there is a siting agreement between the sponsor and the host community. Siting agreements address such matters as facility operations, monitoring, contingency measures, compensation, local benefits, and community consultation procedures (Gerrard, 1995). Compensation and local benefits ensure that the host community is a net beneficiary (Gerrard, 1995; Kasperson et al., 1984). The benefits package can be the driving force of the process (e.g., a bidding procedure).

Sometimes voluntary siting processes and siting agreements include provisions to minimize need (e.g., import controls, reduction, recycling, technological innovation) and to share the burden (e.g., national allocation, dispersed facilities) (Gerrard, 1995; Rabe, 1994). Regulators are fully involved in the process. Government responsibilities and commitments are cleared specified (Gerrard, 1995; Rabe, 1994). Sometimes, special-purpose bodies, with broad and independent management and operating at arm's length from generators and regulators, are created to facilitate the process. The process is more effective if undertaken within an environmental, resource, and waste management policy and program framework. Such institutional reforms can enhance process credibility. Local control and oversight generally extends through operations and post operations. Co-management provisions are commonly included in host community agreements.

Voluntary siting processes, consistent with direct and participatory democratic principles, ensure a high degree of community influence and control. The voluntary nature of the process tends to reduce social and perceived risk concerns (Gerrard, 1995). Communities with cultures conducive to the acceptance of controversial facilities tend to be drawn out (Gerrard, 1995). Procedural and substantive inequities often are ameliorated. Conservation and strategy, plan, and policy formulation can be encouraged. However, environmental conditions can be less suitable with a limited number of voluntary communities (Gerrard, 1995). Equity concerns may arise if communities volunteer that are low income, vulnerable, and already heavily affected (Gerrard, 1995). It is often difficult to define community boundaries (Gerrard, 1995). Few communities may volunteer, perhaps none (Gerrard, 1995). There can be a high environmental cost if there is an urgent need and no volunteer communities at the end of the process. Sometimes, there will be major trade-offs and conflicts among neighboring, access route, and regional communities. Divisions within communities can be exacerbated, especially if there is a close vote.

There have been some notable "successes" with voluntary siting processes, although definitions of "success" can, of course, vary. Certainly the track record of conventional, especially coercive siting approaches has been dismal. But it is far from clear whether voluntary siting is "the" model for siting locally unwanted land uses. Which elements are crucial to the success of the process have yet to be fully demonstrated. The role of local factors could receive greater consideration. It is even less clear whether voluntary siting is a model for a democratic EIA process. How critical, for example, is siting to the process? Locational choices are often very limited for many types of proposals subject to EIA requirements. How important is the scale and type of facility? Usually, much less time and far fewer resources are available than is the case for major energy and waste management facilities. How serious are the negative tendencies of the approach? Is it possible, and how should the negative tendencies of the approach be offset? Voluntary siting experiences should be explored. They display many features that ideally characterize a democratic EIA process. But care should be taken not to assume that what appears to have worked well in a limited range of situations for a narrow range of facilities is necessarily "the model" for instilling democracy into the EIA process.

7.4.3 The People and Communities Affected by the Proposal Decide

A democratic EIA process assumes that people and communities affected by proposed actions are willing and able to make key decisions, alone or in a partnership arrangement. Some might argue that this is not a realistic or appropriate assumption. Table 7.1 highlights key characteristics of several concepts related to how people and communities can become more autonomous and thereby better able to make decisions.

Bioregionalism and communitarian approaches provide visions, principles, and strategies for more autonomous communities. EIA practitioners (at both the regulatory and applied levels) can work with the public and communities in developing, refining, and applying the lessons and insights from such concepts. Bottom-up approaches to community autonomy, such as empowerment, community development, and mobilization, can be instructive to community and environmental activists and organizers. EIA practitioners can encourage, support, and facilitate such efforts. Concepts such as traditional knowledge and lay science contribute to a community-oriented knowledge base—a knowledge base conducive to community participation and empowerment. EIA practitioners can support the development and accommodation of community knowledge. They can conduct, with public guidance and support, community power structure analyses. Community autonomy can be maintained and enhanced by such analyses. EIA capacity building provides skills, knowledge, resources, and institutional reforms needed to further community autonomy objectives.

Each concept starts from the premise that communities should control their own affairs. Knowledge and control flows upward from individuals and communities rather than downward from experts and government. The local community takes on a larger share of the roles and responsibilities conventionally assigned to technical specialists, politicians, and government officials. Specialist, politicians, and government officials actively assist and support this reorientation. The existing distribution of power is challenged. Assumptions regarding scientific and political legitimacy are tested. The primacy of technical, scientific, and rational knowledge and methods is replaced by a far greater emphasis on the knowledge, experiences, and perspectives of local individuals, groups, and organizations. Disciplinary and other conventional categories are crossed and transcended.

Planning and decision-making processes become more social, political, adaptive, informal, subjective, participative, and context-dependent. The public is at the center of rather than peripheral to the process. The process is guided by community values and aspirations. Proposed actions are evaluated as catalysts for or against

Table 7.1   Example of Potentially Relevant Local Autonomy Concepts

Bioregionalism	Focuses on developing self-reliant economic, social, and political systems
	Seeks to develop a territory's fullest potential through reliance on systems of production that draw on local resources, do not degrade the ecosphere, and require consideration of long-term implications as compared with short-term gains
	Requires basic changes in beliefs, attitudes, and values concerning human—natural environmental interactions
	Political power most effective at local level; leads to personal and community empowerment
	Stresses value of interdisciplinary analysis, experiential knowledge, and social learning
	Proposals viewed as catalysts for positive community and regional change
	Built around ecological principles; creates sustainable systems of production
Empowerment	People planning for themselves (self-respect, reliance, and determination); builds on community knowledge and capacity to organize and act
	Multidirectional effort by grassroots groups to secure influence and control over procedures that affect their lives; challenges conventional power relationships; involves appropriating, extending, exercising, negotiating, and mobilizing popular support
	Participation leads to community empowerment and community improvement
	Ladder of empowerment (bottom to top): atomistic (individual unit), embedded individual (individual within larger settings or structures), mediated empowerment (in context of relationship between expert and client), sociopolitical empowerment (links individual to community through collective social action and challenging of oppressive institutional arrangements), political empowerment (community or group the locus of change, operationalized through policy changes and access to community resources)
	Appropriates EIA and SIA to community priorities; facilitates community participation and control; extends into less formal settings where community influence greater; negotiated participation in territorial campaigns for more acceptable local outcomes and mobilization of popular support
	Role of practitioner to establish mechanisms that allow for true participation and influence; responsibility shifted to affected people and community; planning collaborative and political; guided by ethic of empowerment
Community development	Process in which people in a community define their wants and devise means to satisfy them
	Concerned with process and outcomes
	Community building carried out by activists and community-based organizations

Table 7.1 (*Continued*)

	Builds on local distinctiveness, escapes constrictions of local traditions, draws in outside opportunities, and limits potential for domination and exploitation
	Opposes dominating forces, reflects on experiences, overcomes barriers, and builds coalitions of support
	Reshapes boundaries and prevailing models; value-laden and political decision-making process
	Emphasizes social utility
Mobilization	Argues that people can be mobilized for political activities or special-interest associations
	People plan and develop own projects
	People learn from own and other experiences, develop contacts with external institutions for resources and technical advice but control how resources are used
	Counters efforts by industry and government to use EIA to achieve scientific and technical legitimacy; raises questions about scientific legitimacy and environmental rationality
	EIA as a catalyst for political mobilization; at core of political debate and decision making
	EIA a microcosm of a more democratic, culturally diverse, and litigious society; EIA a forum for democratic debate
Communitarian	Focuses on self-management practices of small communities; stress the social side of human nature
	Belief that the human community is the best form of human organization for respecting human dignity, for safeguarding human decency, and for facilitating a way of life open to self-revision and shared deliberation
	Stresses individual human dignity and increased social responsibility
	Each community develops own agenda; not majoritarian but strongly democratic
	Seeks to make government more representative, participative, and responsive to all members of communities
	Supracommunity—a community of communities (the community of humankind)
	Actively maintains institutions of a civil society; communities and polities have a duty to be responsive to their members and to foster participation and deliberation in social and political life
Traditional knowledge	Generally refers to environmental knowledge possessed by indigenous peoples; more a way of life than a knowledge base; knowledge inseparable from people
	Integrates social, ethical, cultural, technical, scientific, and historical in a single process (within own context)
	EIA should accommodate traditional knowledge based on trust, respect, equity, and empowerment
	Holistic; spiritual at the core
	Goal is coexistence rather than integration
	Different from each situation; flexible and fluid
	Often misunderstood and misapplied when viewed outside indigenous contexts

(*Continued*)

Table 7.1 (*Continued*)

Lay science	Interpretation considered a social act; cannot be separated from social context
	Lay research and discovery as an antidote to industry-dominated science
	Environmental professionals realize that there must be interaction between scientist and community; professionals help community volunteers participate in data acquisition and interpretation
	Lay participation in science (e.g., monitoring); basic training provided; residents and workers able to test and defend their own common sense
	Empowers community groups to take part in science and technology policy decisions
	Helps reduce public fears, increases public understanding, and contributes to better science
Power structure analysis	Part of a political approach to EIA and SIA
	Recognizes that proposed actions affect community power structures and stratification systems
	Analyzes effects on local autonomy, power of various factions, vertical linkages, local leadership structure, and processes of distributive politics
	Considers degree roles, institutions, and values transformed and continuing success in providing structure and meaning to community life
	Outcomes complex and dependent on history and context
	Analysis can support management actions
	If community vitality and cohesion undermined, could inhibit ability to cope with and manage change; if enhanced, communities in a better position to bargain with proponents and regulators and to control own fate
Capacity building	Long-term voluntary process of increasing ability of communities (and countries) to identify and solve own problems and risks and to maximize opportunities; aim is self-sufficiency
	Can involve developing and implementing educational and research programs and strengthening educational and research institutions
	At community level involves building up place-based institutional capacity; importance of openness to new relationships, knowledge flow from a wide range of sources, and enhancement of adaptive capability
	Role of EIA and SIA as a learning process that contributes to the ability of communities and societies to change
	With EIA can include developing a library of EIA reports, maintaining databases, establishing practitioner networks, collecting examples of good practice, undertaking demonstration projects, producing newsletters, and inviting guest speakers
	Importance of linking EIA to development planning, programming, and licensing

Sources: Craig (1990), Diffenderfer and Birch (1997), Etzioni (1993, 1995), Gagnon et al. (1993), Healey (1997) Heinman (1997), Lee (2000), McClendon (1993), Novek (1995), Rickson et al. (1990b), Rocha (1997), Thompson and Williams (1992), UNEP (1997), Weaver and Cunningham (1985), Welles (1997).

the satisfaction of community needs and aspirations. The process is broadened to encompass the political roles and activities of community activists and nongovernmental organizations. Governments and practitioners actively foster participation, local empowerment, community self-reliance, and the realization of community goals. Both the EIA process and EIA capacity building are guided by the ethic of community empowerment. EIA becomes a tool for furthering community development. The potential community power and structure implications of proposed actions are analyzed. Communities are assisted, when requested, in their efforts to cope with and manage change, consistent with community ends.

7.4.4 Correcting Power Imbalances

Direct and participative democratic theory presumes that people have equal access to power. Such, of course, is rarely the case. Invariably, there are inequities in access to power among groups and segments of society. There also are perspectives conventionally underrepresented in the EIA process. Table 7.2 highlights key characteristics of several concepts aimed at correcting power imbalances.

EIA practitioners can help offset power imbalances. They can prepare alternative EIA documents, interpretations, and analyses, consistent with advocacy theory. They can integrate social and environmental justice concerns into EIA requirements, processes, and documents. They can, working within government, undertake and support social and environmental equity and progressive planning. The can advocate and support decentralization and deconcentration. They can operate outside government as advocates, activists, and community organizers. They can encourage and support ecological political activities and environmental and social

Table 7.2   Examples of Potentially Relevant Concepts for Correcting Power Imbalances

Advocacy	Argues that planning documents (such as EIAs) reflect the existing distribution of power in society
	Promotes the preparation of alternative plans that reflect the interests and perspectives of underrepresented and vulnerable groups; assumes a pluralistic society and adversarial relationships; envisioned as a means of promoting democratic pluralism
	Practitioner an advocate; advocacy a bridge between political and professional
	Could result in multiple EIAs, shared data but separate interpretation documents, single EIA but client not proponent, or single EIAs but separate peer reviews for each interest group
	Decisions might have to be made by an arbitrator or by means of a quasijudicial procedure
	Many variations (e.g., multiple advocacy: encompasses all major perspectives, ideological advocacy: advocate represents a perspective rather than a client, environmental advocacy: focuses on the development of a consciousness of a shared destiny)

(Continued)

Table 7.2 (*Continued*)

Equity and progressive planning	Government-sponsored progressive planning; information a source of power
	Seeks to redistribute power, resources, and participation away from elites and toward poor and racial and ethnic minorities
	Planning based on substantive, political redistribution goals; aimed at reducing negative social conditions caused by disparities; aims to create justice, fairness, and equity
	Seeks to constrain and modify dominant power sources in process; promotes collaborative governance
	Faces challenge of growing ethnic and racial diversity; seeks coexistent viability of ethnic and racial groups
	Emphasis on citizen participation, power sharing, decentralization, and social self-determination
Political–economic mobilization	Structural (class-based) analysis of inequities in the distribution of resources and power
	Practitioners said to legitimize the existing distribution of power and resources and to perpetuate inattention to incompatibility of democratic political processes with capitalist political economies
	Alternative role for practitioners to identify how inequities created and to define means of facilitating social control
	Equity considerations substituted for efficiency; requires state backing; linked to politics of empowerment, redistribution, and community
	Mobilization is all forms of political action that fall short of revolution; informed by political–economic structural (e.g., class) analysis
Social and environmental movements	Emergent action groups that seek social and environmental transformation; depends on successfully mobilizing social collectivities; relies heavily on the concept of power in numbers
	Source of political energy; a major role in encouraging citizen participation and in keeping issues on political agenda
	Underpinned by perceptions of common purpose and shared grievances; action-oriented
	May opt out of process and resort to protests, civil disobedience, demonstrations, boycotts, etc.
	Can assume an effective role in monitoring decisions and in acting as an ethical watchdog
Social and environmental justice	Refers to the fair treatment and meaningful involvement of all people, regardless of race, color, national origin, or income concerning the development, implementation, and enforcement of laws
	Minority, low-income, and indigenous people should not bear a disproportionate share of negative impacts of government actions
	Analysis determines whether and to what extent injustices occur; may require measures to prevent, reduce, and offset
	Requires a demographic analysis, an impact assessment and community involvement
	Can pertain to both procedural and substantive injustices; draws on theories of justice

Table 7.2   (*Continued*)

	Seeks equal opportunity, affirmative action, and cultural inclusion; looks to examples of democratic citizenship in resisting injustices and in making the boundaries of community life more inclusive; inspiration from citizen activism
Critical social science and theory	Communications seen as political; process reflects systematic patterning of communications—influences community organization, citizen participation, and autonomous citizen action
	Critique of distribution of power in society and how reflected in practice; research is action-oriented; informs practical action; more complex than class—ecological, alienation, and interrelationships crises
	Practice as communicative action; importance of communications in the search for consensus
	Practice should seek to correct communications distortions and to institute communications-enabling rules (e.g., speak comprehensively, sincerely, legitimately, the truth); ideal speech situation
	Seeks to reveal underlying sources of social relations and to empower people, especially the less powerful
Deep ecology and ecological politics	Asserts that humans are but one species among many; no less valuable in own right than humans
	Biocentrism and self-realization at core; identification beyond humanity; favors noninterference with nonhuman parts of biosphere; concerned with quality of human life and health
	Committed to a decentralized and democratic ideology; generally allied with left (only way to give full expression to ecological demands); long-term perspective; driven by a sense of urgency
	Stresses harmony with nature, intrinsic worth of nature, simple material needs, global perspective, and natural limits
	Sometimes a political party; more often direct political action (e.g., protests, boycotts); linked to environmental movements; long-term goal of reestablishment of human society on a sustainable basis
	Some success in transforming traditional decision-making systems, in altering agenda, and in stimulating debate; expanded society's conscious; generally faces major obstacles
	Environmentalists: many differences but share core values; many political approaches; usually campaigns involve a coalition of groups; political action based on belief that necessary to sway public opinion to move politicians
	Environmental lobby groups: large multiissue groups, smaller, more focused groups, education, research and policy development centers, law and science groups and real estate conservation groups
Political SIA	Community conflict the focal point of practice; EIAs engender conflict and are undertaken where there is conflict
	Major social process is one of conflict over resources; interests are the basic elements of social life
	Social order based on manipulation, and control by dominant groups; seeks a more humane society

(*Continued*)

Table 7.2   (*Continued*)

	Central role of conflict analysis and conflict management; stresses openness, participation, empowerment, community development; tendency to be holistic and subjective
	Consistent with critical theory; based on understanding of processes and structures of change
Feminism and eco-feminism	Assumes that women are exploited, oppressed, and devalued by society
	Interest in changing the conditions of women's lives; seeks extension of women's rights
	Importance of gender analysis in impact assessment (different roles, knowledge, and values)
	Eco-feminism: addresses links between androcentrism and environmental destruction; based on similar attitudes
	Seeks ethics-based environmental planning (grounded in responsibility and carrying reciprocity)
	Requires an interactive, face-to-face, democratic decision-making process; needed to counteract the problem of internalized patriarchical structures and values
Institutional advisory groups	Formed on a permanent or ad hoc basis; offers guidance on a range of issues and activities, including policy making
	Legitimacy depends on degree represents and is supported by interest community; since appointed no mandate from interest community; legitimacy determined by source and quality of advice
	Roles: means of testing public reaction, forum for expressing public opinion, places controversial issues into objective opinion arena, involves expert critics in decision process
	Role in EIA includes panels for reviewing EIA documents and providing advice to government (e.g., Canada)
	Can bring valuable insight and criticism into decision-making process; offers analysis and alternative viewpoints at modest expense; in recent years broadened to extend representation (e.g., gender, language, region, sector)
	Dangers: used to placate opposition, co-opt opponents, offer symbolic response to problem, delay action, persuade opponents and provide publicity and patronage instrument
	Increasingly environmental nongovernmental organizations have developed capability of offering equally high caliber advice; sometimes with financial assistance
Participant and intervener funding	Can be provided for participation in review and approvals (e.g., mediation, panel reviews)
	Sometimes provided for participation in earlier stages of EIA process
	Usually based on criteria
	Applied to such costs as peer review, legal services, administration, coordination, and travel

Sources: Bass (1998), Benveniste (1989), Birkeland (1995), Burdge (1994), Checkoway (1994), Craig (1990), Davidoff (1965, 1978), Day (1997), Devall (1985), Fainstein and Fainstein (1985), Filyk and Côté (1992), Forester (1989), Friedmann (1987), Gorz (1980), Greer-Wooten (1997), Halstead et al. (1984), Healey (1997), Hoch (1993), Maynes (1989), Mertzger (1996), Morrone (1992), Newman (1997), Pickvance (1985), Rawls (1971, 2001), Ritzdorf (1996), Rothblatt (1978), Smith (1993), Wilson (1992).

movements. They can undertake and accommodate critiques of power imbalances. These critiques could draw, for example, upon political–economic mobilization, critical social science and theory, deep ecology and eco-feminist theories and analyses. They can be open to perspectives and critiques (e.g., deep ecology, eco-feminism) that fall outside the mainstream of EIA practice.

EIA practice, consistent with political SIA, can focus on the analysis and management of community conflict. The EIA decision-making process can be broadened to establish, support, and give greater weight to the analyses and suggestions of institutional advisory bodies and nongovernmental organizations. Intervener and participant funding can be provided so that unrepresented and underrepresented segments of society can participate more fully in the EIA process. EIA capacity building efforts can identify and ameliorate power imbalances. The net result of these power-corrective actions should be an EIA process consistent with direct and participatory democratic principles. The EIA process and outcomes also are more likely to be fair, just, and equitable (see Chapter 9).

7.5   INSTITUTING A DEMOCRATIC EIA PROCESS

7.5.1   Management at the Regulatory Level

Democratic EIA requirements, as illustrated, in Figure 7.5, should seek to bring government closer to the people. Senior governments should, wherever practical, decentralize and deconcentrate their EIA operations. Senior-level EIA requirements should be delegated to intermediate-level governments wherever appropriate and practical. They should be harmonized where delegation is inappropriate. Intermediate- and local-level EIA systems should be encouraged and supported. Intermediate governments should explore, with local governments, delegation potential and harmonization opportunities. Close and frequent consultation should be maintained with governments closer to the people. Senior levels should justify when delegation is not practical and centralized control is essential. Harmonization and coordination with lower government levels should not be a matter of senior-level discretion.

All governments (the United States, Canada, the European Union, and Australia) actively, but to varying degrees, promote and support the early and ongoing involvement and influence of interested and affected parties and communities. Support and encouragement includes, for example, public participation requirements, accommodating traditional knowledge, EIA capacity building, environmental justice requirements, the use of institutional advisory groups, the provision of intervener and participant funding, and the support of nongovernmental organizations.

Decentralization and Delegation   The National Environmental Policy Act (NEPA) and regulations refer to notifying, involving, and making documents available to states, local agencies, Indian tribes, and the public. State and local agencies (with similar qualifications) can be cooperating agencies. Federal agencies are to

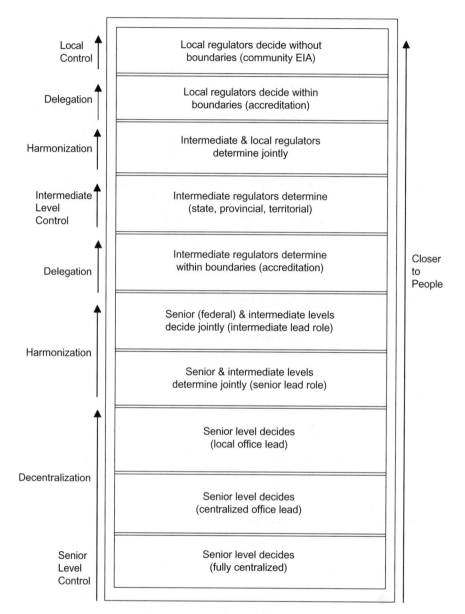

Figure 7.5 Ladder of regulatory control.

cooperate with state and local agencies to reduce duplication. Provisions are made for joint planning processes, joint environmental research and studies, joint documents, and joint public hearings. Federal and state agencies, participating in joint processes, can be joint lead agencies. NEPA authority can be delegated to heads and subsidiary heads of field organizations. States agencies, in certain limited situations involving federal grants to state agencies, can prepare EISs for jointly funded projects. The state agencies must have statewide jurisdiction. Federal agency guidance and independent evaluation are required (Bass and Herson, 1993). Actions by nonfederal parties on federal land or requiring federal funding, regulatory approval, or permits can require NEPA approval. Federal agencies vary in how they apply NEPA requirements to such projects (Kreske, 1996). The NEPA task force established in May 2002 is reviewing intergovernmental collaboration, including joint lead processes. Several states have modeled their EIA requirements after NEPA, although thresholds and procedures vary. The U.S. approach offers a potential for selective decentralization and delegation. Coordination and harmonization are clearly priorities. Delegation and decentralization do not appear to be objectives. They also are not addressed consistently.

In Canada a major effort has been made to harmonize and coordinate federal and provincial EIA requirements. There are both multilateral and bilateral EIA agreements. Coordination is addressed in regulations and guidelines, through various consultation bodies and through regional CEAA (Canadian Environmental Assessment Agency) offices. Amendments to the Canadian Environmental Assessment Act refer to promoting cooperation and coordination between federal and provincial governments. Generally, when both federal and provincial requirements are triggered there is a single process, a single hearing, a single set of documents, and a coordinated review process. It is proposed that a federal environmental assessment coordinator be established for each project.

The results of applying these measures have been mixed. Some provinces prefer either an accreditation system or a system where one level assumes the lead role when both federal and provincial EIA requirements are triggered. Recent proposed amendments to the act stimulate that it does not apply in the Yukon, in the Northwest Territories (NWT), and to Indian bands under the Indian Act. Triparty (federal, territorial, aboriginal) EIA regimes are begin established in the Yukon and in the NWT. EIA substitution agreements are being developed with aboriginal land claims authorities. EIA regimes also are being established under the First Nations Land Management Act. The federal government in Canada, although clearly committed to enhanced EIA coordination, appears ambivalent regarding decentralization and delegation. The variations in approach between the provinces and territories are not wholly explicable in terms of different circumstances.

The European approach creates a framework and addresses broad principles and minimum requirements through the EIA/SEA directives. Member states determine the details. EIA requirements are generally integrated into existing project consent procedures. Legislative disparities are partially addressed by approximating laws. Ample notification and coordination provisions are included. The European Commission provides extensive EIA guidance material and applied research. Considerable discretion remains with member states to adapt EIA requirements to local circumstances. Over time this discretion could be reduced (or not) as the directives are amended and refined to, for example, comply with such European Commission conventions as the Aarhus Convention on Access to Information, Participation in Decision-making and Access to Justice in Environmental Matters. Courts could assume an important role. EIA guidelines could become the "standard" of adequate practice. A delicate balance will be required between ensuring a consistent, minimum standard of good practice across the member states and the need to ensure that each EIA system reflects local needs and priorities. Further moves toward centralization, on the grounds of consistency, should be approached with caution, fully substantiated, and tested for effectiveness and consequences. Implications for reduced local autonomy should receive particular attention.

The Australian EIA approach reflects a clear commitment to delegation. The current commonwealth EIA requirements have been preceded by a long history of intergovernmental cooperation and agreements. The objectives of the Environmental Protection and Biodiversity Conservation Act 1999 refer to strengthening intergovernmental cooperation and minimizing duplication. The commonwealth government clearly defines its environmental interest (i.e., designated matters of national environmental significance, environmentally significant commonwealth projects, environmentally significant projects on commonwealth land). Regulatory or permitting requirements are not EIA triggers. Under the act, state and territorial EIA requirements can be accredited through bilateral agreements. States and territories must be consulted before matters of national environmental significance are declared. A state or territory may refer a proposal to the minister. The minister has to consult with the state or territory regarding the assessment approach to be applied. States and territories also can add approval requirements and conditions. The Australian EIA approach addresses delegation directly and consistently. It clearly defines the national environmental interest and includes specific accreditation procedures. The legislation is new. Its effectiveness, in keeping EIA close to the people, will depend on the contents and application of bilateral agreements.

Local Autonomy   U.S. federal EIA requirements and guidelines make ample provision for involving Indian tribes and local agencies in federal EIA processes. Local agencies (with similar qualifications) can be cooperating agencies. Indian tribes can be cooperating agencies when effects occur on reserves. Local agencies and Indian tribes can participate in joint planning processes, documents, hearings, research, and studies. Federal agencies are to cooperate in fulfilling state and local EIA requirements supplementary to but not in conflict with those of NEPA. Agencies are to identify inconsistencies between a proposed action and approved state, local, or tribal plans or laws. Agencies also are to describe the extent to which they would reconcile inconsistencies. A violation of a local environmental law is a factor in determining impact intensity. The federal government consults with tribal governments on a nation-to-nation basis. Interdisciplinary place-based decision-making approaches that meet community needs and minimize adverse environmental effects could be conducive to community autonomy (US CEQ, 1997a). Some states require that local governments conduct EIA. Numerous municipalities and some tribal governments have their own EIA requirements. Much federal NEPA guidance is relevant to state- and local-level EIA practice. NEPA requirements and guidelines do not directly foster local autonomy. However, place-based decision making and local government participation and plan consistency provisions could help maintain local autonomy. The ample NEPA-related resource materials and training also can be beneficial to community-level EIA.

Canadian federal EIA requirements address local autonomy primarily in terms of aboriginal communities. Aboriginal EIA regimes can be established in accordance with land claims agreements and under the First Nations Land Management Act. The federal government is helping to establish these regimes and is preparing prototype EIA guidelines for aboriginal people. Aboriginal people are a partner in the EIA regimes being created in the territories. Co-management and decentralized decision making are a tradition in the north and in some provinces. EIA requirements and guidelines in the territories emphasize benefits to northerners and consistency with the priorities and values of local residents and communities. Changes to federal EIA requirements refer to considering community knowledge and aboriginal traditional knowledge when preparing EIA documents. The increased emphasis on regional studies is conducive to addressing community-level cumulative environmental effects. Several municipalities integrate EIA requirements (often related to the protection of sensitive areas) and municipal planning. Federal EIA guidelines and applied research are partially applicable at the local level. Consistency with community priorities and plans could receive more attention. The maintenance of local autonomy also could be a policy objective.

European Union EIA requirements are generally directed toward implementation at the member state level. Some European states are implementing EIA through existing procedures, including those related to environmental and land-use planning. Communities assume a major role in land-use planning. The SEA directive is applied largely to town and country plans and programs. This may result in more consistency in addressing environmental concerns in planning documents. But more involvement by state environmental authorities in land-use planning could also contribute to greater centralization. Alternatively, the marriage of planning and EIA requirements could open up public decision making and ensure greater consideration of community autonomy concerns. Local authorities in many European states have EIA requirements and have prepared EIA guideline documents. Some countries (e.g., Norway and the United Kingdom) provide EIA training for municipal planners and local participants and provide guidance and advice to municipal authorities seeking to integrate EIA into land-use planning. The numerous EIA centers scattered across Europe could provide insights of value to municipal governments in establishing and maintaining EIA systems. The planned independent International Commission for Impact Assessment could also provide nonbinding advice on the adequacy of information for decision making. Assistance for establishing the commission is being provided by the Netherlands and by the European Commission.

The Australian EIA legislation refers to involving governments, the community, landowners, and indigenous peoples in a cooperative approach to environmental protection and conservation. It notes the role of Indigenous people in conservation and in the sustainable use of Australia's biodiversity. It promotes the use of indigenous peoples' knowledge of biodiversity and the involvement and cooperation of the owners of the knowledge. Significance to indigenous tradition is a significance criterion for species and ecological communities. The *Administrative Guidelines on Significance* provide criteria for determining the significance of matters of national environmental significance. The *Guidelines and Criteria for Determining the Need for and Level of Environmental Impact Assessment in Australia*, prepared by the Australian and New Zealand Environment and Conservation Council (ANZECC), refer to vulnerable human communities, impacts on community values, land use, and resource use, and effects on the amenity, values, and lifestyle of the community (ANZECC, 1996). The Australian requirements and guidelines (consistent with the ANZECC guidelines) could devote more attention to community autonomy, especially concerning the significance of impacts of commonwealth government projects and projects on commonwealth lands or waters.

Public Influence and Control   Federal U.S. EIA requirements and guidelines make a general commitment to open decision making and public involvement. There are specific notification and document availability requirements. Detailed scoping requirements and guidelines are provided. Public meetings are optional. They are a requirement with agencies such as the Department of Energy. Guidelines have been prepared to facilitate effective public involvement. Judicial review of regulatory compliance for final environmental impact statements can occur. Concerned citizens and public interest groups play an important role in the EIA process and through the courts in enforcing NEPA requirements (Bass and Herson, 1993). Some groups and individuals who participated in the US CEQ effectiveness review of NEPA believe that public involvement is largely a "one-way track" (US CEQ, 1997a). They also feel that they are treated as adversaries. The CEQ report calls for earlier public involvement, a broader interpretation of public involvement, and a true partnership with the community. The U.S. EIA requirements and guidelines provide opportunities for public involvement. More proactive efforts could be made to facilitate enhanced public influence in EIA decision making.

The purpose of the Canadian EIA legislation refers to ensuring an opportunity for public participation in the EIA process. There are notification and document availability requirements. Public review, in the form of mediation or a review panel, is an option. EIA guidelines stress the value of public involvement programs, the importance of providing information to the public, the need to consider public concerns, the potential role of interested parties in scoping, and the desirability of building consensus. Public involvement has been limited for screenings but has been more extensive for comprehensive studies. Responsible authorities generally believe that they are doing a good job of involving the public. Community and environmental groups are often dissatisfied with their EIA public participation experiences. They have expressed concerns with the time provided for review and the extent to which their concerns are reflected in final reports. The five-year review of the Canadian Environmental Assessment Act acknowledges the many concerns expressed regarding opportunities for meaningful participation in screenings and comprehensive studies. Legislative changes call for a greater emphasis on public participation in screenings and comprehensive studies and a formal opportunity for public involvement during scoping. Greater use of mediation and dispute resolution is also promoted. Canadian EIA requirements and guidelines are more oriented toward public information and periodic involvement than to public influence and shared decision making. Some provinces and territories are further down the path toward co-management, decentralized decision making, and community-based consultation.

European EIA requirements provide member states with considerable discretion regarding public notification and involvement. General references are made to transparent decision making, consulting with the public, sufficient time frames, and access to information. The Aarhus Convention on Access to Information, Public Participation in Decision-making and Access to Environmental Justice (1998) (in force October 2001) guarantees the rights of access to information, public participation in decision making, and access to environmental justice. It also includes provisions pertaining to specific activities and to plans, programs, and policies which require that the public be informed early in an environmental decision-making procedure and in an adequate, timely, and effective manner. The Aarhus Convention is being refined and implemented with directives (2003/4/EC) (including possible amendments to the EIA and SEA Directives), an implementation guide and handbooks (Bond and Stewart, 2002; DETR, 2000; ECE, 2000; EU, 2003; Stec, 2003; Teller and Bond, 2002). These measures should further reinforce the opportunities for enhanced public involvement and influence.

European Commission EIA guidelines, especially regarding scoping, provide advice concerning the types of organizations to consult, consultation methods, and good practice consultation procedures. Public consultation requirements vary considerably among the member states. They vary from an opportunity to provide written comments to a requirement for public meetings or hearings. Public consultation procedures have apparently improved in recent years. Most states do more than the minimum. A greater effort could be made to enhance public influence in EIA decision making.

The Australian EIA requirements provide for public notification, public access to assessment documents, and opportunities for public comments. Reference is made to describing the public consultation that occurs, identifying parties affected, and including a statement mentioning any communities that may be affected and describing their views. The ANZECC guidelines identify degree of public interest as a criterion for determining impact significance (ANZECC, 1996). The Australian requirements provide a framework for informing and involving the public. More specific requirements and guidance could be provided to encourage greater public influence and to provide shared EIA decision-making opportunities.

Correcting Imbalances   The United States has instituted extensive environmental justice EIA requirements and guidelines. They extend from individual environmental justice agency strategies. They pertain to potential impacts on minority populations, low-income populations, and Native Americans. The requirements and guidelines address data collection and analysis procedures, impact determination, access to information and effective participation. Reference is made to overcoming cultural, linguistic, institutional, and geographic barriers. Note is made of the U.S. government's responsibilities and treaty obligations to Native Americans. Examples of enhanced outreach methods and procedures are cited.

Numerous basic and advanced NEPA courses are available. Access to the courts has helped ameliorate power imbalances. Inputs from advisory bodies, such as the Citizens' Advisory Committee on Environmental Quality, can broaden the public decision-making basis. The United States has taken significant steps to address power imbalances, primarily by means of environmental justice requirements. Power imbalances also pertain to other segments of the population and many nongovernmental organizations. The public is usually at a disadvantage in being able to participate fully in the EIA process, as compared with government and industry. The United States could take additional steps to address these imbalances.

The Canadian EIA requirements provide for independent advice from an environmental assessment panel or from a mediator. Intervener funding is currently available for participation in panel reviews and in mediation proceedings. Participant funding is proposed for public involvement in reviewing draft EIA guidelines, preparing for and attending scoping meetings, reviewing the EIA, and preparing for and participating in public hearings. Specific criteria govern funding disbursement. The Canadian Environmental Assessment Agency provides numerous EIA training opportunities and sponsors applied EIA research. An aboriginal advisory committee is proposed. The agency is seeking to more effectively incorporate traditional knowledge into the EIA process. In both Canada and the United States it is now possible for nongovernmental organizations to appeal to the Commission for Environmental Cooperation under the North American Agreement on Environmental Cooperation (a subagreement under NAFTA) for failure to enforce environmental laws. The Canadian EIA requirements and guidelines could more explicitly address environmental injustices and related power inequities.

The European EIA requirements do not address power inequities directly. EIA guidelines refer to considering the sensitivity of people. The Aarhus Convention on Access to Information, Public Participation in Decision-making and Access to Environmental Justice (1998)(in force October 2001) guarantees access to justice on environmental matters. Access to an independent and impartial review body is available if questions or concerns are refused or addressed inadequately. Power imbalances could be ameliorated by advice and evaluations from EIA centers and from the planned independent International Commission for Impact Assessment. European EIA requirements and guidelines could go further in identifying and ameliorating power inequities in the EIA process.

The Australian EIA requirements seek to involve indigenous people in biodiversity conservation and enhancement. Reference is made to promoting the use of indigenous peoples' knowledge cooperatively with the owners of the knowledge. The Australian legislation establishes a Threatened Species Scientific Committee, a Biological Diversity Advisory Committee, and an Indigenous Advisory Committee. An independent review of the act is to be undertaken within 10 years of its commencement. One factor considered when deciding whether or not to approve an action is a person's environmental history—a concern frequently raised by nongovernmental organizations. The Australian Commonwealth Government has prepared an EIA training resource manual for developing countries. The ANZECC significance guidelines suggest considering whether the proposal will result in inequities between sectors of the community. The Australian EIA requirements and guidelines could more directly identify and ameliorate power inequities in the EIA process.

The Propensity to Centralize A tendency to centralize EIA requirements sometimes emerges in multilevel EIA systems. Centralization is justified on the grounds that (1) it facilitates consistency among EIA systems, (2) upper-level EIA practitioners have more expertise, and (3) senior-level EIA practitioners are more objective and less susceptible to the pressures of development interests. Although there is some validity to these arguments, they are often overstated. Consistency can be an overrated. The ideal attributes of an EIA regulatory system have yet to be determined. The senior-level characteristics are not necessarily "the model." Diversity can be advantageous. It provides an opportunity to assess the effectiveness of a variety of regulatory approaches. Also, lower-level EIA systems need to be designed to suit local environmental characteristics and priorities.

Often, EIA practitioners, at the intermediate and local levels, have comparable qualifications. They generally have greater local knowledge and experience. Where needed, they can seek (and generally do) advice from senior-level specialists. The argument that intermediate- and local-level EIA practitioners are more susceptible to political pressures is unfair to environmental professionals operating at those levels. If the intent is to make the EIA system more democratic, the result is a more overtly political EIA process. A democratic EIA process means more influence and control by those most directly affected by proposed actions. Also required are measures to ensure that all interests are represented at the table, power imbalances are ameliorated, and community accountability and influence are enhanced.

The downside of centralization is reduced accountability to and control by the people. If EIA requirements and procedures are to be more democratic, centralizing propensities should be resisted unless it can be demonstrated that they are essential. The need for centralization should be fully and publicly justified. Centralization should be kept to a minimum. The value of community knowledge and experience should be emphasized rather than discounted. Any reallocation of EIA responsibilities among government levels should be undertaken with the involvement and consent of those affected most directly.

7.5.2 Management at the Applied Level

Figure 7.6 illustrates an example of a democratic EIA process. The figure and description that follow depict an EIA process that fosters local autonomy and maximizes public influence and control, especially by individuals, organizations, and communities most directly and severely affected by proposed actions. The democratic EIA process also is concerned with identifying and ameliorating power imbalances.

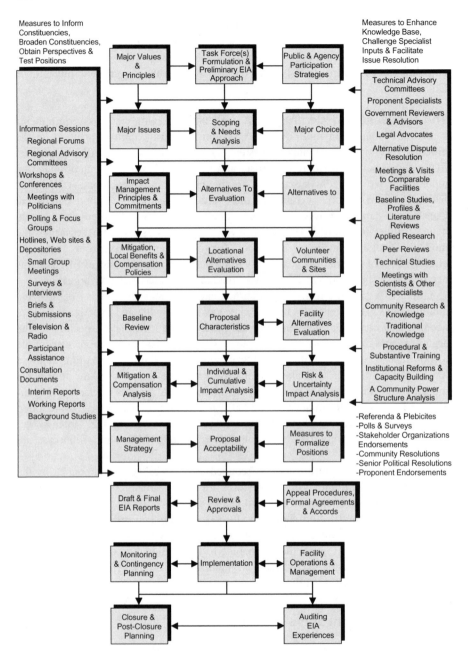

Figure 7.6 Example of a democratic EIA process.

The process is built around the concept of delegated decision making. It is equally well suited to a co-management or shared decision-making approach. The process is an expression and fulfillment of public perspectives. It seeks to build public confidence, trust, and acceptance. Equal and independent parties, with well-defined rights and responsibilities, come together voluntarily to make mutually beneficial decisions. Participants can opt in or out of the process at any time. EIA practitioners act as facilitators and collaborators.

Startup One or more public task forces guide and manage the EIA process. The process begins with meetings or workshops among individuals, groups, and organizations interested in or concerned about a problem and/or a proposed action. These initial meetings identify major issues and perspectives. They also help determine participant expectations. Participants jointly identify potential candidates who are willing and able to be task force members. The members represent collectively a cross section of values, perspectives, and interests relevant to the problem or proposed action. The parties may agree to including respected and credible independent individuals and, where pertinent, proponent representatives. Proponents, if included, are in a minority position. Two or more task forces are established if there are major opinion differences regarding task force membership. These task forces operate in parallel. They meet periodically through the process to reach, where possible, common positions.

The major principles, to guide the process and structure the task force operations, are identified. Measures (e.g., participant assistance) are instituted to ameliorate power inequities and to ensure the full and fair participation of all parties. General ground rules for operating the task force are established. Major process decisions are identified. Measures, analyses, and procedures to provide a decision-making basis are determined. Procedures for involving and testing the support of the public at large and for involving politicians and government officials are established. These decisions are integrated into an overall EIA approach. The approach also addresses resource requirements and schedule. The approach is refined progressively through the EIA process. The problem, need, or opportunity to be addressed is explored thoroughly, both within the committee and with potentially affected communities. Key issues and potentially relevant choices are identified. The EIA process is scoped to focus on major issues, choices, perspectives, stakeholders, and sensitive and significant environmental components.

Pre-approval Decisions Like most EIA processes, the EIA process leading up to review and approvals is built around a series of decisions. Unlike most EIA processes, the public (operating through the task force) assumes the lead role. Specialists, politicians, and government officials are informed, consulted, and involved, as needed. The public controls the process, inputs to the process, and outputs from the process. the task force guides the analyses and undertakes the interpretations and evaluations associated with each decision. A proposed action may be on the table from the outset. Alternatively, there may simply be a problem (or opportunity) and a general sense of ways for solving the problem or taking advantage of the opportunity. In either case the major available and reasonable choices are identified, screened, and compared by the task force. The task force seeks a consensus regarding a preferred course of action. General impact management, including compensation and local benefits, principles, and commitments are formulated.

A voluntary siting approach is applied if location is a choice. A broadly acceptable (e.g., appropriate physical, social, and economic conditions) region or area is identified. Mitigation, local benefits, and compensation policies are formulated. Voluntary communities and, in turn, sites are solicited within the acceptable region and area. Communities volunteer based on municipal council resolutions. They can withdraw from the process at any point. Environmental and social justice considerations are integrated into the analysis. The task force is modified to add members from the voluntary communities. Extensive consultation is undertaken within the voluntary communities. Local advisory committees are established in each area. Referenda are undertaken in each voluntary community to determine the level of public support. The preferred community is selected, taking into account the degree of public support, environmental suitability, social equity, and economic constraints. Further analyses are undertaken within the volunteer community to identify potentially environmentally suitable areas and sites. The site(s) selected also are voluntary (e.g., public land or private property from a willing vendor) within the acceptable areas. Environmental suitability and the degree of support from neighbors surrounding the site and along access routes to the site are considered when making the final determination.

An analysis of baseline environmental conditions at and surrounding the selected site is undertaken. The community helps identify sensitive and significant environmental components and processes. The proposal characteristics, applied in the locational analysis, are refined. Practical design, operations, closure, and post-closure alternatives are identified. The task force screens and compares the alternatives. Community preferences and concerns are central when selecting the preferred alternatives. Consideration also is given to environmental effect, uncertainty, and technical and economic differences.

The task force, supported by technical analyses and ongoing community consultation, identifies, predicts, and interprets individual and cumulative environmental effects. Ways of preventing and offsetting negative effects and of enhancing benefits are determined. The impact significance interpretations take into account mitigation potential. Compensation and local benefits policies and measures are refined, based on proposal characteristics, local environmental conditions, and local preferences. Calculated and perceived risks and uncertainties and their potential implications are explored. Uncertainties are reduced by supplementary analyses, where practical. Community preferences regarding risk and uncertainty acceptability and management play a key role. Individual impact management measures are consolidated into an overall impact management strategy. The impact management strategy includes monitoring, contingency measures, compensation, local benefits, post-approval consultation, and preferred closure and post-closure options.

Periodic workshops and meetings are held if there is more than one task force. The task forces compare analyses and attempt to come to common positions. Where practical, sensitivity analyses explore differences. Conciliation, facilitation, and mediation help identify and expedite discussions. An arbitrator, an independent commission, or a review panel may be used when a consensus position cannot be achieved. Referenda are used to determine community positions on the acceptability of the proposed action (or actions, if task forces come to more than one conclusion). Polls or surveys are undertaken in areas directly affected. Resolutions, endorsing or opposing the proposed action, are sought from elected representatives (at each relevant government level) and from each nongovernmental organization involved in the process.

Knowledge Base   The task force draws upon a diverse knowledge base: baseline studies, literature reviews, applied research, and reviews of comparable facilities. It determines its own research and advice requirements. It has the final say regarding terms of reference and in selecting specialist advisors. Consistent with advocacy theory, it supports analyses from multiple perspectives. Analyses are tested by peer review. Legal advice is provided where appropriate. Support analyses are structured around the knowledge requirements associated with each decision in the EIA process. The support analyses include an assessment of community power structure implications.

Heavy reliance is placed on community involvement in baseline analyses, on community knowledge, and where pertinent, on traditional knowledge. The community actively participates in data collection, analysis, and interpretation, both at the site and through visits to comparable facilities. Technical training, institutional improvements, and capacity building measures are instituted where needed to facilitate community understanding, involvement, and control.

Technical advisory and government agency review committees are constituted where needed. These committees provide advice regarding the technical soundness of analyses and the likelihood of regulatory compliance. Training in effective group methods is provided to task force members, as needed. A concerted effort is made to minimize communications misinformation, distortions, and barriers.

Validating the Decisions   In a democratic EIA process the activities of the task force are "legitimate" only when they reflect the perspectives, interests, and preferences of the overall public. The process links task force activities to broader constituencies. Close contact is maintained with elected officials at all pertinent government levels. Ongoing political participation and support is seen as essential to community understanding and acceptance.

The participation activities extend well beyond the conventional meetings, open houses, and other measures commonly employed in EIA to inform and to obtain feedback from the public. Techniques such as small group meetings, workshops, forums, and conferences provide a more interactive, personal, and collaborative approach to addressing process-related issues and choices. Polling, surveys, interviews, and focus groups are employed to obtain more in-depth and structured feedback. Advisory committees provide continuity of involvement. The extent of involvement is broadened through hotlines, Web sites, television, and radio.

Referenda are used at key decision points (e.g., whether a community should volunteer, proposal acceptability) to obtain a comprehensive community response. Inputs from groups, organizations, and segments of society traditionally underrepresented in the EIA process are actively solicited. Participant assistance is provided to facilitate involvement.

Numerous consultation (adjusted to the needs of varying constituencies) documents are prepared and circulated broadly. Interim, background, and working reports are prepared to provide a sound basis for each decision. The public is provided with ample opportunities to comment on draft and final documents. Several community document repositories are established to ensure that documents are widely and readily available. Documents also are available at a project Web site. Briefs and submissions from individuals and nongovernmental organizations are actively solicited. Responses are provided to all comments and suggestions received.

Approvals and Post Approvals The proposed action proceeds to review and approvals only when there is clear community acceptance and preferably, support. An appeal is available to an independent review body for participants dissatisfied with the process or its outcomes. Analyses and consultations are detailed in draft and final EIA documents. The ongoing involvement of government agencies minimizes the likelihood of unforeseen regulatory concerns. The public, through the task force and an array of consultation procedures, is a full participant in determining approval conditions. If the proposed action is approved, commitments to the community and to neighbors are formalized first in draft accords and then in final agreements. The agreements address such matters as measures to reduce need, local benefits, facility operations, monitoring, contingency measures, compensation and closure, and post-closure planning and management.

The agreements provide the foundation for continued public influence and control right through implementation. A neighborhood committee is established to co-manage the facility. Community representatives have a major say in facility operations and in monitoring and contingency planning. They also contribute to closure and post-closure planning and management. The experiences associated with the process are documented and made available to others wishing to apply a similarly democratic EIA approach.

7.6 ASSESSING PROCESS EFFECTIVENESS

Table 3.5 (in Chapter 3) presents criteria and indicators for assessing the positive and negative tendencies of the EIA processes described in Chapters 3 to 10. In this section we summarize examples of positive and negative tendencies of democratic EIA processes.

Scientific *rigor* is rarely a high priority with democratic EIA processes. The incorporation of scientific analyses is highly discretionary. It is selectively and not always consistently applied. Technical and scientific specialists are sometimes viewed with suspicion. More emphasis is placed on group consensus than on explicit and substantiated assumptions, findings, and interpretations. Conclusions and recommendations are usually well supported. Scientific contribution is generally a low priority.

The public commonly favors a *comprehensive* definition of the environment. More attention is devoted to local environmental knowledge and less to specialist knowledge. Problems and opportunities are broadly defined. A holistic perspective is common. Interrelationships and cumulative effects are not always assessed consistently and thoroughly.

Democratic EIA processes are *systematic* in some areas and less so in others. Decision making is open. But more emphasis is placed on procedural principles and outcomes (e.g., consensus) than on ensuring explicit and traceable decision making. The process direction is clear, but objectives evolve. Methods are not always identified or applied systematically. The process identifies and evaluates major choices. It is less suited to identifying and assessing complex and technical suboptions. Major impacts and management measures are usually assessed thoroughly. Less well-known and subtle impacts often receive less attention. This can be problematic if potentially significant impacts are not widely known or understood.

Democratic EIA processes are *practical.* They seek public acceptance—often a prerequisite to project implementation. They focus on major issues and trade-offs, as defined by the public. They generally ensure clear and succinct documents. However, democratic EIA processes also rest on several fragile assumptions. They presume that the public is willing and able to manage a complex process. They expect that diverse perspectives and interests can be accommodated and reconciled sufficiently to advance the process through each decision. They anticipate that there will be environmentally suitable volunteer communities and sites. The process can break down if one or more of these assumptions turn out to be unrealistic. In addition, the government occupies a peripheral position. Regulatory compliance is not always a priority. The process is often open-ended, which can stretch out schedules and consume additional resources.

Democratic EIA processes are highly conducive to maintaining and enhancing stakeholder *democratic* influence. There is a high degree of public control. Community and traditional knowledge are readily accommodated. The process is explicitly political. Informed public acceptance is more likely than with technical approaches. Community acceptance is less likely if the process breaks down, if no options are considered acceptable, if there are fundamental value conflicts, or if equity rather than community control concerns predominate.

Democratic EIA processes are inherently *collaborative*. They facilitate stakeholder understanding and involvement. They are conducive to dialogue, consensus building, and conflict resolution. Participants jointly define roles and responsibilities. The process suffers if the knowledge, insights, and perspectives of other parties (e.g., proponents, industry, regulators, specialists) are not fully appreciated and considered.

Fairness tends to be a major public concern with democratic EIA processes. Measures to facilitate procedural fairness are generally integrated into the process.

Substantive fairness is commonly a public priority. It is sometimes difficult to maximize both voluntarism and outcome equity. Democratic EIA processes are general guided by broad *ethical* principles. More precise ethical imperatives, standards, and decision rules are not usually formulated. Multiple ethical perspectives can be accommodated. Rights and responsibilities are commonly recognized.

Democratic EIA processes are designed and adapted to suit the social and cultural context. Group processes are often *flexible* and creative. An open approach to defining the problem, seeking out opportunities, and identifying possible solutions is common. Scanning ahead and reconsidering past decisions does not generally occur. Risks and uncertainties are usually a major public concern. Perceptions may dominate. Calculated risks are sometimes discounted.

Democratic EIA processes are well suited to *integrating* diverse values, forms of knowledge, perspectives, and ideals. The holistic perspective, associated with democratic EIA processes, can facilitate bridging and transcending disciplinary, professional, and EIA barriers. Links to proposal planning, related decisions, and to related environmental management forms are not generally priorities.

7.7 SUMMING UP

In this chapter we describe a democratic EIA process, a process that shifts power from specialists and politicians to the public. The people most directly affected by proposed actions have a major say in whether and how actions proceed. Groups, segments of society, and perspectives, commonly excluded or underrepresented, assume a more prominent position in the EIA process. Table 7.3 is a checklist for formulating, applying, and assessing a democratic EIA process.

The three stories provide contrasting pictures of the role of democratic values in an EIA process. In the first example, a stakeholder group went outside a narrowly defined SIA process to obtain the knowledge and insights required to enhance their influence within the SIA process. In the second example, an EIA process collapsed, largely because of a failure to provide even a moderate level of public participation and influence, especially in early EIA process decisions. In the third story, local influence was enhanced but based on disinformation and a nonrepresentative role for the public. The stories provide an initial sense of the potential benefits associated with a greater level of public influence and the dangers when the public has minimal influence in the EIA process or when the process is distorted or compromised.

Although the value of public participation in widely acknowledged, the public too often has a minor role in the EIA process. Members of the public frequently are or believe themselves to be powerless in major decisions that affect their lives. This problem is exacerbated by imbalances in the distribution of power. The most vulnerable segments of society tend to be the least influential. The solution is an EIA process that delegates or shares decision-making authority with the public.

Democracy is rule by the people. The EIA process should be an expression and fulfillment of direct and participatory democratic concepts and principles. The

Table 7.3   Checklist: A Democratic EIA Process

The response to each question can be yes, partially, no, or uncertain. If yes, the adequacy of the process taken should be considered. If partially, the adequacy of the process and the implications of the areas not covered should be considered. If no, the implications of not addressing the issue raised by the question should be considered. If uncertain, the reasons why it is uncertain and any associated implications should be considered.

DEFINITIONS AND DISTINCTIONS

1. Is the EIA process consistent with direct democratic principles?
2. Is the EIA process consistent with participatory democratic principles?
3. Do politicians, government officials, and the courts involve the public, share responsibility with the public, and are they responsive to the public in the EIA process?

THE PEOPLE DECIDE

1. Is the EIA process designed to facilitate a high degree of direct public influence and control?
2. Do the local politicians assume a prominent role in the EIA process?
3. Does the public have a major say in each decision in the EIA process?
4. Is a clear rationale provided for how the public is defined?
5. Is a clear rationale provided for how the public is to participate in each decision?
6. Is a clear rationale provided for what represents a choice by the public?
7. Is a clear rationale provided for any bounding of the public's role in decision making?
8. Does the process provide opportunity(s) for public control?
9. Does the process provide opportunity(s) for shared decision making?
10. Does the process provide an opportunity for voluntary siting?
11. Does the process draw on the lessons of resource co-management and of voluntary siting experiences?

THE PEOPLE AND COMMUNITIES AFFECTED BY THE PROPOSAL DECIDE

1. Is the process based on visions, principles, and strategies for more autonomous communities?
2. Does the process facilitate community empowerment, development, and mobilization?
3. Does the process promote and accommodate community and traditional knowledge?
4. Are the implications for community power analyzed?
5. Are community EIA capacity building and institutional reforms promoted and facilitated?
6. Is a concerted effort made to shift, to the extent practical, the roles and responsibilities traditionally assigned to technical specialists, politicians, and government officials to the public?
7. Is the public at the center of rather than peripheral to the EIA process?
8. Is the process guided by community values and aspirations?
9. Are proposed actions evaluated as catalysts for and against the satisfaction of community needs and aspirations?
10. Does the process actively seek to involve community activists and nongovernmental organizations?

(Continued)

Table 7.3   (*Continued*)

CORRECTING POWER IMBALANCES

1. Is provision made for the preparation of alternative EIA documents, interpretations, and analyses?
2. Are social and environmental justice concerns integrated into EIA requirements, processes, and documents?
3. Are social and environmental equity and progressive planning objectives promoted?
4. Are decentralization and de-concentration advocated and supported?
5. Are community advocates, activists, and organizers encouraged and supported?
6. Are social and ecological political activities and movements encouraged and supported?
7. Are power imbalances critiqued, drawing upon pertinent political, economic, social, ecological, and feminist theories and analyses?
8. Is the process open to perspectives and critiques that fall outside the mainstream of EIA practice?
9. Does the process focus on the analysis and management of community conflict?
10. Are the analyses and suggestions of institutional advisory bodies and nongovernmental bodies fully integrated into the process?
11. Are participant and intervener funding provided to underrepresented and unrepresented groups and segments of society so that they can participate fully in the process?
12. Do EIA capacity building efforts devote greater attention to identifying and ameliorating power imbalances?

MANAGEMENT AT THE REGULATORY LEVEL

1. Do senior governments, wherever practical, decentralize and deconcentrate their EIA operations?
2. Are senior-level EIA responsibilities delegated to intermediate-level governments wherever appropriate and practical?
3. When delegation is not possible or appropriate are EIA systems harmonized?
4. Does the senior level clearly demonstrate when delegation is not practical and when centralized control is essential?
5. Do senior levels promote and support the autonomy of local EIA systems?
6. Do EIA systems promote and support early and ongoing public participation and influence?
7. Do EIA systems provide opportunities for shared and delegated decision making?
8. Do EIA systems ameliorate power imbalances?
9. Do EIA systems resist the propensity to centralize?

MANAGEMENT AT THE APPLIED LEVEL

1. Is the EIA process guided and managed by one or more public task forces?
2. Are EIA practitioners process facilitators and collaborators?
3. Does the EIA process incorporate such startup activities as:
   a. Public task force formulation?
   b. Major values and principle identification?
   c. Public and agency participation strategy formulation?

Table 7.3 (*Continued*)

---

d. Preliminary EIA approach formulation (including process design, procedures for establishing the knowledge base, and procedures for validating the decisions)?

e. Scoping and needs analysis (incorporating the identification of major issues and choices)?

4. Does the public control the process, inputs to the process and outputs from the process?

5. Does the EIA process incorporate such preapproval decisions as:

   a. Impact management principles and commitments determination?
   b. Broad analysis of alternatives to the proposal?
   c. Voluntary siting approach?
   d. Formulation of mitigation, local benefits, and compensation policies (extending from community concerns and preferences)?
   e. Baseline review (with community and traditional knowledge prominently featured)?
   f. Proposal characteristics formulation?
   g. Community evaluation of facility design, operations, closure, and post-closure alternatives?
   h. Individual and cumulative impact analyses (prominent role for community and traditional knowledge)?
   i. Mitigation and compensation analyses (community concerns and preferences stressed)?
   j. Risk and uncertainty analyses (community concerns, perceptions and preferences stressed)?
   k. Impact management strategy formulation?
   l. Application of measures to come to a common position if more than one public task force?
   m. Application of measures to formalize community and stakeholder positions on proposal acceptability?

6. Does the public, operating, for example, through task forces, determine its own research and advice requirements?

7. Does the EIA process establish and make effective of a diverse knowledge base by:

   a. Making effective use of baseline studies, literature reviews, applied research, and reviews of comparable facilities?
   b. Supporting analyses from multiple perspectives?
   c. Applying peer review strategically?
   d. Making effective use of proponent and specialist analyses and knowledge?
   e. Drawing effectively on the skills of specialists in alternative dispute resolution?
   f. Working closely with government officials with regulatory responsibilities?
   g. Drawing upon legal advice, where appropriate?
   h. Assessing community power structure implications?
   i. Making effective use of community and traditional knowledge?
   j. Fostering community participation in data collection, analysis, and interpretation?
   k. Instituting technical and procedural training, institutional improvements, and capacity-building measures to facilitate community understanding, involvement, and control?

8. Do task force actions and decisions reflect the perspectives, interests, and preferences of the overall public by:

   a. Maintaining close and ongoing contact with politicians?
   b. Making effective use of;
      (1) Involvement methods such as meetings and open houses?

(*Continued*)

Table 7.3   (*Continued*)

    (2) Collaborative techniques such as small-group meetings, workshops, forums, and conferences?

    (3) Techniques, such as polling, surveys, interviews, and focus groups, to obtain more detailed and structured feedback?

    (4) Continuous involvement methods, such as advisory committees?

    (5) Procedures to broaden involvement such as hotlines, Web sites, television, and radio?

    (6) Methods, such as referenda, to obtain a comprehensive community response, at key decision points?

    (7) Methods, such as intervener and participant funding, to involve underrepresented groups and segments of society?

  c. Ensuring that interim, background, and working reports provide a sound decision-making basis and:

    (1) Are broadly circulated and are readily available (e.g., community depositories, Web site)?

    (2) Are adapted to the needs of varying publics?

    (3) Offer sufficient time for public comments and suggestions?

    (4) Incorporate and respond to public suggestions, advice, briefs, and submissions?

9. Does the EIA process only proceed to review and approvals when there is clear community acceptance and preferably support?

10. Does the EIA process incorporate such approval and post-approval activities as:

  a. Preparation of draft and final EIA documents (with ample public and agency involvement)?

  b. Appeal procedure to an independent review body?

  c. Public involvement in determining conditions, if the proposed action is approved?

  d. Preparation of accords and agreements to formalize commitments to neighborhood and community?

  e. Co-management of facility?

  f. Community influence in closure and post-closure planning?

  g. Documentation of community experiences with process?

PROCESS EFFECTIVENESS

1. Does the process balance rigor and public control, while ensuring an adequate level of applied scientific performance?

2. Does the process balance specialist and community knowledge, maintain a holistic perspective and thoroughly and consistently assess individual and cumulative effects?

3. Does the process systematically and explicitly address major choices, impacts, and trade-offs?

4. Does the process concentrate on major issues, community acceptance, and clear documentation to facilitate implementation?

5. Is the process highly conducive to maintaining and enhancing community influence and control?

6. Does the process facilitate public understanding, involvement, and collaboration?

7. Is the process procedurally fair, and does it effectively balance substantive fairness with community influence and control?

Table 7.3 (*Continued*)

8. Is the process suited to the community and environmental context, and does it adapt flexibly to changing conditions?
9. Does the process effectively balance predicted and perceived risks and uncertainties?
10. Does the process accommodate and link a diversity of values, forms of knowledge, perspectives, and ideals?
11. Is the process linked effectively to proposal planning, related decisions, and related environmental management forms?

courts, politicians, and government officials should involve, delegate power to, share power with, and be responsive to the public.

The EIA process should be designed to facilitate a high degree of direct public influence and control. The public, working closely with local politicians, should assume the lead role in shaping and guiding the process. There should be an influential public role for each decision within the process. There should be a clear rationale for how the public is defined, what represents a public choice, and any bounding of the public's role. The EIA process should draw on the principles, insights, and experiences of community resource co-management and voluntary siting approaches.

The EIA process should help make people and communities more autonomous and better able to make decisions about matters that affect their lives. Lessons should be derived from visions, principles, and strategies for more autonomous communities. The EIA process should facilitate community empowerment, development, and mobilization. It should promote and accommodate community and traditional knowledge. It should assess community power structure implications. It should facilitate community EIA capacity building and institutional reform. Roles conventionally assumed by politicians, government officials, and technical specialists should be shifted to the public, wherever practical. The public should be at the center of the process. Community activists and nongovernmental organizations should be featured prominently. The EIA process should be guided by community values. Proposed actions should be assessed as catalysts for or against the realization of community needs and aspirations.

The EIA process should be designed and managed to minimize power imbalances. This may necessitate preparing alternative EIA documents, interpretations, and analyses. Social and environmental justice concerns should be integrated into the process. Imbalances should be ameliorated by decentralizing and de-concentrating power. Environmental equity and progressive planning concepts and principles can be instructive. The EIA process should encourage and support community advocates, activists, and organizers. Social and ecological political activities and movements also should be encouraged and supported. Identifying and characterizing power imbalances should be aided by drawing upon critiques from pertinent political, social, ecological, and feminist theories and analyses. The process should be opened up to perspectives and critiques outside mainstream EIA practice. The analysis and management of community conflict should receive particular attention. The EIA process should draw on the analyses and suggestions of institutional advisory bodies and nongovernmental organizations. Financial and other assistance should be provided to underrepresented and unrepresented segments of society so that they can more effectively participate in the process. EIA capacity-building efforts should devote greater attention to identifying and rectifying power imbalances.

Senior governments should, wherever practical, decentralize and de-concentrate their EIA operations. EIA responsibilities should be delegated to intermediate government levels where appropriate and practical. A clear rationale as to why central control is essential should be provided whenever delegation does not occur. The autonomy of local EIA systems should be promoted and supported. EIA systems should promote and encourage early and ongoing public participation and influence. Opportunities for delegated or shared decision making with the public should be provided where practical and appropriate. EIA systems should ameliorate power imbalances. They also should resist the propensity to centralize.

A democratic EIA process seeks to maximize public influence and control, foster local autonomy, and correct power imbalances. Decision-making authority is delegated to or shared with the public. The process is designed and managed to build public confidence trust and acceptance.

One or more public task forces guide and manage the example democratic EIA process. EIA practitioners act as facilitators and collaborators. The process begins with the task force formulation, the identification of major values and principles, the formulation of public and agency participation strategies, and the development of a preliminary EIA approach. Major issues and choices are identified. The process is scoped. The need for action is determined. The balance of the process, leading up to review and approvals, is built around decisions. The task force, drawing on an extensive knowledge base and maintaining close contact with the broader public, makes each decision. Major available alternatives to the proposed action are identified, screened, and compared. Impact management principles and commitments are determined. Proposed facilities are located using a voluntary siting approach. Proposal characteristics are formulated. Community and traditional knowledge are featured prominently in the baseline analyses. Community concerns and preferences provide the basis for mitigation, local benefits, and compensation policies and measures. Facility design, operations, closure, and postclosure options are evaluated. Individual and cumulative impact analyses, taking into account mitigation potential, compensation measures, risks, and uncertainties, are undertaken. An overall impact management strategy is formulated. Differences among task forces are reconciled, where practical. Measures such as referenda, council resolutions, and organizational endorsements are used to gauge public and stakeholder acceptability of proposed actions.

The public task force draws on an extensive knowledge base. Reference is made to technical studies, baseline studies, community profiles, literature reviews, applied research, peer reviews, and visits to comparable facilities. Community and traditional knowledge is supported and accommodated. Community power structure implications are explored. The task force works closely with technical specialists, government officials, legal advocates, and procedural specialists. It receives technical and procedural training, as needed. Institutional reforms and capacity building occur where needed to support the process. The task force makes an ongoing effort to reflect the perspectives, interests, and preferences of the overall public. Close contact is maintained with politicians. A variety of methods are applied to communicate with, involve, and collaborate with the public. Interim, background, and working documents provide a sound decision-making basis. The documents are readily available, are adapted to the needs of different publics, and fully integrate public concerns and preferences. The EIA process only proceeds to review and approval when there is clear community acceptance and preferably support. There is ample public and agency involvement in preparing draft and final EIA documents. An appeal procedure to an independent review body is available. The public participates in determining approval conditions. Commitments to individuals and communities are formalized in accords and agreements. Proposed facilities are co-managed. The public has a major influence in closure and post-closure planning. Community experiences with the process are documented.

Scientific rigor is not commonly a priority with democratic EIA processes. It is addressed selectively. Democratic EIA processes generally employ a broad definition of the environment and adopt a holistic perspective. Local environmental knowledge may be favored over specialist knowledge. Analyses tend to focus on major issues and choices. They are not always systematic, especially in addressing subtle differences, complex alternatives, and potential impacts not widely known or understood. The focus on public acceptance can facilitate implementation. Democratic processes can break down in dealing with complex issues or where there is major opposition or divisions within the community. The process can be time-consuming and costly. Democratic EIA processes are highly conducive to maintaining and enhancing stakeholder influence. They also are inherently collaborative. The process suffers if all relevant perspectives and interests are not integrated. Democratic EIA processes generally stress procedural and substantive fairness. Substantive fairness is not always consistent with public control. The process is often flexible and creative. It does not always scan ahead or reconsider past decisions. Perceived risks and uncertainties tend to be favored over predicted risks and uncertainties. The process integrates diverse values, forms of knowledge, and perspectives. It often bridges and transcends disciplinary, professional, and EIA barriers. Links to proposal planning, related decisions, and other environmental management forms are not usually priorities.

CHAPTER 8

HOW TO MAKE EIAs MORE COLLABORATIVE

8.1 HIGHLIGHTS

In Chapters 7 and 8 we pose two choices concerning how EIAs might become more democratic. The first choice, delegating or sharing decision-making authority with the public, was addressed in Chapter 7. This chapter is concerned with the second choice, taking steps to ensure that the public is an active and ongoing participant in a collaborative EIA process. Final decision-making authority, however, continues to reside with the proponent and with the regulators.

Collaboration is defined broadly to encompass all forms of public participation short of delegation or shared decision making. Collaboration implies a joint endeavor of the public and of other stakeholders. Therefore, forms of public participation, not fully collaborative, are included in the analysis but as prerequisites to or subsets of a collaborative EIA process. Nonparticipation, either warranted or not warranted, is not considered. Also not addressed, except in the sense of dangers to guard against, is illegitimate participation (e.g., deliberately incomprehensible, insincere, untruthful) (Forester, 1989). The differences between collaborative and democratic EIA processes are largely a question of degree. Both processes seek to enhance the role and influence of the public in decision making.

- The analysis begins in Section 8.2 with three applied anecdotes. The stories describe applied experiences associated with efforts to make EIA practice more collaborative.

Environmental Impact Assessment: Practical Solutions to Recurrent Problems, By David P. Lawrence
ISBN 0-471-45722-1   Copyright © 2003 John Wiley & Sons, Inc.

- The analysis in Section 8.3. then defines the problem, which is the gulf between the potential benefits of collaborative EIA processes and the more modest benefits achieved by public participation approaches commonly evident in EIA requirements and practices. The direction is exploring the potential for and means of making EIA processes more collaborative.
- In Section 8.4 we first consider the possibility that valid and significant disadvantages and constraints largely preclude a collaborative EIA approach to public participation. This explanation, although partially valid, is found wanting. The second possibility is that there is an extensive foundation of sound analysis and good practice, which could provide the basis for collaborative EIA processes. But the relevant source materials are immense, of varying quality, and scattered across numerous related fields. What is required is a succinct presentation and analysis of the major building blocks of a collaborative EIA process. Major distinctions drawn in the analysis include (1) principles and practices, (2) consultation, (3) communications, (4) mutual education, (5) negotiations, and (6) collaboration.
- In Section 8.5 we apply the insights, distinctions, and lessons identified in Section 8.4 and describe the properties of a collaborative EIA process at both the regulatory and applied levels. In Section 8.5.1 we explore how EIA requirements could facilitate collaborative EIA practice and in Section 8.5.2 demonstrate how a collaborative EIA process could be expressed at the applied level.
- In Section 8.6 we evaluate how well the collaborative EIA process presented in Section 8.5 satisfies ideal EIA process characteristics.
- In Section 8.7 we highlight the major insights and lessons derived from the analysis. A summary checklist is provided.

8.2 INSIGHTS FROM PRACTICE

8.2.1 Public Participation in Strategic Environmental Assessment

The third revision of the zoning plan of the municipality of Weiz was subject to a voluntary strategic environmental assessment (SEA), in accordance with the European Union's SEA Directive Proposal of December 1996 [COM (96) 511 final]. Weiz has a population of approximately 9200 and is located in the northern part of Austria's southeastern province Styria. The approach undertaken for preparing the assessment was to integrate the SEA provisions, including participation of stakeholders, into the procedure for revising the zoning plan for the time period 1999–2004, in accordance with the Styrian Spatial Planning Act. The SEA identified, analyzed, described, and assessed the environmental effects of the no-action alternative and of two plan alternatives.

Unfortunately, two separate public hearings were held for the SEA: one for the drafted zoning plan and the other for its environmental report. The environmental report describes the environmental and socioeconomic impacts of the no-action alternative and of the two plan alternatives. The evaluation method presented in the environmental report involved the application of a weighting procedure. It also integrated public and agency suggestions and incorporated mitigation measures. A total of 150 visitors attended the first hearing. Most of the visitors were inhabitants of Weiz. Only a few visitors attended the latter meeting, even though a short and easily understand explanation paper had been prepared and published in the local official newspaper. The explanation paper was distributed, free of charge, to every household and included a simplified map of the drafted plan and a nontechnical summary of the environmental report. All citizens and many actors (e.g., various departments of the provincial government, neighboring municipalities, chamber of commerce, chamber of labor) had the opportunity to comment on both documents (the drafted plan and the environmental report) over an eight-week period.

Public involvement was probably constrained because an environmental impact assessment (EIA) of a plan is a new instrument and to some members of the public is somewhat abstract. Consideration should be given to the adaptations needed to foster public consultation for planning-level documents. It may be desirable to begin early in the process with some form of public education concerning the nature of planning documents, the role of SEA in plan evaluation, and the importance of and opportunities for public involvement in the plan making and evaluation process. Having two public hearings also was clearly problematic. Separating planning from evaluation, both in the sense of the planning process and in terms of public involvement opportunities, is clearly artificial and confusing to the public.

For future SEAs it seems more appropriate to organize only one public hearing for both the drafted plan and for its environmental report. A larger issue, which may need consideration, is the extent to which planning and SEA documents, and their associated public consultation procedures, should be separated, partially integrated, or fully integrated. Regardless of the approach adopted, public involvement methods more suited to a broad level of public participation in the preparation and review of general planning documents will be needed. Particular consideration should be given to how best to facilitate a high level of early, ongoing, and meaningful public involvement in the planning/SEA process. It may also be helpful if a responsible and qualified individual or team coordinates the public consultation procedures and creates and maintains a professional and transparent communication/involvement structure.

RALF ASCHEMANN
Austrian Institute for the Development of Environmental Assessment

8.2.2 Effective Listening and Adapting to Individual Needs and Perspectives

The public is understandably uneasy when the prospect is raised of establishing an extrahigh-voltage 500-kilovolt electrical transmission lines in close proximity to their homes and their community. Reaching consensus on the preferred location for a 100-kilometer line through southwestern Ontario's rich farmland provided a uniquely successful demonstration of the value of collaborative decision making within an EIA process. In the mid-1980s, the former Ontario Hydro required additional high-voltage transmission lines to obtain power out of its Bruce Nuclear Power Development and to provide stability for the bulk transmission system in the southwestern part of the province. Ontario Hydro, predominantly an engineering organization, was always supportive of social science research related to transmission-line routing and accepting of innovative approaches to community relations and public participation.

Over a seven-year period that included system planning and later, transmission route planning, the community relations staff implemented a public information-feedback program and joint planning initiatives with key stakeholders. Public information centers, multistakeholder working groups, and ongoing liaison with key community stakeholders succeeded in obtaining the understanding and support necessary to proceed with the construction of the transmission lines through numerous municipalities from Bruce County to London and east to the Hamilton area.

Respect for several important public consultation principles was key to a positive outcome. As the EIA process allowed for the selection of any route alignment within a broad series of corridors, the public participation program stressed listening and obtaining feedback as opposed to persuading or convincing the public of the merits of a particular route. Early and ongoing involvement was essential. Staff spent time understanding the situation of individual farmers and their respective community organizations. Changes were made to the public consultation program to accommodate their need to work on the farm in the spring, summer, and fall. Many working group meetings were held during the winter months after harvest and prior to spring planting. Key decisions were transparent and open to public influence. Frequent and appropriate communications with municipal politicians and staff helped assure local politicians that their voters were being treated fairly and respectfully at all times. Good ideas, arising from public suggestions, were considered and where appropriate, implemented. Finally, Ontario Hydro realized that the process of public acceptance would take time. By the end of the formal public participation process, it was apparent that there were few issues that could not be addressed.

This story demonstrates that effective collaborative involves a blending of communications, mutual education, and consultation mechanisms that enable joint problem solving among many competing interests. It underscores the need for patience, transparency, an open mind, and effective listening. It reinforces the importance of beginning from the concerns, needs, and perspectives of each individual, group, and organization interested in and potentially affected by a proposed activity and of skillfully adapting and applying a diversity of public participation methods.

DAVE ABBOTT
Ontario Power Generation

DAVE HARDY
Hardy Stevenson Associates

8.2.3 Building Credibility, Trust, and Consensus

A major steel company that uses auto bodies as part of its raw materials required a landfill to handle what is referred to as *shedder fluff*: the fiberglass, plastic, glass, and foam parts of an automobile. This fluff is considered nonhazardous, solid industrial waste appropriate for landfill disposal. The company considered possible locations for a landfill. It concluded that a landfill on its own property was preferable. The facility abuts Lake Ontario. The area bordering the lake is used extensively for fishing. There also are walking trails along part the lakeshore. The proposed landfill would occupy about 40 acres, would have an operating life of approximately 20 years and would entail the erection of a 50-foot-high three-sided berm extending up to 1 mile in length between the lakeshore and the current operating facility.

The EA review parties included the company, the planning departments of the Durham regional government and the town of Whitby, the Thickson Point residents (about 30 households adjacent and to the east of the facility along the lakefront), the corridor area ratepayers (approximately 3000 households in the area), and the Ministry of Environment. Some nearby residents were initially opposed to the proposed facility. The company was naturally in favor of the proposal. The town, the region, and the Ministry of Environment did not take a position either in favor or against the proposal.

After some initial public consultation events at the beginning of the EA work, the company, on the advice of the regulators (Ministry of Environment) and its public consultation consultants, decided to establish a public liaison committee (PLC) to attempt to resolve outstanding issues. This decision followed from considerable internal discussion within the company because it meant full and open disclosure and collaboration. Ultimately, the company decided that the risk was worth taking.

A retired deputy police chief was selected as independent chair. The chair, assisted by a social impact/public consultation firm, guided the overall process and ensured open information flows and the addressing of all issues. The town, the region, and the ministry attended as observers. The PLC meet monthly for the first year to identify contentious issues and to form an agenda for resolving issues. As time proceeded, it became clear that this process was having some success. The process continued for about four years (through 24 formal meetings). During that time issues were addressed and solutions acceptable to all parties identified. Ultimately, the stakeholders all agreed to the design, location, and structure of the proposed shedder berm as well as to the measures to protect and preserve the walking trails and lake shoreline.

The EA report, submitted with the support of the PLC, indicated that all significant outstanding issues had been addressed successfully. The Minister of Environment approved the EA and praised the company, the stakeholders, and the open, inclusive, and collaborative approach for identifying and resolving issues. The expense and potential delays from a hearing process were avoided because the EA was submitted without any organized or stakeholder opposition. The PLC continues to meet regularly to monitor implementation and facility operations and to deal with issues as they arise. As of the spring 2002, 38 PLC meetings held been held.

This case example demonstrates the potential procedural and substantive rewards of a collaborative EIA process. It illustrates that mutually beneficial solutions require on the part of all parties, patience, flexibility, and a willingness to invest the time and energy to make the process work for the benefit of all. Collaborative EIA processes also tend to be more effective when supported by independent and credible advice and assistance.

PETER HOMENUCK
IER Planning, Research and Management Services

8.3 DEFINING THE PROBLEM AND DECIDING ON A DIRECTION

The preceding stories illustrate some of the complexities and subtleties associated with collaborative EIA processes. The first story demonstrates that public participation, at the SEA level, requires consultation methods adapted to the characteristics of strategic planning processes. Particular attention needs to be devoted to the extent to which strategic planning and SEA and associated consultation activities can and should be merged. The second story illustrates that complex EIA processes require multiple involvement methods and opportunities, considerable patience, open minds and an open planning process, and a concerted effort to understand, adapt to, and integrate local needs and perspectives (i.e., effective listening). The third story demonstrates that each party must accept the risks associated with a process that requires considerable time and effort and will involve solutions likely to depart markedly from initial positions. Credibility, trust, and consensus are built incrementally and iteratively and can be facilitated by independent and credible advice and assistance.

Numerous benefits, as highlighted in Figure 8.1, have been ascribed to effective public participation in public and private decision making. Public participation is intrinsically beneficial to participants and to society (Day, 1997; Nagel, 1987; Pateman, 1970). It is consistent with human nature. It is ethically just to involve the public in decisions that could affect their lives (Praxis, 1988).

Effective public participation has considerable developmental value (Nagel, 1987). It can facilitate a greater level of interest and involvement in public life (Morgan, 1998; Pateman, 1970). It can enhance the confidence of and political skills of participants (Day, 1997; Dunning, 1999). Citizens can more ably articulate their preferences and demands (Day, 1997). The application of these skills can empower people, further community identity and development, foster environmental sensitivity, and contribute to a more democratic and responsive political system (Barrow, 1997; Day, 1997).

Public goals can be advanced by effective public participation. The public can have more direct access to decision making prior to final determinations (Hyman et al., 1988; Shepherd and Bowler, 1997). Public values, perspectives, and preferences are incorporated into decision making (Morgan, 1998). Unrepresented people are able to present their views. The public is able to examine expert knowledge and

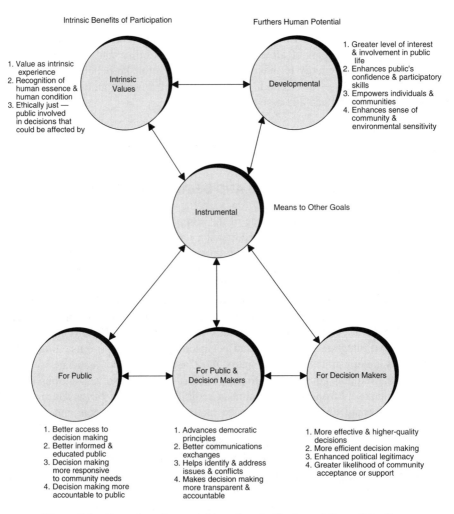

Figure 8.1   Examples of potential benefits of effective public participation.

to weigh and test the positions and decisions of elected representatives (Pateman, 1970). Public alienation and marginalization is less likely to occur. The public has a better understanding of environmental conditions, project characteristics, local issues, and potential impacts (Lee, 2000). Decision making is more likely to reflect and be responsive to stakeholder and community needs (Shepherd and Bowler, 1997; US EPA, 1998b). Decision making is fairer and more accountable to the public (Barrow, 1997). A check on government and private action is provided. There is a greater potential for individuals, groups, and communities to use EIA and other decision-making tools to help solve their own problems and to better influence their own futures (SERM, undated).

Better decisions and more effective and efficient decision making can result from effective public participation (Howell et al., 1987; Lee, 2000). A means is provided for obtaining local and traditional knowledge and for determining local issues, perspectives, and values (Bisset, 1996; Glasson et al., 1999; Morgan, 1998). The public can help diagnosis problems, formulate and evaluate alternatives, identify, predict, and integrate impacts, interpret impact significance, determine appropriate mitigation, compensation, and monitoring measures, and decide on proposal acceptability (Barrow, 1997; Greer-Wooten, 1997; Hughes, 1998; US EPA, 1998b). Public contributions to decision making can make it easier to establish priorities. Management expertise can be enhanced (Praxis, 1988). Decision makers are better able to plan for and adapt to change (Day, 1997; Lee, 2000). Costs and delays associated with public opposition are less likely (Glasson et al., 1999). Project management can focus on key public issues (Praxis, 1988). Approvals and implementation tend to be less complex and confrontational (SERM, undated). Decision makers are viewed as more credible and the decision-making process is perceived as more legitimate (Barrow, 1997; Creighton et al., 1983; Smith, 1993). There is likely to be less hostility and a greater level of community trust, acceptance, and sometimes support for both the process and the decisions resulting from the process (Hyman et al., 1988; Lee, 2000; Shepherd and Bowler, 1997).

The public and decision makers can benefit jointly from a decision-making process based on popular sovereignty and political equality principles. All parties benefit when misunderstandings are clarified and when information and knowledge is shared effectively (Hughes, 1998). Effective public participation provides a means of identifying and a forum for resolving issues (Bisset, 1996; Hughes, 1998). It offers a mechanism for building consensus and for avoiding and resolving conflicts (Greer-Wooten, 1997; Praxis, 1988). It can contribute to more open, transparent, and democratic planning and decision making (CCMS, 1995; Shepherd and Bowler, 1997). Narrow technical biases can be ameliorated. A countervailing force is established to offset administrative and political power concentrations. Broad public involvement and support also can facilitate sustainability initiatives (Barrow, 1997; SERM, undated).

It would be difficult to realize most of these ascribed benefits in an EIA process characterized by late public involvement and/or through periodic public involvement events intended largely to inform the public. Instead, early public involvement and a more continuous and interactive EIA process seem more in order. Effective two-way communications and mutual education seem essential. Mechanisms to anticipate, avoid, and resolve disputes, to build consensus, to collaboratively plan, to solve problems, and to take advantage of opportunities all appear necessary. As illustrated in Figure 8.2, these elements of effective public participation need to be guided by general principles, goals, and good practices, structured by integrative frameworks and bounded by limits of acceptable practice.

It is far from clear if or how well these elements of effective EIA public participation are being or are likely to be satisfied. There are both positive and negative patterns and trends. There also is considerable variability in the quality of EIA practice. EIA requirements and practices have significantly increased public information

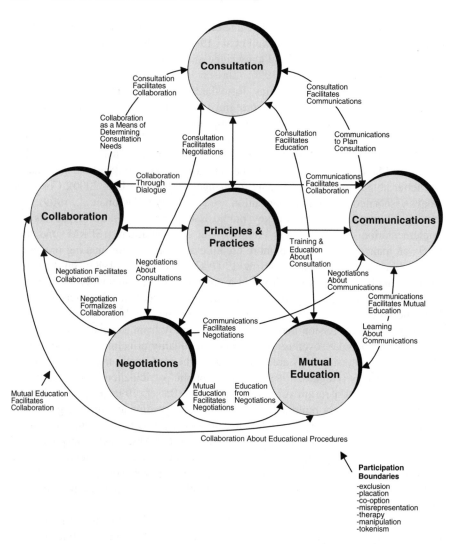

Figure 8.2 Examples of effective public EIA participation.

and input into agency decision making (US CEQ, 1997a). Earlier and more continuous public consultation is being emphasized to a greater extent. There are numerous examples of sincere, creative, and effective approaches for involving the public, for resolving disputes, and for collaboratively solving problems (Carpenter, 1991; Creighton et al., 1983, 1999; Desario and Langton, 1987; Gray, 1989; Susskind and Cruikshank, 1987; Susskind et al., 1999). A concerted effort has been made to provide detailed guidance and to identify and interpret the lessons of public involvement, alternative dispute resolution, and collaborative planning practice (Creighton et al., 1983, 1999; CSA, 1996; PCSD, 1997; Praxis, 1988; US DOE, 1994, 1998; US EPA, 2001a,b). In recent years there has been a greater emphasis on two-way communications, on transparent and accountable decision making, on outreach to traditionally underrepresented groups and nongovernmental organizations (NGOs), on facilitating procedural and distributive justice, and on community empowerment. These trends, analyses, and guidance materials, although pertaining to many forms of public and private environmental decision making, also are largely applicable to EIA practice.

Concurrent with these positive trends and developments, there has been a tendency for public agencies to opt for forms of EIA (e.g., mitigated FONSIs and environmental assessments in the United States and screenings in Canada) that preclude or severely restrict public involvement (Shepherd and Bowler, 1997; Soloman et al., 1997). The trends toward deregulation, the application of business principles and concepts to public administration, and privatization could further inhibit public involvement in EIA practice (Bisset, 1996; Sinclair and Diduck, 2001). The treatment of public participation in EIA guidelines is highly variable. Too frequently, guidelines are confusing, lacking in practical guidance, and weak in proactively advocating public involvement (Hughes, 1998). The public role in EIA practice is often poorly defined (Harrop and Nixon, 1999). Much of the time, public involvement begins after major decisions have been made and occurs at only two or three key decision points in the EIA process (Freudenberg, 1983; Shepherd and Bowler, 1997; Soloman et al., 1997). Often, public consultation is limited to disseminating information and gathering public comments, frequently in poorly structured processes. The range of public consultation methods employed in practice remains narrow and there is a tendency to overemphasize quantitative methods and biological and physical impacts (Solomon et al., 1997). Public participation rarely extends into the post-approval stage (Harrop and Nixon, 1999).

The gulf in perspectives between proponents and the public regarding the need for and role of public participation is still considerable (Fell and Sadler, 1999). The levels of public distrust and cynicism remain high concerning the motives of decision makers and the weight they attach to public comments and suggestions (Mittelstaedt et al., 1997). Citizens sometimes are frustrated and feel that they are being treated as adversaries rather than as welcome participants (US CEQ, 1997a). In some cases the role of citizens is substantial and influential. But in others it is largely symbolic. There is a particular need for reforms to enhance the involvement and influence of indigenous peoples and governments (Mittelstaedt et al., 1997; Paci et al., 2002; US CEQ, 1997a).

The messages from EIA effectiveness reviews are equally mixed. Effectiveness reviews clearly demonstrate the benefits of public participation for improved project design, environmental soundness, and social acceptability (Hughes, 1998). Effective public participation increases the likelihood of approval and reduces the potential for costly delays during approvals and implementation (Bisset, 2000). Overall, the benefits of public participation generally exceed the costs (Glasson et al., 1999). Nevertheless, agencies such as the World Bank acknowledge that they have not yet achieved their public consultation goals (World Bank, 1997).

The effectiveness surveys, undertaken as part of the *International Study of the Effectiveness of Environmental Assessment* (Sadler, 1996), point to considerable room for improvement in the institutional frameworks for public scrutiny and participation, in the performance of public participation activities, and in addressing public consultation in procedural and technical guidelines. The role of the public is largely limited to scoping and to the review of EIA documents (Sadler, 1996). Many do not see assessment as a learning process that results in significant changes. Only about one-fourth of respondents indicated that EIA had resulted in a significant improvement in the acceptance of public involvement and input (Sadler, 1996).

Despite the many benefits of effective public participation and the array of instructive analyses and guidance materials, the reality of public participation still falls well short of the promise (Lawrence et al., 1997). This discrepancy between potential and performance raises several questions. (1) Are there major disadvantages, which largely offset the benefits of effective public participation? (2) Are there unresolved and perhaps irresolvable issues, which generally preclude an enhanced level of public participation in EIA practice? (3) Is the problem simply one of applying insights and lessons from the available literature and guidance materials? (4) Are the relevant prescriptive materials too scattered and in need of consolidation and succinct presentation? (5) Is it necessary to integrate additional concepts and frameworks? In Section 8.4 we answer each of these questions. We also seek to maximize the benefits of effective public participation (as highlighted in Figure 8.1) and to address and integrate the elements of effective EIA public participation (as displayed in Figure 8.2). The Section 8.4 analyses provide the foundation for the collaborative EIA process presented in Section 8.5.

8.4   SELECTING THE MOST APPROPRIATE ROUTE

8.4.1   Disadvantages and Issues

Public participation is a generic term for all types of activities designed to include the public in the decision-making process, prior to and after a decision. The role of the public is direct and acknowledged (Lee, 2000). Members of the public influence or attempt to influence decision-making outcomes (Nagel, 1987). Most parties generally accept that public participation in EIA practice is desirable. However, qualifications and exceptions are rapidly identified. There also are varying definitions of what represents public participation (e.g., involvement or influence) and when it is most appropriate for it to occur. "When" concerns both the circumstances under which public participation is appropriate and when in the EIA process. Table 8.1 lists examples of reasons commonly advanced for why public participation is undesirable or why it should be severely limited. Comments regarding the validity of the reasons also are provided. Public participation may be viewed as undesirable by proponents and regulators or by members of the public, albeit for different reasons.

Table 8.1 Analysis of "Disadvantages" of Public Participation

Proponents and Regulators

It costs too much.	The public participation program can be designed to stay within available resources.
	Public participation is likely to be a tiny fraction of overall project costs. It is a cost of good practice.
	The public often contributes insights that lead to cost savings.
	Public opposition is more likely with a closed process. The additional costs associated with approvals, litigation, and implementation are likely to be far greater than the costs of public participation.
It will lead to delays.	Public participation can extend timelines but is an essential facet of democratic decision making.
	A properly designed and managed public participation program should not result in significant delays.
	The time associated with preparing for and participating in lengthy, adversarial hearings can be considerable. Hearings can often be avoided or greatly abbreviated with effective public participation.
	The implementation timetable, even if approval occurs, will probably be extended because the stage has not been set through effective public participation.
It will make decision making less efficient.	Short-term efficiency gains are likely to be more than offset by the costs and delays associated with not consulting the public.
	Public knowledge and experience can provide insights that lead to greater efficiencies.
	Public issues can help focus the EIA process (i.e., scoping).
	The public has a right to be involved in decisions that affect them.
	One of purposes of EIA is to broaden the decision making to encompass more than efficiency concerns.
It is divisive. It will result in a partisan process.	Not involving the public can be even more divisive. Conflict is deferred and usually exacerbated.
	Effective public participation can help build consensus and avoid, ameliorate, and resolve conflicts.
	Public participation is an effective check against partisan technical analyses and interpretations.
The public lacks the knowledge and skills to contribute to the process.	Part of effective EIA participation involves enhancing public understanding.
	The public usually possesses valuable local knowledge and experiences, which can enhance the process.
	Much of EIA practice involves interpretations of significance and acceptability. The public can and should contribute to such interpretations.

(Continued)

Table 8.1   (*Continued*)

The public will adopt a "not in my backyard" position.	NIMBY is a natural, reasonable, and appropriate response to a potential intrusion into the community.
	Early public involvement can mean that the public is a partner in identifying and comparing choices.
	There are generally multiple publics with multiple perspectives.
	Perspectives often change through the course of the public participation program.
	NIMBY is less likely with a voluntary siting approach.
It will raise the project profile and empower opponents.	The public has a right to be informed about decisions that might affect them.
	An effective public participation program can clarify misunderstandings that are often the basis for conflict. A worthwhile project should be able to stand up well to public scrutiny.
	Effective public participation can reduce opposition and lead to a greater level of community acceptance and support. It could help scope or even avoid legal action.
	A demonstrably inadequate public participation can be a source of even greater power for opponents.
The public only has one point of view (e.g., environmentalists), which we know.	There are multiple publics.
	The public has multiple values, perspectives, and interests.
	Public perceptions, attitudes, and positions often change through the process.
	Even if positions are understood, they should still be expressed. It is unlikely that the bases for positions will be fully understood without effective public participation.
Short-term local interests may have to be overruled by regional, long-term needs.	Agencies and regulators can retain final decision-making authority.
	Local participation is essential to fully understand the trade-offs involved.
	Local participation can help in addressing mitigation and compensation options and measures.
	The dichotomy can be a false one. Sometimes local participation makes it possible to identify win–win solutions.
We will lose control of the process.	Proponents and regulators can always retain final decision-making authority.
	The delegation or sharing of control is a choice. It can increase decision-making credibility and legitimacy.
	There are many forms of public participation that do not involve power sharing or delegation.
We lack the necessary participation skills.	There are numerous specialists in public participation and alternative dispute resolution.
	There are many user-friendly public participation resource materials, which are readily available.

Table 8.1   (*Continued*)

	Public participation training programs are widely available.
The process will be high-jacked by activists.	There is a danger that a few individuals can dominate a public meeting.
	There are many other public participation methods that are less prone to such problems.
	Activists usually have something to contribute and should be consulted.
	Approaching multiple publics using a variety of methods can largely offset such problems.
The participants are not "representative" of the public.	Public participation is a voluntary process. Some element of self-selection is inevitable.
	Ensuring that the full range of relevant values and interests are integrated into the process is usually more important than how representative an individual is of a larger constituency.
	A valid concern or suggestion is of value regardless of the level of support.
	The use of multiple consultation methods and ample opportunities for stakeholder representatives to consult with their constituencies can further ameliorate the issue of representation.
It will lead to land speculation.	Land speculation is an issue only during site selection.
	Land can be purchased or expropriated at values prior to site announcements. Land can be optioned.
	The validity of the choice can be subject to public scrutiny before final decisions are made.
It will result in the release of confidential materials.	This is rarely a valid issue.
	It is usually possible to protect proprietary information in an EIA process.
	Confidentiality issues tend to arise more frequently as part of hearings and court cases. Effective public participation can reduce the likelihood and scope of such proceedings.
It will confuse the process. It will be less rational. There will be multiple perspectives and possibly errors.	There should be minimal confusion with a well-designed and well-managed public participation program.
	Data can be checked for accuracy. Facts and values are mixed in practice.
	Multiple perspectives simply reflect the value context. Different perspectives can facilitate interpretations, address uncertainties, and help resolve problems.
	Complex projects require more, not less, participation to ensure that potential impacts and uncertainties are adequately identified and interpreted. Effective public participation planning is essential.

(*Continued*)

Table 8.1 (*Continued*)

It will create expectations that we cannot fulfill. The results could be inconclusive.	The scope and limits of the public participation program can be clearly defined from the outset.
	Public participation provides a means of determining and of transmitting commitments to the community.
	Inconclusive public participation results could simply reflect multiple perspectives and divisions within the community. This does not preclude taking the comments and suggestions provided into consideration.
	Sometime public participation can help build consensus and identify and narrow differences.
Public issues will dominate the process. The environment will suffer.	Public issues generally closely parallel environmental issues and uncertainties.
	The public is often a useful source for identifying and interpreting potential environmental impacts and uncertainties. Perceptions of impacts and uncertainties are real social impacts.
	The public participation program provides a means of addressing public misconceptions.
	The assessment of public issues does not preclude considering other issues and concerns. The public is usually supportive of addressing impacts, uncertainties, and concerns identified by others.
The public will lose interest or will not be interested in policies or programs.	Participation is voluntary.
	A focused and well-planned participation program is more likely to maintain interest.
	Participation for plans, policies, and programs requires alternative approaches (e.g., the involvement of national and regional NGOs, surveys) rather than no participation.
The project is too urgent.	If the need is genuine the public is often supportive of an accelerated project schedule. Urgent timelines are frequently artificial.
	Public participation can usually be designed to meet a project schedule.
	Even when the need is urgent, it is still essential to minimize adverse impacts and uncertainties. The public can contribute to such efforts.
It is too early in the process. We don't yet have a proposal.	Involving the public when there is a need or opportunity and no clear proposal is the best time to begin public participation. In this way the public can fully participate in the decisions leading up to the proposal determination.
	One major public objection tends to be that the decisions have already been made.
The project is too small. There are no or negligible impacts.	Small projects with no to negligible impacts are not (or should not be) subject to EIA requirements.

Table 8.1 (*Continued*)

Such projects can be addressed with categorical or class assessments or by means of a streamlined EIA process. There should be a bump-up provision for significant impacts. Some level of public participation can still occur.

There is still an argument for decision-making openness and transparency regardless of the impact scale.

Negligible is a matter of interpretation. The public can contribute to such interpretations.

Nongovernmental Organizations and Individual Members of the Public

The decisions have already been made. What's the point?	There may be cases of tokenism and placation where decisions have already been made. Where this is clearly the case, it may be appropriate not to participate.
	Sometimes participating can lead to a reconsideration or reversal of a decision.
	Often, decisions are tentative. Public scrutiny can lead to reversals.
The public will be co-opted or manipulated.	The public can participate while making it clear that they do not endorse either the process or the outcome.
	As with proponents and regulators, members of the public should not make up their minds from the outset.
	Changing positions based on new knowledge and thoughtful deliberations is not co-option or manipulation.
	It is the proponent and regulator's responsibility to avoid bias and misrepresentation. The public, with adequate support, can test data and interpretations (e.g., independent peer reviews).
We don't have the time.	The public participation program should be designed so that the time requirements for individuals and groups to participate are no more than absolutely necessary for effective participation.
	The use of a range of methods to involve various publics should reduce the time burden on any one group or individual.
	The timing and duration of public participation activities should respect the other demands on the time of participants (e.g., planting season). Outreach methods (e.g., kitchen table meetings) can sometimes be helpful.
We don't have the resources.	Sometimes participant or intervener funding is warranted. The criteria for funding should be clearly specified.
	Payments for expenses can often be helpful, as well as the provision and sharing of resources.
	Public participation programs can be designed to ensure that resource constraints do not preclude or seriously inhibit public participation, especially for traditionally underrepresented groups.

Sources: Barrow (1997), Canter (1996), Day (1997), Glasson et al. (1999), Nagel (1987), Priscoli (1982), SERM (undated), Shepherd and Bowler (1997), UNEP (1997).

The case against public participation, detailed in Table 8.1, is usually either dubious or overstated. If public participation is accepted as necessary (i.e., a right) and generally desirable, then arguably the burden of proof should be on those seeking to prevent or curtail public participation. Moreover, to realize the benefits of public participation (as highlighted in Figure 8.1), it generally is more desirable for public participation to start early in the EIA process and to occur either continuously or at frequent intervals throughout the EIA process. Circumstances do vary, however. Occasionally, there may be valid reasons for precluding or limiting public participation activities. At a minimum, consistent with the goal of decision-making transparency, a clear rationale should be provided for such limitations.

The concerns listed in Table 8.1 can be valid when a public participation program is poorly designed or executed. Care should be taken to ensure that avoidable public participation "disadvantages" do not occur. The public participation program should be appropriate (i.e., suited to the situation), efficient (i.e., time and other resources are not wasted), and effective (i.e., achieves the shared objectives of the participants). The scope and limits of the program should be jointly determined and should be clearly specified from the outset. The program should evolve in conjunction with the EIA process. It should include an appropriate blending of consultation (e.g., information exchange, continuous involvement), communications (e.g., publicity, dialogue), education (e.g., stakeholder, proponent, mutual), negotiations (e.g., to identify, avoid, and resolve disputes), and collaboration (e.g., to build consensus, to solve problems jointly, to create win–win solutions) elements.

Except in very special circumstances, the "disadvantages" of public participation in the EIA process do not appear to be valid or can be avoided or minimized through good public participation practice. Therefore, public participation disadvantages do not provide an adequate explanation for the discrepancy between the potential and the performance of public participation in the EIA process.

Public participation issues can be more of a challenge. Figure 8.3 highlights some issues often encountered in EIA public participation practice. Clearly, the management of public participation activities in the EIA process require numerous complex, difficult, and subjective decisions. These public participation issues do not provide a rationale for not undertaking or severely restricting public participation activities. Quite the opposite! The issues are not resolvable through technical analysis or political expediency. Instead, they should be addressed jointly with interested and affected parties. A breadth of perspectives can make it easier to identify and explore issues. A sounder foundation also can be established for reaching and substantiating interpretations and decisions.

Public participation has been a component of EIA practice for more than 30 years. The quality and effectiveness of practice (both in EIA and in related fields) has advanced rapidly, especially over the past 10 to 15 years. Thus there are ample good practice examples and guidance materials to help in designing and executing public participation and dispute resolution programs, for anticipating and managing problems and dilemmas, for making difficult judgments, for reconciling or accommodating conflicting perspectives, for ameliorating obstacles, and for managing

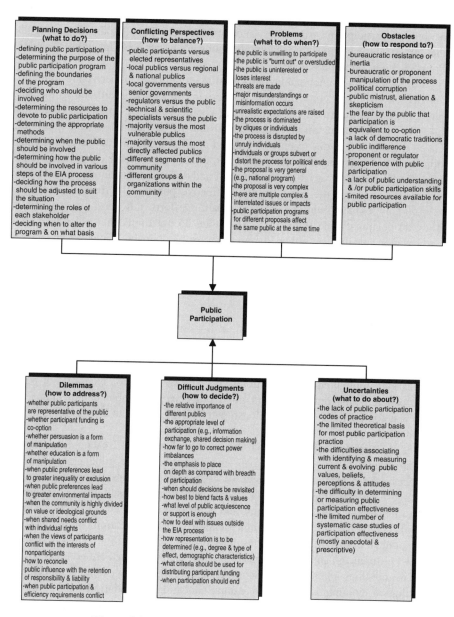

Figure 8.3 Examples of public participation issues.

uncertainties. Dilemmas, obstacles, and problems, which appear impossible in the abstract, can generally be worked through in practice, especially by effectively drawing upon the knowledge, experience, and judgment of the public. The difficult issues that emerge in a public participation program reflect the complexities of decision making in a pluralistic society. There are no quick fixes. A sufficient record of public participation successes and failures has been amassed to suggest that practical resolutions or accommodations to issues such as those listed in Figure 8.3 can often be reached.

At the same time, much public participation literature is not derived from theory, nor does it provide a coherent basis for deriving theory. It is difficult to measure public participation effectiveness. Codes of good conduct are largely lacking. Analyses of public participation effectiveness tend to be qualitative and anecdotal. Given the difficult issues often encountered and the uncertain conceptual foundation, it is not surprising that the quality and effectiveness of EIA public participation efforts are highly variable. The limits of and difficulties sometimes encountered in practice reinforce the need for more effective public participation. They may occasionally explain why public participation efforts fall short of aspirations. They do not justify the status quo. Also, they do not imply that potential public participation benefits are either inappropriate or unattainable. They do underscore the need to plan and execute public participation programs thoughtfully and jointly, to draw upon the lessons of public participation practice, and to anticipate and effectively address the many types of issues that often emerge in practice.

8.4.2   Principles and Practices

Public participation in EIA practice has advanced to the point that a core body of prescriptive knowledge is emerging. Table 8.2 identifies general, consultation, communications, mutual education, negotiations, and collaboration examples of public participation goals, principles, and good practices. The table demonstrates that there is a considerable knowledge base potentially relevant to EIA process design and management.

It is apparent from Table 8.2 that the various elements of EIA public participation are highly interdependent. Goals and principles guide good practice. Good practices extend from and contribute to goals and principles. The general goals, principles, and practices provide a framework for the consultation, communications, mutual education, negotiations, and collaboration elements. Effective consultation is conducive to effective communication. Mutual education is more effective when built on a base of effective consultation and communications. Negotiations (to address differences) and collaboration (to build shared visions) are complementary. Both negotiations and collaboration are enhanced when they extend from effective consultation, communications, and mutual education. Negotiations and collaboration can foster more effective consultation, communications, and mutual education. Sensitivity to these interdependencies is essential to effective public participation in the EIA process.

Public consultation and negotiations (especially alternative dispute resolution) have received the most attention in the EIA literature and in the literature of related forms of environmental management. Communications is commonly characterized as communicating to the public. Education has tended to be defined as public education. Public education is sometimes equated with persuading or

Table 8.2   Examples of EIA Public Participation Goals, Principles, and Good Practices

Goals	Principles	Good Practices
	General	
To produce better decisions	Be honest, open, inclusive, and responsive	Undertake community or social profiling
To earn the trust of public participants	Explicitly identify public participation objectives; clearly identify decisions to be made	Select methods to match objectives, context, issues, publics, and stage in EIA process
To increase decision-making transparency and accountability		
To promote fairness	Explicitly identify public participation limits	Focus on issues as identified by the public
To build credibility with those who might be affected	Design public participation efforts to match situation	Design the process to accommodate stakeholder values
To integrate public participation and decision making	Ensure that all interests are represented	Clearly define roles and responsibilities
To contribute to more environmentally responsible decisions	Ensure adequate resources, including time, for effective public participation	Ensure sufficient time and flexibility for adequate public participation
	Recognize that the people have a right and a responsibility to manage their own affairs	Seek to make interactions informal and personal
		Recognize and ameliorate barriers to participation
	Design process to be responsive to community needs	Inform public of process progress clearly and frequently
	Recognize public contribution to process	Evaluate, with public, by stage, effectiveness of public participation measures; adjust and supplement as needed
		Employ qualified and unbiased participation specialists
	Consultation	
To involve interested and affected parties in the EIA process	Involve members of the public in decisions that might affect them	Identify relevant interest groups
To identify public values and concerns	Work for broad participation	Prepare a public involvement plan
To provide to the public relevant information regarding the proposal, possible options and potential impacts	The public has a right to information relevant to potential decisions that might affect them	Involve the interested and affected publics in formulating the public involvement plan
To obtain feedback from the public concerning values, perspectives, preferences, and suggestions	Provide an opportunity for those otherwise unrepresented to express their views (outreach); provide the resources necessary to ensure their effective participation	Select and adapt involvement methods to stakeholder characteristics
To provide in-depth involvement opportunities		Design consultation for the convenience of the public
		Interview representatives of each group to identify potential concerns

(Continued)

Table 8.2   (*Continued*)

Goals	Principles	Good Practices
To involve traditional unrepresented and underrepresented groups and segments of society	Obtain and accommodate local and traditional knowledge The public should have an opportunity to comment prior to each decision in the process Responses should be provided to all public comments and suggestions Involve the public early in the process (e.g., problem definition, alternatives identification, criteria identification, public identification) Involve regulators from the outset	Share information openly Clearly explain how public input will be used; provide explanations if input rejected; provide prompt responses Maintain the visibility of the public consultation program Identify and ameliorate barriers to information flow (e.g., lack of awareness, legal, financial, technical) Place greater emphasis on interactive formats, such as workshops or coffee klatches, in preference to public hearings or large public meetings Fully document record of public involvement Involve the public in approvals and implementation

Communications

Goals	Principles	Good Practices
To enhance public, proponent, and regulator understanding To facilitate the interchange of ideas among citizens To encourage respectful speaking and listening To establish and maintain a dialogue between those responsible and those affected by possible actions To minimize communications distortions To ensure that information is accurate, relevant, and unbiased To provide an opportunity for those otherwise unrepresented to express their views (outreach) To employ effective communications skills	Minimize inaccuracies Communications materials should be adapted to the needs of each participant group Provide the public with accurate, timely, pertinent, and understandable information Interpretations, ideas, options, and management measures should be substantiated and open to reasoned criticism Recognize that feelings equal facts Facilitate interagency communications and cooperation Recognize that listening is a critical element of participation	Learn to speak the public's language Recognize that process communicates content Use professional expertise to create opinions not to kill them off Look at the range of values, not just the numbers Ensure that documents are well planned, organized, edited, and presented Minimize or explain technical language where must include; avoid jargon Design documents to suit audience Use third-party mechanisms when there are arguments over facts

Table 8.2   (*Continued*)

Goals	Principles	Good Practices
	Recognize that communications is two-way Provide channels for receiving, evaluating, and responding to individual, group, and societal public concerns and suggestions Establish a working rapport with all stakeholders Ensure that documents clearly communicate local sentiments to decision makers	Be proactive in communicating with the public Inform public of communications channels to EIA team, to regulators, and to decision makers Ensure that communications is clear, concise, and noncondescending Make effective use of visual techniques (e.g., photo-simulation)

Mutual Education

Goals	Principles	Good Practices
To enhance public knowledge about possible actions, environmental conditions, and possible impacts To enhance proponent and regulatory knowledge about local conditions, values, needs, and concerns To promote mutual, social, collaborative, and transformative learning To foster cognitive enhancement (the acquisition of knowledge) and moral development (growth in the ability to make judgments about right and wrong) To promote critical EIA education (education about and through EIA)	Seek out and make use of public knowledge Treat traditional knowledge as a valid form of knowledge Recognize learning as a step to conflict avoidance or resolution Successful stakeholder involvement requires agency staff training or expert assistance Facilitate learning about facts, values, and social identities Ensure learning is free from coercion and distortion Be open to alternative perspectives Ensure the free expression of attitudes, feelings, and intentions Treat the EIA process as a learning process (e.g., contributes to ability of communities and societies to learn and change) Seek to improve the intelligence capacity of government agencies and of communities	Provide for local capacity building (to participate more effectively) where needed Provide for participant training Plan educational programs/ activities in partnership with stakeholders Seek to integrate personal/ experiential/contextual knowledge with processed knowledge Distinguish between cognitive (knowledge dominant) and social learning (responsive communications leading to policy reframing) Reflect critically about presuppositions Play close attention to fairness and competence Use dialogic and argumentative processes to promote learning Foster and recognize interactions among critical listening, reflection-in-action, and constructive argumentation Integrate learning from practice stories

(*Continued*)

Table 8.2 (*Continued*)

Goals	Principles	Good Practices
	Negotiations	
To avoid and reduce conflict	Ensure that information to support process is complete and accurate	Identify potentially controversial issues and seek resolution with the appropriate parties
To develop decisions that are mutually acceptable to interested and affected citizens	Ensure that the full range of interests are represented and that all are free to negotiate with other stakeholders	Plan conflict resolution process, especially pre-negotiations; ensure agreement on rules and procedures
To search for new conceptions of values	Correct power imbalances	Start with joint fact-finding
To meet a mix of people's substantive, procedural, and psychological interests	Ensure that third parties (e.g., mediator) have adequate training and experience and are acceptable to participants	Highlight underlying assumptions
To reduce the risk of subsequent misunderstanding		Seek to identify low-cost trades
To ensure a just and equitable process	Ensure that all parties have sufficient resources and authority	Design the process to suit the type of conflict
To ensure just and equitable outcomes	Provide for early and ample opportunities for conflict resolution	Understand the role of interpersonal dynamics and help people to move on
To further advancement toward social, environmental, and sustainability ends	Negotiate over interests not positions	Define measures of success (e.g., products, acceptance, interests protected, responsibilities defined, relationships established and maintained)
	Consider a wide range of alternatives that reconcile differences	Stave off angry confrontation
	Agree on principles or criteria to evaluate alternatives	Seek points of mutual agreements; focus on options for mutual gain
	Document the agreement	Provide sound technical data and support to process and stakeholders
	Agree on the process by which the agreements are to be revised	Employ practical approaches for dealing with disruptive behavior
	Ensure commitments are observed	Clarify the presumed liability of participants, confidentiality agreements, legal agreements, and extent to which precedents are or are not being set
	Ensure outcomes are monitored and enforced	Visibly isolate extremes
		Keep public informed of progress
		Conduct post hoc evaluations of effectiveness

Table 8.2 *(Continued)*

Goals	Principles	Good Practices
	Collaboration	
To build consensus To build and sustain trust To build support for and acceptance of decisions To ensure procedural and outcome fairness To foster collaborative and creative explorations of problems and opportunities To obtain tangible environmental and sustainability outcomes (i.e., goodness of decision)	Ensure information to support process is complete and accurate Involve the public in idea generation and problem solving Do not substitute compromise for good problem solving Treat analysis as a joint effort rather than a battle over facts Define the problem rather than propose solutions or take positions View the situation as an opportunity for collaboration, not competition Recognize the interdependence of process and substance Seek to define common goals and shared visions of the future (community and environment)	Be clear regarding boundaries, who invited to participate, expectations of contributions by participants, how facilitators chosen, how information generated will be used, and who owns Separate people and their personalities from the problem Ensure that process is flexible and where appropriate, experimental Provide sound technical data and support to process and stakeholders Keep public informed of progress Undertake documentation in partnership with community leaders Be attentive to the distribution of power by stakeholders and facilitators Adopt activist mediation model (process and outcome) Make effective use of methods for creatively redefining problems and for generating, selecting, and evaluating ideas Conduct post hoc evaluations of effectiveness

Sources: Bauer and Randolph (2000), Bisset (2000), Clark (1994), Creighton et al. (1999), Daniels and Walker (1996), Diduck and Sinclair (1997), Fell and Sadler (1999), Forester (1999), Glasson et al. (1999), Healey (1997), Howell et al. (1987), Interorganizational Committee (1994), Lauber and Knuth (1998), Manring et al. (1990), March (1998), Maser (1996), Maynes (1989), Moore (1986), Morgan (1998), Praxis (1988), Priscoli (1982), Priscoli and Homenuck (1986), Rickson et al. (1990), SERM (undated), Smith (1993), Smith et al. (1997), Susskind (1999), Susskind and Cruikshank (1987), UNEP (1997), US EPA (2001b), Webler et al. (1995), Weiss 1989).

even manipulating the public. Collaboration is often seen as an extension of negotiations (i.e., building on win–win solutions to conflicts). More attention should be devoted to principles, concepts, and methods of two-way communications, mutual education, and creative and substantive collaboration (see subsequent subsections of this chapter).

Additional effectiveness reviews could help derive, refine, and test public participation principles and practices. Such analyses could demonstrate which practices contribute the most and the least to achieving public participation goals. They could illustrate critical interdependencies. They could identify when principles or practices are complementary and when they operate at cross-purposes. They could contribute to more effective public participation planning and management and to more effective integration of public participation into the EIA process. They could also demonstrate how the EIA process could be reformed and adapted to foster more effective public participation.

Public participation concepts and categories of methods often are displayed as continua, as illustrated in Figure 8.4. Continua are useful for grouping methods. They can illustrate which categories of methods are best suited to achieving alternative citizen participation goals (e.g., citizen control, citizen autonomy, citizen influence, citizen involvement). They can indicate which groupings are more appropriate to situations characterized by varying mixes of cooperation and conflict. They can assist in role definition for public agencies, the public, and third parties. They can provide a general sense of major methods' characteristics (e.g., degrees of formality, continuity, and intensity). There are some inconsistencies in the placement of various categories along the continua. These differences reflect varying definitions of categories (e.g., one- versus two-way communications and education) and varying role interpretations.

A continuum clearly and succinctly displays major differences. But only differences for a single criterion can be displayed at a time. The impression can be created that only one category can be used (it is possible to use several in an EIA process) and that categories further along the continuum are somehow better (it is more often a case of matching the methods to the context). Public participation methods can be classified in ways that do not involve continua. They can, for example, be categorized by function (e.g., information dissemination, information collection, initiative planning, reactive planning, decision making, participation process support) or by operational characteristics (e.g., large group meetings, small group meetings, organizational approaches, media, community interaction, legal mechanisms). Matrices can display differences along more than one dimension. Table 8.3, for example, clusters methods by public participation element (e.g., consultation) and by role (e.g., information exchange, continuous involvement, formal involvement).

Public participation texts and manuals generally describe the characteristics, advantages, and disadvantages of numerous individual methods. Sometimes, connections are drawn between the methods and public participation goals. Ideally, a consistent set of criteria (based on public participation goals and principles) would be applied to each method grouping and/or each method. The application of scaling

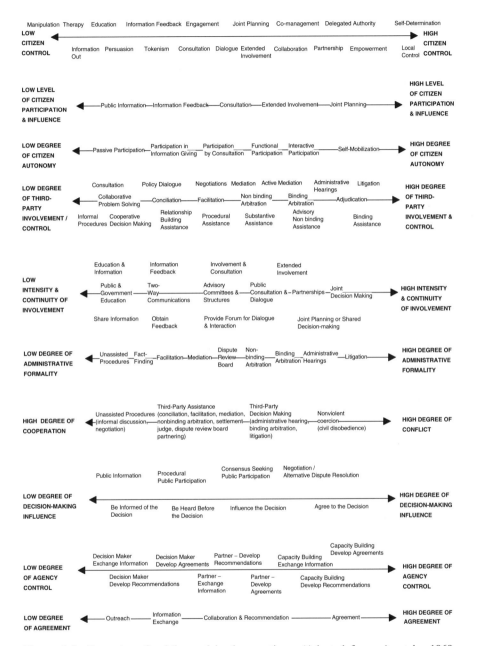

Figure 8.4 Examples of public participation continua. (Adapted from: Arnstein, 1969; Bisset, 2000; Creighton et al., 1999; Fell and Sadler, 1999; Health Canada, 2000b; Hughes, 1998; Parenteau, 1988; Praxis, 1988; Smith, 1993; Susskind and Madigan, 1984; Westman, 1985.)

procedures could help ensure that differences are consistently addressed. Such analyses would be further strengthened if linked to EIA process activities (e.g., scoping), if assessed for varying contexts (e.g., third-world countries), and if supported by systematic reviews of EIA effectiveness analyses.

Table 8.3   Examples of Methods

	Consultation	
Information Exchange	Continuous Involvement	Formal Involvement
Information in (e.g., interviews, surveys, polling, focus groups, public comments, community profiling, call in television, direct e-mail, hot lines, mail-in response forms, door-to-door canvassing, responsiveness summaries, briefs, submissions, content analysis, cumulative brochures, letters to the editor)	Advisory committees, councils, groups, and boards	Hearings
	Task forces and groups	Litigation/adjudication
	People's panels	Referenda and plebiscites
	Citizens' review board	Commissions
	Breakfast meetings	Inquiries
	Community impact committee	
	Community planning council	
	Citizens action committees	
Information out (e.g., briefings, exhibits, displays, contact person, telephone network/phone tree, computer bulletin boards, community liaison officer, political preview, demonstration projects, document circulation, feature articles)		
Town meetings, open houses, and workshops		
Conferences, roundtables		
Contests		
Stakeholder meetings		
Ombudsperson or representatives		
Televoting, 1–800 numbers, media-based issue voting		
Workbooks and community mapping		

Table 8.3 *(Continued)*

Communications		
Publicity	Dialogue	Enhanced Dialogue
Traditional publications (e.g., newspaper inserts, information kits, brochures, newsletters, fact sheets, mail-outs, paid advertisements, plain language communications)	Coffee klatches Kitchen table meetings Search and consensus conference Constituent assembly Roundtables Retreats	Relationship building assistance (e.g., counseling/therapy, conciliation, team building, informal social activities) Search and consensus conference
Audio/visual (e.g., film presentations, video, slide presentations, tape)	Computer-assisted participation and interactive www/ e-conferencing	Issue conference Capacity building and outreach Technical assistance and participant funding
Media (e.g., radio and television interviews, Web sites, media releases, public service announcements, press kits, newspaper inserts, news conferences)	Online discussion groups Participatory television/cable television Community-sponsored meetings	Structured workshops Citizen employment Citizen honoraria Coordinator or coordinator catalyst
Information fairs/exhibits Translations Group presentations	Field offices Advice and argumentation	Discourse ethics Combating misinformation (critical theory) Procedural justice

Mutual Education		
Community Education	Proponent, Regulatory, and Specialist Eduction	Mutual Education
Technical and financial assistance	Procedural training Substantive training	Storytelling Relationship building
Citizen training	Networking	assistance (e.g., counseling/
Lectures and workshops	Comparable proposal and	therapy, conciliation, team
Computer-based programs	environment review	building, informal social
Publications and translation	Citizens' juries and panels	activities)
Site visits, depositories, and resource materials	Community profiling Traditional knowledge	Coaching/process consultation Participatory research
Formal education, integration into existing curricula	Citizen surveys	Study circles and study groups Participatory drama
Simulation exercises and photo-simulation		Social and collaborative learning
Citizen training programs		Transformative learning
Seminars, discussions, and position papers		Deliberative learning
Media campaigns		
Speaker's bureau and panels of experts		
Technical advisors and peer reviewers		
Demonstrations and demonstration projects		

(Continued)

Table 8.3   (*Continued*)

Negotiations

Unassisted	Third-Party Assistance	Third-Party Decision Making
Informal discussion	Fact-finding	Dispute prevention
Negotiation	Conciliation and facilitation	Advisory nonbinding assis-
Conciliation	Mediation	tance (e.g., nonbinding
Information exchange meet-	Conflict anticipation	arbitration, summary jury
ings	Conflict assessment	trial)
Interest-based negotiation	Technical advisory board	Administrative hearing
Policy dialogue	Minitrial and nonbinding	Binding arbitration
	arbitration	Med-arb
	Settlement judge and dispute	Mediation, then arbitration
	review board	Dispute panels (binding)
	Settlement conference	Private courts/judging
	Negotiated rule making	Litigation/adjudication
	Community dispute	
	resolution centers	

Collaboration

Joint and Collaborative Planning	Joint Management	Creative Collaboration
Roundtables, conferences,	Co-management boards and	Community visioning and
and working groups	councils	shared vision planning
Cooperative/collaborative	Partnering and partnership	Brainstorming,
problem solving	agreements	brainsketching, and
Role-planning	Co-jurisdiction	brainwriting
Joint planning	Steering committees	Delphi process
Coalition building	Public authorities	Nominal group process
Strategic choice	Community representatives	Lateral thinking methods
Large-group response	on boards	Think tanks
technique	Citizen assemblies	Active mediation
Consensus building	Community forums	Simulation, modeling, and
Collaborative planning		scenario writing
Charrette		Creative problem solving
Niagara process		
Trade-off games		
Samoan process		
Multicriteria group		
decision-making models		
Constructive engagement		

Sources: Canter (1996), Creighton et al. (1983, 1999), Daniels and Walker (1996), De Bono (1992), Forester (1989, 1999), Friend and Hickling (1997), Glasson et al. (1999), Health Canada (2000b), Howell et al. (1987), Morgan (1998), Praxis (1988), Saarikoski (2000), SERM (undated), Sinclair and Diduck (2001), Smith (1993), Susskind et al. (1999), US DOE (1998), US EPA (2001a,b).

8.4.3   Consultation

With public consultation or involvement the public is informed about proposals. They also express their views about proposals. These interactions occur prior to decision making (Parenteau, 1988). Public concerns and suggestions are taken

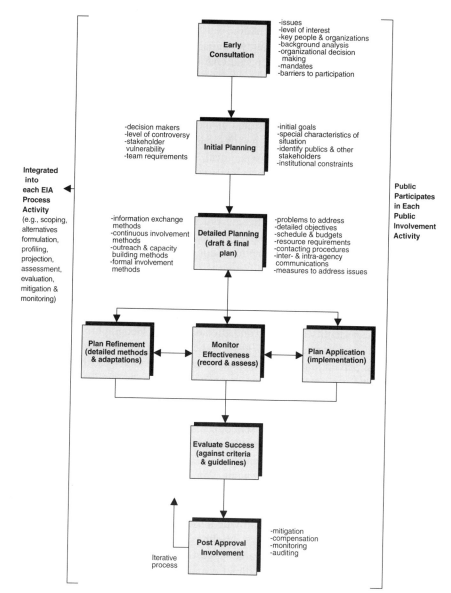

Figure 8.5   Examples of an EIA public consultation process. (Adapted from Burdge and Robertson, 1994; Creighton et al., 1999; Praxis, 1988; US DOE, 1998; US EPA, 2001a,b; Wolfe et al., 2001.)

into account by decision makers (US EPA, 1988b). Public inputs inform but do not dictate decisions. Proponents and regulators retain final decision-making authority (Smith, 1993).

Figure 8.5 illustrates how a public involvement process might unfold. Early consultation provides a general sense of such matters as issues, levels of interest, key people and organizations, organizational mandates and decision-making procedures, and barriers to participation. Early consultation activities provide a basis for initial consultation planning. Initial consultation planning establishes preliminary goals, determines the special characteristics of the situation, indicates study team requirements, highlights institutional constraints and stakeholder vulnerabilities, and identifies decision makers, the various publics, and other stakeholders.

Stakeholder identification is a critical element of public involvement planning. Individuals, groups, organizations, and segments of society can be differentiated based on, for example, location (e.g., local, regional, national), interests (e.g., industry, environment, community service), and characteristics (e.g., social, cultural, economic, political) (Bisset, 2000; Hughes, 1998; Praxis, 1988). Further distinctions can be drawn among types of individuals (e.g., community leaders, local informants, local experts, politicians, practitioners), groups (e.g., professional, environmental, cultural, recreational, service), and organizations (e.g., government agencies, businesses, institutions, media, labor unions) (Canter, 1996; Priscoli and Homenuck, 1986). There will be differences among stakeholders in the extent to which they are involved or not involved, informed or not informed, organized or not organized, united or divided and supportive, opposed to or apathetic to the proposed action (Praxis, 1988; Priscoli and Homenuck, 1986). Stakeholders, third parties, and staff can identify interested and affected parties. Geographic, demographic, historical, and comparative analyses also can help in stakeholder identification. The characteristics, perceptions, and positions of each stakeholder can be determined, appreciating that positions and alliances change, sometimes rapidly.

Detailed consultation planning generally involves preparing a draft and then a final public involvement plan. The plan is likely to characterize problems, determine objectives, establish schedules and budgets, allocate resources, establish contacting procedures, identify communications channels, and determine specific procedures for addressing identified issues (Praxis, 1988; UNEP, 1997; US EPA, 1998b). A public involvement plan can include information exchange, continuous involvement, outreach and capacity building, and formal involvement methods.

Information exchange methods provide a useful means of transmitting information to the public (e.g., newsletters, briefings, displays, background papers), of receiving information, comments, and suggestions from the public (e.g., surveys, public comments, response forms, direct e-mail), and for agencies and the public to exchange information and viewpoints (e.g., open houses, meetings, workshops). Information exchange methods can be geared to large or small audiences. If undertaken effectively, they can reach a major proportion of the population. They generally occur periodically in an EIA process.

Continuous consultation methods (e.g., an advisory committee) involve a small number of stakeholder representatives. The committee meets frequently throughout the EIA process. The committee can address issues, analyses, options, and trade-offs in much greater depth than is possible in information-exchange sessions (Maynes, 1989; Praxis, 1988; US EPA, 1998b). Continuous involvement is more effective when aided by facilitation and when adequately supported by influential agencies. It is ineffective if there is a lack of trust or if the viewpoints expressed through the committee are not taken seriously (Priscoli, 1982). Continuous involvement and information exchange methods can be complementary. Continuous involvement methods can explore issues and concerns identified in information-exchange sessions. Information exchange stresses breadth of involvement. It provides a mechanism for committee representatives to test positions and concerns with constituency groups. Continuous involvement emphasizes depth of involvement.

Outreach and capacity-building methods help bring into the EIA process traditionally unrepresented or underrepresented groups and segments of the population. Outreach and capacity building can take the form of technical or financial assistance. It can entail the supply of technical resources such as phone conferences and e-mail support and the provision of translation and facilitation services (US EPA, 2001b). Such methods can enhance the capacity of organizations and groups to participate effectively in the EIA process. Communities also can be empowered to leverage additional resources and to capitalize on existing civic assets (US EPA, 2001b). Outreach and capacity building can support and supplement both information exchange and continuous involvement methods.

Formal involvement methods, such as hearings, commissions, and inquiries, tend to assume the characteristics of judicial procedures. Such methods can provide a useful way of presenting and testing evidence. They are often adversarial, however, and can be intimidating to the public, especially if technical and financial resources are not made available to public groups and organizations (Maynes, 1989). Referenda and plebiscites provide a formal mechanism for testing agreements obtained through continuous involvement procedures or for obtaining public feedback on major proposals and options. Such procedures can be costly, can oversimplify complex choices, and are occasionally divisive.

Public involvement plans are not simply implemented. Refinements and adjustments occur throughout the EIA process, based on an ongoing assessment of changing circumstances and of methods effectiveness (Howell et al., 1987). Public involvement procedures and the public involvement plan are merged with rather than distinct from the EIA process (Burdge and Robertson, 1994). Separate consultation objectives and methods are selected and adapted to meet the requirements of each EIA process activity (e.g., scoping, alternatives formulation, impact assessment) (Burdge and Robertson, 1994). The overall effectiveness of the public involvement procedures is assessed prior to approvals. This helps identify supplementary involvement measures, which should be instituted to address identified gaps and weaknesses (Wolfe et al., 2001). Public involvement extends into the post-approval period to ensure that public concerns associated with project implementation, mitigation, compensation, and monitoring are fully considered. Public involvement, in common with the overall EIA process, is highly iterative (Praxis, 1988).

Public involvement procedures can establish a strong foundation for a collaborative EIA process. Although necessary, they tend not to be sufficient. Although partially addressed through good practice guidance, insufficient attention tends to be devoted to the specific mechanisms by which effective two-way communications and mutual education can occur. Public involvement methods, by themselves, tend to be ineffective in avoiding, managing, ameliorating, and resolving conflicts. They also tend to neglect the development and application of specific techniques and procedures for building consensus, for creative problem solving, and for collaboratively contributing to the realization of substantive environmental objectives.

8.4.4 Communications

Communications involves interactions among people. It is the bridge between environmental analysis and decision making (Holling, 1978). The communications act includes the parties involved (who), the message (what), the form by which the message is encoded (how), the audience (to whom), and a result (with what effect) (Bishop, 1975, 1983). Effective communications can facilitate understanding, conflict resolution, consensus building, and decision making. Ineffective communications can lead to a lack of understanding or to misunderstandings. It also can undermine consensus, exacerbate conflict, and inhibit decision making.

References to communications in EIA literature tend to revolve around communications in EIA documents. Stress is place on facilitating understanding through clear, focused, and consistent document presentation and organization (Morgan, 1998). Documents should focus on the needs and concerns of and be readily understandable to the target audience. General references also are made to developing and refining the verbal and written communications skills of EIA practitioners (Daniels and Walker, 1996).

Table 8.4 highlights the characteristics of several communications concepts relevant to EIA process management. These concepts underscore the central role of dialogue in EIA practice. EIA, in common with planning, is a collective, dialogic, practical, and moral activity. It is, therefore, essential to provide conditions conducive to open, unencumbered, undistorted, and noncoercive dialogue. This may require the formulation and application of mutually acceptable communicative ethical principles. Steps may need to be taken to correct power imbalances. Ideal speech characteristics could be explicitly recognized. A concerted effort could be made to identify, avoid, and minimize communications misinformation and distortion.

The characteristics of effective and ineffective advice giving could be considered. EIA practitioners (and related procedural specialists such as facilitators) could help establish and maintain the conditions required for effective and appropriate argumentation, persuasion, and storytelling. They could provide and derive important insights from EIA practice stories. They could help participants explore the

Table 8.4   Examples of Potentially Relevant Communications Concepts

Dialogue	A form of communications in which understanding and respect are goals; intentions include maintaining social contacts and affiliation, eliciting and gaining information, promoting relationships, changing the environment, and others
	In dialogue, participants present their own perspectives, listen carefully to the perspective of others, remain open to change, speak for selves and from personal experience, allow others to express their perspectives safely, learn significant new things about selves and others, find shared concerns with people holding different perspectives, explore doubts and uncertainties, ask questions based on true curiosity, explore the complexity of issues without polarization, and collaborate to create better futures
	Communications process models include diffusion processes (to public), collection processes (from public), and diffusion–collection processes (information disseminated with intent of obtaining response)
	Messages can be received at the perceptual, the cognitive, and the judgmental levels
	Roles in dialogue: sender of message (determine own believes, motives, and beliefs); receiver of message (listening for decisions, listening with empathy, nonverbal communications)
Communicative planning and action	Sees planning as an exercise in collective, participatory action
	Argues that the best window onto planning practice is planning discourse; discursive interaction the most important element of planning practice
	Recognizes that planning process may manipulate citizen action and lead to systemic distortions; systemic distortions are avoidable
	Seeks to facilitate informed, open, unforced, and unmanipulated citizen action
	Seeks sincere, comprehensive, and appropriate communications; self-consciously chooses actions to overcome institutional barriers and to become more egalitarian
	Seeks a deliberative style of debating issues and interests; an open dialogue among equals
	Principles: all important interests (identified and articulated) at table; all stakeholders fully and equally informed and able to be represent their interests; all equally empowered in discussion; power differences from other contexts must not influence who can speak or who is listened to or not; allows all claims and assumptions to be tested and all constraints tested; all participants able to assess the speaker's claims; all must speak sincerely, honestly, and comprehensively; groups should seek consensus
	Can be obstructed by social inequities and tensions and conflicts among groups; criticized as politically naive

(Continued)

Table 8.4 (*Continued*)

Discourse or communicative ethics	Procedural approach to moral justification; procedural morality
	Assumes that the basic unit of meaning is the speech act, that meaning is inseparable from the role of language in structuring practices and social interactions and that truth and normative rightness are essentially discursive matters
	Seeks to engender ideal speech situation: freedom of access, equal rights to participate, truthfulness on the part of participants, and absence of coercion in taking positions
	Endeavors to ensure that all relevant voices get a hearing, the best arguments given the present state of knowledge are brought to bear, and only the unforced force of better arguments determine the yes or no of participants
	Requires that all members be prepared to listen for differences not only in interests but in values and cultural references
	Sets itself the tasks of deriving argumentation rules for discourse in which moral norms can be justified
	Role of discourse ethics: to examine the normative validity of public action-guiding norms, to examine not just whether all affected participants might accept a norm but whether the norm deserves to be accepted by them, given the process in which they might consider them
Misinformation	Various forms of misinformation impede and distort communications
	Managing comprehension (e.g., deliberate ambiguity, jargon, ideological language, obscure messages)
	Managing trust (e.g., false assurances, symbolic decisions, marshalling respectable personage to gain trust, ritualistic appearance of openness)
	Managing consent (e.g., decisions reached without legitimate representation of public interest, arguing technically when acting politically, appeals to adequacy of participation, not addressing systemic failures)
	Managing knowledge (e.g., decisions that misrepresent actual possibilities to the public before a decision is made, misrepresenting costs, benefits or options, ideological or deceptive presentation of needs)
	Managing control (e.g., withholding information, misleading information or judgment, inconsistencies in what is being said, gaps in argumentative chain, undue persuasion, professionalization of debate)
	Need to address and combat the effects of unequal power relations and misinformation; most misinformation avoidable even when systemic
Advice, argumentation, and persuasion	Requirements of advice: relationship of persons of trust and truth, a basis in the world through knowledge and experience, expressed in reasonable and justifiable stories and a public understanding of who we are as a community

Table 8.4 (*Continued*)

	Planning as a dialogic and argumentative process; involves marshaling evidence and giving reasons, minimizing the exclusion of relevant information, encouraging the testing of conjectures and welcoming rather than punishing value inquiry
	Planning as action in a flow of persuasive argumentation; expressed in awareness of differing or opposing views
	Need to meander skillfully: arguing for visions, constructing inclusive processes, negotiating the meaning of key concepts, responding to unexpected events, taking existing rules and prior decisions into account (while seeking to change problematic ones), relying on own substantive knowledge (while being open to other forms of knowledge and expertise), configuring arguments (in the face of contestable configurations), and arguing persuasively in diverse media and forums
	Argumentation affected by conflicts with others over meaning, media in which persuasive efforts occur, events that create new opportunities and constraints, institutional rules and previous decisions, legalistic procedures that inhibit understanding and innovation, social and institutional factors, and opposition to open and inclusive processes
	Rhetorical frame (persuasive use of story and argument in policy debate) as distinct from action frame (frames that inform policy practice)
Story telling	Planning arguments are characteristically expressed as stories
	Stories describe events, provide explanations, warn of dangers, identify benefits, report relevant details, search for others' meanings, confess mistakes, justify recommendations, and prepare others
	Need to understand the significance of the very messiness, complexity, detail, and moral entanglement of living stories
	Stories are accounts of value and identity, of abiding concern, and of complexities; ignored at practical risk
	Stories are morally thick, politically engaged, and practical
	The discursive process needs to be designed to explore different storylines about possible actions
	Suggested convergence strategy when varying stories: a pluralistic strategy; embrace rather than seeking to resolve or ignore controversy; consistent with an open moral community

Sources: Bishop (1983), Fischler (2000), Forester (1989, 1999), Habermas (1993), Healey (1997), Huxley and Yiftachel (2000), Innes (1998), Kreiger (1981), Lauria and Soll (1996), Mandelbaum (1991), Patton et al. (1989), Sager (1994), Schön and Rein (1994), Taylor (1998), Throgmorton (2000).

characteristics of and potential for accommodating diverse arguments and stories within the EIA process. The institution of effective EIA communications measures can contribute to enhanced public consultation. Such measures also are conducive to a collaborative EIA process. Additional building blocks, concerned with mutual education, conflict resolution, and consensus building, however, are still required.

8.4.5 Mutual Education

Education in EIA practice is conventionally depicted as using information dissemination and general instruction to create public awareness of proposed actions and issues, to encourage more responsible environmental stewardship, and to facilitate informed decision making through enhanced understanding (Morgan, 1998; Praxis, 1988; SERM, undated). It is generally recognized that educational activities and programs should be jointly planned with stakeholders.

Education is sometimes acknowledged as a necessity and a precondition for advanced levels of public involvement, conflict resolution, and collaboration (Diduck and Sinclair, 1997; Maser, 1996). The EIA process has been characterized as a learning process—a process that can help communities and societies change and that can improve the intelligence capacity of government agencies (Rickson et al., 1990b). The assumption tends to be made that learning is one way (i.e., to the public). It is sometimes recognized that proponents, regulators, and practitioners also need to receive training (e.g., in consensus building and conflict resolution techniques) and can learn much from the public. Increasingly reference is made to the mutual learning that occurs through dialogue and debate among stakeholders (Daniels and Walker, 1996; Diduck and Sinclair, 1997; Webler et al., 1995).

Table 8.5 highlights the characteristics of various mutual education concepts relevant to EIA process management. These concepts demonstrate that there are many forms of mutual learning (e.g., cognitive, social, practical, collaborative, transformative, critical, traditional knowledge) possible in EIA practice. They illustrate how mutual learning integrates the cognitive, the moral, and the practical. They show how learning can be approached from multiple perspectives, how it varies depending on the historical, social, and cultural context, and how it integrates and transcends such distinctions as personal and processed knowledge, facts and values, and people and the environment.

Mutual learning is interactive, social, reflective, critical, practical, affective, holistic, and democratic. It facilitates learning about facts, values, issues, decision-making processes, and the participants in the process. It can further democratic values. Participants in mutual leaning are transformed by the experience. Mutual learning is conducive to learning about and through the EIA process. It is more likely to occur when supported by accurate information and a noncoercive environment. Third parties, such as facilitators, can help participants adapt and apply mutual learning.

Mutual learning concepts, coupled with more conventional educational methods, such as the training of participants, can contribute to more collaborative EIA processes. Education in and through EIA broadens and reinforces the base established through public involvement and communications measures. Additional measures, however, are needed to address conflicts, to build consensus, and to advance substantive environmental goals.

Table 8.5 Examples of Potentially Relevant Educational Concepts

Mutual education	No single party, organization or discipline holds the key to understanding; therefore, mutual learning critical
	Types of learning: about what is (facts and explanations), about what should be (values) and about participants
	In mutual learning, personal experiential and processed knowledge are integrated
	Ideal conditions for learning: accurate and complete information, freedom from coercion, openness to alternative perspectives, ability to reflect critically upon presuppositions, equal opportunity to participate, and ability to assess arguments in a systematic manner and to accept rational consensus as valid
	Learning involves various thinking modes (e.g., concrete experience, reflective observation, abstract conceptualization, active experimentation); combined to form learning dialectics
Social learning	Social learning: the process of framing issues, analyzing options, and debating choices in the inclusive deliberation
	Occurs when citizens involved in working out mutually acceptable solutions mature into responsible democratic citizens and reaffirm democracy
	Distinction between cognitive learning (where knowledge is a dominant variable) and social learning (based on responsive communications leading to the reframing of a policy issue)
	Two general component of social learning: cognitive enhancement (i.e., the acquisition of knowledge) and moral development (i.e., growth in the ability to make judgments about right and wrong)
	For social learning to occur there must be a free expression of attitudes, feelings, and intentions
	No predetermined outcomes; supported by information from multiple perspectives, citizens add value, there are serious and substantive discussions, and discussions are supported by neutral facilitators
Practical and deliberative learning	Practitioners learn and reflect as they act with others in practical situations
	Practitioners reflect in action, make moves, evaluate results of moves, and reconsider working theories; practice can lead theory; theory and practice integrated
	Practitioners learn alone or from or with others; can learn from systematic studies and by listening to practice stories from thoughtful practitioners; deals explicitly in the everyday language of practical life
	Double visioning: ability to act from one perspective while holding awareness of other possible perspectives
	Reflective transfer: the process by which patterns detected in one situation are carried over as projective models to other situations where used to generate new causal inferences and are subjected to new, situation-specific validity tests

(Continued)

Table 8.5 (*Continued*)

Collaborative learning	Sees EIA as a learning and civic discovery process where people act together and find new solutions
	Designed to address complex and controversial issues; combines elements of systems methods, mediation/dispute management with experiential learning theory
	Process: introduction to process, identify situation to be improved, share situation perceptions and description, dialogue about interests and concerns, develop transformative models, compare models with reality, and collaborative arguments about desirable and feasible change
	Emphasizes learning and negotiation interaction as the means through which learning and progress occurs
	Attributes: stresses improvement (rather than solution), situation (rather than problem or conflict), concerns and interests (rather than positions), systems thinking (rather than linear thinking); recognizes that considerable learning about science, issues, and values will have to occur before implementable improvements are possible
Critical and transformative learning/critical EIA education	Critical pedagogy: accepts the transformative possibilities of willed human action on an individual and social level; student centered with emphasis on democratic dialogue
	Major descriptors of critical pedagogy: participatory, situated (in student thought and language), critical, dialogical, desocializational (students desocialized from passivity in classroom), multicultural, research oriented, activist (classroom is active and interactive), and affective (interest in broad development of human feelings)
	Transformative theory of social learning: explores not only how our arguments change in dialogue and negotiations but how we change as well; transforming ends, ideas, and ourselves
	Transformative learning: a comprehensive theory of how adults learn; focuses on learning process and accommodates social context; describes how individuals improve instrumental (how to control and manipulate the environment) and communicative competence (trying to understand what someone means when he or she communicates)
	Learning not just through arguments, reframing of ideas, and critiques of expert knowledge; also through transformations of relationships, responsibilities, networks, competence, and collective memory and memberships
	Critical EIA education encompasses both education about EIA and education through EIA; includes education about project, environment, how decision-making processes and project decisions can be challenged, and how members can work together to pursue their own goals
	Critical EIA education: contributes to human democratic liberation, to assessment activities, and to fostering of critical consciousness (enables public to evaluate dominant discourse and to present forceful counterarguments)

Table 8.5   (*Continued*)

Traditional knowledge	Cumulative, dynamic body of knowledge and beliefs about the relationship of living beings with one another and their environment handed down through generations by cultural transmissions; biophysical, cultural, and cosmological; represents a cognitive spiritual awareness based on the relationship of indigenous people and their environment
	Built up over time and continuing into the present, by people living in close contact with the natural environment
	Attribute of societies with historical continuity of resource use practices (generally, indigenous or tribal); is unique to each tradition and is closely associated with a given territory; varies among different Indigenous societies
	Usually, linked to a belief system that stresses respect for the natural world; takes a holistic perspective which stresses the place of humans with the natural system; four perspectives: taxonomic, spatial, temporal, and social
	Oral communications; taught through observation and experience; explained based on spiritual and social values
	Can assist with building relationships between proponents and indigenous peoples; barriers—perceptual, skepticism of scientific community and political obstacles; needs to be controlled at the community level

Sources: Berkes (1993), BC EAO (2001), Brascoupé and Mann (2001), Daniels and Walker (1996), Forester (1999), Gadgil et al. (1993), Healey (1997), Johannes (1999), Mezirow (1994), Paci et al. (2002), Saarikowski (2000), Schön and Rein (1994), Sinclair and Diduck (2001), Sköllerhorn (1998), Webler et al. (1995).

8.4.6   Negotiations

As a form of public participation in the EIA process, negotiation is based on a conflict and interest-oriented view of society. Negotiations can be aided or unaided. Aided negotiations can follow the route of litigation through the courts or can employ alternative dispute resolution (ADR) mechanisms. This analysis focuses largely on the potential roles of various forms of ADR in the EIA process. It does not preclude unaided negotiations. It also recognizes that ADR tools can be applied for purposes other than avoiding, managing, and resolving conflict.

ADR is based on the theory that the people involved in a controversy, because they know their own needs and interests, are best able to develop reasonable and lasting solutions (US EPA, 2000a). ADR is voluntary and flexible (Bingham and Langstaff, 1997). It involves stakeholders discussing differences and working together as a group to solve problems or to address issues (SERM, undated). Neutral third parties (e.g., a facilitator, a mediator) often assist the parties in reaching mutually acceptable accommodations. Third parties (e.g., active mediation) are not always neutral. They can help ensure equitable procedures and fair, enduring, and environmentally sound outcomes (Susskind and Madigan, 1984). Authorities retain final decision-making authority with some forms of ADR (e.g., facilitation, mediation) but not with others (e.g., binding arbitration) (Susskind, 1999). Parties to the process are not contractually liable for their actions during negotiations (McGlennon and Susskind, undated). ADR seeks to avoid, mitigate, and resolve conflict, without resorting to litigation and where existing administrative procedures are ineffective (US EPA, 2001a).

ADR has been applied in many situations (e.g., adjudication, rule making, policy development, enforcement actions, permit issuance, contract administration, EIA) (US EPA, 2001a). The types of conflicts, which can be addressed through ADR, are many and diverse. They can, for example, concern resource allocation, policy priorities, jurisdiction, environmental quality standards, data, values, interests, and relationships (Campbell and Floyd, 1996; Moore, 1986; Priscoli, 1999). Conflicts can be perceived or latent (whether parties conscious of), manifest or potential (whether taking place), real or displaced (whether actors correctly conceive), system dependent or independent (whether generated internally or externally), zero or variable sum (win lose or win–win potential), cooperative or noncooperative (information exchange and coalition building potential), means or ends oriented (objectives or instruments), formal or informal, institutionalized or ad hoc (Sager, 1994). Conflicts can also be characterized based on the number of parties involved, the types of parties involved, and the extent to which there is agreement regarding goals and problem definition (Priscoli, 1999). Outcomes from ADR can include pacification, settlement (i.e., procedural accommodation), or solutions (i.e., substantive improvements) (Sager, 1994).

ADR is not always appropriate. It should be possible to identify and include all relevant parties. All parties must be willing and able to come to the table voluntarily, to negotiate in good faith, and to reach a settlement on behalf of their constituents. Each party should formally accept ADR as potentially preferable to either not participating or to litigation (i.e., they gain some value) (Amy, 1987; Bingham, 2001; Susskind et al., 1999). The agency with final decision-making responsibility should support the process. The participants should be prepared to accept the negotiating ground rules and the negotiating structure (Bingham, 2001). The issues should have crystallized or "ripened" to the point that a common purpose can be agreed to, alternative courses of action can be determined, trade-offs and compromises can be identified, and solutions capable of joint acceptance are possible (Amy, 1987). Agreements reached through negotiations should be reasonable and capable of implementation (Moore, 1999). Legal challenges should be unlikely (Rodwin, 1982). ADR is not a good idea if policy precedents are likely to be set or if unacceptable environmental conditions could result (Bingham, 2001; Moore, 1999). ADR is very difficult, but not impossible, when there are fundamental clashes of values or principles.

It should be possible to ensure a relative balance of power among the parties (Amy, 1987; Armour and Sadler, 1990). Adequate resources and relevant data should support the process. There should a deadline and some urgency for a decision (Bingham, 2001; Moore, 1999). There must be sufficient time for consensus building (Susskind et al., 1999). Third-party support (e.g., facilitation, mediation) and training for participants (if needed) should be available (Emond, 1990; Susskind et al., 1999). It should be possible to address both technical and nontechnical issues. Information should be freely shared among the parties (Bingham, 2001). It should be possible to create a clear map outlining how consensus is to be built (Susskind et al., 1999). Communications with broader interests should be maintained throughout the process (Bingham, 2001). Final agreements should be written and signed by each participating party representative. The resulting document should be legally binding and enforceable. Opinions vary as whether the process should be transparent or confidential (Bingham, 2001; Emond, 1990).

Figure 8.6 highlights some potential characteristics of an ADR process. The process starts with an overview of the factors (e.g., issues, range of parties) that determine whether ADR is possible, appropriate, and timely compared with the available alternatives (such as litigation or conventional administrative procedures). It may for desirable to formalize this review in a conflict assessment (Susskind, 1999). Parties associated with each interest are identified. Credible representatives for each party are determined and recruited (Susskind and Madigan, 1984). Funding commitments are obtained (Susskind, 1999). Assurances are sought that decision makers will take the process outcomes seriously and will allow sufficient time to ensure a sincere consensus-building effort. Procedural rules for the process are drafted and refined in consultation with parties. An overall strategy or plan is formulated and refined (Moore, 1999). The strategy addresses such matters as schedule, timing, resource requirements, training needs, communications methods, third-party assistance needs, roles and responsibilities, contacting procedures, decision-making links, and procedures for maintaining communications with constituents (Susskind, 1999). A mediator or facilitator is identified, together with a recorder (Susskind and Madigan, 1984). An agenda for the initial negotiations session is prepared (Moore, 1999).

The negotiations process is highly iterative but appears to coalesce into four overlapping steps: (1) initial deliberations, (2) focusing, (3) detailed deliberations, and (4) final refinements. During initial deliberations, the underlying interests of each party are identified, background data are obtained and exchanged, the negotiations skills of participants are enhanced (where necessary), the committee structure is determined, supplementary data collection, analysis, and review (i.e., joint fact-finding) takes places, the concerns and priorities of each party are identified, initial concept statements are formulated, and general efforts are made to build trust, rapport, and cooperation among the parties (Moore, 1999; Susskind, 1999). Focusing involves identifying key issues, determining points of agreement and disagreement, establishing the scope and boundaries for the negotiations' packages, identifying possible negotiations packages, generating texts to focus discussions, and excluding clearly unacceptable packages (Moore, 1999; Susskind, 1999). In detailed negotiations, packages are presented, concessions and commitments are advanced and traded, the consequences of the packages are determined, the packages are evaluated and possibly combined, solutions are sought which are mutually acceptable and which will maximize joint gains, measures are identified to prevent and offset

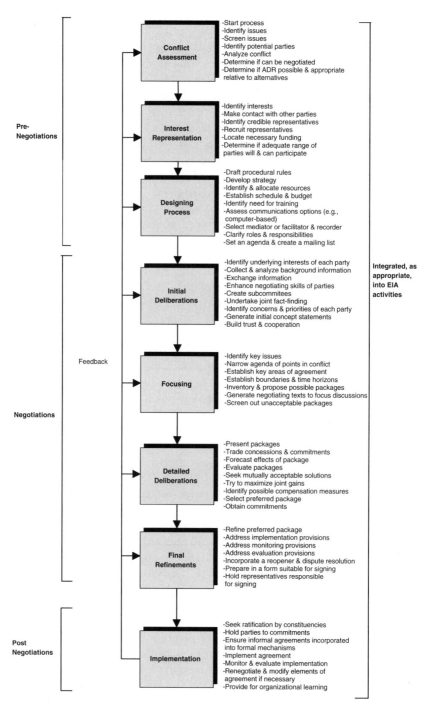

Figure 8.6 Example of an ADR process. (Adapted from: Carpenter and Kennedy, 1988; Harashina, 1995; Moore, 1999; Sadler, undated; Smith, 1993; Susskind, 1999; Susskind and Cruikshank, 1987; Susskind and Madigan 1984; Susskind et al., 1999.)

negative features, a preferred package or package combination is identified, and commitments are obtained from each party to the preferred package (Susskind and Madigan, 1984). Final refinements elaborate on the preferred package to add implementation, monitoring, compensation, and post negotiations evaluation provisions. A reopener or dispute resolution mechanism is often included (Susskind, 1999). The final package is prepared in a form suitable for signing by the parties. The parties are held responsible for signing the agreement.

During negotiations the parties are expected to be responsible (act in good faith), open (all concerns explicit), respectful (all heard and taken seriously), trustworthy (nothing held back, no hidden agendas), fair (power inequities offset), flexible (discussion based on interests rather than predetermined positions), and constructive (search for outcomes that meet and further the interests of all) (Innes, 1996; Innis and Booher, 1999; Praxis, 1988; Susskind, 1999). The parties should strive for but not insist on consensus (Nagel, 1987). Consensus can be either unanimity (a good idea but not always practical) or overwhelming agreement (in contrast to a bare majority) (Susskind, 1999). During negotiations, authority does not have to be given up (any party can walk away at any time), and principles and interests need not be abandoned (the process seeks solutions that respect and further the interests of each participant) (Susskind et al., 1999). A third party, such as a facilitator or a mediator, can assist in consensus building, conflict resolution, and joint problem solving (Smith, 1993). A record of the process should be kept (Susskind, 1999). As the process unfolds it may be necessary to revisit earlier stages.

Implementation can involve ratification by constituents. The parties are held to their agreements (Susskind and Madigan, 1984). It may be necessary to ensure that informal agreements are incorporated into formal mechanisms. The provisions of the agreement need to be implemented. Implementation should be monitored and evaluated. Monitoring results may necessitate renegotiations or modifications to elements of the agreements. The lessons and insights obtained through negotiations and monitored should be incorporated into the organizational learning mechanisms of the process participants (Susskind, 1999).

Conflicts or potential conflicts, which might warrant ADR, can arise at any stage in the EIA process where interests, values, or perspectives might clash (e.g., scoping, significance interpretation, impact management). They can concern the design and execution of public participation activities. They can occur both prior and subsequent to approvals. They can pertain to procedure or to substance. ADR procedures can help avoid conflict and scope issues when instigated near the outset of the EIA process. They can help keep the process on track when dealing with difficult interpretative issues such as alternatives evaluation and impact significance. They can help bring the process to a successful conclusion when dealing with troublesome issues surrounding mitigation, compensation, monitoring, and implementation. ADR can represent an alternative to or can scope a hearing or court action.

Various ADR methods may be appropriate at different stages in the EIA process. Table 8.6 lists examples of negotiations and ADR methods and mechanisms. The methods vary in the roles they can perform (e.g., conflict avoidance, fact-finding, determining if ADR is practical, expediting meetings, problem solving, dispute

Table 8.6   Examples of Negotiations Methods and Mechanisms

Unaided negotiations	Contending parties work out differences without help
	Distinction between position (each side argues from positions) and interest (alternative solutions that meet interests or needs presented) negotiations
	Differences resolved based on compromise or interest-based principles
	Informal: serves to soften hard positions, explore underlying interests, develop options, and reach a mutually acceptable resolution
Conflict anticipation and convening/ conflict assessment	Dispute prevention: third party identifies potential disputes before opposing positions fully identified
	Involves communications, building personal relationships, and establishing procedures for addressing issues before they become disputes
	Conflict assessment: a document that spells out what the issues are, who are the stakeholders, where they disagree, and where they find common ground; usually prepared by a neutral outsider based on confidential interviews with key stakeholders
	Allows the assessor to explore the parties' incentives and willingness to negotiate in good faith; creates an opportunity for the assessor to educate stakeholders about what it takes to bring consensus process to successful conclusion
	Produces recommendations regarding who has a stake, what issues are important to stakeholders, and whether makes sense to proceed given constraints and circumstances
	Phases: introductions, information gathering, analysis, process design, report writing, and report distribution
Information exchange and joint fact-finding	Information exchange meetings: parties share data and check out perceptions of each other's issues, interests, positions, and motivations in an effort to minimize unnecessary conflicts over facts
	Used to establish a shared framework for analyzing a dispute, resolving disputes on matters of fact, and clarifying disagreements of fact
	Fact-finding can be used in scientific, technical, or business disputes in which knowledge is highly specialized
	Third-party subject matter expert is chosen by the parties to act as a fact-finder or independent investigator
	Can identify facts and areas of agreement and disagreement; expert then submits a report or presents the findings
Conciliation	Neutral party, generally with no stake in the dispute, to informally communicate separately with disputing parties for the purposes of reducing tensions, build trust, and agreeing on a process for resolving the issue
	Attempts to assist negotiators in searching for accommodations
	Helps establish a basis for direct negotiations
Facilitation	Facilitator: a nonpartisan or neutral-trained specialist who helps people design effective meetings and problem-solving sessions and then acts as the meeting leader on behalf of the group; does not have the authority to make substantive decisions on behalf of the group

Table 8.6 (*Continued*)

	Focuses on process and uses available tools to create and foster an environment conducive to joint problem solving
	Assists parties in coming together to exchange views, share information, and clarify differences; helps promote meetings that are purposive, efficient, productive, and civil
	Examples of responsibilities: taking care of meeting logistics, reminding parties of ground rules, intervening when someone violates the ground rules, keeping discussion on track and schedule, summarizing and focusing discussion, identifying key points, clarifying issues and interests, orienting the group to objectives, promoting effective communications, eliciting creative options, and maintaining a nonthreatening environment that encourages people to participate
Mediation	Involves the intervention of an acceptable, impartial, and neutral third party, who has no decision-making power, to assist contending parties in voluntarily reaching their own mutually acceptable settlement of issues
	Mediator is only concerned with process issues, works hard to ensure that process is fair and unbiased in the eyes of the parties at the table; process is voluntary, informal, and confidential
	Assists in clarifying issues, facilitating information collection, exchange and evaluation, identifying key issues, uncovering hidden interests, designing an effective negotiations process, developing options for dispute resolution, and helping to identify and formulate areas of agreement
	Mediation process: decision to start, mediator contacts parties and outlines process and logistics, mediation meetings held, mediator's report is prepared and report signed by parties and submitted to approving authority
	Conventional mediator: no stake in outcome; active mediator: works hard to ensure that process is fair, unbiased, and open to all parties affected by the outcome, whether they sit or do not sit at negotiating table, seeks to ensure that the outcome is viewed as fair by the community at large, is reached efficiently, and remains stable after bargaining
Minitrial, dispute review board, or disputes panel	Minitrial: expedited presentation of positions and evidence to a panel composed of senior decision makers representing each participant and a third party; authorized representatives hear case and negotiate agreement; parties can present summary proofs and arguments; third parties can advise, mediate, or make advisory opinion
	Voluntary, expedited, nonjudicial, informal, and confidential procedure; used to address complex technical issues where litigation costs would be high and senior decision makers want maximum control of terms of settlement
	Dispute review board or dispute panel: provides the parties with an objective evaluation of the dispute by fully qualified experts; opinion of the board is advisory, with the parties negotiating a final resolution

(*Continued*)

Table 8.6   *(Continued)*

Inquiries, public hearings and adjudication	Formal judicial or quasijudicial proceeding; parties meet in an adversarial setting before an impartial judge or panel
	Can ensure thorough presentation and testing of evidence; can address issues not suitable for or remaining after ADR
	Can be intimidating, time-consuming, and expensive
	Can exaggerate conflicts; encourages people to take positions rather than share fundamental interests or to engage in problem solving; sets up a win–lose situation
Arbitration	Parties select a neutral individual or panel with expertise on issues at dispute and set rules or norms to apply
	Parties (or counsel for each party) present their case (facts, positions, and formal arguments); arbitrator recommends a basis for settlement
	Nonbinding arbitration: parties are not bound to submit to arbitrator's decision; but advise normally carries a great deal of weight
	Binding arbitration: parties agree to live by arbitrator's decision even if they are unable to reach a voluntary decision themselves
	Each party seeks to design most reasonable outcomes for presentation to arbitration; emphasis on design rather than argument
	Results in an unbiased judgment, avoid problems of litigation, prevents loss of face, and encourages a decision
Negotiated rule making	Process of bringing together representatives of various interest groups and a government agency to negotiate the text of a proposed rule; goal is for the committee to reach consensus on the text of a proposed rule
	Parties need to perceive that transaction costs of developing, implementing, and enforcing regulations in the usual fashion are high and that significant costs savings are possible from a different approach
	Involves convening the appropriate parties, clarifying roles and responsibilities regarding consensus-based decision-making process, reaching and testing the scope of the agreement (joint problem solving and fact-finding, option development and evaluation, selection of a preferred option), and binding parties to their commitment (draft rule published and subject to comment period)

Sources: Bingham (1986, 2001), Bingham and Langstaff (1997), Campbell and Floyd (1996), Creighton et al. (1999), De Bono (1992), Forester (1999), Lowry et al. (1997), Morgan (1998), Moore (1999), Saarikoski (2000), Sager (1994), Smith (1993), Susskind and Madigan (1984), Susskind et al. (1999), Weber (1998), Westman (1985).

resolution). They also vary in the role of participants, in the formality of proceedings, in the degree of confidentiality, and in the types of situations into which they are applied. The application of ADR methods in EIA should take into account these differences, general ADR characteristics, strengths and limitations, the specific characteristics, strengths and limitations of individual ADR methods, and the match between methods and context.

ADR methods have had considerable success in contributing to settlements (Creighton, 1999; Sipe and Stiftel, 1995). Participants tend to be very satisfied with the process (Sipe and Stiftel, 1995). ADR has been effective in identifying the rationale for settlements (Creighton, 1999). It is often faster and less costly than litigation (Campbell and Floyd, 1996; Harashina, 1995). It is credited with building and enhancing relationships, facilitating higher-quality decisions, identifying and solving conflicts and problems, furthering procedural and substantive equity, enhancing the likelihood of approvals, implementation, and compliance, and contributing to stakeholder empowerment (Campbell and Floyd, 1996; Creighton et al., 1999; Harashina, 1995; Harrop and Nixon, 1999; Innes and Booher, 1999; Smith, 1993; Smith et al., 1997). Managers often favor ADR because it is voluntary, nonjudicial, confidential, and does not necessitate control delegation or sharing (Creighton et al., 1999). Stakeholders may prefer ADR because it is informal, nonintimidating, and conducive to joint problem solving.

There are, however, many situations, as noted earlier, where ADR is inappropriate or impractical. Some suggest that agreements reached through ADR tend to be vague and general (Neuman, 2000). It is feared that by operating on the fringes of institutional structures, ADR will either be ineffective or will undermine representative democracy and state intervention (Fischler, 2000; Neuman, 2000). ADR, it is argued, could lead to ethical, democratic, or environmental sacrifices if objectives, principles, or interests are compromised in the quest for consensus (Smith, 1993). Citizens may find themselves at a disadvantage because of imbalances in training and expertise (Smith, 1993; Smith et al., 1997). They may find the time commitments too onerous (Canter, 1996; Smith, 1993). They may resist the need for coalitions with other interests. They may fear co-option (Canter, 1996). Agencies could be reluctant to participate because ADR is inconsistent with their conventional operating procedures (Manring et al., 1990; Smith, 1993). They may hesitate to involve others in decision making on the grounds that they will lose control or that they will be opening up their decision-making processes to scrutiny and legal challenges (Bingham and Langstaff, 1997). They too may lack negotiating skills and experience (Sachs, 1982). ADR could run counter to the financial interests of outside counsel. Some argue that it is difficult to enforce implementation and monitoring requirements obtained through ADR (Smith, 1993).

The many ascribed benefits of ADR are, according to some, overstated (Bingham, 2001). While acknowledging that ADR has a high settlement rate, they suggest that it does not necessarily result in a higher compliance rate (Sipe and Stiftel, 1995; Sipe, 1998). Each ascribed ADR benefit, it is suggested, should be treated as a success measure and should be systematically tested against experience (Bingham, 2001). ADR brings parties together to resolve disputes and to solve problems. Disputes are frequently resolved. Most ADR disadvantages and constraints can, with judicious application and a heightened awareness of potential pitfalls and limitations, be avoided or ameliorated. It is less apparent whether and how ADR generates creative solutions to problems or proactively generates win–win options and opportunities, which move beyond the reconciliation of interests (De Bono, 1992). The specific techniques for creative collaboration are not well developed in ADR. It is possible that by relying excessively on the skills of facilitators and mediators, groups will be slow to develop or to apply their own collaborative skills (De Bono, 1992). ADR, applied appropriately and in the right circumstances, can assume a vital role in EIA public participation. But alternative approaches and a wider range of creative methods are probably needed before an EIA process can be said to be fully collaborative.

8.4.7 Collaboration

Collaboration is about people cooperatively working together in a joint endeavor with substantive aspirations. The orientation, with collaboration as compared with negotiations, shifts from interests and positions to perspectives, from problems to visions and opportunities, from conflict management to creative exploration, and from negotiations to collaboration. Negotiations may (or may not) be an element of or precede collaboration. Collaboration can build on a base of but is more than the sum of effective consultation, communications, mutual education, and negotiations. Effective collaboration transcends the other elements of EIA participation.

Table 8.7 highlights the characteristics of several collaborative concepts relevant to EIA process management. Some concepts, such as joint fact-finding, active mediation, and consensus building, also are used in negotiations. Others, such as procedural justice, closely parallel such related communications and educational concepts as discourse ethics. These commonalities underscore the many interactions among EIA participation elements. Collaboration encompasses, integrates, and transcends involvement, communications, mutual education, and negotiations.

The concepts exhibit numerous ideal collaborative process characteristics. Collaboration, for example, should be inclusive. It should include multiple interests, issues, values, and perspectives. The participants should bring to the table a diversity of relevant knowledge and experience. A collaborative process should be jointly undertaken and owned by the participants. Power and responsibility should be shared. Technical specialists (e.g., fact-finders) should assume a support rather than a lead role. A collaborative EIA process should be directed toward and guided by substantive environmental management, environmental justice, and sustainability ends. The process should be positive and purposive. Third parties (e.g., active mediators) should support the realization of stable, efficient, and "good" outcomes.

The participants should generate and then pursue visions, goals, objectives, and opportunities. They should also rectify and ameliorate problems. The process should be open, voluntary, informal, flexible, cooperative, and consensus seeking. It should be guided by procedural, democratic principles such as clarity, honesty, commitment, mutual recognition, mutual respect, trust, open-mindedness, confidence, fairness, and attention to detail. The process should be procedural just. Power imbalances should be offset. The process should be creative and synergistic. Participants should jointly discover and explore new ideas and solutions. They should make effective use of methods (e.g., models, scenarios, role-playing simulation, brainstorming, lateral thinking procedures, nominal group process) and group

Table 8.7 Examples of Potentially Relevant Collaboration Concepts

Joint fact-finding	Information exchange: parties share data and compare perceptions, perspectives, and motivations
	Fact-finder is an independent expert chosen by the parties to conduct investigations
	Fact-finders have technical expertise; they investigate and analyze issues
	Joint fact-finding can provide a shared factual and analysis basis for collaboration
Joint planning	Representatives of the interested parties participate in a committee with the power to make a binding decision
	Involves dialogue, shared responsibility, multiple perspectives, and in-depth deliberations
	Tends to work best when serious environmental impacts and trade-offs, a wide range of complex issues, many concerned public and agency groups, sufficient time for planning, public strongly desires formal involvement program, and agency has capacity to support
	Works best when mutual respect and trust, recognition of knowledge, experience and respect of each participant, recognition of individual and joint rights, and responsibilities and agreement on meeting procedures
Partnering	Formal or informal means to improve and build the relationship between government and another party and/or to work with one or more parties to achieve a common goal
	Used primarily during contract performance; goal is a more cooperative, team-based approach
	Partners share some level of responsibility, planning, and decision making and ownership of process and product; resources, expertise, energy, and risks shared
	Parties jointly define a clear vision, goals, and action items and then work together to achieve; process is working when sharing, clear expectations, trust and confidence, commitment, responsibility, courage, understanding, and respect, synergy (outcome more than sum of partners), and excellence
Group problem solving and opportunity seeking	Problem solving: group identifies problem, analyzes problem, identifies and evaluates possible solutions, and develops a plan for implementing the problem solution agreed to by the group
	Opportunity seeking: begins with search for positive possibilities rather than reaction to something going wrong
	Seeks to adhere to certain virtues of group inquiry such as clarity, honesty, openmindedness, and attention to detail
	Various characterizations of group development process (e.g., forming, storming, norming, performing) and of group maintenance behavior (e.g., harmonizing, gate keeping, encouraging, compromising, standard setting and testing and relieving tension)
	Numerous individual and small-group methods for redefining and analyzing problems, for generating ideas, for evaluating and selecting ideas, and for implementing ideas (e.g., brainstorming, nominal group, process, forced relationships or free associational, related or unrelated stimuli, lateral thinking techniques, charrettes, simulation games) *(Continued)*

Table 8.7 (*Continued*)

Coalition building/ networking	Networking refers to linking stakeholders through formal or informal channels so as to bring about plan formulation and implementation
	Networks exist over time, are invitational, are numerous, have a limited capacity, are only as good as their members, depend on exchanges and incentives, tend to focus on selected actions, are channels of action, are a source of power, and take place in a symbolic context
	Network tasks: map the terrain, gain information and identify actors; identify the relevant leverage points; select the tentative coalitions of support; float the initial image to symbolize the possibility of action; adapt the technical argument to the requirements of support and opposition; organize the coalition to trigger the multiplier and maintain and feed the coalition moves toward implementation; can be extended to collaboration in institutional design
Active mediation	Challenge to move beyond role as process people; mediators are nonpartisans but concerned with representation of affected parties in the mediation process and the efficiency, stability, and well-informed character of potential mediation outcomes
	Moves beyond search for acceptable agreements within a given space of interests; the space is altered and the participants are transformed and empowered
	Searches for new possibilities and agreements; concern with the decision's goodness and the quality of agreements
	Involves a broader conception of political life, the public interests, deliberation and debate, mutual recognition and discussion, learning, and civic discovery
Procedural and environmental justice	Assumes that the procedures used to arrive at decisions are significant determinants of satisfaction separate from the effect of outcomes; procedures perceived as unfair might reduce satisfaction with what are otherwise judged as objectively fair decisions
	Practice of fairness: talk to trust, to outside selves, to mutual respect, and to joint search for mutually acceptable reasons and measures
	Fairness rules: decision consistency, suppression of bias, accuracy, error correction, representativeness of groups of affected people, ethically compatible with fundamental moral and ethical values and correction of power imbalances
	Also necessary to address environmental justice issues
Collaborative planning	Process through which parties who see different aspects of a problem can constructively explore their differences and search for solutions that go beyond their own limited visions of what is possible
	Embraces sustained dialogue, stresses common ground, and promotes shared vision of the future
	Involves shared power, open discussions, and shared values
	Characterized by face-to-face dialogue, mutual learning and voluntary participation; critical listening, reflection-in-action, and constructive argument all interact
	Fosters the inclusion of all members of political communities while acknowledging their cultural diversity

Table 8.7 (*Continued*)

Collaborative public participation process	Characteristics: shared vision and objectives and measurable outcomes; process is equally managed by stakeholders; involves up-front planning, conflict resolution and open communications among participants; balanced and inclusive stakeholder participation; strong leadership; capacity created for stakeholders to understand information; possible help from facilitators
	Sound informational base, rests on sound democratic participation principles, honors a full spectrum of values, and holds everyone responsible for success
	Begins with mutual education; no one leader; no one excluded from table; works together with community to define shared visions that sustain the community and the environment
	Requires advance planning, management support, funding for a facilitator and other expenses, and sufficient time to reach useful results; crucial to maintain a balance of power among participants
Co-management	Government and stakeholders work cooperatively to undertake integrated management of the environment and natural resources in a sustainable way consistent with goals of parties; applicable primarily to indigenous communities; allows parties with an interest in the ownership and management of natural resources to power share; various mechanisms (e.g., a cooperative, a co-management council)
	Makes it possible to integrate local community interests with third-party and government interests
	Principles: public ownership and government responsibility, cooperate as partners, stewardship of natural resources and environment and integration of environment/natural resources, economic development and social well-being, inclusive process
	Lessons: government commitment, open debate about long-term direction, meaningful third-party agreement, a reflective and evolving process, real teeth and shared decision making, direct community input to resource inventory and planning, community and local government staff, formal agreements, a broad range of interests, coalitions of interests, multiresource framework, interim measures, co-management plan and implementation, action linked to information and understandability
Consensus building	Process in which people agree to work together to resolve common problems in a relatively informal, cooperative manner; two meanings of consensus: unanimity and positive support from large proportion of participants
	Good consensus-building process: includes all relevant and significantly different interests, is driven by a purpose and tasks that are real, practical, and shared by the group, is self-organizing, allows participants to decide on ground rules, objectives, tasks, working groups, and discussion topics, engages participants, keeps participants at the table, interested and learning through in-depth discussion, drama, humor, and informal interaction, encourages challenges to the status quo and fosters creative thinking, incorporates high-quality

(*Continued*)

Table 8.7 (*Continued*)

	information of many types and assures agreement on its meaning, seeks consensus only after discussions have fully explored the issues and interests and significant effort has been made to find creative responses to differences
	Represents a way to search for feasible strategies to deal with uncertain, complex, and controversial planning and policy issues of common concern; properly designed can produce results that approximate the public interest
	Creative efforts can be enhanced by techniques such as scenario formulation and role-playing simulation
Shared visions planning	Way to use computers to help stakeholders to participate in rigorous planning analyses; developed by Institute for Water Resources, U.S. Army Corps of Engineers
	Shared vision models are built using new, user-friendly graphical simulation models; built with decision-maker and stakeholder involvement; models used to evaluate alternative plans according to decision criteria
	Marries systems engineering, public policy, and public involvement; similar to adaptive environmental assessment and management
	Because experts and stakeholders build models together, conducive to developing a consensus view of how system works as a whole and how it affects stakeholders and the environment
	Model flexibility makes it easy to analyze sensitivity of conclusions to errors in data, changed forecasts, or conflicting assumptions
	Other more general visioning approaches combine team building with an alternative futures planning process to produce shared visions; factors provide a basis for themes, which are, in turn, built into scenarios and strategies which are compared, which then form the basis for short-medium-terms goals and action plans; process completed with assignments and target dates
Constructive engagement	Approach that brings groups together to establish and monitor a facility's environmental activities through a cooperative, nonadversarial partnership
	Takes many forms (e.g., citizen advisory groups, stakeholder negotiations, formal mediations, "good neighbor agreement" processes, oversight committees, independent organizations)
	Have dealt with issues such as site location, facility operations, emissions and waste controls, worker health and safety, regulatory relief, site cleanup, and pollution prevention
	Offers an approach to improving communications among stakeholders and for finding creative solutions about facility activities

Sources: Bauer and Randolph (2000), Benveniste (1989), Creighton et al. (1999), De Bono (1992), Forester (1999), Gray (1989), Healey (1997), Innes (1996), Innes and Booher (1999), Kreiger (1981), Lawrence et al. (1997), Laws (1996), Maynes (1989), Moore and Woodrow (1999), Mosley et al. (1999), Nagel (1987), Patton et al. (1989), Praxis (1988), PCSD (1997), Saarikoski (2000), SERM (undated); SIFC (1996), Susskind and Cruikshank (1987), Susskind and Madigan (1984), Susskind et al. (1999), US EPA (2000a,b, 2001a,b), Westman (1985), Witty (1994).

development and maintenance techniques, conducive to fostering and applying creativity.

Participants in an effective collaborative process are altered and empowered by the experience. The process should not be confined to the group. Ongoing contacts should be maintained with constituents. The process, to be more than an interesting experience, must be practical and real. Outcomes should be formalized (e.g., a plan, a strategy, an agreement, a contract, a rule, an EIA, facility operations) and capable of implementation. The collaborative process, to be effective, needs to be supported by management (e.g., resources, time) and by the public. Implementation may necessitate networking and coalition building and maintenance. Sometimes, institutional design and reform may be required.

8.5   INSTITUTING A COLLABORATIVE EIA PROCESS

8.5.1   Management at the Regulatory Level

Consultation   US EIA requirements and guidelines make numerous references to open decision making, public notification, early public involvement, and public involvement at key decision points. Documents must be made available to other governments (states, Indian tribes, local agencies) and to the public. Comments must be invited. There are opportunities for public involvement during scoping and during EIA document review. There is the potential for judicial review of the final EIS. Several federal departments [e.g., Department of Energy (DOE), Environmental Protection Agency (EPA)] have public participation policies, public participation guidelines, and additional public participation requirements (e.g., DOE requires a public meeting). Some guidelines refer to the participation of indigenous groups and tribal citizens. U.S. environmental justice requirements are conducive to broadening the range of publics involved in the EIA process. The US CEQ effectiveness review notes that some groups and individuals continue to remain concerned about when public consultation begins, the limited number of involvement opportunities during the EIA process and the extent to which public comments and suggestions are taken seriously by decision makers (US CEQ, 1997a).

Public participation in the EIA process is a purpose of the Canadian Environmental Assessment Act. The Canadian EIA requirements include specific notification and document availability requirements and provisions for an electronic registry. Intervener funding is currently available. It is proposed that participant funding also be provided. The extent of public involvement in screening reports and comprehensive studies has been criticized. Some environmental groups, for example, suggest that public involvement occurs too infrequently (especially with screening reports), occurs too late in the process, provides insufficient time for the public to participate effectively, and does not adequately integrate public concerns and suggestions. Responsible authorities tend to believe that they are doing a good job of involving the public. The need for additional public involvement opportunities is acknowledged in the five-year review of the Canadian EIA requirements. Steps are being taken to enhance public participation in both screenings and in comprehensive studies (e.g., public involvement during scoping). Legislative changes refer to considering community and aboriginal traditional knowledge and to establishing an aboriginal advisory committee. Some Canadian provinces (e.g., Alberta and British Columbia) have more experience with continuous involvement procedures (e.g., community advisory committees) and with community-based EIA.

European Union EIA project and SEA directives include general references to informing the public, to making documents available, to providing opportunities for the public to express opinions on draft documents, and to allowing sufficient time for consultation and transparent decision making. Individual states have considerable discretion in determining the timing and extent of public involvement. The Aarhus Convention on Access to Information, Public Participation in Decision-Making and Access to Environmental Justice (1998) (into force October 2001) includes more detailed requirements regarding public access to information and early and effective public participation in the decisions associated with specific activities and with plans, policies, and programs. The Aarhus Convention, together with the directives, guides, and handbooks for implementing the convention, should enhance the potential for effective EIA public participation. States tend to do more than the minimum required.

The Australian EIA requirements include provisions regarding notification procedures, document availability, and the soliciting of public comments in response to draft EIA documents. The legislation refers to a cooperative approach involving governments, communities, landowners, and indigenous peoples. EIA documents are required to identify affected parties, to indicate how the communities may be affected, and to describe the views of the public regarding the proposed action. Several provisions are included (e.g., the promotion of the use of indigenous peoples knowledge, the consideration of species and ecological communities of particular significance to indigenous traditions, an indigenous advisory committee) to help involve indigenous peoples.

The four jurisdictions generally include measures to ensure that information is provided to the public and that comments and concerns are obtained from the public. Some progress has been made to further the involvement of traditionally unrepresented and underrepresented segments of society. There are very limited provisions for continuous, in-depth public involvement. There remains room for improvement in ensuring early and frequent public involvement, in involving the public in EIA screening decisions, and in demonstrating how public concerns and preferences have influenced decision making.

Communications   The area of communications is addressed directly in U.S. EIA requirements through document content requirements. Two-way communications, dialogue, and improved communications are recurrent themes in EIA guidance documents. Efforts to facilitate enhanced communications are evident from the preparation of stakeholder directories (US DOE, 2002) and from the soliciting of advise from stakeholder groups and advisory committees. These efforts have been only partially successful. Some groups and individuals still maintain that communications is often one way and sometimes that they are treated like adversaries (US CEQ, 1997a).

Effective communications is a common theme in Canadian EIA requirements and guidelines. A citizen's guide to the EIA process has been prepared. References are made to understandable and focused documents and to communications among government agencies, with other governments and with aboriginal peoples. Specific guidance to facilitate more effective communications is not provided. The same general pattern is evident in Europe and Australia. There is a requirement in Europe to prepare a nontechnical summary. The Aarhus Convention includes specific provisions concerning the collection and dissemination of environmental information. These provisions address such matters as when information is to be provided, how and in what form. Both Europe and Australia refer to clear and understandable documents and to effective two-way communications with the public. The Australian EIA requirements refer to special arrangements for affected groups, with particular communications needs having adequate opportunity to comment on proposed actions. Reference also is made to taking into account the proponent's environmental history.

The four jurisdictions generally address EIA communications in terms of communicating with the public either directly or through documents. General references are made to effective two-way communications. Some scattered references are made to adapting communications strategies and documents to meet the needs of communities and of various segments of society. More specific advice could be provided concerning communications principles, communications skills, offsetting communications distortions, and applying measures to foster and enhance dialogue among EIA process participants.

Mutual Education In the United States, numerous basic and advanced NEPA courses are available to enhance agency, practitioner, and public education about EIA requirements and procedures. EIA process participants can draw upon a host of EIA guidelines, handbooks, checklists, fact sheets, and outlines. Several federal agencies have compiled public participation case studies, methods guidance, and innovative practice materials to enhance the knowledge base of EIA practitioners. Individual agencies provide public participation training programs, sponsor research, and provide technical assistance. Reports preparing by advisory and working groups, councils, and committees have helped educate government officials and others about stakeholder perspectives and knowledge. Community outreach is a major focus of environmental justice requirements and guidelines. The importance of instituting measures to overcome linguistic, cultural, institutional, geographic, and other participation barriers is stressed.

Canada provides considerable EIA guidance, some oriented to EIA practitioners and some geared to a more general audience. Public participation guidelines have been prepared by and for the Canadian Environmental Assessment Agency (CEAA). Various federal agencies have prepared public participation policies and guidelines. CEAA provides EIA training oriented to overall requirements, to specific types of EIA documents, and to cumulative effects assessment. It also has an annual research program and sponsors numerous workshops and seminars. Changes to the EIA legislation refer to considering community and aboriginal traditional knowledge. Guidelines for integrating traditional knowledge into the EIA process are being prepared by the agency and have been prepared by certain individual federal agencies.

A proposed European Commission amendment to the EIA directive, to ensure conformity with the Aaarhus Convention, should contribute to a sounder foundation for mutual EIA education. European EIA guidelines describe the potential roles of public participation in screening, scoping, and EIA document review. Numerous EIA centers have been established across Europe. These centers provide independent advice and undertake applied research. Several individual European states train municipal planners and local participants in integrating EIA into land-use planning. Plans are under way to establish an International Commission for Impact Assessment. The commission is hoped to have a global outreach. It is to provide nonbinding advice (with the consent of government) on the adequacy of information for decision making.

Australia provides EIA guidelines and specific guidelines regarding "matters of national environmental significance." Degree of public interest is one of the significance criteria suggested by the Australian and New Zealand Environmental Conservation Council (ANZECC) in its guidelines and criteria for determining the need and level of EIA in Australia. The Australian EIA legislation creates a series of independent advisory committees (e.g., Threatened Species Scientific Committee, Biological Diversity Advisory Committee, Indigenous Advisory Committee). Australia also has prepared an EIA training resource manual for use in developing countries.

The four countries have devoted considerable attention to educating EIA practitioners and the public about EIA, both in general and with specific reference to public participation principles and practices. There also are numerous, albeit scattered efforts to integrate community and traditional knowledge into EIA requirements and practices. Very little attention, except for general references, has been devoted to mutual learning concepts, methods, and practices.

Negotiations   Negotiations, or more specifically, alternative dispute resolution (ADR), have received considerable attention in the United States. Highly controversial is a significance factor under NEPA. ADR has been used extensively in the United States by federal agencies to negotiate rules, develop policies, plans and strategies, issue permits, adjudicate disputes, administer contracts, and grants and involve stakeholders. Various federal agencies have prepared ADR guidelines, have derived lessons from ADR case studies, and sponsor ADR training programs. They also have prepared resource guides for evaluating environmental conflicts in communities. The role of ADR in EIA practice appears to be less extensive. However, much ADR guidance material is relevant to EIA practice. In addition, many dispute resolutions centers have been established across the United States, which could provide the advice needed to integrate EIA and ADR.

ADR has received much less attention in the other jurisdictions. A voluntary negotiations process, with an independent and impartial mediator, is an option under Canadian EIA legislation. Changes to the legislation refer to promoting the greater use of mediation and dispute resolution and to a greater role for the agency in building consensus and in resolving disputes. ADR, as an option for resolving EIA disputes, does not appear to have received much attention in either Europe or Australia.

The logical departure point for assessing the potential role of ADR in EIA is to draw on the extensive U.S. knowledge base. The experience of Canada with mediation as an option under EIA requirements could also be monitored. Additional attention could be devoted to the specific adjustments needed to both EIA and ADR requirements and practices to link and integrate these two related environmental management fields more effectively.

Collaboration   Collaboration, as reflected in EIA requirements and guidelines in the four jurisdictions, assumes many forms. In the United States, EIA requirements and guidelines include numerous references to partnering with or working jointly or cooperatively with states, local agencies, and Indian tribes. Environmental justice EIA requirements probably facilitate collaboration. Various federal agencies have devoted considerable attention to collaborative planning, resource management, program administration, and facility management. The President's Council on Sustainable Development has detailed lessons learned from collaborative approaches (PCSD, 1997). Less attention appears to have been devoted to the potential roles of collaboration in EIA requirements and practices.

Collaboration in Canadian EIA requirements largely refers to working cooperatively with provincial and territorial governments and with aboriginal governments and peoples. Numerous formal and informal mechanisms (e.g., regional EA committees, EA agreements, joint panels) have been introduced to facilitate cooperation and joint planning. Federal intervener and participant (proposed) funding provisions are likely to be conducive to collaborative EIA approaches. Federal EIA guidelines include general references to cooperation and consensus building. The tri-party EIA approach (federal, territorial, indigenous people) adopted in the north is a form of joint planning. The federal government (through the Department of Indian Affairs and Northern Development) and several provinces and territories have considerable experience with resource co-management with indigenous peoples and with decentralized decision making.

Collaboration in EIA requirements and guidelines in Europe pertains largely to collaboration among European states. Access to environmental justice requirements in Europe will probably facilitate collaborative planning approaches. EIA centers also could assist such efforts. The Australian EIA requirements emphasize a cooperative approach to environmental protection and conservation among governments, communities, landowners, and indigenous peoples. The legislation addresses cooperative arrangements between the Commonwealth and states and territories in considerable detail. The legislation also includes environmental and resource co-management provisions.

Collaboration appears to be well developed in the EIA requirements of the four jurisdictions regarding interactions among governments. Some attention has been devoted to mechanisms for facilitating stakeholder involvement and collaboration (e.g., participant funding, environmental justice). Considerable experience has been acquired in applying collaborative approaches in such related fields as resource, environmental, and facility planning and management. Especially in the United States, some attention, has been devoted to general collaborative planning approaches. Practical collaborative planning methods and frameworks and adaptations to EIA requirements needed to facilitate collaborative EIA practice could receive more attention.

8.5.2   Management at the Applied Level

Figure 8.7 illustrates a collaborative EIA process. The figure and the description that follows depict a process consistent with EIA public participation goals, principles, and practices. Specific consultation, communications, mutual education, negotiations, and collaboration concepts, methods, and processes are incorporated into the process. The process seeks to enhance collaborative planning and decision making by EIA stakeholders. Communications, mutual education, negotiations, and collaborative elements are grafted onto and integrated into the EIA process. The process is broadened to encompass numerous publics. It is supported by a sound knowledge base. The process is informal, open, inclusive, interactive, and people-centered.

Startup and Planning   The collaborative EIA process begins with initial consultation. Issues are identified. Level of interest is determined. An overview of environmental characteristics is undertaken. Historical grievances are noted. Remedial actions are taken where practical. Key people and organizations are identified. Pertinent organizational mandates, characteristics, and constraints are described. This initial context scanning provides the basis for a conflict assessment. The conflict assessment considers issues, potential parties, and potential conflicts. It then decides whether alternative dispute resolution is possible and appropriate. Key parties that might be interested in or potentially affected by the proposed action are identified. The parties are contacted. Appropriate representatives are identified and recruited as members of an advisory committee. The committee includes a diversity of proponent, government, and public stakeholder representatives. The committee does not have final decision-making authority. However, because of the membership breadth and the in-depth deliberations anticipated, the findings and recommendations that emerge from the committee are expected to have considerable decision-making "weight."

The committee establishes and agrees to procedural rules and principles. It identifies an appropriate range of subcommittees. EIA and public participation planning are integrated. The overall EIA/public participation plan addresses such matters as

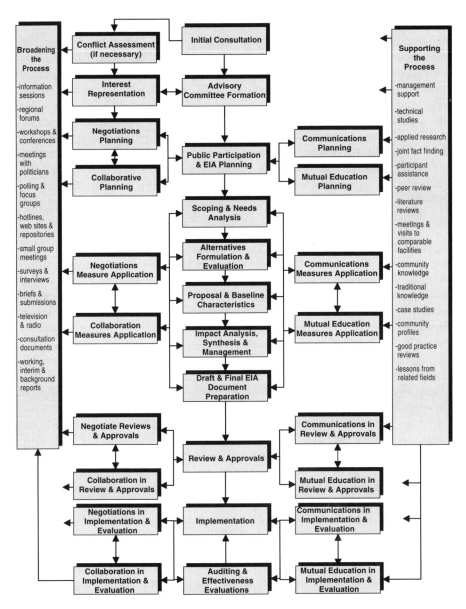

Figure 8.7 Example of a collaborative EIA process.

general principles and goals, issues, and problems to address, activities and tasks, schedule, resource requirements, technical specialist input requirements, budgets, roles and responsibilities, and public involvement procedures. The plan includes communications (e.g., information exchange procedures, communications goals, principles and methods, measures to overcome communications barriers, plans for groups and organizations with special communications needs), mutual education (e.g., community, proponent and regulator training and education requirements, mutual education goals, principles and methods, plans for groups and organizations with special educational needs), negotiations (e.g., third-party assistance, measures to offset power imbalances, negotiations goals, principles and methods, conflict identification, and management procedures), and collaboration (e.g., core values and preliminary visions, collaboration goals, principles and methods, outreach and capacity building methods, participant assistance requirements, procedures to foster creative collaboration) elements. A draft plan is first prepared. The draft plan is modified based on comments from committee members and from a broader audience. Further refinements and adjustments to the plan occur as the EIA process unfolds. An evaluation of the effectiveness of the startup and planning activities is undertaken.

Application The core application activities largely mirror those commonly associated with EIA processes. The overall process is scoped. The need for action is assessed. Alternatives, of varying types and at different levels of detail, are systematically generated, screened, and compared. Baseline and proposal characteristics are determined. Individual and cumulative impacts, risks, and uncertainties are identified and predicted. Impact significance, with and without mitigation, is interpreted. Compensation, monitoring, and contingency measures are determined. An overall impact management strategy is formulated. Techniques such as sensitivity analyses test the robustness of conclusions. Draft and final EIA documents are prepared. Modifications to the draft EIA documents reflect both regulatory and public concerns and preferences. The process is highly iterative. Unlike most conventional EIA processes, it is the advisory committee that takes the lead, and it is the EIA and other specialist practitioners who assume a support role.

The committee, with the assistance of specialists, guides both the technical and the procedural (i.e., public involvement, communications, mutual education, negotiations, collaboration) activities. Public involvement includes numerous procedures to provide information out to the public, to obtain information, comments, and concerns from the public and to exchange information and perspectives with the public. Communications involves interactions both within the committee and between the committee and constituents. It includes activities such as preparing clear, succinct, accurate, and readily understandable documents, identifying and offsetting misinformation, distortions and communications barriers, facilitating and maintaining dialogue, ensuring that appropriate adjustments are made to reflect the varying communications requirements of different groups and segments of society (e.g., based on language and cultural differences), establishing and keeping communications channels open, ensuring that EIA documents are known about and are readily available and ensuring that local perspectives and concerns are accurately reported and effectively integrated into EIA documents.

Mutual education extends well beyond community education. Public knowledge of proposed actions, options, environmental conditions, potential impacts and management measures is enhanced. Training and education opportunities are provided to public committee members and, where appropriate, to other members of the public. Mutual education also entails educating the proponent, the regulators, and the specialists, ensuring that the process fully accommodates community and traditional knowledge, facilitating learning about facts, values, issues, decision-making processes and participants, integrating learning from practice stories, ensuring that learning is free from distortion or coercion, and promoting mutual, social, transformative, and critical learning.

Negotiations are tailored to the conflicts that emerge and to the characteristics of individual EIA activities. Concerns, priorities, issues, and points of conflict are identified. Conflicts are, where practical, avoided, "staved off," and ameliorated. Remaining conflicts are, wherever possible, resolved. Third parties (e.g., facilitators, mediators) assist the negotiations. Possible conflict resolution packages are identified, screened, and compared. Commitments are obtained. Where warranted, the packages selected are incorporated into agreements, suitable for signing by stakeholder representatives. Provisions are included to ensure the monitoring of agreement implementation and to permit mutually acceptable adjustments to suit changing circumstances. Collaboration builds rapport, trust, consensus, and support. Common goals and shared visions of the future are formulated. Procedural inequities are offset. The creative and collaborative exploration of problems and opportunity is fostered. Effective use is made of synergistic methods for generating novel approaches to complex issues and concerns. The process is focused on achieving environmental and sustainability improvements without sacrificing outcome equity, especially for disadvantaged groups and segments of society.

The customary EIA activities are all undertaken in accordance with regulatory requirements and good practice, but in a manner that integrates all elements of effective collaboration. The effectiveness of the EIA and public participation process and outcomes are evaluated both during and subsequent to the completion of the application activities. The evaluation leads to both procedural and substantive adjustments and refinements.

Review, Approvals, and Post Approvals   A sound decision-making basis is provided for all parties that should have a say in whether and how the proposed action proceeds. Communications and mutual education ensure that all parties are fully appraised, in a form suited to their needs, of all matters relevant to their deliberations. Care is taken to avoid and offset communications breakdowns and knowledge deficiencies. Negotiations and collaboration enable the parties to avoid, reduce, and resolve disputes and to generate creative approaches and solutions that serve both the interests of the parties and the broader public interest. Effective negotiations and collaboration eliminate the need for or dramatically scope formal public review proceedings (e.g., hearing, litigation before the courts).

Collaboration activities extend into the post-approvals stage. Communications and mutual education ensure that all parties are fully informed of implementation activities. Proponents and regulators are made aware of community concerns and preferences as they emerge. Communications among the parties is facilitated. Prompt action is taken to correct misinformation and communications distortions.

Knowledge deficiencies are corrected. Conflicts are avoided where practical. Negotiations serve to ameliorate and resolve residual conflicts. Collaboration ensures that impact management measures are efficient, focused, and effective. Local benefits, rapport, trust, and cooperation are maintained and enhanced. The effectiveness of all elements of the EIA/public participation process is evaluated immediately following approvals. Additional effectiveness evaluations are undertaken periodically through implementation and whenever major unforeseen circumstances and remedial actions.

Broadening and Supporting the Process   Confining participation to an advisory committee composed of stakeholder "representatives" places an impossible burden on committee members. Committee members can only be assured that they are representing the views, interests, and concerns of their constituents if there are ample opportunities for the broader public to be involved in the process. The process provides for such opportunities prior to each major decision. A variety of involvement procedures inform and obtain input from the public. The public is provided with an ample range of different types of involvement opportunities (e.g., information sessions, small group meetings, television, radio, Web sites). Involvement procedures are tailored to meet the needs of various groups, organizations, and segments of society. A particular effort is made to involve those groups and segments of the community likely to be the most directly affected and which are especially vulnerable to change. Positions adopted by the committee or by subcommittees are tested with such techniques as surveys, polling, and interviews. Committee members carefully compare their perspectives and positions against those contained in briefs and submissions. The broader public is provided with an opportunity to respond to background, interim, and draft reports. Expanding the base of public understanding and involvement contributes to an enhanced level of comfort for both committee members (that they are effectively representing the views of their constituents) and the public (that their concerns, interests, and preferences are being adequately represented).

A collaborative EIA process is highly dependent on an adequate level of support. There must be a strong management commitment to the approach—a commitment reflected in adequate resources and sufficient time for the process to proceed at its own pace. The activities of the committee are supported by sound technical studies, applied research to address areas of uncertainty, community knowledge, and where applicable, traditional knowledge. Community involvement in establishing a sound knowledge base is supported, where needed, by participant funding and by procedural and substantive training. Specialists assist joint fact-finding. Committee members draw upon the insights and lessons obtained through visits to comparable facilities, community profiles, case studies, literature reviews, reviews of experiences in related fields, and good practice reviews. Peer reviewers test technical analyses. Good communications practice is reflected in how documents are structured, presented, and edited. The support activities reflect the needs, preferences, priorities, and expectations of the committee members, of regulators, and of the broader public.

8.6 ASSESSING PROCESS EFFECTIVENESS

Table 3.5 (in Chapter 3) presents criteria and indicators for assessing the positive and negative tendencies of the EIA processes described in Chapters 3 to 10. In this section we summarize examples of positive and negative tendencies of collaborative EIA processes.

Scientific *rigor* tends to be a secondary consideration with group processes. The adequacy of the scientific and technical analyses and their application are highly dependent on effective links between the committee members and scientific and technical specialists and resources. Scientists can play a useful support role. The committee is likely to press for the explicit presentation of major assumptions, methods, findings, and interpretations. Methodological details may receive less attention.

A diverse stakeholder committee is likely to support *comprehensiveness* in a broad definition of the environment, of problems, of alternatives, and of potential direct and indirect effects. The committee is likely to exhibit a holistic perspective. Groups tend to undertake and review analyses at a broad level of detail. Complex and detailed support analyses may not always be reviewed thoroughly or fully incorporated into decision making.

The committee is likely to be *systematic* in making explicit the rationale for all major interpretations and conclusions. The route to decisions is sometimes difficult to trace with group processes. Major issues, trade-offs, and impact management measures are generally considered systematically. Option comparisons and impact analyses are not always undertaken and reviewed consistently. The committee may have difficulty dealing with multiple options and a complex array of suboptions. Consensus building and conflict resolution tend to be higher priorities.

Stakeholder groups tend to stress *substantive* environmental improvements and sustainability. Environmental values and ideals are usually priorities. Collaborative processes seek both procedural effectiveness and substantive objectives. Stakeholder groups are usually open to multiple perspectives and to a broad definition of environmental knowledge. There is a danger that substance may be sacrificed in the quest for consensus. Difficult decisions sometimes mean proceeding despite opposition from some interests. The interests of future generations and of other unrepresented parties are not always fully considered with group processes.

A collaborative EIA process is *practical*. It focuses on major issues and trade-offs. Collaboration is consistent with good EIA practice. Implementation is facilitated if a consensus among major stakeholders is reached and maintained and if major disputes are resolved. The process produces succinct and readily understandable documents. Process practicality is highly dependent on maintaining effective links to regulators and constituents. The process can be costly and time consuming because of the resources required to build consensus, to resolve conflicts, and to maintain external links. These front-end costs can be more than offset by a less adversarial review and approvals stage. Care must be taken to ensure that the committee does not become a bottleneck. There is also the danger that the process could break down because of major value divisions.

Collaboration can enhance *democratic* local influence. Major stakeholders have committee representatives. The extent of local influence depends on the committee structure, how well the representatives reflect local priorities and the effectiveness of the links between the committee and the broader public. A collaborative EIA process approaches but stops short of power sharing or delegated decision making. Some stakeholders may be reluctant to participate because the committee is advisory. There must be a clear management commitment to the process and a clear demonstration that early and interim committee outputs strongly influence and shape decision making. The process is well suited to soliciting and accommodating traditional and community knowledge. Stakeholder committees are generally sensitive to political implications. Provided that external links are maintained, effective collaboration can contribute to community acceptance.

A collaborative EIA process is, by definition, *collaborative*. It is highly conducive to stakeholder understanding and involvement. Roles and responsibilities are defined jointly. The stakeholders design and execute the process jointly. Considerable stress is placed on consensus building and on conflict resolution. Effective and ongoing links back to constituents are critical if understanding, agreement, and support are to extend beyond the committee. External resistance and opposition can result in committee members altering their positions or even dropping out of the committee. One or a small number of committee members who refuse to look beyond positions to individual and shared interests can potentially derail the process.

Procedural fairness is usually a priority with group processes. Procedures are generally put in place to offset power inequities and to facilitate procedural fairness. Stakeholder committees often raise numerous ethical concerns. Environmental justice is commonly a priority. Rights and responsibilities are frequently noted as concerns. Substantive equity, although probably a concern, may or may not be considered thoroughly or consistently. Although many *ethical* issues and concerns are raised, group processes sometimes have difficulty addressing ethical issues, implications, trade-offs, and dilemmas systematically. The process is not likely to be shaped and guided by well-defined ethical imperatives and standards.

Collaborative EIA processes are generally *adaptive* because they are informal and people-centered. Committees are usually willing to reconsider previous analyses and to alter the process when issues arise and when circumstances change. The burden on the committee, even when well supported and managed, is considerable. It is unlikely that the committee will scan forward or reconsider past analyses unless there is external pressure. Stakeholder committees are generally open to broadly defining problems and opportunities. The presence of a diversity of stakeholder representatives tends to result in a good match between process and context. Group processes can be creative with adequate procedural support and methodological advice. Effective external links are essential if problems and opportunities are to be characterized and explored creatively. The process is less conducive to individual creativity. Risk and uncertainty are usually major concerns with some stakeholders. The stakeholders generally ensure that risk and uncertainty perceptions and alternative management perspectives are not overwhelmed by risk and uncertainty analyses. Systematic analysis of risks and uncertainties is more likely if some committee members are familiar with the fields and if there is sound technical support.

Collaborative EIA processes tend to be conducive to *integrating* diverse values, forms of knowledge, perspectives, and ideals. Groups tend to be open to and supportive of perspectives and frameworks that adapt, integrate, and transcend individual disciplines, professions, and EIA types. They generally expect that proposal planning and the EIA process should be integrated. Links to and from related decisions and related environmental management forms and levels do not tend to be priorities.

8.7 SUMMING UP

In this chapter we describe a collaborative EIA process, a process in which the public is an active and ongoing participant. The three stories illustrate some of the complexities and subtleties associated with collaborative EIA processes. The first story demonstrates that alternative participation approaches are needed at the SEA level, especially in terms of early involvement and concerning if and how best to link and merge planning and evaluation-related consultation activities. The second story shows that complex projects, in particular, require multiple methods and opportunities, patience, openness, effective listening, and adaptations to reflect the needs and perspectives of individuals and organizations. The third story demonstrates that each party must be willing to take the risks and invest the time and energy to participate in a process aimed at generating mutually beneficial solutions. Credibility, trust, and consensus are built incrementally and iteratively and can be facilitated by independent and credible advice and assistance.

Collaboration encompasses all forms of public participation short of delegation or shared decision making. Stakeholders jointly undertake a collaborative EIA process. Noncollaborative forms of public participation are prerequisites or subsets. The problem is the gulf between the potential benefits of collaborative EIA processes and the more modest benefits achieved by public participation approaches commonly evident in EIA requirements and practices. The direction is exploring the potential for and means of making EIA processes more collaborative. Six complementary elements of effective public participation practice are identified: (1) core principles and practices, (2) consultation, (3) communications, (4) mutual learning, (5) negotiations, and (6) collaboration. Table 8.8 is a checklist for formulating, applying, and assessing a collaborative EIA process.

Effective public participation has intrinsic benefits. It furthers human potential and offers numerous individual and joint benefits for the public and for decision makers. EIA public participation practice often fails to realize these benefits fully. The possibility is raised that the shortfall occurs because of disadvantages associated with public participation or because of largely irresolvable issues associated with public participation practice. The public participation disadvantages are generally either dubious or overstated. They can generally be avoided, overcome, or largely ameliorated. Restrictions to public participation in the EIA process should

Table 8.8   Checklist: A Collaborative EIA Process

The response to each question can be yes, partially, no, or uncertain. If yes, the adequacy of the process taken should be considered. If partially, the adequacy of the process and the implications of the areas not covered should be considered. If no, the implications of not addressing the issue raised by the question should be considered. If uncertain, the reasons why it is uncertain and any associated implications should be considered.

GOALS, PRINCIPLES, AND GOOD PRACTICES

1. Does the EIA process realize the potential benefits of effective public participation?
2. Does the EIA process incorporate all elements of effective public participation?
3. Does the EIA process effectively avoid and reduce public participation disadvantages?
4. Is public participation effectively integrated into the EIA process?
5. Does the EIA process effectively manage public participation issues?
6. Is the EIA process based on well-defined and consistently applied public participation goals, principles, and good practices?
7. Are the public participation methods selected appropriate to the situation and process objectives, and are they applied effectively?
8. Is public participation in the EIA process planned effectively?

CONSULTATION

1. Is the EIA process based on well-defined and consistently applied consultation goals, principles, and good practices?
2. Are the consultation methods selected appropriate to the situation and process objectives, and are they applied effectively?
3. Does public consultation occur early in the EIA process?
4. Is the full range of interested and potentially affected parties identified?
5. Are public consultation opportunities provided prior to each decision?
6. Is public consultation planned and executed effectively?
7. Is information to and from the public exchanged effectively?
8. Are outreach and capacity-building methods used to bring into the EIA process traditionally unrepresented and underrepresented groups and segments of the population?
9. Is public consultation effectively integrated into the EIA process?
10. Are continuous involvement opportunities provided for major stakeholders?
11. Is public consultation effectiveness assessed?

COMMUNICATIONS

1. Is the EIA process based on well-defined and consistently applied communications goals, principles, and good practices?
2. Is communications effectively integrated into the EIA process?
3. Are the communications methods selected appropriate to the situation and process objectives, and are they applied effectively?
4. Are EIA documents clear, focused, understandable, and targeted to meet the needs and characteristics of varying audiences?
5. Are steps taken to enhance the communications skills of EIA process participants?
6. Does the EIA process facilitate open, unconstrained, undistorted, and noncoercive dialogue?

Table 8.8   (*Continued*)

7. Are effective steps taken to prevent and counter communications misinformation and distortion?

8. Is the EIA process conducive to effective and appropriate argumentation, persuasion and storytelling?

9. Is communications effectiveness assessed?

MUTUAL EDUCATION

1. Is the EIA process based on well-defined and consistently applied mutual education goals, principles, and good practices?

2. Is mutual education effectively integrated into the EIA process?

3. Are the mutual education methods selected appropriate to the situation and process objectives, and are they applied effectively?

4. Is the EIA process treated as a learning process?

5. Is the process conducive to community learning, proponent/regulator learning, and mutual learning?

6. Does the process accommodate community and traditional knowledge?

7. Is the EIA process conducive to social, collaborative, transformative, and critical learning?

8. Does the EIA process facilitate practical and deliberative learning?

9. Is mutual education effectiveness assessed?

NEGOTIATIONS

1. Is the EIA process based on well-defined and consistently applied negotiations goals, principles, and good practices?

2. Is negotiations effectively integrated into the EIA process?

3. Is alternative dispute resolution applied in appropriate situations?

4. Is a conflict assessment undertaken?

5. Are interests effectively represented in the process?

6. Is the negotiations process effectively planned, undertaken, and implemented?

7. Do third parties assume an effective role in the process?

8. Are the results of negotiations formalized in signed agreements, which can be monitored and adjusted during implementation?

9. Are the negative tendencies of negotiations effectively prevented and offset?

10. Are the negotiations methods selected appropriate to the situation and process objectives and are they applied effectively?

11. Is negotiations effectiveness assessed?

COLLABORATION

1. Is the EIA process based on well-defined and consistently applied collaboration goals, principles, and good practices?

2. Is collaboration effectively integrated into the EIA process?

3. Is the collaborative process directed toward the achievement of substantive environmental visions and objectives?

4. Is the EIA process inclusive, jointly undertaken, positive, and purposive?

(Continued)

Table 8.8 *(Continued)*

5. Do third parties assume an effective role in the process?
6. Is the process guided by procedural democratic principles, and is it procedurally fair?
7. Is the process conducive to creative problem solving and opportunity seeking?
8. Are the participants altered and empowered, and does the process extend to external coalition building?
9. Is the process supported by management and built into decision making?
10. Are the collaboration methods selected appropriate to the situation and process objectives, and are they applied effectively?
11. Is collaboration effectiveness assessed?

MANAGEMENT AT THE REGULATORY LEVEL

1. Do EIA requirements and guidelines facilitate and ensure early, frequent, continuous (where appropriate), and effective public notification, information exchanges, and involvement?
2. Do EIA guidelines provide adequate guidance regarding public participation goals, principles, and good practices?
3. Do EIA requirements and guidelines facilitate the involvement of traditionally un-represented and underrepresented segments of society?
4. Are there adequate public involvement requirements for each type of EIA document?
5. Do EIA requirements stipulate that the public's contribution to decision making needs to be specified?
6. Do the EIA requirements and guidelines foster effective communications to the public and through EIA documents?
7. Do the EIA requirements and guidelines foster effective two-way communications, including adaptations for different audiences?
8. Do the EIA guidelines provide specific advice concerning communications principles, skills, measures to offset distortions, and measures to foster and enhance dialogue among EIA process participants?
9. Do the EIA requirements and guidelines help educate EIA practitioners and the public about EIA?
10. Do the EIA requirements and guidelines facilitate accommodating community and traditional knowledge into EIA practice?
11. Do EIA guidelines provide advice regarding mutual learning concepts, methods, or practices?
12. Do the EIA requirements and guidelines provide opportunities for alternative dispute resolution?
13. Is good practice ADR guidance provided?
14. Do the EIA requirements and guidelines foster collaboration in the EIA process among governments?
15. Do the EIA requirements and guidelines foster collaboration in the EIA process with and among stakeholders?
16. Is good practice collaboration guidance provided?

MANAGEMENT AT THE APPLIED LEVEL

1. Is the EIA process guided and assisted by one or more advisory committees?

Table 8.8 (*Continued*)

2. Does the EIA process incorporate such startup and planning activities as:
    a. Initial consultation?
    b. Advisory committee formation?
    c. Conflict assessment?
    d. Interest representation?
    e. Overall public participation planning?
    f. Integration of communications, mutual education, negotiations, and collaborative planning into public participation and EIA planning?

3. Does the EIA process incorporate consultation, communications, mutual education, negotiations, and collaboration application activities into:
    a. Scoping and needs analysis?
    b. Alternatives formulation and evaluation?
    c. Proposal and baseline characteristics?
    d. Impact analysis, synthesis, and management?
    e. Draft and final EIA document preparation?

4. Does the EIA process incorporate consultation, communications, mutual education, negotiations, and collaboration measures into review and approvals?

5. Does the EIA process incorporate communications, mutual education, negotiations, and collaboration measures into implementation?

6. Are evaluations undertaken during and subsequent to the EIA process of the effectiveness of public consultation, communications, mutual education, negotiations, and collaborations measures?

7. Are effective actions taken to broaden public participation in the EIA process?

8. Are effective actions taken to support the public participation measures instituted in the EIA process?

PROCESS EFFECTIVENESS

1 . Does the process balance collaboration with an adequate level of applied scientific performance?

2. Does the process maintain a holistic perspective, broadly define the environment, options, and effects, and ensure that analyses are undertaken at an adequate level of detail?

3. Does the process systematically and explicitly address major choices, impacts, and trade-offs?

4. Does the process contribute to substantive environmental improvements?

5. Is the process focused on major issues and trade-offs, and does it facilitate approvals and implementation efficiently and effectively?

6. Is the process conducive to enhanced local influence?

7. Does the process facilitate public understanding, involvement and collaboration?

8. Is the process procedurally fair, and does it balance collaboration effectively with outcome fairness?

9. Is the process suited to the context, and does it anticipate and adapt flexibly to changing conditions?

10. Does the process accommodate and link a diversity of values, forms of knowledge, perspectives, and ideals?

11. Is the process effectively linked to proposal planning, related decisions, and related environmental management forms?

always be justified. There are numerous issues associated with EIA public participation practice. These issues pertain to conducting public participation activities,
balancing conflicting perspectives, treating problems, obstacles, dilemmas, and
uncertainties, and making difficult judgments. Rather than justifying restrictions
to public participation, the issues underscore why issue identification, exploration,
and resolution should be collaborative. They also point to the need to learn from
public participation practice.

Examples of EIA public participation goals, principles, and good practices are
identified. Public participation methods are often placed along continua to demonstrate
major methods characteristics, their suitability for achieving different objectives,
and their match to varying situations. Public participation methods can also be characterized by function or by operational characteristics. The characteristics, advantages, and disadvantages of categories of and individual methods should be
considered before they are applied. Methods should be consistent with EIA process
goals and should be appropriate to the context.

Public consultation or involvement includes informing the public, integrating the
views of the public, and interacting with the public, all prior to decision making.
A public consultation process includes early consultation, initial and detailed
planning, plan application and refinement, the monitoring of plan effectiveness
during and subsequent to plan application, and post-approval involvement. Public
involvement should begin early in the EIA process and should be integrated into
each EIA process activity. Stakeholder identification is especially important. Information exchange, continuous involvement and outreach, and capacity building
methods can be applied. Formal involvement methods sometimes occur near the
end of the process to present and test evidence. Public involvement plans are usually refined and adjusted through the process.

Communications is concerned with clear, focused, and understandable documents, the communications and advice-giving skills of EIA practitioners, and
undistorted and noncoercive dialogue among EIA participants. Applying procedural
ethical principles, insights from EIA practice stories, and effective argumentation
could enhance communications in the EIA process.

Education in the EIA process is conventionally portrayed as educating the community. *Mutual education or learning* works both ways. It involves the parties
learning together and potentially being transformed. The EIA process should be
treated as a learning process, an opportunity for all parties, individually and collectively, to enhance their knowledge and their intelligence capacity. The application
and accommodation of traditional knowledge and concepts such as social, collaborative, practical, critical, and transformative learning could facilitate learning
about and through the EIA process.

The process of *negotiation* is concerned with avoiding, resolving, and ameliorating conflict in the EIA process. Alternative dispute resolution (ADR) provides the
EIA process with a means of negotiations not limited to conventional administrative
procedures or to litigation through the courts. Third parties such as mediators or
facilitators should generally assist the process. ADR is not always appropriate.
The appropriate conditions must be satisfied. ADR processes generally involve

pre-negotiations (e.g., conflict assessment, interest representation, designing the process), negotiations (e.g., initial deliberations, focusing, detailed deliberations, final refinements), and post-negotiations (i.e., implementation). ADR can potentially be integrated, in different forms, into several EIA process activities. ADR methods range from unaided negotiations through procedural assistance (e.g., conciliation, facilitation, mediation) to quasijudicial mechanisms (e.g., minitrials, public hearings, arbitrations). Methods should be selected to suit the process objectives and the situation. ADR advantages and disadvantages should be considered when determining if, how, and when in the EIA process ADR methods are to be applied.

Consultation, communications, mutual education, and negotiations, individually and collectively, can contribute to *collaboration* in the EIA process. Collaboration is about people working together in a joint endeavor with substantive aspirations. Collaboration emphasizes perspectives, visions, opportunities, and creative joint exploration by EIA process participants. Effective collaboration is inclusive and open, involves multiple perspectives and forms of knowledge, is jointly undertaken by stakeholders, and is directed toward and guided by substantive environmental management, environmental justice, and sustainability ends. Collaboration extends from a sound knowledge base and extends to interested and affected parties.

The EIA administrative arrangements in the four jurisdictions are partially conducive to collaborative EIA processes. They help ensure information exchanges with the public and outreach to less involved groups and segments of society. Additional emphasis could be placed on earlier, frequent, and continuous involvement, on involvement in screening documents, and on decision-making links. Communications to the public through EIA documents is partially addressed. More advice could be provided regarding communications principles, communications skills, communications distortions, and facilitating dialogue. Much attention is given to educating the public about EIA participation opportunities and practitioners about participation approaches. More attention could be devoted to integrating community and traditional knowledge and to applying mutual learning concepts, methods, and practices. Considerable guidance regarding ADR is provided in the United States. Mediation is an EIA option in Canada. More consideration could be given to the potential roles of ADR in the EIA process. Collaboration among governments is well developed but in need of further refinement. Measures to facilitate stakeholder collaboration have received some attention. More consideration could be given to collaboration planning methods and frameworks and to facilitating collaborative EIA processes.

The example collaboration EIA process begins with an initial round of consultation, a conflict assessment, the formation of an advisory committee (the focal point of the process), interest representation, and public participation and EIA planning. The EIA process and public participation planning and execution are fully integrated. The various elements of effective public participation (involvement, communications, mutual education, negotiations, and collaboration) are integrated into EIA and public participation planning, into major EIA activities (e.g., scoping and needs analysis, alternatives formulation and evaluation, proposal and baseline characteristics, impact analysis, synthesis, and management, draft and final EIA document preparation), into review and approvals and into implementation. The effectiveness of public participation actions is evaluated during and subsequent to each stage in the process. The process is built on a solid knowledge foundation and includes frequent and numerous links to the broader public.

Collaborative EIA processes are particularly effective in fostering local influence and in facilitating public understanding, involvement, and sometime acceptance. They usually ensure procedural fairness and integrate diverse perspectives and forms of knowledge. They are holistic in perspective and tend to employ a broad definition of problems, the environment, choices, and effects. They adapt well to context and to changing circumstances. They usually focus on major issues and trade-offs. They generally seek to achieve substantive environmental ends and equitable outcomes, effectively integrate community knowledge, and are sensitive to community perceptions of risk and uncertainty. Scientific rigor, detailed and consistent technical analyses, and links to related decisions and forms of environmental management tend to be lower priorities. Collaborative EIA processes can break down over major value divisions. There is a risk that substantive environmental ends, outcome equity, and unrepresented interests will be compromised in the quest for consensus. Effective links to constituents and to the broader public are essential. Management, procedural, and technical support is critical. The process tends to be costly and unfolds at its own pace prior to approvals. But it can result in substantial cost savings during approvals and implementation because of less adversarial relationships.

CHAPTER 9

HOW TO MAKE EIAs MORE ETHICAL

9.1 HIGHLIGHTS

In this chapter we respond to the challenge to make EIA processes and outcomes more fair, equitable, and just. We also seek to identify and advance rights (especially those of disadvantaged groups) and to ensure that duties are fulfilled. These concerns are all ethical. More precisely, they fall under the umbrella of normative (what ought to be), applied (directed toward the resolution of practical problems), and practical (moral questions, the answers to which involve commitments to action) ethics. We also illustrate how ethical concerns can be integrated into EIA processes.

- The analysis begins in Section 9.2 with two applied anecdotes. The stories describe applied experiences associated with efforts to make EIA practice more ethical.
- The analysis in Section 9.3 then defines the problem, which is an insufficient effort at explicit and systematic integration of ethical considerations into EIA processes and process outcomes. In this section we demonstrate the ubiquitous nature of ethical concerns in EIA practice.
- In Section 9.4 we define key terms, describe relevant ethical concepts, and highlight major distinctions. We also present examples of procedural fairness principles, distributional fairness principles, and ethical rights and duties.

Environmental Impact Assessment: Practical Solutions to Recurrent Problems, By David P. Lawrence
ISBN 0-471-45722-1   Copyright © 2003 John Wiley & Sons, Inc.

These concepts, principles, and distinctions provide the basis for defining an ethical EIA process.

- In Section 9.5 we detail how an ethical EIA process could be implemented at the regulatory and applied levels. In Section 9.5.1 we infuse ethical concepts and perspectives into EIA regulatory requirements and guidelines and in Section 9.5.2 integrate an ethical perspective into an applied EIA process.

- In Section 9.6 we assess how well the ethical EIA process presented in Section 9.5 satisfies ideal EIA process characteristics.

- In Section 9.7 we highlight the major insights and lessons derived from the analysis. A summary checklist is provided.

9.2 INSIGHTS FROM PRACTICE

9.2.1 A Collaborative Approach to Achieving Net Environmental Benefits

Western Australia has a three-phase assessment process for water allocation. In phase 1 a regional water allocation plan is prepared by the state's water allocation agency, the Water and Rivers Commission (WRC). These plans or strategies are broad and may embrace more than one water resource system and are prepared following an extensive community consultation process that includes a strategic review of environmental issues. The Environmental Protection Authority of Western Australia (EPA) usually provides advice under the Environmental Protection Act. In phase 2 a subregional water allocation plan is developed by the WRC determining the limits for potential water resource development of a particular source. The EPA may choose to assess the plan formally or merely to provide strategic advice. Phase 3 involves preparation by a proponent of a detailed source development proposal in accordance with the subregional allocation plan. The proponent prepares an environmental review and the EPA conducts its assessment in accordance with its procedures for environmental impact assessment under the act if the EPA requires a formal assessment.

In the late 1990s, in response to declining stream flows, the state's major water supplier, the Water Corporation, became interested in redevelopment of the Stirling–Harvey water supply scheme. Located in an agricultural area approximately 140 kilometers south of Perth (population 1.3 million), the scheme would supply water to the nearby town of Harvey, the local irrigation district, as well as to the Perth metropolitan water supply system via a 106-kilometer pipeline. It included the planned construction of a new 56-gigaliter Harvey Dam downstream from the much smaller Harvey Weir.

In 1998 the WRC prepared a phase 2 subregional allocation plan (i.e., an environmental impact assessment) of the proposed redevelopment. The WRC's EIA focused on determining water allocations for public supply, irrigation, and the environment but covered a wide range of environmental, social, and economic issues and examined a number of alternative redevelopment schemes. From the earliest planning stages extensive community involvement occurred with all landowners and other stakeholders in the Harvey Basin. Although a time-consuming and intensive process, it resulted in significant alterations to the original concept for the project. This included a shift away from a largely engineering perspective to one that attempted to reconcile conflicting objectives through direct negotiation with stakeholders.

Construction of the new Harvey Dam would result in the inundation of approximately 370 hectares of land, including the existing Harvey Weir and 18 farm properties. In addition to negotiating compensation packages with landowners directly affected (as opposed to nonvoluntary acquisition), equity issues arose at the community level. The impetus for scheme redevelopment was to supply water to a metropolitan area more than 140 kilometers away. However, in the original proposal there was little benefit to the local community. In fact, as originally conceived, the proposal would have resulted in a number of existing recreation uses (e.g., fishing), existing land uses (e.g., agriculture, forestry), and future land uses (e.g., tourism) having to be modified significantly so as to ensure the maintenance or improvement of water quality and thereby avoid costly water treatment.

Over the course of the EIA, considerable thought was given to finding ways to have the proposal result in a net social and environmental benefit to the local community. A range of commitments were made in the allocation plan, including the development of a water quality protection policy in consultation with affected landowners and a reservoir recreation plan addressing the existing and potential recreation values of the area. These two elements formed the basis for the subsequent development of a watershed management plan for the Harvey Basin. The Environmental Protection Authority (EPA) of Western Australia approved the EIA in 1998. A phase 3 EIA by the Water Corporation for scheme redevelopment followed shortly. That EIA built on the numerous environmental and social commitments in the WRC's allocation plan. Due to the phased or tiered nature of the process, the EPA set the level of assessment at a public environmental review, which is a lower level of assessment.

To offset losses of habitat due to inundation and to ensure a net environmental benefit, a commitment was made to rehabilitate 200 hectares of land near the reservoir. In addition, two parcels of native bush with high conservation values were purchased by the proponent and then vested in the Crown as A class reserves. The Ngalang Boodja Nursery, an aboriginal-run enterprise, undertook some of the rehabilitation work. This business was created and funded jointly by the proponent and the federally funded Noongar Employment and Enterprise Development Aboriginal Corporation and provided socioeconomic benefits to a traditionally disadvantaged sector of the population in the region. The Water Corporation also committed $750,000 over five years to support the establishment and activities of the Harvey River Restoration Trust. The community-based trust would distribute money to land-care groups and individual land managers to restore waterways within the Harvey Basin.

The recreation plan developed in conjunction with the local government was designed to boost ecotourism in the area. The plan included an amphitheatre for major concerts and cultural events, a boardwalk and walk trails, grassed picnic areas with gazebos and barbecues, and a viewing platform. Existing recreational activities such as fishing and canoeing would be able to continue. Accommodating recreation and tourism values meant that the drinking water supply would require a higher and more costly level of treatment than originally planned by the proponent.

The Water Corporation submitted its PER in March 1999 and received EPA approval in November 1999. Construction of the $275 million Stirling–Harvey Redevelopment Scheme began in early 2000 and was completed in November 2002. This story points to the value of a concerted effort to ensure net environmental and community benefits at the local level. Concentrated costs and dispersed benefits are not inevitable, but it takes considerable thought and collaboration if an imbalance is to be corrected.

JO ANN BECKWITH
Department of Resource Development
Michigan State University

9.2.2   The Benefits of Procedural Equity[*]

Participant funding provides financial resources to facilitate public participation in large-scale EIAs. Participant funding can contribute to a more substantive dialogue. It allows participants to develop and obtain independent technical expertise in specific EIA issues (see, e.g., Lynn, 2000; Lynn and Watherin, 1991). Money can be used to prepare and participate in scoping meetings, to review draft assessment guidelines, to review the proponent's environmental impact statement (EIS), and/ or to prepare and participate in public hearings. The Sable Gas Panel Review case illustrates the benefits of participant funding programs.

The Sable Gas Panel Review, an assessment of a natural gas project situated in the Canadian Maritimes, was undertaken between 1996 and 1997. A total of $125,000 was disbursed to nine interveners in the hearings. The interveners included the Conservation Council of New Brunswick, the Union of New Brunswick Indians, the Native Council of Nova Scotia, the Ecology Action Centre, the Maritime Pipeline Landowners Association, the Clean Nova Scotia Foundation, the Citizens Coalition for Clean Air, the Nova Scotia Salmon Association, and the Allergy and Environmental Health Association. The money provided offset the administrative costs associated with the EIS. It enabled the various interest groups to devote the time and effort required to research key issues related to the EIS and to participate in the hearings process. The research undertaken by the interveners addressed a host of ecological, economic, heritage, health, and social issues associated with various aspects of the proposed project. Approximately 50 reports were prepared pertaining to such project-related issues as corridor selection procedures; terrestrial and aquatic ecosystems impacts; easement and property rights; sea

---

[*]The research participants have been assigned code names.

ice conditions; impact management; compensation and monitoring procedures; risk and health concerns; energy conservation and global warming; employment and economic opportunities; fisheries, mining, recreational, and petroleum development implications; archaeological and land claims effects; air, water, and groundwater pollution; and cumulative and sustainability effects.

Participants in the Sable Gas Panel review expressed a range of opinions about whether their participation in the hearings was effective. While some believed that participation had an impact on the panel decisions ["I bet there are a few people who think a little bit differently, even if it is only about one thing, one aspect" (Andrew)], others were more skeptical ["I don't feel that the panel—it didn't change anything. It didn't seem that the panel took much into consideration from the presentations that we did" (Marie)].

Some panel recommendations can be linked directly to the research, testimony, and questioning of public interveners. The interveners' perspectives were reflected, for example, in panel recommendations concerning information requirements related to the impacts of the subsea pipeline on the valued ecosystem components identified in Betty's Cove (recommendation 5); regarding the development of contingency plans to focus on spill prevention, response, and strategies for cleaning up the marine and terrestrial environments (recommendation 16); and pertaining to a written protocol detailing the proponent–aboriginal roles and responsibilities for cooperation and monitoring (recommendation 45).

Although participants expressed concerns related to the participant funding, including the (small) amount of money and the (late) timing of the disbursement of funds, it is clear that public participation was facilitated by this resource, and in turn, this participation influenced the project design. "Thank god for intervener funding because the little that we did have, we put to really good use" (Dana). This story demonstrates that measures such as participant funding, which offset procedural inequities, can have both procedural (more effective public participation) and substantive (reduced adverse environmental and enhanced positive effects) benefits.

PATRICIA FITZPATRICK and JOHN SINCLAIR
University of Manitoba

9.3 DEFINING THE PROBLEM AND DECIDING ON A DIRECTION

The two stories point to the procedural and substantive benefits of ensuring procedural equity. They also demonstrate the value of moving beyond avoiding and minimizing adverse effects to a concerted effort to ensure net ecological, economic, and social benefits at the local and regional levels. The insights provide a partial and preliminary sense of how EIA processes and outcomes can become more ethical. A more detailed exploration of the potential role of values and ethics in EIA requirements and practice is required, however.

Some critics suggest that EIA practitioners and potentially affected groups and individuals are often "talking a different language." The EIA practitioners tend to take great pains to demonstrate how the procedures they employ are systematic and consistent, to highlight the many opportunities for public involvement, and to show how overall adverse impacts are minimized. The public stresses that the EIA process is unfair. They argue that benefits and adverse impacts are unfairly distributed. They insist that their rights have been ignored or diminished. They suggest that proponents and regulators have not made clear, verifiable, and enforceable commitments to the public and to the environment. Both parties are frustrated because their message is not getting through.

Both EIA practitioners and the public in EIA processes grapple with the values associated with human conduct (Jiliberto, 2002). These values provide the principles and standards applied by each party to assess whether the proposed action and the EIA process are "good" or "bad," "right" or "wrong." Ethics is a branch of philosophy that addresses whether actions are moral (i.e., good, bad, right, or wrong). Neither party tends to view their conflicting positions, perspectives, and interests as elements of an ethical debate. However, by acknowledging the ethical nature of EIA, the first step is taken toward establishing a framework for accommodating perspective and interest differences.

EIA is an inherently ethical activity. It seeks to advance environmental values (Hettwer, 1991; Jiliberto, 2002). It is subjective, moral, and value-full (Finsterbusch, 1995; Mostert, 1996). Value-based interpretations and judgments are made in every EIA activity (Enk and Hornick, 1983). Often, EIA is perceived as biased, sometimes with good reason (Beder, 1993). The ethical basis for interpretations and judgments is occasionally presented explicitly. Too often, it is not. As detailed in Chapters 3 and 4, EIA practice tends to be shrouded in the language and pretence of objectivity.

EIA is prescriptive, predictive, and interpretative. Uncertainties, ambiguities, and alternative interpretations abound, especially when predicting future conditions (with and without a proposed action) and when determining impact significance. Notwithstanding the inevitable ethical uncertainties and dilemmas, EIA seeks to provide a sound decision-making basis. This leaves EIA practitioners at both the regulatory and applied levels with considerable administrative discretion. Consequently, they have an ethical obligation to justify their positions and actions. They also have a responsibility to seek out and respond to the values and ethical positions of other participants in the EIA process.

Issues of procedural fairness are inherent to the EIA process. The EIA process must be perceived as fair, from multiple perspectives, if it is to be accepted as legitimate (Firth, 1998; Laws, 1996). Procedural fairness is both an end in itself (consistent with democratic decision-making values) and a means of reducing public dissatisfaction and of enhancing the potential for public acceptance (Kasperson et al., 1988; Lawrence et al., 1997). The proposals assessed through EIA result in temporal (e.g., exacerbating historical inequities, adverse effects on future generations), spatial (e.g., inequities in the distribution of costs and benefits and of services and facilities), and social group (e.g., disadvantaged groups bearing a greater share of the burden of adverse impacts) inequities (Interorganizational Committee, 1994). They also contribute to changes in the distribution of political power. The potential for distributional inequities tends to be a particular concern when siting LULUs (locally unwanted land uses) and when assessing the social and environmental justice implications of proposed actions (Liu, 1997; Morell, 1984).

EIA is a form of applied research. Hence there is a need for EIA practitioners to apply ethical research standards and to consider the ethical dimension of different forms of social inquiry (Chase, 1990). Procedural ethical principles and standards come to the fore in public consultation and in joint efforts with stakeholders to negotiate mutually acceptable solutions. EIA practitioners, as environmental professionals, should also comply with the ethical standards of professional organizations such as the National Association of Environmental Professionals (NAEP) and the International Association for Impact Assessment (IAIA).

EIA is one among many instruments for advancing sustainability and for furthering the cause of environmental and social justice. Social equity has been identified as a key element (some would say a prerequisite) of social sustainability (Boyce, 1995; Gardner and Roseland, 1989; Leith, 1995). The unequal distribution of environmental hazards has become a major public policy concern (Albrecht, 1995; Weinberg, 1998). The recognition that the proposals assessed through EIA requirements can exacerbate such inequities has resulted in initiatives to integrate environmental justice concerns into U.S. EIA requirements.

EIA does not operate in a vacuum. It is inevitably influenced by the "rights revolution," by debates concerning the role of justice in public policy, and by alternative characterizations of human and natural environmental relationships (Chase, 1990; Etzioni, 1995; Ignatieff, 2000; Rawls, 2001). Often, these debates are or could be framed in ethical terms. Applied fields such as EIA, environmental management, and planning draw increasingly upon ethical theory to explore and apply ethics in public policy more systematically and explicitly (Beatley, 1989; Finsterbusch, 1995; Harper and Stein, 1992).

It is evident from the above that ethics is and should be a central attribute of EIA practice. The question then is how best to proceed from the recognition of the role of ethics in EIA to its full integration into the EIA process.

9.4 SELECTING THE MOST APPROPRIATE ROUTE

9.4.1 Definitions

Ethics is a branch of philosophy concerned with the moral rules, principles, and standards that govern conduct. Ethics depend on values. EIA is a prescriptive field of practice. Therefore, *normative ethics* (which seeks to arrive at moral conduct standards) and *applied or practical ethics* (which study specific practical problems and involve a commitment to action) are especially relevant. In this chapter we focus on integrating ethical principles and standards into the EIA process.

Judging from the criticisms of EIA practice, the ethical concepts of equity, fairness, justice, rights, and duties seem especially pertinent. *Equity* concerns treating people impartially (i.e., treating everyone in the same way). *Fairness* involves treating people reasonably, consistent with moral rules or standards. *Justice* is concerned with moral rightness (an end). Justice also involves determining rights and administering rewards and punishments (a means). *Rights* are the expression of values to which people have a moral and sometimes legal claim. *Duties* or responsibilities represent a moral and sometimes legal obligation from one person to another.

As illustrated in Figure 9.1, these five ethical concepts are highly interrelated. Each concerns moral principles and standards of human conduct. Each involves judgments regarding right or wrong behavior. Equity, justice, and fairness are commonly used interchangeably. Although their meanings clearly overlap, there also are distinct differences. For example, equity could be viewed as a subset of fairness (i.e., equity is not the only standard of fairness). Fairness, in turn, could be considered a subset of justice (i.e., fairness is not the only standard of justice). Justice determines and enforces rights and duties. It also represents a means to achieve equity and fairness. Rights can be a precondition to fairness, justice, and equity. Duties implement rights, fairness, justice, and equity. There are equity, justice, and fairness rights and duties.

Integrating ethical concerns into EIA practice involves considering the potential role of equity, fairness, justice, rights, and duties in both the EIA process (i.e., a procedural focus) and in outcomes from the process (i.e., a substantive or distributional

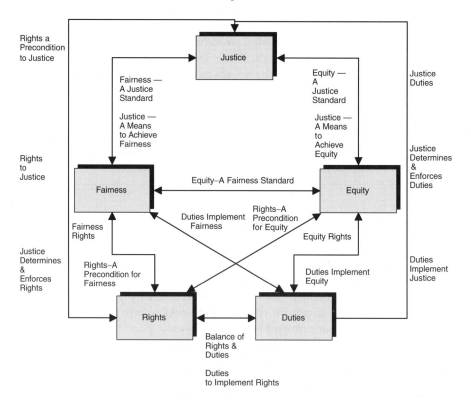

Figure 9.1   Examples of interactions: justice, fairness, equity, rights, and duties.

focus). Ethical concerns can take many forms in EIA practice. They can be issues, objectives, principles, criteria, standards, decision rules, or requirements. They can be integrated into methods, into planning and decision-making processes, and into organizational structures and procedures.

9.4.2 Distinctions

Even when limited to normative applied ethics, ethics is a diverse field of theory and practice. Therefore, it is necessary to be selective regarding potentially relevant ethical distinctions, subfields, and concepts. Table 9.1 lists several examples of potentially relevant ethical concepts, subfields, and distinctions. Key characteristics and potential EIA process implications are identified. Table 9.1 demonstrates that

Table 9.1 Potentially Relevant Ethical Concepts

Concepts	Key Characteristics	EIA Implications
Practical ethics	Addresses specific moral questions; answering a question involves making a commitment to action	EIA must grapple with moral questions and involves a commitment to action
	Can entail an appeal to a relevant moral rule, the questioning of the relevant moral rule, the justification of general moral rules, or the resolution of moral dilemmas (when the same action falls under two different but acceptable rules)	The decision rules that guide EIA decision making, because they are value-based, are arguably relevant moral rules
		EIA decision rules (moral rules) can be appealed to, questioned, and justified
	Moral reasoning normally involves clear definitions, the evaluation of arguments, the analysis of social institutions, the collection of historical data, and assent to a series of prescriptions for action that the reasoning supports	EIA often involves conflicting values and moral positions; frequently it is necessary to choose between or seek to reconcile conflicting moral rules (i.e., resolving moral dilemmas)
		The EIA process is arguably a form of moral reasoning; hence EIA practice can benefit from the insights and lessons of practical ethics
Deontological ethics	Applies absolute or foundational normative standards, duties, or principles of moral conduct (irrespective of consequences)	Participants in EIA processes often judge proposed actions based on absolute standards (i.e., nuclear power or clear-cutting unacceptable)
	Duty based (e.g., duties applicable to every situation)	Absolute standards can be useful in screening
		Also helpful in understanding basis for stakeholder positions and in determining proponent duties

(Continued)

Table 9.1   (*Continued*)

Concepts	Key Characteristics	EIA Implications
Teleological or consequentialist ethics	Normative principles of choice Rightness or wrongness of an action depends on the consequences of the action (total good consequences outweigh the total bad consequences) Utilitarianism is a form of teleological ethics (the public good is the sum of all preferences or the greatest good for the greatest number)	Implicit in much of EIA practice Tendency in EIA to focus on minimizing the negative rather than on comparing total good versus total bad or on maximizing benefits Once explicit, can recognize limitations with utilitarian approach and potential benefits of applying other normative standards
Rawlsian ethics	Right to extensive system of basic liberties (restrictions to liberty for sake of liberty of others) Protect resources for future (just savings) Greatest benefit to the least advantaged Lowest cost to least advantaged	Explicit consideration of distribution of costs and benefits by social group A moral rule consistent with focus on environmental and social justice (maximize utility of worst off) Combines consideration of social justice, liberty, and resource protection
Libertarianism	Elevates individuals and their rights above all others	Tendency, especially for public proposals, to assume that greater public interest should always prevail over individual rights Points to need to consider, and, to the extent practical, minimize losses of individual rights and freedoms
Discourse or communicative ethics	Seeks to counteract misinformation Seeks to ensure procedural fairness	Recognizes that EIA is a dialogue among interested and affected parties Recognizes need to minimize communications distortion and to facilitate procedural fairness
Procedural fairness or equity	The fairness of consultation and choice procedures Seeks to enhance democratic decision-making processes	Provides a basis for determining when procedures unfair and for formulating and applying rules and measures to prevent and offset May require additional measures to facilitate the involvement of traditionally underrepresented groups and organizations

Table 9.1 (*Continued*)

Concepts	Key Characteristics	EIA Implications
Critical ethics	Focuses on inequalities in the distribution of power and knowledge as a catalyst for action	Recognizes that power inequities a component of an ethical analysis; inequities can be exacerbated by proposed action (e.g., centralization of authority) Provides a basis for efforts to reduce and offset political inequities
Communitarian ethics	Focuses on the normative values of and control by local communities Normative values arise from the community	EIA proposals can inhibit or enhance local empowerment Local control (e.g., voluntary communities) one approach to the siting of "locally unwanted land uses"
Egalitarian ethics	Stresses the need to treat people equally and for those who receive the benefits to accept the burdens Merit of action dependent on whether the process distributes basic rights and duties justly and equitable	The unequal distribution of benefits and burdens is a recurrent issue Provides a basis for identifying and, where practical and appropriate, preventing or offsetting inequities
Distributional or outcome equity or fairness	Focuses on the distribution of resources, benefits, and costs over time, over space, and among social groups	Distributional inequities is a recurrent issue Provides a basis for identifying and, where practical and appropriate, preventing or offsetting inequities Consistent with social and environmental justice
Research ethics	Concerned with the ethical standards applied in natural and social science research	EIA is a form of applied research Many research guidelines available Particular concern when undertaking research involving indigenous peoples
Professional ethics	Concerns conduct of professionals in practice Codes of conduct applied by specific professions such as planners, engineers, scientists, environmental managers, and EIA specialists	Many professions involved in EIA process Professional codes of conduct facilitate ethical behavior Codes of conduct for environmental professionals conducive to good environmental practice

(*Continued*)

Table 9.1   (*Continued*)

Concepts	Key Characteristics	EIA Implications
Environmental ethics	Concerned with the moral basis of environmental responsibility Extends ethical rights to other organisms and to ecological communities Ecocentric perspective	Ethical responsibilities to environment a central attribute of EIA practice Helpful perspective in assessing environmental and impact significance
Feminist ethics	Focuses on women's issues and women's moral reasoning Emphasizes responsibility, obligation, and care more than rights, rules, and justice	Consistent with a discursive, inclusive, relational, nonexploitive, and nonmanipulative EIA process Helpful model for integrating ethical with technical Useful perspective on balancing rights and duties in fair processes
Sustainability ethics	Responsibility of current generations to future generations Ethics on a global scale Distributional equity a key element of social sustainability	Provides a framework for integrating ethical with other decision-making considerations Extends distributional analysis to include rights of and responsibilities to future generations Addresses social equity and sustainability links Broadens EIA temporal and spatial perspectives
Ethical pluralism	Addresses ethics from multiple perspectives using multiple methods and standards	Range of ethical perspectives, methods, and standards could be used to assess options Conflicting ethical standards and perspectives are possible and can be addressed through the EIA process

Sources: Beatley (1989), Etzioni (1995), Finsterbusch (1995), Forester (1989), Gardner and Roseland (1989), Harper and Stein (1992), Hendler (1994), Howe (1990), Kasperson et al. (1984), Lawrence et al. (1997), MacNiven (1982), Patton and Sawicki (1993), Rawls (1971, 2001), Taylor (1986).

the EIA process is a forum within which practical ethics are expressed and applied. The EIA process applies (or should apply) both professional and research ethics. Multiple ethical standards, principles, and decision rules are available for assessing proposed actions and for conducting EIA processes. It will sometimes be helpful to apply a plurality of ethical principles, standards, and decision rules. The preferences for and the manner in which ethics are applied will vary depending on the value systems of process participants. As with values, ethical perspectives and positions change and evolve. Making the evolving ethical perspectives and principles of EIA process participants explicit can reduce confusion and sometimes ameliorate conflict.

Ethical trade-offs and dilemmas are highly likely with multiple perspectives, values, participants, and potential ethical principles and standards. An ethical analysis should seek to identify and address ethical issues, trade-offs, and dilemmas. Ethical principles and standards can be applied to both procedures and proposed actions. They can also be applied to individual process activities (e.g., research, significance interpretation, consultation, communications). Ethical principles and standards are likely to vary by discipline (e.g., social, political, ecological, sustainability) and by perspective (e.g., feminist, traditional knowledge). Substantive ethical principles and standards can be determined only after analyzing the potential distribution of effects over time, over space, and among social groups. Measures will often be necessary to prevent and offset procedural and substantive distributional inequities. Efforts to address substantive fairness and equity issues are likely to be inhibited if participants perceive the EIA process to be unfair.

Part of an ethical analysis involves making the rights and duties of participants explicit. Rights extend beyond process participants to future generations and to the environment. Rights will often conflict. It is therefore necessary to identify and assess the implications of conflicting rights. Duties are not limited to proponents. They extend to regulators and to all process participants. Interpretations of appropriate duties will probably vary among participants. These varying interpretations also need to be explored.

An ethical analysis can build from ethical codes of practice, applied research ethical principles, and efforts to integrate social and environmental ethics into corporate planning. Practice-based precedents such as environmental justice initiatives and the application of substantive equity principles in siting waste management facilities could be particularly relevant. Many useful concepts, principles, and distinctions can be culled from applied ethics literature, especially efforts in directly related fields of practice such as environmental management and planning. Varying conceptions of the role of ethics within broader integrative frameworks and concepts, such as sustainability, could be especially instructive to EIA process management.

9.4.3   Procedural Fairness

Procedural fairness is concerned with the fairness of the EIA process. It includes both how consultation with interested and affected parties is undertaken and how choices are made (Kasperson et al., 1988; Lawrence et al., 1997). Procedural fairness principles and standards can pertain to the rights of participants, to the duties of the proponent and EIA team members, and to the responsibilities of process participants.

All interested and affected parties have a right to participate effectively in the EIA process. They may also see it as their right to be involved in designing and adapting the EIA process. They are likely to be particularly concerned with timely access to all relevant information and analysis and to timely (e.g., prior to major decisions) and adequate (e.g., sufficient time to formulate, review, and respond) involvement provisions. Rights also concern the ground rules for participating in and withdrawing from the process. They can extend to how participants are treated and to how their knowledge is incorporated into the process. Procedural rights can vary among social groups. Specific provisions to offset procedural inequities may be necessary for disadvantaged groups, traditionally underrepresented groups, and indigenous peoples. It will sometimes be necessary to address historical inequities at the outset before some parties will accept the process and its outcomes as potentially legitimate.

Proponents and EIA team members have a duty to establish a clear and understandable EIA process. Time frames should be reasonable. The need for the proposed action should be established. Reasonable alternatives should be considered. Assumptions, interpretations, conclusions, and recommendations should be explicit and substantiated. The proponent and team members, together with the EIA process, should be sufficiently flexible to adjust to changing circumstances and to adapt to language and cultural variations. They should respond to the concerns and suggestions of other process participants. They should respect the rights and values of other participants. They are obliged to minimize bias, to provide accurate data and analysis, to correct errors, to identify uncertainties and their implications, to comply with regulatory requirements, and to record and fulfill commitments. They should seek to remove barriers to understanding and participation. The EIA team members should comply with applicable ethical codes of conduct. The EIA process and methods should be consistent with good practice standards. The EIA process should be monitored and efforts made to enhance its effectiveness.

In exchange for the fulfillment of rights and duties, such as those cited above, all parties are commonly expected to participate in good faith. They also are accountable for their actions and should maintain contact with and be accountable to their constituents. They should not engage in rhetoric or make sensational charges. Depending on the process, all parties could attempt to reach a consensus or accommodate conflicts.

It may sometimes be advantageous to formalize procedural rights and duties through written agreements. These agreements will probably evolve in conjunction with the EIA process. Procedures may also be necessary to address situations where rights conflict or where there are conflicting interpretations of duties. Appeal procedures may be needed for such matters as the timely provision of all relevant information. A commitment could be made to institute an open, fair, impartial, and independent review at the end of the EIA process.

The choice and application of rights and responsibilities will vary among EIA processes. A suite of good practice ethical procedural principles could evolve over time. These principles could be adapted to suit individual proposal and environmental circumstances and to meet the needs and expectations of process participants.

9.4.4 Distributional Fairness

Distributional fairness pertains to the distribution of risks, costs, and benefits over space (community/regional/state/provincial distribution, fair/unfair locations), over time (historical inequities, current, future), and among social groups (income, ethnic, indigenous peoples, class, age, other susceptible populations). It can refer to the allocation of services and resources (generation/receipt, fair/unfair distribution, opportunities). It can concern the extent to which individual liberties and local decision-making powers are reduced or enhanced. Distributional fairness takes into account the fairness of cumulative effects and the carrying capacity and vulnerability to change of social, economic, and ecological systems.

The aggregation of distributional fairness concerns can take many forms. Total impacts and costs can be compared or net benefits to society can be determined. Both these approaches are consistent with a utilitarian ethical approach. Alternatively, net benefits by social group and by geographic area can be determined. The latter approach is more conducive to identifying and addressing social and environmental injustices. Fairness and acceptability determinations can be influenced by the availability of reasonable alternatives, by the potential for avoiding and mitigating inequities, and by applying equity compensation measures and local benefits.

Distributional fairness, as with procedural fairness, involves rights and duties. Distributional rights could pertain, for example, to avoiding unnecessary adverse effects, to reducing and mitigating adverse effects, and to compensating for significant adverse effects, which cannot be mitigated to acceptable levels. The rights of potentially affected populations to receive benefits and to avoid impacts could vary depending on the degree of potential harm, on the extent of adverse impacts already incurred, and on the degree to which the potentially affected population is socially and economically disadvantaged.

The proponent and EIA study team could have a duty to determine the distribution of costs and benefits and to ascertain the vulnerability of various groups. They might be expected to identify and redress historical and current burdens and hazards, to protect the interests of current and future generations, to accept the burden of proof, and to bear the full costs of the EIA process. There may be an expectation that benefits and unavoidable risks are to be shared and that there should be fair access to compensation based on clearly defined and consistently applied criteria. Meeting societal needs could be viewed as a shared responsibility. Some parties could suggest that risks should only be imposed voluntarily, that the greatest benefit should accrue to the least advantaged, and that a community that once has been selected should be assured that it will not be selected for future facilities.

9.4.5   Rights and Duties

A right is a claim on others that a person or a group of persons has and that is enforced by law, custom, or education (MacNiven, 1982). Rights express and often give a legal meaning to values. They tend to highlight some injustices (e.g., barriers to access to information) and to devote less attention to others (e.g., economic inequities). They protect our right to be equal (e.g., equal protection under the law) and to be different (e.g., minority rights). Privileges are possible within a rights system (e.g., affirmative action). Rights help determine what is right. Rights often conflict. However, rights systems tend to provide a means of adjudicating rival claims. Rights and duties have a reciprocal relationship. Each right entails an obligation (Ignatieff, 2000). The appropriate balance between rights and duties is often highly contentious. The EIA process is one among many forums within which rights and duties are expressed and applied.

An ethical EIA process determines and applies rights and duties. As highlighted in Table 9.2, there are various types of rights (rights about). Rights are possessed by different segments of the population (rights of). Rights address a range of concerns (rights to). Different parties exercise duties (duties by). The duties concern specific subjects (duties about). There are those who benefit from the conscientious application of duties (duties to).

Table 9.2 lists examples of rights and duties. Rights can be possessed by, for example, proponents, communities, indigenous people, consumers, workers, landowners, and governments. Rights can be extended to the environment and to future generations. There are basic or fundamental human rights and freedoms. Rights can, for example, concern such matters as health and safety protection, the application of democratic principles, compliance with legal requirements, social and economic concerns, the equality of access or treatment, responsibilities to minority populations, the continuation of traditional activities, the mitigation of and the provision of compensation for adverse effects, the protection of renewable and nonrenewable resources, and the maintenance and enhancement of environmental quality and ecological integrity. Rights can apply to such matters as how basic human needs are to be fulfilled, information is to be shared, consultation activities are to be conducted, decisions are to be made, personal freedoms and privacy are to be maintained, continued resource use is to occur, and safety is to be assured. Rights might concern how parties are to be treated fairly, how their languages and culture are to be protected, how they are to coexist, and how they are to continue to determine their own futures.

Many parties could have duties in an EIA process, including, for example, proponents, governments, professionals, researchers, nongovernmental organizations (NGOs), and individuals. The duties could be directed toward governments, communities, workers, NGOs, constituents, the environment, and future generations. The duties could concern how health and safety is to be determined and

Table 9.2   Examples of Rights and Duties

Rights		
Rights to	Rights About	Rights of
Basic needs	Fundamental freedoms	Proponents
Information	Health and safety	Regions and communities
Consultation	Democratic principles	Indigenous peoples
Liberty	Legal requirements	Consumers
Continued	Social concerns	Workers
resource use	Economic concerns	Landowners
Coexistence	Equality of access and	Interested and affected parties
Privacy	treatment	Governments
Safety	Minority populations	Future generations
Fair treatment	Traditional activities	
Self-determination and	Resource use	
consent	Ecological integrity	
Language and	Environmental quality	
culture	Mitigation and compensation	

Duties		
Duties to	Duties About	Duties of
Other governments	Health and safety	Proponents
Indigenous peoples	Environmental stewardship	Governments
Regions and communities	Planning and decision	Professionals
Workers	making	Researchers
Knowledgeable individuals	Information access	Current generation
Nongovernmental	Consultation	Nongovernmental
organizations	Legal liabilities	organizations
Constituents	Research procedures	Participants in the EIA
Environment	Institutional controls	process
Future generations	Training and employment	
	Compensation	
	Local benefits	
	Risk and uncertainty	
	management	
	Treatment of rights	
	Treatment of historical	
	grievances	
	Respect of culture and values	

protected. They could relate to environmental objectives and performance standards. They might pertain to the design and application of the EIA process, including, for example, information generation and sharing, public consultation, research procedures, and respect for culture and values.

Duties often extend to organizational obligations regarding such matters as legal liabilities, training and employment procedures, compensation and local benefits procedures and policies, social and environmental performance standards, and risk and uncertainty management standards and procedures. Additional duties are likely to be needed when indigenous peoples are involved in the EIA process. The latter duties could involve such matters as respecting self-determination goals and aspirations; respecting rights; seeking to preserve the culture, identity, and way of life of indigenous people; adapting planning, decision making, and research procedures; and providing compensation and local benefits to help indigenous people advance their own goals.

9.5   INSTITUTING AN ETHICAL EIA PROCESS

9.5.1   Management at the Regulatory Level

The four jurisdictions (the United States, Canada, the European Union, and Australia) all tend to spell out (in varying levels of detail) the EIA process-related responsibilities of the proponent and of government reviewers. The jurisdictions generally address public procedural rights to the extent of including minimum public notification, access to information, and public involvement requirements. These requirements are not generally portrayed as rights. There has been a general move in each of the jurisdictions to facilitate the involvement of disadvantaged groups and to take into account the rights, knowledge, culture, and traditional activities of indigenous people.

Various approaches have been taken to address distributional fairness concerns. The United States has integrated detailed environmental justice requirements into federal EIA requirements. EIA requirements in the northern territories of Canada stress the need for proposed actions to optimize benefits for northern residents and communities. European and Australian EIA stress the need to consider intergenerational equity.

Table 9.3 provides four examples of regulatory initiatives that address ethical considerations. The U.S. EIA environmental justice requirements address the potential for disproportionately high and adverse impacts on minority populations, low-income populations, and Indian tribes. The European Commission has instituted requirements to facilitate public access to information, public participation in decision making, and public access to justice. The Canadian EIA guideline document stresses the need to assess proposals in terms of the community and regional social and economic benefits provided. The Australian EIA legislation contains several requirements to accommodate the traditions, needs, and knowledge of indigenous people. It also addresses the rights (including ecological rights and the rights of future generations) and duties of various parties.

Overall, the regulatory approaches in the four jurisdictions fall well short of the measures described in Section 9.4. There remains some latitude for more specific procedural and distributional fairness provisions. Such provisions might have a

Table 9.3   Examples of Regulatory Ethical Approaches

US ENVIRONMENTAL JUSTICE EIA REQUIREMENTS

(BASS, 1998; US EPA, 1995, 1998a; WILKINSON ET AL., 1998)

Focuses on environmental justice

Addressed through guidelines from the U.S. Environmental Protection Agency and the U.S. Council on Environmental Quality

Considers disproportionately high adverse human health or environmental effects on minority populations, low-income populations, and Indian tribes

Points to need to determine population and community composition, to consider demographic, economic, risk, cultural, ethnic, historic, and policy characteristics and issues; and to identify patterns of natural resource consumption

Also addresses community role in EIA process (e.g., consultation procedures, alternatives analysis, mitigation analysis) and adaptations and methods for determining disproportionately high and adverse effects

Includes consideration of cumulative or multiple adverse exposures from other environmental hazards

EUROPEAN COMMISSION CONVENTION OF ACCESS TO INFORMATION, PUBLIC PARTICIPATION IN

DECISION MAKING, AND ACCESS TO JUSTICE IN ENVIRONMENTAL MATTERS

(UN ECE, 1998, DETR, 2000; ECE, 2000; EU, 2003; STEC, 2003)

Convention refined and applied by directives, guides and handbooks

Addresses basic requirements regarding public access to information, public participation in decision making, and access to justice regarding environmental matters

Outlines duties of governments to promote environmental education and awareness, to recognize and support environmental organizations, and to provide relevant information to the public promptly when there is imminent human health or environmental threat

Includes public participation provisions for early and effective public involvement in specific activities and for plans, policies, and programs

Includes provisions regarding information access appeal rights (e.g., review by court of law or another independent and impartial legal body, provision of information to public regarding administrative and judicial review procedures)

Provides for the establishment of appropriate assistance mechanisms to remove or reduce financial and other barriers to access to justice

NORTHWEST TERRITORIES, CANADA—BENEFITS GUIDELINES ASSOCIATED WITH MINERAL

EXPLORATION IN THE INUVIALUIT SETTLEMENT REGION

(ENVIRONMENTAL IMPACT SCREENING COMMITTEE, 1999)

Developed under the authority of a land-claims agreement

Identifies principles that companies engaged in mineral exploration are to comply with (e.g., preservation of Inuvialuit cultural identity and values, equal and meaningful participation of Inuvialuit, protection and preservation of arctic wildlife, environment and biological productivity)

Describes objectives of benefits guidelines (e.g., local business support and encouragement, Inuvialuit economic participation and self-reliance) and defines industrial benefits (e.g., procurement policy to optimize benefits to regional businesses, working with other parties to identify business opportunities)

Includes specific provisions regarding employment, training, and consultation

Provides for annual report to track program costs, wages, employment, purchases of goods and services, consultations undertaken, and benefits from relevant programs

(Continued)

Table 9.3   (*Continued*)

AUSTRALIAN ENVIRONMENTAL PROTECTION AND BIODIVERSITY CONSERVATION ACT (1999)

Establishes indigenous advisory committee

Recognizes biodiversity and sustainability roles and knowledge of indigenous people in objectives of act

Includes special rules to protect indigenous interests in the planning process for commonwealth reserves (e.g., traditional uses of reserves, procedures to resolve disagreements between director and indigenous people's land council, role in preparation of management plan, rules to protect indigenous interests)

Principles of ecologically sustainable development in act refer to integrating long- and short-term equity considerations and to intergenerational equity (i.e., present generation should ensure that the health, diversity, and productivity of the environment is maintained or enhanced for the benefit of future generations)

Act contains very specific descriptions of obligations of minister, proponent, and commissioners; identifies rights and obligations of witnesses; also identifies liability of executive officers for corporations

Provisions regarding listed threatened species and ecological communities and concerning whales and other cetaceans a recognition of an environmental right; liabilities also detailed

Regulations to act provide for special arrangements for affected groups with particular communications needs

"harder edge" if they were described in terms of rights and duties, perhaps along the lines of and extending from the Australian legislation. It could be worthwhile to adopt a more formalized approach to access to information, decision making, and justice rights as has occurred in Europe. It might be advantageous to formalize the requirement to undertake a fairness distributional analysis as has occurred in the United States. Such distributional analyses could be more broadly defined, as described in Section 9.4.4. As EIA requirements move toward a greater emphasis on sustainability, it could be necessary to introduce specific provisions concerning the rights of future generations. The stress placed on local benefits in northern Canada seems to have the potential for broader application.

On first inspection, the broader application of such measures could be very appealing. But there is a danger in too much precision at the regulatory level. The interested and affected parties vary among proposals and settings. EIA processes frequently involve a negotiation of procedural and distributional rights and duties. These negotiations occur both between proponents and regulators and among interested and affected parties. It could be worthwhile, in many cases, to formalize such negotiations. In this way, confusion can be minimized and conflict contained. The establishment of general ethical ground rules at the regulatory level could expedite proposal specific discussions and negotiations. However, the parties must also have sufficient latitude to come to agreements and accommodations which best match local circumstances and are consistent with the needs and aspirations of the participants. The auditing of proposal-specific experiences in treating ethical concerns could help identify recurrent issues where direction and guidance from the regulatory level did or could facilitate the EIA process.

9.5.2 Management at the Applied Level

Figure 9.2 is an example of an EIA process. Figure 9.2 and the process description that follows incorporate many ethical EIA elements. EIA process managers and participants can "pick and choose" the relevant and appropriate elements.

Startup The process begins with an overall study design. The study design incorporates a preliminary public and agency consultation plan. This step ensures that the EIA process is structured and focused. Consideration is given to redressing past grievances and injustices (an historical equity issue). A concerted effort is made to identify ethical issues and conflicting ethical perspectives and positions. The analysis is based on both secondary source reviews and discussions with interested and affected parties. An initial overview of applied ethical literature is undertaken to identify pertinent concepts, theories, and distinctions. These analyses contribute to study design refinements and to scoping the EIA process.

Ethical Foundation Once the startup activities are completed, the emphasis shifts to identifying procedural fairness principles (to guide and structure interactions with stakeholders) and distributional fairness principles (to guide the analysis of distributional effects). The procedural fairness principles address such concerns as timely and complete access to information, the fair treatment of participants (including assistance to disadvantaged groups), the right to participate fully in planning and decision making, the removal of participation barriers, and access to an open, fair, impartial, and independent review process. The distribution fairness principles concern such matters as undertaking a distributional analysis (with a particular emphasis on adverse effects on and benefits to minority, low-income, indigenous, and other susceptible populations), assessing the fairness of cumulative hazards (including the consideration of social and ecological carrying capacity), and instituting measures to manage equity-related impacts (mitigation, compensation, local benefits, monitoring).

Methods for determining distributional differences are formulated. The rights and duties of each major participant in the EIA process are identified. The principles, methods, rights, and duties are refined and adjusted based on stakeholder discussions. Ethical research rules and professional standards are formulated for environmental and ethical specialists. These rules and standards refine and adapt professional and disciplinary codes of practice. Measures (e.g., participant funding, expert advice) are developed to offset procedural inequities. These measures focus on the needs of disadvantaged groups, organizations, and communities. Ethical considerations are built into the EIA process goals and objectives.

The procedural fairness principles, in combination with the measures to offset procedural inequities, lay the groundwork for procedural fairness rules. The procedural fairness rules are determined jointly with stakeholders. They ensure

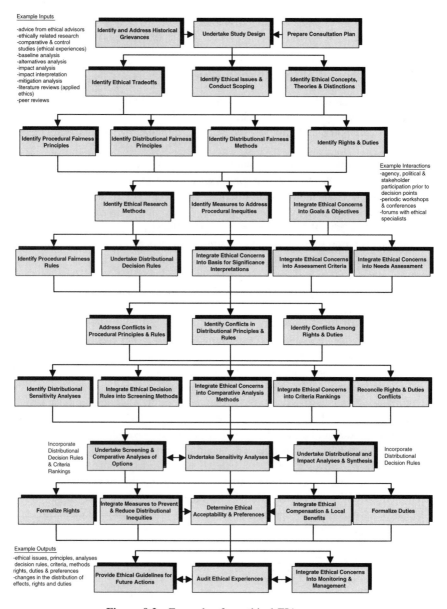

Figure 9.2 Example of an ethical EIA process.

that the dialogue and debate minimize distortion and are fair to all participants. The distributional fairness methods determine the magnitude and nature of distributional inequities over time, over space, and among social groups. Distributional fairness decision rules link the distributional analysis to decision making. They address such matters as minimizing overall adverse impacts to society (especially to the least advantaged), maximizing benefits to society (especially to the least advantaged), and minimizing undue burdens on future generations. Ethical concerns are integrated into such impact analysis activities as needs determination, assessment criteria formulation, and impact significance interpretation factors.

Refinements Conflicts among procedural principles and rules generally occur. Rules and principles are adjusted to resolve or at least to accommodate conflicts. A concerted effort is made to reach a consensus among parties regarding changes to procedural principles and rules. Conflicts are also likely among distributional principles and rules and among rights and duties. Again, a consensus among parties is sought. Residual differences are addressed by sensitivity analyses. Methods such as mediation are applied to address conflicting perspectives. A composite list of distributional sensitivity analyses is formulated. Ethical considerations are integrated into screening, option comparison, and criteria ranking methods.

Integrating ethical considerations in assessment methods includes formulating ways to explore ethical concerns, trade-offs and dilemmas; procedures for assessing choices against ethical goals and consequences (including future generational implications); approaches for assessing individual and cumulative effects from multiple perspectives (a pluralistic approach); methods for combining or modifying options to enhance ethical benefits; measures to prevent and offset inequities and to recognize and reinforce rights; procedures for testing outcomes against varying principles, theories, and methods; and methods for testing choices against ethical regulatory policies, laws, standards, guidelines, positions, and preferences.

Application and Decision Making The options are screened using distributional decision rules. Options remaining after screening are compared, taking into account the distributional analysis as well as the ethical inputs to criteria selection and rankings. The ethical distributional analysis extends and refines the impact analysis. Distributional principles and decision rules help determine the need for mitigation and the acceptability of the proposed action. Sensitivity analyses address uncertainties regarding distributional decision rules, the allocation of duties, and the probable effectiveness of ethically oriented mitigation, compensation, and local benefit measures.

Ethically preferred options are selected. Whether the proposed action is ethically acceptable is determined. Net benefits to society are determined. The distribution of costs and benefits among social groups (especially the disadvantaged), the extent to which rights are infringed upon, whether spatial and temporal inequities are ameliorated or reinforced, and whether political inequities are exacerbated or reduced are all considered. Whether the anticipated allocation of duties is appropriate and is likely to result in the adequate management of potential injustices and inequities also is taken into account.

The duties and rights associated with implementing the proposed action (if approved) are formalized and built into terms and conditions. Measures are instituted to monitor the actual distribution of effects, the effectiveness of measures to address inequities, the extent to which rights are maintained and the extent to which duties are fulfilled. The effectiveness of the EIA process in addressing ethical concerns is reviewed. Ethical guidelines are prepared for future related actions.

Inputs, Outputs, and Interactions The process is supported by advice from ethical advisors and peer reviewers and by ethically related research and literature reviews. The experiences of comparable and control communities in addressing ethical concerns are considered. Ethical analyses are combined with other planning and decision-making factors. Ethical considerations assume a pivotal role within broader planning and decision-making activities.

Agencies, elected representatives, and stakeholder groups are highly involved in the ethical EIA process, consistent with procedural fairness principles. A variety of involvement approaches are applied. Ethical specialists could formulate the principles, rules, rights, and duties. Modifications could be made based on the comments received. Alternatively, interested and affected parties could take the lead, with ethical specialists providing a support function. A possible middle ground entails the proponent, ethical specialists, and stakeholders jointly integrating ethical concerns into the EIA process. Forums and workshops are convened to broaden the range of ethical perspectives. Both specialists and nonspecialists participate in such forums. Depending on the location of proposed actions, it could be especially important to accommodate the ethical perspectives associated with traditional knowledge. Interpretations and conclusions are tested from multiple ethical perspectives.

There are numerous interim documentary outputs (e.g., ethical issues, principles, distribution analyses, decision rules, methods, rights, duties). A clear documentary trail is provided of how and why ethical concerns were addressed in the EIA process. The ethical analyses are expected to contribute to positive changes in the distribution of effects, rights, and duties within affected communities and among affected populations (both human and nonhuman).

9.6 ASSESSING PROCESS EFFECTIVENESS

Table 3.5 (in Chapter 3) presents criteria and indicators for assessing the positive and negative tendencies of the EIA processes described in Chapters 3 to 10. In this section we summarize examples of positive and negative tendencies of the ethical EIA process.

Scientific *rigor* is not likely to be a priority with this process. Procedural fairness and the distribution of effects, however, are priorities of social scientists and of ethical specialists. Independence and explicit presentation are consistent with procedural fairness principles.

An ethical EIA process is likely to address *comprehensively* all matters pertaining to fairness, equity, justice, rights, and duties. Nonethical concerns are likely to receive less attention. The integration of ethical concerns into more conventional

EIA processes would tend to broaden the definition of problems and opportunities. It could also help ensure that multiple perspectives are applied to interpretations and judgments.

Options are likely to be addressed *systematically* for all ethical matters. Other concerns would receive less attention. The stress on procedural fairness should ensure that procedures are open and that documents are succinct and understandable.

An ethical EIA process is *substantive* because it considers the distribution of benefits and adverse effects systematically. The management and acceptability of distributional effects are also considered. An ethical EIA process is guided by explicit values. Equity is a key component of sustainability. Substantive environmental concerns, unrelated to equity, fairness, and justice, tend to receive less attention.

An ethical EIA process is *practical*. It defines and applies rights and duties explicitly and systematically. It focuses on key ethical issues and trade-offs—issues and trade-offs that tend to be critical in terms of stakeholder opposition and acceptance. A heavy emphasis on ethical matters may not be a wise use of resources given the need to address many other concerns. It could also inhibit implementation to the extent that ethical matters are not the priority of all parties, including regulators, associated with implementation.

A strong emphasis on procedural fairness, on distribution fairness, and on defining and supporting rights tends to be conducive to a greater level of *local democratic influence* in the EIA process. Political implications are addressed when considering distribution of power implications and when defining and applying rights and duties. Stakeholder acceptance could be facilitated if equity, fairness, and justice concerns are stakeholder priorities. The emphasis on multiple perspectives could be helpful in accommodating traditional knowledge.

The emphasis on procedural fairness is conducive to a *collaborative* EIA approach involving all interested and potentially affected parties. A particular effort would be made to involve traditionally unrepresented and underrepresented groups and organizations. Government involvement is not likely to be a priority, except in ensuring that affected parties have access to relevant government agencies. The explicit identification of rights and duties could reduce the potential for misunderstandings. Consensus building and conflict resolution are facilitated only to the extent that ethical issues are the predominate concerns of stakeholders.

Ethical concerns are a priority with this EIA process. Both procedural and substantive fairness would be assessed thoroughly. Rights and duties would be identified and applied. Fairness, justice, and equity would be addressed from multiple perspectives.

Adaptability is defined quite narrowly with this EIA process (i.e., flexible in identifying and responding to ethical concerns). The bargaining process, associated with addressing procedural fairness, rights, and duties, can be quite adaptable. Risks and uncertainties are addressed in a distributional sense. Other risk and uncertainty aspects tend to be secondary considerations.

The *integration* of diverse values, perspectives, and knowledge is generally consistent with an ethical EIA process. Implications for and from related decisions are priorities only when they pertain to ethical concerns. The same is the case for links across disciplines, professions, and EIA types to proposal planning and to related environmental management forms. An ethical EIA process can be adapted to match the temporal (e.g., historical inequities, impacts on future generations), the spatial (e.g., spatial inequities), and the social and cultural context (e.g., inequities among population groups). It is linked to the institutional and political context because it considers political distributional effects and defines and applies rights and duties.

9.7 SUMMING UP

In this chapter we respond to the challenge to make EIA processes and outcomes more fair, equitable, and just. The discussion is concerned with identifying and advancing rights and ensuring that duties are fulfilled. We describe, with two stories, the benefits that can accrue from procedural and distributional equity initiatives. We provide the conceptual underpinning for an ethical EIA process and describe an ethical EIA process as it might be applied at the regulatory and applied levels. We then assess the ethical EIA process against ideal EIA process characteristics. Table 9.4 is a checklist for formulating, applying, and assessing an ethical EIA process.

Table 9.4   Checklist: An Ethical EIA Process

The response to each question can be yes, partially, no, or uncertain. If yes, the adequacy of the process taken should be considered. If partially, the adequacy of the process and the implications of the areas not covered should be considered. If no, the implications of not addressing the issue raised by the question should be considered. If uncertain, the reasons why it is uncertain and any associated implications should be considered.

CONCEPTUAL DISTINCTIONS

1. Does the process address equity, fairness, and justice concerns explicitly?
2. Does the process address rights and duties explicitly?
3. Does the process identify and apply relevant ethical concepts?
4. Does the process address procedural fairness concerns systematically?
5. Does the process address distributional fairness concerns systematically?
6. Does the process address the rights and duties of major process participants systematically?

MANAGEMENT AT THE REGULATORY LEVEL

1. Do the regulatory requirements:
   a. Detail the process duties of the proponent and of government reviewers?
   b. Identify minimum public procedural rights?
   c. Facilitate the involvement of disadvantaged groups?
   d. Take into account the rights, knowledge, culture, and traditional activities of indigenous people?
   e. Integrate environmental and social justice concerns?

Table 9.4 (*Continued*)

f. Facilitate the maximization of local benefits?

g. Ensure that intergenerational equity concerns are addressed?

h. Identify the duties and liabilities of all parties?

i. Formalize public access to information?

j. Formalize early and effective public involvement in decision making?

k. Formalize public access to justice in environmental matters?

l. Address the potential links between equity and sustainability?

MANAGEMENT AT THE APPLIED LEVEL

1. Does the process incorporate such startup activities as:

    a. Measures to redress past grievances and injustices?

    b. Identification of ethical perspectives, positions, and issues?

    c. Identification of ethical trade-offs?

    d. Identification of pertinent ethical concepts, theories, and distinctions?

2. Do the foundational EIA process activities:

    a. Identify procedural fairness principles?

    b. Identify distributional fairness principles?

    c. Identify methods to determine distributional fairness?

    d. Identify ethical research rules and professional standards?

    e. Identify measures to offset procedural inequities?

    f. Integrate ethical concerns into EIA process goals and objectives?

    g. Integrate ethical considerations into the determination of need?

    h. Integrate ethical considerations into assessment criteria?

    i. Integrate ethical considerations into significance interpretation factors?

    j. Establish procedural fairness rules?

    k. Establish distributional fairness decision rules?

3. Do the refinement EIA process activities:

    a. Identify conflicts in procedural principles and rules?

    b. Adjust the procedural rules and principles to resolve and accommodate conflicts?

    c. Identify conflicts among distributional principles and rules?

    d. Identify conflicts among rights and duties?

    e. Seek to address distributional principle and rule conflicts with sensitivity analyses?

    f. Seek to resolve conflicting perspectives regarding rights and duties?

    g. Integrate ethical considerations into option screening methods?

    h. Integrate ethical considerations into option comparison methods?

    i. Integrate ethical considerations into criteria ranking methods?

4. Do the application and decision-making EIA activities:

    a. Screen options against distributional decision rules?

    b. Compare options taking into account distributional impacts and ethical inputs to criteria and criteria rankings?

    c. Apply sensitivity analyses that address ethical uncertainties?

(*Continued*)

Table 9.4 (*Continued*)

---

  d. Determine ethically acceptable options?

  e. Determine ethically preferred options?

  f. Formalize rights?

  g. Formalize duties?

  h. Integrate measures to prevent and reduce distributional inequities?

  i. Integrate compensation and local benefits to address distributional inequities?

  j. Provide ethical guidelines for future actions?

  k. Integrate ethical concerns into monitoring and management?

  l. Audit the EIA process experience in addressing ethical concerns?

5. Do the EIA input, output, and interaction EIA activities:

  a. Provide for advice from ethical advisors and peer reviewers?

  b. Draw upon ethically related research and literature reviews?

  c. Consider the experience of comparable and control communities in addressing ethical concerns?

  d. Integrate ethical concerns with other planning and decision-making factors?

  e. Effectively involve elected representatives and stakeholders in addressing ethical issues in the EIA process?

  f. Convene forums to broaden the range of ethical perspectives?

  g. Accommodate, where pertinent, the ethical perspectives inherent in traditional knowledge?

  h. Test interpretations and conclusions from multiple ethical perspectives?

  i. Fully document how and why ethical concerns were addressed in the EIA process?

  j. Demonstrate that the proposed action will make a positive contribution in the distribution of effects, rights, and duties within affected communities and among affected populations (both human and nonhuman)?

PROCESS EFFECTIVENESS

1. Is the process rigorous in the treatment of both ethical and nonethical concerns?

2. Is the process comprehensive, appreciating that ethical EIA processes tend to treat some activities more comprehensively than others?

3. Is the process environmentally substantive, recognizing that environmental concerns unrelated to ethics will tend to receive less attention?

4. Is the process practical, recognizing that ethical concerns will receive a larger share of available resources?

5. Does the process facilitate a greater level of local influence, assuming that ethical concerns are a stakeholder priority?

6. Is the process collaborative, assuming that ethical concerns are a stakeholder priority?

7. Does the process thoroughly address procedural fairness, distributional fairness, rights, and responsibilities?

8. Is the process flexible, appreciating that the emphasis is on anticipating and responding to ethical issues and trade-offs?

9. Does the process facilitate integration, acknowledging that the focus in any integrative efforts will be on ethical concerns?

Establishing a foundation for an ethical EIA process begins with two stories that illustrate how procedural equity measures and a focus on net benefits at the local level can enhance an EIA process. An overview of the major shortcomings identified by critics is then presented. The critics emphasize that insufficient attention is being devoted to the fairness of the EIA process, to the distributional consequences of proposals subject to EIA requirements, to the rights of participants in the process, and to the duties of proponents, regulators, and other process participants.

The major concerns raised by the critics (fairness, equity, justice, rights, and duties) are all concerned with the moral rules, principles, and standards that govern human conduct (i.e., ethics, or more specifically in this case, normative, applied, practical ethics). These terms are each defined. Interconnections are highlighted. Several key ethical concepts are described briefly, together with implications for EIA process management. The concepts largely concern situations in which ethics might be applied and alternative ethical standards for judging behavior.

An overview of procedural fairness is presented. Procedural fairness is concerned with both how consultation takes place and how decisions are made. It includes principles and rules pertaining to the rights and duties of process participants. Several examples of procedural fairness rights and duties are presented. Examples of distributional fairness distinctions, principles, and duties are identified. Distributional fairness pertains to the distribution of risks, costs, and benefits over space, over time, and among social groups. It can refer to the allocation of services and resources and to impacts on individual liberties and on local decision-making powers. It considers the fairness of cumulative effects and the relationships to social, economic, and ecological carrying capacity and to vulnerability to change. A description is provided of possible rights and duties. Rights express and give legal meaning to values. Duties entail obligations. Rights and duties can be expressed and applied through the EIA process. There are various types of rights and duties. They apply to different population segments, and they concern a range of subjects. Some rights and duties are established through regulatory requirements. Others are determined, often through discussions and negotiations, during individual EIA processes.

EIA regulatory requirements in the four jurisdictions identify some proponent and government review duties. They identify minimum public notification and involvement rights. More consideration is being given to measures to facilitate the involvement of disadvantaged groups and to accommodate the rights of indigenous peoples, knowledge, culture, and traditional activities. Varying approaches are being taken regarding such matters as environmental justice requirements, access to information, the provision of local benefits, and the treatment of intergenerational equity. More could be done to address ethical concerns at the regulatory level. However, a balance should be maintained between greater structure and guidance and the need to make proposal and setting specific adaptations and refinements.

An example of an ethical EIA process is described. The process begins by considering historical grievances and by identifying ethical issues and trade-offs. Relevant literature and experiences are canvassed. These overview analyses provide the basis for identifying procedural and distributional fairness principles and methods.

Rights and duties, ethical research methods, and measures to address procedural inequities are addressed. Ethical concerns are integrated into EIA process goals and objectives.

Procedural and distributional fairness principles are refined into decision rules. Ethical concerns are integrated into significance determination factors, assessment criteria, and the needs analysis. Conflicts among procedural principles and rules, among distributional principles and rules, and among rights and duties are identified. The conflicts are resolved or accommodated to the extent practical. Residual conflicts are addressed through sensitivity analyses. Ethical concerns are integrated into screening methods, comparative analysis methods, and criteria ranking.

Distributional analysis, distributional decision rules, and criteria rankings are incorporated, where applicable, into the screening of options, the comparison of options, and the impact analysis. Ethical uncertainties are addressed by sensitivity analyses. Ethically acceptable and preferred options are selected. Ethical concerns are built into mitigation, compensation, and local benefit measures. Rights and duties are formalized. Ethical concerns are incorporated into monitoring and management. Ethical guidelines are prepared for future actions. The EIA experience in addressing ethical concerns is audited.

The EIA process is supported by advice from ethical advisors, applied research, and reviews of comparable situations. Ethical concerns are integrated with other planning and decision-making activities. Agencies, elected representatives, and stakeholders participate in identifying and applying ethical concerns. Efforts are made to broaden the basis of involvement and to ensure that multiple ethical perspectives test interpretations and conclusions. The role of ethical concerns in the EIA process is fully documented.

The ethical EIA process thoroughly addresses fairness concerns. It tends to support stakeholder collaboration and enhanced local influence. It is selectively conducive to making EIA more rigorous, comprehensive, substantive, practical, flexible, and integrative. It addresses each objective, but only to the extent that ethical concerns are involved.

CHAPTER 10

HOW TO MAKE EIAs MORE ADAPTIVE

10.1 HIGHLIGHTS

In this chapter we address how EIA processes can adaptively anticipate and respond to the uncertainties associated with difficult problems in chaotic and complex environments. It is commonplace in EIA literature and practice to emphasize that the EIA process should be adaptive, flexible, and iterative. Specific means for accomplishing this aim are less evident. The major approaches advanced for managing uncertainties are controversial and only partially or indirectly connected to EIA process management. In this chapter we provide a systematic, integrated approach to managing uncertainties in the EIA process.

- The analysis begins in Section 10.2 with two applied anecdotes. The stories describe applied experiences associated with efforts to make EIA practice more adaptive.
- The analysis in Section 10.3 then defines the problem. The problem is a failure to characterize and manage uncertainties in the EIA process adequately. In this section we explain why it is necessary to formulate EIA processes that adaptively manage uncertainties.
- In Section 10.4 we explore the characteristics of the problems and environments that generate uncertainties, identify uncertainty types and sources, describe general adaptation strategies and tactics, and characterize four major uncertainty management approaches: (1) risk assessment and management,

Environmental Impact Assessment: Practical Solutions to Recurrent Problems, By David P. Lawrence
ISBN 0-471-45722-1 Copyright © 2003 John Wiley & Sons, Inc.

(2) human health impact assessment, (3) the precautionary principle, and (4) adaptive environmental assessment and management.

- In Section 10.5 we detail how an adaptive EIA process could be implemented at the regulatory and applied levels. In Section 10.5.1 we explain how regulatory requirements and guidelines can facilitate uncertainty management and adaptation and in Section 10.5.2 describe how the uncertainty concepts, strategies, tactics, and approaches can be linked and combined in practice.
- In Section 10.6 we assess how well adaptive EIA processes satisfy ideal EIA process characteristics.
- In Section 10.7 we highlight the major insights and lessons derived from the analysis. A summary checklist is provided.

10.2   INSIGHTS FROM PRACTICE

10.2.1   Adapting an EIA Process to Both Changes and Flaws

An EIS was prepared for the siting of a new high-security correctional facility near Minneapolis, Minnesota. Siting of a correctional facility is a difficult process, as it is generally an unwanted land use by potential neighbors. The Minnesota Department of Corrections (DOC) needed a location for a new $80 million high-security facility but did not want to place the facility in a community where it was not wanted. To address this concern, the DOC put out a request for proposal (RFP) to rural communities that wanted the proposed facility as an economic development tool. Nine communities responded to the RFP. A committee made up largely of corrections and state building staff made security, economic development, and provision of infrastructure the primary criteria for selecting the preferred location. It was not until the preferred location had been selected and endorsed by numerous politicians that an EIS process was initiated.

The EIS evaluated all nine alternative locations. The EIS also included an effective public consultation process involving the use of interagency scoping meetings, newsletters, individual interest-group meetings, and public comment meetings. Public consultation went beyond the state requirements to ensure that a broader range of concerns was heard than had been captured during the DOC siting process. This included one meeting at the farmhouse of an elderly opponent and her neighbors across the road from the preferred site to hear their views in an interactive and nonthreatening environment.

The draft EIS revealed a number of important findings. First, the proposals submitted by the nine communities were prepared and endorsed largely by the business segment of the local population. However, there was a large segment of the population that was not involved or aware of the proposals and that had a very different perspective on the amount and type of growth that was healthy for their community. Second, environmental features were not given careful attention in the DOC selection process. As a result, the preferred site contained a larger percentage of wetlands than most of the other nine sites. The Clean Water Act required that the DOC demonstrate consideration of avoidance as the first criterion in comparing sites. This factor left the preferred site open to a high risk of losing a contested case hearing. Third, there had been little interaction with state and federal agency with environmental responsibilities.

These findings led to a series of interagency meetings culminating in a meeting of six state agency commissioners where it was decided to move the preferred location to a community that was able to demonstrate greater local support across the community as a whole and greatly reduced potential for environmental effects. This EIS process was started too late in the process to avoid unnecessary siting effort. The process did ultimately serve its purpose well by bringing issues to the table not initially considered by the DOC so that an informed decision could be made in the best interest of the public as a whole.

Adaptation generally is seen as making adjustments to the EIA process to address events, issues, and preferences that emerged during the process, which would have been difficult to anticipate at the outset. However, adaptability also means modifications to overcome flaws in the process. In this case some of the difficulties that arose appear to have been, at least in part, foreseeable. The broadly based and effective consultation program failed to identify the perspectives and concerns of those segments of the population that were unaware of or had not responded to the proposal. Perhaps a more collaborative consultation process might have identified those concerns earlier in the process. Or perhaps some segments of the population become actively involved only when they perceive a direct and immediate threat to their way of life. If the latter, this suggests that an EIA process should be sufficiently flexible to adapt to the views and perspectives of different segments of the population as they become more or less involved in different stages of the process. The failure to identify significant environmental features would appear to have been avoidable with earlier and more systematic agency consultation, greater attention to regulatory requirements, and a more comprehensive treatment of substantive environmental concerns. Still, with any EIA process, agency concerns and requirements can change and supplementary analyses may need to be undertaken to overcome analysis gaps and errors. The EIA process must be sufficiently flexible to make such adjustments.

This story suggests that a highly collaborative approach can be an effective way of making the necessary adjustments to address unforeseen (both avoidable and unavoidable) circumstances, even quite late in the process. However, major late "course correction" is a risky business. More attention to preventing and containing avoidable process-related flaws can make it easier to design and manage an adaptable EIA process that can address changing, unforeseeable circumstances rapidly and systematically.

DARRYL SHOEMAKER
HDR Engineering

10.2.2   Operating Within and Reshaping a Complex and Chaotic Jurisdictional Environment

For a major capital project, an EIA is a complex undertaking at the best of times. It can become especially complicated when a large geographic area is involved, the proponent has no political power, and approval procedures are overlapping and ill- or undefined. The Greater Toronto Area (GTA) of the province of Ontario encompasses five regional governments. Each regional government has authority over several local municipalities, each of which offers a range of lower-tier municipal services to the total population of close to 4 million residents. In the mid-1990s, the province of Ontario downloaded many of its responsibilities to regional and local municipalities.

Two areas of approval are particularly significant for large public capital works projects: (1) environmental approvals and (2) land-use approvals. Regional and local governments are able to provide environmental approvals for routine projects, with generally known effects. These are called *class environmental assessments*. They also have the responsibility of identifying when a full environmental assessment study is required, known as an *individual EA*. The provincial Ministry of the Environment, through the minister, retains ultimate authority for approving an undertaking subject to an individual EA.

Regional and local governments are also given the responsibility to approve or reject land-use changes that may be associated with an undertaking. Furthermore, regional governments can approve or reject land-use decisions made at the local level by local municipalities. The provincial government through the Ontario Municipal Board has ultimate approval over land-use changes.

The York region, located to the north of Toronto, is landlocked and dependent on other regions for potable water for its residents. Obtaining a new secure source of water meant obtaining land-use and environmental assessment approvals and constructing facilities to Lake Ontario through its sister jurisdiction—the Durham region. Following a strategic EA, the Long Term Water Supply Master Plan, they decided to proceed with a $600 million undertaking involving a water intake tunnel, a raw water pump station, a water filtration plant, a storage reservoir, and water transmission mains (pipelines). The right-of-way required obtaining approvals for crossing their neighbor, the Durham region, and the city of Pickering, a lower-tier municipality within the region of Durham.

The York region concluded that an individual EA was required. York had approval authority over its own lower-tier municipalities but no authority over its neighboring municipalities. Provincial legislation was silent on whether York Region's approval authority extended beyond its borders. If left unresolved, a conflicting set of approval and rejection requests could be submitted to the province. Accordingly, it was necessary to sort out roles informally.

Fortunately, the region of Durham saw benefits in the project. It agreed not to complicate or interfere with the EA approval process but to instead, to support the York region. The region of Durham, however, was an active commentator. In this way they continued to exercise authority over the local municipality and to be sensitive to the needs of local residents.

The York region staff was accustomed to consulting its own residents. Consulting with directly affected city of Pickering residents was seen to be so important that they helped local residents to become organized and helped the residents to provide effective input to the EA by paying for independent peer review consultants. Although local residents continued to oppose the project, they were generally supportive of the EA process, and their comments led to improvements to the definition of the undertaking throughout the EA. Formal discussions between the York region and the city of Pickering occurred on a staff-to-politician/staff basis.

The York region consulted with organizations and agencies responsible for review and approval under the Environmental Assessment Act. The formal approval body, the Provincial Ministry of the Environment, coordinated the input of other ministries, such as for agriculture and transportation. Exercising their authority to assess land-use implications of proposed changes and approve or reject those changes, the city of Pickering opposed the water supply facilities subject to the York region and the city completing a satisfactory community compensation agreement. With those agreements in place, the York region would formally submit the EA, including land-use conditions and agreements, to the Minister of the Environment for approval. The undertaking was Ontario's first individual EA to be submitted under its newly revised Environmental Assessment Act. It was ultimately withdrawn after a third region entered into a more cost-effective agreement with the York region to provide water.

This story demonstrates that undertaking complicated projects in complex jurisdictional environments often will necessitate careful attention to formal and informal institutional arrangements. Often, new roles will need to be defined and existing roles adapted. New and modified relationships will need to be established and modified as circumstances change and evolve. Care will be required in defining and coordinating the multiple, often overlapping approval requirements. Operating in a complex and chaotic jurisdictional setting necessitates a highly adaptive EA process. It can also mean reshaping and, in some instances, creating a new jurisdictional environment. The boundaries among process, institutional arrangements, and context are necessarily fluid. Flexibility and adaptability are especially important when identifying alternatives and when defining the proposed action. The ultimate decision to withdraw the project demonstrates that EA studies can extend information gained through strategic environmental assessments by identifying socioeconomic and natural environmental costs of the preferred project against the costs and benefits of other options.

DAVE HARDY
Hardy Stevenson and Associates Limited

10.3   DEFINING THE PROBLEM AND DECIDING ON A DIRECTION

The two stories illustrate the issue of coping with uncertainty in different ways. In the first story, adaptations are required as the process unfolds, in part because of unanticipated circumstances and in part because of process flaws. EIA processes are generally too complex to anticipate fully what might occur. Some mistakes and flaws are inevitable. EIA processes vary in their ability to adapt to such eventualities. The second story demonstrates that it is sometimes necessary to reshape and refine institutional arrangements when requirements, roles, and procedures are ill- or undefined. These is especially the case with large, complex undertakings that involve multiple jurisdictions. In such cases an adaptive approach to process management, to establishing and refining institutional arrangements, and to contextual variations is essential. Both stories demonstrate that a collaborative approach can help deal with complex and evolving circumstances. The two stories provide only a superficial and selective sense of the role of uncertainty in the EIA process and how it might be managed.

Uncertainty is about not knowing and about not being sure (Yoe, 1996). EIA practice has been faulted both for being overly deterministic (i.e., unsupportable precision) and for being overly vague (i.e., a lack of precision). In the former case uncertainty is not acknowledged. In the latter case vagueness stems more from a lack of effort to be precise than from an acknowledgment of uncertainty and its implications. EIA practice tends to assume that a single number can represent the range of values potentially associated with a measured or predicted parameter (Carpenter, 1995). Such thinking fails to acknowledge or account for natural variation, knowledge gaps, or indeterminacy (Power et al., 1995). It operates under the illusion that present and future conditions can always be measured, predicted, and controlled readily and precisely (Hodgson and White, 2001). EIA analyses and decision making, based on such thinking, neither acknowledge nor explicitly consider uncertainty (Reckhow, 1994). Equally unacceptable are vague, unsupported, qualitative statements about current or future conditions (Culhane et al., 1987; Malik and Bartlett, 1993). Such statements provide the reader with minimal insight regarding what is known (or knowable) or unknown (or not known with precision). False assurances of certainty are misleading. Vague and unsupported "musings" are similarly uninformative. EIA effectiveness reviews demonstrate that accurate forecasts, the use of confidence limits (as a means of acknowledging uncertainties) and monitoring (as a means of testing the accuracy of forecasts and the effectiveness of mitigation measures) are still more the exception than the rule (Culhane et al., 1987; Sadler, 1996).

Uncertainty is ubiquitous in environmental decision making and in EIA practice (Gibson, 1992; Tonn, 2000). Uncertainty is not well understood. As Tonn points out, there can be (1) too much uncertainty (inadequate effort to reduce), (2) too much certainty (a failure to consider the consequences of inaccurate predictions), (3) conflicting perspectives on certainty and uncertainty, (4) misrepresented certainty, (5) misunderstood uncertainty, (6) the confounding of uncertainty and values, and (7) a lack of foresight (Tonn, 2000). Uncertainties occur throughout the EIA process (Gibson, 1992). Not acknowledging uncertainties or addressing uncertainties with simplistic and unsupported "safety factors" can impair decision making and contribute to inequities (e.g., increased uncertainties in the lives of the weak) (Cardinall and Day, 1998; Marris, 1996). Explicitly addressing uncertainties and, by extension, follow-up is essential to good EIA practice (Government of Canada, 2001; Sadler, 1996; Yoe, 1996). Paralysis is not inevitable (Gibson, 1992). Uncertainty analyses help hedge away from large losses and aid in avoiding and reducing the potential for nasty and tragic surprises (Gibson, 1992; Reckhow,

1994). An uncertainty-oriented EIA process is necessarily flexible, adaptive, and iterative. Flexibility (anticipating and adjusting rapidly to changing conditions) and iteration (linking EIA activities and stages with feedforward and feedback loops) are commonly identified as good practice principles (IAIA and IEA, undated; Sadler, 1996).

Accepting the need to address uncertainties and for adaptive EIA processes is only a beginning. Good practice uncertainty management principles need to be identified (Sadler, 1996). Dubious assumptions (e.g., equating vagueness with flexibility, assuming that reducing uncertainty will increase certainty) need to be avoided (Hodgson and White, 2001). Identifying uncertainty types and sources, characterizing uncertainty concepts, and formulating, adjusting, and applying adaptation approaches, methods, and concepts all require further attention. EIA literature and practice often acknowledge the uncertainties associated with difficult issues and problems and with complex and chaotic environments. Reference is frequently made to matching the EIA approach to the problem and the environment. Less attention is devoted to characterizing those problems and environmental conditions most prone to uncertainty and to exploring EIA management implications.

Uncertainty is commonly coupled with risk as alternatives to certainty. Arguably, risk is a subset of uncertainty (i.e., a form of uncertainty to which probabilities can be attached). Risk, however, goes further in considering potential negative implications. Risk combines probabilities with harmful outcomes—to people, to property, and to ecological systems. The treatment of risk in EIA guidelines and practice is often either superficial or highly variable (Eccleston, 1999b; Malik and Bartlett, 1993; Sadler, 1996). Increased attention has been given to if and how risk and more particularly risk assessment and management could be linked and integrated with EIA (Barrow, 1997; Canter, 1993b; Carpenter, 1995; Erickson, 1994; Harrop and Nixon, 1999; Ugoretz, 1993; Westman, 1985). Although the need for EIA to consider risk is broadly acknowledged, the merits of elements of risk assessment and management and whether and how the two fields might best be linked, integrated, or modified have been debated intensely. The debates extend to comparisons with alternatives to risk assessment and management (e.g., performance standards, semiquantitative hazards assessment). In recent years it has broadened to encompass alternative risk, uncertainty, and health effects management approaches (e.g., human health impact assessment, the precautionary principle, adaptive environmental assessment and management)—approaches that could provide a framework for, be subsumed within, or represent an alternative to risk assessment and management.

The precautionary principle has been identified as a sustainability principle (Sadler, 1996). It is integrated into EIA requirements in some jurisdictions (Australia and the European Union, for example), although implications have yet to be determined. There are, however, numerous definitions, a host of positions concerning potential EIA and decision-making roles and a lengthy list of ascribed advantages and disadvantages. Adaptive environmental assessment and management (AEAM) blends scientific, ecological model building with adaptive, heuristic group planning and decision making. It has been applied widely in environmental

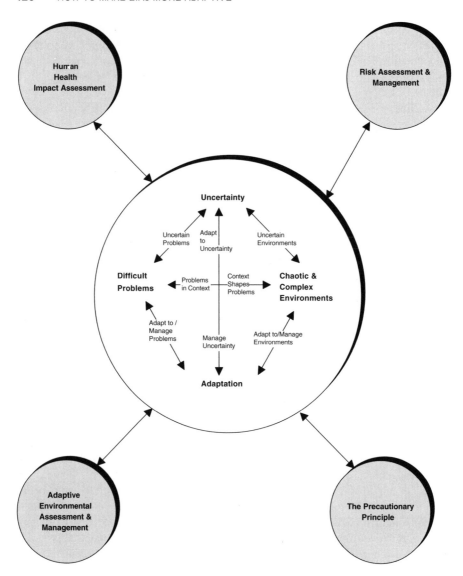

Figure 10.1   Examples of uncertainty management elements.

and resource management, although frequently only partially and sometimes with mixed results. AEAM has been identified as an effective approach to uncertainty management in EIA (US CEQ, 1997a). As with the precautionary principle, there is an intense debate surrounding AEAM. Its potential EIA practice roles and attributed strengths and deficiencies are often overstated.

Harm, in the sense of human health effects, is a component of risk. Although human health risk is invariably a major public concern, human health effects are often not addressed or are addressed only superficially, partially, inconsistently, and inadequately (Arquiaga et al., 1994; BMA, 1998; Canter, 1990; Davies and Sadler, 1997; Ortolano and Shepherd, 1995; Steinemann, 2000). EIA guidelines tend to devote little (often, no) attention to health concerns or define health issues very narrowly (Sadler, 1996). The propensity to equate health-related concerns with environmental standards ignores the health implications of substances and processes not covered by standards (Arquiaga et al., 1994). A major response to these types of deficiencies has been health impact assessment (HIA). Several jurisdictions have issued HIA guidelines in recent years (EnHealth Council, 2001b; Health Canada, 2000a; IPHI, 2001; WHOROE, 1987, 2000a,b). The effectiveness of these guidelines has yet to be determined. HIA institutional relationships, the relationship between HIA and other forms of impact assessment, and the relationships between HIA and other risk and uncertainty management approaches also require additional attention.

The problem, then, is a combination of confusion regarding the nature of uncertainty and the related concepts of risk and health effects and ambivalence concerning the most appropriate approach (or combination of approaches) for managing uncertainties in the EIA process. As illustrated in Figure 10.1, the direction is (1) to describe uncertainty sources, types, and concepts; (2) to classify the types of problems commonly associated with high levels of uncertainty; (3) to identify the relevant properties of chaotic and complex environments and systems; (4) to provide an overview of general adaptation strategies and tactics; (5) to describe risk assessment and management and potential EIA process relationships; (6) to describe human health impact assessment and potential EIA process connections; (7) to describe the precautionary principle and potential EIA process roles; (8) to describe adaptive environmental assessment and management and potential EIA process links; and (9) to explore interconnections among the uncertainty management concepts and approaches. Collectively, these analyses provide the basis for adaptive EIA process management at the regulatory and applied levels.

10.4   SELECTING THE MOST APPROPRIATE ROUTE

10.4.1   Uncertainty

Uncertainty, broadly defined, is any situation where we are not absolutely sure (i.e., the opposite of certainty) (Yoe, 1996). There is doubt, incertitude, or lack of clarity. There may be an absence of knowledge (something is not known or knowable), knowledge may be partial, or knowledge may be imprecise. Uncertainty, narrowly defined, focuses on situations where the direction or system characteristics are known but the nature of the outcome or its probability is unknown (Carpenter, 1995; Dearden and Mitchell, 1998; Hyman et al., 1988). Risk is excluded where probabilities can be ascertained. This analysis (i.e., risk as a subset of uncertainty) applies the broader definition. With both risk and uncertainty, it is largely a question of degree (of certainty or uncertainty), with overlapping or highly permeable boundaries between the two concepts. Interpretations of when there is uncertainty and how much uncertainty exists are subjective, social, and political (Gullett, 1999a).

There are many *uncertainty forms*. Uncertainty can be quantitative or qualitative, objective or subjective (CEC, 2000; Dearden and Mitchell, 1998; US ACE, 1992). There can be scientific or methodological uncertainties concerning the choice of parameters, the measurements made, the conditions of observation, the samples drawn, the models used, and the causal relationships employed (Carpenter, 1995; CEC, 2000; CRAM, 1993; Rowe, 1994; US EPA, 1998c, 1999). Perceptions of uncertainties often vary between scientific and technical specialists and lay observers (Grima et al., 1986). There can be substantive knowledge or epistemological uncertainties regarding organizational or environmental systems (Cardinall and Day, 1998; Friend and Hickling, 1997; Mostert, 1996; US EPA, 1998b,c). Knowledge uncertainties can sometimes be reduced through additional analysis but can also be inherent (i.e., fundamental limits to our knowledge of the world) (Tonn, 2000). Uncertainties can pertain to the past, to the present, or to the future (Rowe, 1994). There can be uncertainties concerning guiding values and desires, especially when values, perspectives, and interests conflict, interact, and change (Cardinall and Day, 1998; Friend and Hickling, 1997; Tonn, 2000). Uncertainties crop up in every aspect of and activity within and linked to the EIA process.

Many *sources* can contribute to uncertainty. There can be a lack of data, knowledge, experience, or understanding (Carpenter, 1995; US ACE, 1992; US EPA, 1998c; Yoe, 1996). Theories, explanatory paradigms, methods, and models can be inadequate (Carpenter, 1995; US ACE, 1992; US EPA, 1998c; Yoe, 1996). Time, expertise, and other resources can be insufficient (Carpenter, 1995). Analyses can lack focus because of an absence of direction or poor management (US ACE, 1992). Institutional capacity constraints or deficiencies in EIA requirements and guidelines can contribute to uncertainty (Mostert, 1996). Uncertainties can be exacerbated by poor communications, errors, bias, conflict, and dubious judgments (Carpenter, 1995; Rowe, 1994; Treweek, 1999; US EPA, 1998c; Yoe, 1996). Uncertainties can result from inherent variations, changing proposal characteristics, randomness, and the multiplicity of intervening variables associated with complex systems (Carpenter, 1995; US EPA, 1998c; Yoe, 1996). Novel situations and new technologies, materials, and methods tend to be especially uncertain and prone to surprise (Carpenter, 1995). Uncertainty is generally heightened as analysis scales are increased, as time horizons are extended and as study schedules are abbreviated.

Table 10.1 describes briefly several key uncertainty-related *concepts*. Recognizing ignorance or lack of knowledge can be humbling. It acknowledges inevitable knowledge gaps. It can stimulate efforts to reduce knowledge deficiencies. Errors and bias will always occur in EIA practice. It can be helpful to understand the different types of errors that can occur, to be sensitive to the conditions that contribute to errors and bias, to focus on those types of mistakes likely to have the most serious implications, and to proactively anticipate, minimize, and correct mistakes and bias. Indeterminism and inconclusiveness underscore the limits to uncertainty reduction and the need to ensure that conclusions are not more definite than supporting analyses. Science offers potential insights into the nature of indeterminism

Table 10.1   Examples of Uncertainty Concepts

Ignorance/Incomplete Knowledge

Lack of knowledge; not all outcomes are known; also ignorance of own ignorance (don't know what don't know)

Scientists surprised by the outcome; they do not know, but with hindsight can usually explain it

Two faces: positively—a humble admission that we don't know what we don't know; negatively—the practice of making decisions without considering uncertainties

Cure: obtain knowledge (e.g., education, training, talking to experts, acquiring experts through hiring, contract, or coordination)

Errors/Mistakes/Bias

Type I errors (false positives—concluding that there is an effect when, in fact, there is none); type II errors (false negatives—concluding that there is no effect when, in fact, there is); type III errors (wrong problem)

Errors of measurement, calculation, and judgment

Bias in data acceptance (can treat research too leniently or too harshly) and bias in data interpretation (e.g., overemphasis on avoiding type 1 errors)

Measures to address errors should provide new information, should not destroy the experimented, and should not cause irreversible environmental change; when errors occur it should be possible to learn from error (a source of new information) and to start over

Uncertainty Principle (Heisenberg)

Places an absolute, theoretical limit on the combined accuracy of certain pairs of simultaneous, related measurements

Specifically gives a theoretical limit to which a particle's position and momentum can be measured simultaneously

Has been elevated by some to the status of a philosophical principle, called the principle of indeterminacy, which has been taken to limit causality in general

Indeterminism/Inconclusiveness

Means that the uncertainties are of such magnitude and variety that they may never be significantly reduced

Scientific knowledge is inadequate; causal chains and networks are open and not understood

Potentially relevant concepts from new physics (e.g., new ideas of time, space, and causation evident in theories of relativity and quantum indeterminism)

Inconclusive means information that cannot lead to conclusive or definitive results

Fuzziness/Vagueness

Fuzziness: vagueness; haze at the edges; degrees of truth; arguable that probability is a special case of fuzziness

Fuzzy thinking is not precise; it reflects truths, not facts or statistics; a convenient way to approximate nonlinear systems

Asks if a particular conclusion, which is always tentative, is more true than untrue, or more untrue than true; by progressive steps, it backs and fills its way, merely reflecting the observed phenomena; a sliding scale

(Continued)

Table 10.1   *(Continued)*

Admits the possibility of partial membership in a class, generalizing what might be otherwise crisp sets into ones where class boundaries are, or cannot, be defined clearly

Reflects judgments that permeate all scientific inquiry and decision making; fuzzy set theory addresses nonspecificity and fuzziness

Potential bridge between probabilistic risk assessments and qualitative assessments; quantifies the qualitative while preserving imprecision; also can be integrated into EIA simulation models (e.g., fuzzy cross-impact simulation)

Ambiguity/Nonspecificity

Having more than one possible meaning; intentionally or unintentionally, obvious (patent) or hidden (latent)

Also pertains to vague, uncertain, or doubtful meaning or interpretation

When faced with ambiguity about rules, obligations, promises, mandates, and duties, practitioners tend to look for precedent, tradition, a source of legitimacy, a consensually based interpretation, and an appropriate and fitting response

Approximations

Simplifications of complex real systems: four types—(1) can be solved exactly but do not know correct equation; (2) to solve problem is impossible, so resort to approximation; (3) simplify equations (a further abstraction from reality); (4) solution too complicated to understand (approximate to make result understandable)

Doubt

Occurs in more complex decision-making contexts; issues and problems typically have no exact precedent or involve several parties with divergent or conflicting interests; also insufficient or unreliable data, disagreement over the importance of variables, and the fact that some variables may not be quantifiable

Generally, handled through the rules and structures of the procedure within which the parties interact

Confusion/Linguistic Imprecision/Dissonance

Dissonance: pure conflict (one statement is true and its rivals are false); addressed by probability theory

Confusion: pure and potential conflict; addressed by possibility theory

Procedural confusion: the complexity and uncertainty of the situation exceeds the problem-solving capacity of existing decision-making techniques, procedures and institutions

Linguistic imprecision: imprecise communications

Surprise

Manifestation of uncertainty that cannot be predicted; a qualitative disagreement between observations and expectations

Typology: local (created by broader scale processes for which there is little or no previous local knowledge); cross-scale (similar to local surprise but larger-scale fluctuation intersects with slowly changing internal variables to create an alternative stable, local system state) and novelty (something truly unique, in which new variables and processes transform the system into a new state)

Table 10.1 (*Continued*)

Revenge effects: ironic, unintended consequences of mechanical, chemical, biological, or medical ingenuity

Surprise generating mechanisms and effects: logical tangles (paradoxical conclusions), catastrophes (discontinuities from smoothness), chaos (deterministic randomness), uncomputability (output transcends rules), irreducibility (behavior cannot be decomposed into parts), and emergence (self-organizing patterns)

Uncertainty Analysis

Analysis of information about risks that are only party known or unknowable; describes the degree of confidence in the assessment

Quantitative uncertainty analyses explicitly describes the magnitude and direction of uncertainties

Qualitative descriptions of uncertainties avoid false sense that know precisely extent of risk, help identify uncertainties with the largest impacts, explains differences in risk estimates generated by different stakeholders, and suggests research opportunities

Sources: Benveniste (1989), Calow and Forbes (1997), Cardinall and Day (1998), Carpenter (1995, 1997), Cartwright (1991), Casti (1994), CEC (2000), Constanza et al. (1992), Coveney and Highfield (1995), Crossley (1996), Dearden and Mitchell (1998), Gunderson (1999), Hodgson and White (2001), Holling (1978), Homer-Dixon (2000), Jones and Greig (1988), Lein (1992), McNeil and Freiberger (1994), Parashar et al. (1997), PCCRARM (1997b), Rothman and Sudarshan (1998), Rowe (1994), Stern and Fineberg (1996), Suter (1993), Tickler and Raffensperger (1998), Treweek (1999), US ACE (1992), US EPA (1998c), Westman (1985), WHOROE (2001c), Yoe (1996).

and into its implications for understanding and action. The uncertainty principle (also from science) illustrates the limits of measurement and causality. Fuzziness or vagueness demonstrates that boundaries often are permeable, blurred, and overlapping (i.e., degrees of truth). Methods (e.g., fuzzy set theory, fuzzy cross-impact simulation) can apply this conceptual insight to bridging the qualitative and the quantitative. Ambiguity and nonspecificity point to vague or multiple meanings and to the need to scrutinize the meanings associated with statements and observations. Approximations illustrate how complex real systems are simplified. Doubt can be both a healthy attitude (consistent with good scientific practice) and an acknowledgment that stakeholders (often with good reason) tend to be skeptical of "experts," specialist analyses, and EIA processes. Confusion can arise because of miscommunications, conflict, and poorly adapted problem-solving approaches. Perceptions of confusion frequently vary among EIA process participants. Surprises cannot be predicted. The generating mechanisms for surprises can be characterized. Uncertainty analysis, a common stage in risk assessment, illustrates that uncertainties can be assessed systematically. There is a danger in confining the analysis of uncertainties to a single stage in the EIA process.

10.4.2 Difficult Problems

The impetus for an EIA process is often a desire to solve a problem or take advantage of an opportunity (the opposite of a problem). Problems and opportunities also arise during the EIA process. A problem has the following general properties: (1) a question, an issue, or a situation triggers the problem; (2) it has negative connotations (e.g., it is perplexing, vexing, or distressing); and (3) it needs to be dealt with, solved, or addressed. Perceptions of the incidence and nature of problems vary among stakeholders (i.e., problems are subjective) (Cartwright, 1973). Problem-solving processes often begin by identifying, describing, defining, bounding, and stating the problem (Bardwell, 1991; VanGundy, 1988). The initial problem statement is refined progressively through the process. EIA practice frequently assumes that the problem is "obvious," that the action proposed will "solve" the problem, and that additional problems will not arise. More attention to problem delineation in EIA could reduce such recurrent mistakes as solving the wrong problem, stating the problem so it cannot be solved, solving a solution, stating the problem too generally, and trying to obtain agreement on the solution before there is agreement on the problem (International Associates, 1986).

EIA, planning, and environmental management problems tend to fall within four broad, overlapping categories: (1) simple or tame problems, (2) compound or semi-structured problems, (3) complex or ill-structured problems, and (4) crises or metaproblems (Cartwright, 1973; Miller, 1993; VanGundy, 1988). Uncertainty progressively increases from level 1 up to level 4. The adverse consequences of uncertainty are particularly acute at level 4. EIA practice has tended to focus on or to assume level 1 and 2 problems. Conventional EIA approaches are poorly suited to addressing level 3 and 4 problems—defined here as difficult problems. Table 10.2 outlines several concepts relevant to difficult problems. An adaptive EIA process should be designed to cope with difficult problems.

Table 10.2   Examples of Difficult Problems

Real Problems

Problems that exist in the real world
Principal characteristics: large size, high spatial and temporal variability, not conducive to experimental control, ambiguous, and poorly defined

Complex Problems

Occur in systems where there are multiple interactions among numerous variables, there are many unknown variables, and relationships are hard to identify and understand
Absence of deterministic and complete information about the options, impacts, and interest groups; also multiple interests and multiple and often conflicting objectives and perceptions of problem
Implications: only superficial control over problems, character of problem often misunderstand, not possible to address problems through training, past nor a good guidepost for future, and many social and economic hierarchies unworkable

Transboundary Problems

Effects cross jurisdictional boundaries (within or among countries) or affecting global commons

Table 10.2 (*Continued*)

Multiple jurisdictions, each with different priorities

Need to create new institutional mechanisms to address; tension because of fears regarding loss of sovereignty

Trans-scientific Problems

Crosses and transcends disciplinary boundaries

Not amenable to analytical scientific methods

Requires intelligent scanning; succession of judicious nudges

Paradoxes, Dilemmas, and Contradictions

Paradox: variety of meanings— (1) something that appears contradictory but which is true; (2) something that appears true but is contradictory; (3) a series of deductions from a self-evident starting point that leads to a contraction

Both visual and linguistic paradoxes

Example: Arrow impossibility theory—demonstrates that no method of combining individual preferences to produce a social choice that meets all democratic choice conditions

Impossibilities/Insoluble/Intractable Problems

Cannot discover all truths

Types: incompleteness, undecidability, logical and practical impossibilities, and technological, cosmological, human, and deep limits

Wicked/Messy Problems

No definitive formulation; no stopping rule; solution are not true or false but good or bad; no immediate and no ultimate test of a solution; every solution is a one-shot opportunity, because there is no opportunity to learn by trial and error, every attempt counts significantly; do not have an enumerable or an exhaustible describable set of potential solutions, nor is there a well-described set of permissible operations; every wicked problem is essentially unique; can be considered a symptom of another problem; the existence of a discrepancy representing a wicked problem can be explained in numerous ways—the choice of explanation determines the nature of the problem's resolution; the practitioner has no right to be wrong

Involves complex and dynamic situations of changing and interdependent problems

Analysis and solutions cannot be standardized into general laws or theories; cannot be managed through traditional analytical science

Latent Time Bombs/Catastrophes/Crises

Latent time bombs: potentially major, sudden disasters such as earthquakes, droughts, floods, or financial collapse; can be interpreted spatial and temporally and can take many forms (e.g., physical, ecological, social, and economic)

Concerns major events, predictions about them are credible, early intervention is understood to be possible and potentially beneficial, costs associated with advance preparation are significant and highly visible

Catastrophes: as it moves through a family of functions, a stable fixed point of the family loses its stability; this change of stability forces the system to move abruptly to the region of a new stable fixed point

(*Continued*)

Table 10.2   (*Continued*)

Tendency of governments to take action after, rather than before, threatening events occur because of the need to engage in cost distribution

Ingenuity Gaps

Shortfall between the rapidly rising need for ingenuity and inadequate supply

Problems intrinsically harder to understand and knowledge slow to develop; result critical time lag between problem recognition and delivery of sufficient ingenuity; converging complexities and connections result in need for high-speed decision making and associated management difficulties

Human knowledge and ingenuity progress at different rates in different domains; impeded by human cognitive limits, intrinsic complexity of subject matter, nature of scientific institutions, and slow and unwieldy economic, social, and political systems

Roles for markets, science, and democracy but failures and constraints associated with each (e.g., market failures, cognitive limits, varied rates of scientific progress in different domains, rising costs, political gridlock and corruption, social turmoil)

Sources: Allen and Gould (1986), Barrow (1998), Casti (1994), Dery (1997), Homer-Dixon (2000), Patterson and Williams (1998), Rittel and Webber (1973).

Simple or tame problems are well defined. Ends can be established readily. Much is known about environmental conditions, technologies, methods, and available alternatives. Simple problems can generally be resolved with standardized, often quantitative procedures and methods. Some but not all the parts are known with compound or semistructured problems. There may be varying perspectives regarding ends. There are likely to be a mix of calculable variables, uncertainties, knowledge gaps, and surprises. Routine procedures will not suffice. Additional analyses are required to fill data gaps. Experimentation, innovative approaches, and practical procedures are needed to deal with new, emerging, and unanticipated issues. Frequent and ongoing stakeholder consultation, mediation, and bargaining are required to cope with varying and conflicting perspectives, values, and interests. Well-defined and managed good-practice EIA procedures are generally adequate for compound or semistructured problems.

High levels of uncertainty and variability and low levels of understanding and control characterize complex or ill-structured problems. Complex problems are dynamic, interdependent, "messy," ambiguous, unique, and real. They involve multiple variables, interactions, and interdependencies. They often defy simplification. Models of complex systems and problems frequently fail to capture critical components and interrelationships. Complex problems are often less amenable to quantification. The past is of limited value either in understanding the present or as a basis for prediction. Ends are not agreed to and means are not known. There are usually multiple perspectives concerning which methods best suit complex problems. Many uncertainties cannot be understood with additional analysis nor managed effectively with good practices. Complex problems transcend disciplinary boundaries. They often extend over broad geographic boundaries, involve multiple jurisdictions, and are long term. Traditional hierarchical institutional structures and analytical methods rarely cope well with complex problems. Complex problems can be ameliorated but not solved. They require creative approaches tailored to their unique and changing characteristics. Flexibility and adaptability are essential to anticipate and accommodate change and surprise. Conventional EIA approaches are not well suited to managing complex problems.

Metaproblems or crises are more than difficult or even intractable—they are deadly or "wicked." Crises take many forms and are often interdependent. Efforts to address metaproblems frequently encounter paradoxes, dilemmas, and contradictions. Metaproblems are impossible to fully understand or manage. Untended, they can rapidly become disasters or catastrophes. No experimental intervention is consequence free. Incremental adjustments and other adaptive behaviors can exacerbate the problem or even trigger an irreversible chain of deadly consequences. Crises often emerge or occur because of a widening gap between the need for and supply of ingenuity. Early intervention is possible, potentially beneficial, but costly. Markets, science, and democratic institutions can all contribute to avoiding, ameliorating, and staving off crises but all have limitations, some profound. Crises require a unique mix of sustained ingenuity, commitment, institutional reform, capacity and network building, leadership, high-speed decision making, and precaution. The likelihood and severity of crises can be reduced, sometimes delayed, and occasionally reversed. Major uncertainties will remain, notwithstanding best efforts. Conventional and even good-practice EIA approaches tend to fare poorly when coping with metaproblems and crises.

10.4.3 Chaotic and Complex Environments

It is often stated that EIA processes should match the environment or the context. Environment or context generally encompasses ecological, political, social, economic, institutional, and technological components and systems. Some context types are more uncertain than others. Classification systems for environments, contexts, situations, systems, or futures generally involve a continuum from the simple to the highly complex (Barrow, 1998; Hodgson and White, 2001; Trist, 1980). Simple systems have a limited number of variables and interactions and a slow and usually predictable pace of change. Such systems are not commonly constrained by human or natural limits. Command and control management approaches tend to work well in such situations. Intermediate levels have greater complexity, more interactions and interdependencies, and a higher level of uncertainty. Decision making is more constrained by environmental conditions. Operating in such environments requires effective planning, consultation, and coordination. Turbulent, complex, and interdependent systems are very difficult to understand, predict, or influence. Decision making in such environments is more effective when oriented toward social learning, judicious experimentation, and the proactive anticipation, review, and selective management of risk, error, and uncertainty.

Context classification systems closely parallel those for problem types. Not surprisingly, the most difficult situations encountered in EIA practice often involve a combination of difficult problems in complex and chaotic environments. In such cases the problems and the context are poorly defined and overlapping. Practitioners and decision makers face the double dilemma of not knowing when to begin (i.e., separating the problem from the context) and not knowing when to end (i.e., no "stopping rule" for determining that the problem response has been adequate). The related concepts of chaos and complexity are highly relevant to EIA practitioners seeking to operate in highly uncertain environments. Table 10.3 outlines

Table 10.3   Examples of Characteristics of Chaotic and Complex Environments and Systems

Chaos: General

Order without predictability; deterministic randomness

Characteristics: outputs transcend rules, local rather than system order, self-referential, sensitive to initial conditions (the "Butterfly effect"), loss of information about initial conditions; basic cause–effect processes still operate among system components but interactions over time and large-scale behavior unpredictable; some systems flip back and forth between chaos and order

Chaotic systems not always complex (chaos can be observed in simple systems)

Analysis and interpretation implications: impossible to know a system's exact initial state (incomputable), prediction logically impossible, errors of measurement and calculation inevitable, amplifies uncertainties, impossible to infer from its present state how it got there, can never be fully understood, surprise inevitable

Planning and management implications: ensemble of forecasts and simple models, sensitivity analyses applied to initial state, look for patterns of system behavior, local and incremental predictions, planning as a succession of judicious nudges rather than a step-by-step recipe, unlike natural systems humans can learn and can change behavior to avoid chaos

Complexity: General

Intricate tangles of shifting and often opposing forces that unfold in unpredictable and frequently surprising ways

Complex systems not always chaotic but common in complex systems

As complexity rises, precise statements lose meaning and meaningful statements lose precision

Characteristics: multiple variables, interactions, and feedback and feedforward loops; absence of deterministic information, acausal, diffuse authority, new laws come into play when the level of complexity increases, sensitive to the smallest changes and perturbations, behavior can flip from one mode to another suddenly and dramatically, openness to outside environments, global behavior outlasts behavior of component parts, and exhibits different characteristics at different scales

Types: disorganized complexity (millions or billions of variables only approached by statistical mechanics or probability theory), organized complexity (moderate number of variables but all variables interrelated)

Critical processes: coevolutionary diversity (competition and interdependence of system entities), structural deepening (individual entities become steadily more sophisticated to improve performance) and capturing software (systems take over or task simpler systems)

Table 10.3 (*Continued*)

Analysis and interpretation implications: counterintuitive behavior of system, as complexity increases not only limited but self-limiting (theory predicts cannot predict), current events heavily influence the probabilities of many kinds of later events

Planning and management implications: unable to predict or manage the behavior of complex systems, results in confusion and sometimes fear, importance of ability to switch between different modes of behavior as environmental conditions are varied, resulting flexibility and adaptability introduces notions of choice and of collective or social learning

Self-Organizing

A self-organizing system produces complex organization from randomness without external intervention; self-organizing systems use feedback to bootstrap themselves into a more orderly structure

In self-organizing systems, orderly patterns emerge from lower-level randomness; opposite of chaotic systems where unpredictable behavior emerges out of lower-level deterministic rules

Complex systems have a tendency to organize themselves into critical states that are optimally sensitive

Self-organizing/self-learning; intelligence builds from bottom up; macrointelligence (system learning) and adaptability derive from local knowledge

Properties (e.g., the whole is greater than the sum of the parts, self-controlled within larger-scale constraints, they evolve)

Aim of management should be to enhance the capacity of the system for self-management, with active management being used to steer it away from large discontinuities

Emergent

Properties of a system that the separate parts do not have; the idea that simple elements that are governed by a few simple rules and operate through trial and error with interaction and feedback can produce persistent and systematic patterns that are quite unlike the original elements

Complex structures seem to display thresholds which, when crossed, give rise to sudden jumps in complexity

Rarely a smooth, steady increase in the consequences of similar changes in complexity

Emergent system-design principles: more is different (critical mass), ignorance is useful (better to build from simple elements), encourage random encounters, look for patterns and pay attention to neighbors (local information can lead to global wisdom)

A top-down analytical approach (dissecting whole into parts) will miss emergent or synergistic properties

Making an emergent system more adaptive generally entails tinkering with different kinds of feedback (positive and negative)

Emergent organizational systems (e.g., a more cellular, distributed network of small units) tend to be more innovative and adaptable to change than hierarchical models

Turbulent/Unstable/Dynamic

Turbulence: the apparently random eddying and twisting of the flow; a special case of chaotic behavior

A dynamic system is in constant flux; the higher its variety, the greater the flux

Instability and commotion are common

(*Continued*)

Table 10.3   (*Continued*)

Nonlinear

A change in a system can produce an effect that is not proportional to its size

Does not obey the laws of addition; generally produces complex and frequently unexpected results

Small changes can produce large effects; large changes can produce small effects

Often a consequence of positive feedback, which tends to amplify small perturbations

Characterized by multiplicative or synergistic relationships among components or variables

Irreducible/Synergistic/Irreversible/Antagonistic

Nonreductionist; behavior cannot be decomposed into parts

Synergy: whole more than sum of parts

Antagonism: whole less than the sum of parts (offsetting)

Irreversible: the one-way time evolution of a real system

Variable/Random/Heterogeneous

The number of possible states of a system is called its variety; a measure of complexity in a system

Inherent randomness or variability (stochasticity); difficult to reduce because an inherent characteristic of system being assessed

A population's natural heterogeneity or diversity, particularly that which contributes to differences in exposure levels or in susceptibility to the effects of chemical exposures

In risk assessment arises from differences in the nature and magnitude of a population's exposure to hazards and from variations in people's susceptibility to hazardous exposures; quantities vary from time to time and place to place

The only thing that control variety is more variety; variety absorbs variety (Ashby's law—the law of requisite variety)

Randomness: uncertainty that is impossible to reduce

Incoherent

Aspirations and activities do not integrate with one another

Do not cohere conceptually, operationally, linguistically, or socially

Function of forecasts and planning: to enhance focus, direction—coherence—for whatever ends

Unpredictable/Surprise

Chaotic systems are unpredictable (lack of predictability inherent rather than situational); starting situation is never the same between two circumstances; outcome can never be predicted and solution to achieve the desired outcome will need to be created in each new situation

Complex systems are deterministic but not predictable or manageable

Interdependent

An increasingly interdependent world

A dense web of causal connections among components

Table 10.3 (*Continued*)

The density, intensity, and pace of interactions sharply increases with complex systems
Positive (reinforces or amplifies initial change) and negative (counteracts the initial change) feedback among system components

Resilience and Stability

Resilience determines the persistence of relationships within a system in the face of sharp and unexpected external pressures
Stability is the ability of a system to return to an equilibrium state after a temporary disturbance; it is quantified in terms of return time
Other definitions emphasize conditions with more than one stable equilibrium, where instabilities can flip into a system into another regime of behavior
A concept that relates to a system's ability to absorb, cope with, and benefit from change, without losing its basic integrity
Originally developed in an ecological context but since applied to economic, social, and political systems
Policy design criteria: maintain different distinct modes of behavior because of, rather than despite variability; the more that variability in partially known systems is retained, the more likely it is that both the natural and management parts of the system will be responsive to the unexpected

Sources: Axelrod and Cohen (1999), Barrow (1998), Calow and Forbes (1997), Cardinall and Day (1998), Carpenter (1995), Cartwright (1991), Casti (1994), Coveney and Highfield (1995), Dearden and Mitchell (1998), Gleick (1988), Greene 1999: Hodgson and White (2001), Hollick (1993), Homer-Dixon (2000), Innes and Booher (1999), Jasanoff (undated), Johnson (2001), Michael (1989), Nicolis and Prigogine (1989), PCCRARM (1997b), Radford (1988), Rothman and Sudarshan (1998), Rowe (1994), Stern and Fineberg (1996), Suter (1993), Treweek (1999), Trist (1980), US ACE (1992), US EPA (1998c), Yoe (1996).

some key properties of chaotic and complex environments. Simply put, chaos involves lower-level order (i.e., system apparently governed by a small number or rules) that evolves into higher-level disorder or randomness (i.e., rules are transcended at the higher level). Chaotic systems are highly sensitive to initial conditions. Complexity begins with disorder or randomness, but order emerges. Such self-organizing behavior results from feedback mechanisms. Complex systems can be organized or disorganized. They involve multiple variables, interactions, and interdependencies. Chaotic systems are not always complex. Complex systems are often chaotic. Both chaotic and complex systems evolve, often abruptly, in unpredictable ways. Both tend to be irreducible, incomputable, irreversible, incoherent, unstable, dynamic, and nonlinear. Errors and surprise are inevitable with complex and chaotic systems.

There are no standardized approaches to operating in chaotic and complex environments. It is prudent to be sensitive to initial conditions, to behavioral patterns at the local level, and to interdependencies. Confusion, fear, and surprise should be expected. An ensemble of simple models in combination with local, incremental predictions are likely to provide more insights than a single, grand model and long-term system-level forecasts. Variety should be matched to variety. Multiple sensitivity analyses (i.e., tinkering with positive and negative feedback mechanisms) can reveal critical interdependencies and potential thresholds. Adaptability and creativity are essential. A cellular network of small organizational units is usually more innovative and flexible than hierarchical models. Organizational and social learning, synthesis, and the capacity to respond quickly as conditions change are attributes to be fostered. Limits, errors, risks, and uncertainties should be priorities. Approaches that intervene selectively to enhance the self-management capabilities of systems, that maintain and reinforce resilience, and that steer systems away from large discontinuities are often more appropriate. Approaches that evolve and change in parallel with complex and chaotic systems are more likely to be effective in coping with uncertainty.

10.4.4 Managing Uncertainty in the EIA Process

Addressing uncertainty in EIA should begin with an attitude or perspective change. Uncertainty is a fundamental process attribute rather than "a distasteful transition to attainable certainty" (Holling, 1978). Priorities shift from prediction and control to adaptability and responsiveness. It is necessary to learn from error, live with and obtain benefits from uncertainty, avoid the unwarranted appearance of certainty, and address uncertainties throughout the EIA process (Canter, 1993b; Dickman, 1991; Hollick, 1993; Mostert, 1996; US ACE, 1992). The EIA process becomes an ongoing investigation rather than as a one-time prediction of impacts (Holling, 1978). The process is iterative (anticipatory scanning and feedback loops), open, and adaptive (Gibson, 1992; Mulvihill and Keith, 1989). It evolves with and selectively and proactively influences both the problem and the context. This perspective shift is necessary at both the regulatory (e.g., performance oriented requirements and guidelines) and applied (e.g., EIA process management) levels. It should also be present in each EIA activity.

An adaptive EIA process begins with a thoughtful, open (to divergent perspectives and interests), and systematic search of the problem space or situation. Uncertainties in the problem definition and in governing norms, values, and interests are identified explicitly (Mostert, 1996). Care is taken to use uncertainty language consistently (Toon, 2000). The EIA process is carefully bounded (US EPA, 1998c). Constraints are openly acknowledged (Cardinall and Day, 1997). Risk and uncertainty issues are identified, objectives are formulated, and methods are determined (US ACE, 1992). A resilient mix of reliable solutions is identified—each treated as a case study. The proposed action(s) encompasses components intended both to prevent failure (i.e., fail-safe) and aimed at responding and surviving if failure occurs (i.e., safe-fail) (Holling, 1978). The proposed action (or more likely, suite of actions) is adapted and refined as circumstances change, both during and subsequent to approvals (Hollick, 1993). Ideally, the action is suited to staged approval (i.e., self-contained components) and implementation. In this way the monitoring results can lead to modifications to and, where warranted, termination of the action (Hyman et al., 1988).

The process involves multiple parties and perspectives in a creative and heuristic search for reversible, low-magnitude, flexible, simple, error-friendly, proven reliable, safe–fail, and harm-reducing options that hedge away from large losses or catastrophic effects, provide benefits even if problems are less serious than feared (i.e., no regrets), involve simple, known, and predictable environmental conditions, have minimum potential for synergistic effects, protect and enhance environmental integrity and sustainability, and can be harmonized with surrounding natural and social systems (Gibson, 1992; Hollick, 1993). Option evaluation criteria reflect these types of properties. Alternative criteria and criteria rankings and multiple sensitivity analyses test varying assumptions and perspectives. The evaluation narrows the list of options, but several potentially acceptable options and option combinations are carried forward into the process as far as practical. Retaining multiple options enhances action and process flexibility (Hollick, 1993). The preferred options are those best able to adapt to changing conditions, pose the least threat to the vulnerable environmental components and systems (assuming flawed predictions and ineffective mitigation), and make the most positive contribution to sustainability (Gibson, 1992; Homer-Dixon, 2000).

Uncertainty is a central consideration in baseline and impact analysis, interpretation, and management. Vulnerable (to impact, change, and surprise) environmental components, interactions, and systems are identified. The analysis focuses on change processes and identifies key variables and processes likely to amplify fluctuations (Gibson, 1993; Hollick, 1993). Major uncertainties are identified once data are obtained (Yoe and Skaggs, 1997). Supplementary analysis and research reduce the uncertainties. Multiple models are developed, refined, and applied to characterize the system. Sensitivity analyses, wide error margins, and confidence ranges test assumptions, assess the consistency of relationships and bound uncertainties (Hyman et al., 1988). Considerable uncertainties remain despite such measures. Predictions are difficult, sometimes impossible. More emphasis is placed on understanding the system than on prediction (Holling, 1978). The future is addressed by exploring planned and unplanned alternative futures using such techniques as scenario analysis (Hollick, 1993). Allowance is made for a wide range of errors and outcomes. More stress is placed on avoiding type II errors (predicting no impacts when impacts occur) than on avoiding type I errors (predicting impacts when no impacts occur) (Interorganizational Committee, 1994). Impacts are assessed over the life cycle of the proposed action under normal and abnormal conditions (Tonn, 2000). Extreme and worst-case scenarios are formulated. Broad safety margins and conservative assumptions are employed. Predictions are expected to be inaccurate. Mitigation measures are assumed to be ineffective (Gibson, 1992). Experiments and pilot projects (both at the site and for comparable undertakings and settings) help refine and test the analysis (Treweek, 1999). Uncertainty, vulnerability to change, reversibility, resilience and adaptability, and consequences of error and failure are major impact significance factors.

The mitigation analysis stresses emergency and contingency planning, early warning systems, reversibility, adaptability, and the availability of fallback positions and damage control systems (De Bono, 1993). Natural mitigation approaches, which require minimal intervention and which recognize and cultivate the self-organization capacity of systems, are favored over methods reliant on a high degree of intervention, control, or "engineering" (Hollick, 1993). A risk and uncertainty analysis identifies, analyzes, interprets, and determines appropriate management measures for risk and uncertainty types and sources (Reckhow, 1994; Toon, 2000; US ACE, 1992; Yoe and Skaggs, 1997). Uncertainty management includes targeted research and error reduction procedures. Uncertainties are presented in a form suitable for decision making and monitoring (Glasson et al., 1999; Holling, 1978). Monitoring begins early in the EIA process by assessing comparable environments, comparable undertakings, and pilot projects. It continues during and following a staged review and approval process. Monitoring focuses on maintaining and enhancing the health of vulnerable environmental components, on detecting emerging discontinuities, and on contributing to environmental integrity and sustainability (Axelrod and Cohen, 1999). Monitoring uncertainties are acknowledged explicitly. Actions taken in response to monitoring err on the side of environmental protection.

Public attitudes toward uncertainty, including value differences regarding uncertainties, are identified (US ACE, 1992). All potentially affected parties are involved in addressing uncertainties (Mostert, 1996). The EIA process is open and collaborative. Reciprocity facilitates trust. Trust among social groups is acknowledged as essential for ameliorating complex, collective action problems (Ostrom, 1998). Consultation is a social learning opportunity, where the knowledge limits of all parties are recognized. What is and is not certain, what is and is not being done about uncertainties, and the rationale for all actions taken or not taken in response to uncertainty are communicated to all parties (Hance et al., 1990). Interpretations are open to challenge and comment and subject to independent peer review (Yoe, 1996).

Study teams are selected and managed with uncertainty in mind. Study team leaders identify productive areas of uncertainty and confusion and lead the team toward opportunities (Hodgson and White, 2001). The study team is willing and able to explore ambiguities, handle uncertainties, tackle difficult and unknown problems, readily adapt to changing situations, span boundaries, focus on essentials, scan ahead, communicate effectively, and accommodate conflict (Hodgson and White, 2001). EIA documentation explicitly identifies risks, uncertainties, limitations, and constraints. Simplifying assumptions and subjective interpretations and choices are fully justified (Mostert, 1996). The limits of data, technologies, methods, and procedures are acknowledged (US EPA, 1998c). Key uncertainty issues and how they were addressed are described (US ACE, 1992). Aspects of uncertainty most likely to affect decision making are identified (Reckhow, 1994).

Decisions are supported by uncertainty-related techniques (e.g., fuzzy set analysis, bounding analysis, subjective probability analysis, expert panels, scenario formulation, sensitivity analyses, life-cycle analysis, simulations, comparative analysis, decision and event trees) (Gibson, 1992; Tonn, 2000; Treweek, 1999; US ACE, 1992; Yoe, 1996). Proposed actions with high uncertainties and potentially grave and likely irreversible consequences are generally rejected (Gilpin, 1995; Hyman et al., 1988). Actions are deferred if short-term studies can reduce high uncertainties and can manage the remaining uncertainties (Gilpin, 1995). Actions are generally approved if experience elsewhere suggests low-magnitude impacts and uncertainties that can be addressed adequately by conditions and monitoring (Gilpin, 1995). Staged approval is applied, where practical, to maximize the opportunity to monitor, adjust, defer, or even terminate proposed actions. Decision making is risk aversive. It hedges decisions away from large losses (e.g., no or least regrets) (Gibson, 1992). The effectiveness of adaptive methods and procedures is continually assessed. Institutional constraints and implications are identified. Adaptive organizations provide for rapid and continuous knowledge acquisition, have effective information flow and communications networks, have regenerative–restructuring capability, have a bias toward action, preventive planning and monitoring, and are vertically and horizontal integrated. They tend to be collaborative, experimental, flexible, creative, reliable, and evolving. They have error-detecting and error-correcting mechanisms, encompass varying critical and systems perspectives, are open to scrutiny, and are responsive to interested parties and diverse interests (Homer-Dixon, 2000; Michael, 1989; Mulvihill and Keith, 1989). Adaptive organizations are much like self-organizing, emergent, complex systems.

Table 10.4 identifies several potentially relevant adaptation concepts. Design confronts complexity with positive visions, pragmatic, tested concepts, and the

Table 10.4   Examples of General Adaptation Concepts and Methods

Design

Idea of a growing whole; operates at many levels in many different ways; object is to incrementally produce wholeness or coherence

Establishes a context for future actions; visionary—experienced and then expressed as a vision; a vision that can be communicated to and felt by others; positive—aim to create a positive character based on urban, ecological, and sustainability visions and principles

Begins from a few well-tested concepts that pragmatically respond to prevailing conditions (initial ideas often from analogies, analogs, and models); consequences explored and reevaluated; ongoing refinement, adjustment, and embellishment

Ingenuity

Sound sets of instructions; minimum ingenuity requirement: shortest set of instructions to solve problem

Ideas that can be applied to solve practical technical, social, and environmental problems

Amount of ingenuity dependent on intrinsic difficulty of achieving goal and kinds and amounts of available resources

Innovation (truly new ideas) and application of known ideas in different ways and in different contexts

Technical (helps solve problems in physical world) and social (well-functioning markets, institutions, social arrangements) ingenuity; social a prerequisite to technical ingenuity

Within social ingenuity can distinguish between structural ingenuity (used to create or reform institutions) and policy ingenuity (for actions pursued within an existing institutional framework)

Measure of ingenuity: quantity (number of instructions) and quality (how well works in practice)

(Continued)

Table 10.4 (*Continued*)

Can distinguish between ingenuity applied to short- and long-term problems; entangled with social and political processes (context)

Creativity

Key characteristic of brain; the ability to improvise in novel situations; intelligence implies flexibility and creativity; capacity for analogy and metaphor especially important; value of affect or emotion in higher-level integrative brain functions

Concerned with the changing of concepts and perceptions and with the generation of new concepts and perceptions

Creativity operates on more than one plane and is liberating (defeats habit by originality), visionary, nonlinear; involves both differentiation, and integration; also transcends logic (thinking aside or laterally), rationality (i.e., the extrarational), language, and science

Creative thinking; can be fostered deliberately with approaches (e.g., lateral thinking, synectics) and by specific techniques (e.g., brainstorming); facilitated by training

Individual and group creativity; better in combination; creativity involves preparation, incubation, illumination, and verification

Potential role in improvement, in problem solving, in realizing value, and in taking advantage of opportunities; particular need for creativity in order to generate future possibilities and to devise ways of coping with multiple possibilities

More than a way to make things better; approaches and techniques seek to break out of old structures, patterns, concepts, and perceptions

Stresses value of employing analogs, metaphors, imagery, illusion, simulation, storytelling, and games, of exploring apparent contradictions, of searching for patterns and interconnections, of transcending dualisms, of identifying hidden assumptions, of formulating theme variations, and of connecting concepts

Strategic Choice

Choosing in a strategic way; stresses interconnections among decisions; focus on planning under pressure; seeks opportunities for managing uncertainties through time

Links technology, organization, process, and product in a process that involves shaping, designing, comparing, and choosing

Creatively manages multiple uncertainties (about the working environment, guiding values and related decisions)

Employs various concepts concerning problems (e.g., current and modified decision problems, broader planning problems, problem focus or foci), options (e.g., option bars, option graphs, exploratory options, composite options), comparisons (e.g., comparison area, relative assessment, advantage comparison, working shortlist, evaluation framework), decisions (e.g., decision areas, decision links, decision schemes, immediate decisions, future decision space, action scheme, commitment packages), and interactions (e.g., lateral connections, switching, looping, coalescing decision areas)

Systematically explores uncertainties (e.g., eliciting limits of surprise, identifying uncertainty areas, linking uncertainties to decision areas, reformulating composite uncertainty areas, comparing alternative responses to uncertainty, weighing uncertainty against urgent decision making, accommodating uncertainty in future decision spaces)

Applies in a flexible and iterative group decision-making process; provides skill development and practical advice

Table 10.4 (*Continued*)

Consilience

Consilience: proof that everything in our world is organized in terms of a small number of fundamental natural laws that comprise the principles underlying every branch of learning

Argues for fundamental unity of knowledge; entails the interlocking of facts and fact-based theory across disciplines and branches of learning (e.g., biology, social science, ethics, environmental policy) to create a common groundwork of explanation (i.e., a "jumping together of knowledge")

Argues for extending the habits of thought (e.g., reductionism, integration, competing hypotheses, no claim accepted as final) that have worked so well in material world into social sciences and humanities; natural sciences already has constructed a webwork of causal explanations ranging from quantum physics to brain sciences and evolutionary biology—already converging

Consilience (i.e., units and processes of a discipline that conform with solidly verified knowledge in other disciplines) a criterion for theoretical quality; others include parsimony, generality, and predictiveness

The greatest challenge is the accurate and complete description of complex systems

Holistic Science

Interconnectedness and interdependence of all living and nonliving systems; social and ecological systems coevolve

Emphasizes: complexity, surprise, nonlinearity, and emergence and on the need for creative, intuitive, adaptive, integrative, normative, trans-scientific, and pluralistic approaches; suspends the constraints of analytical thought

Broadens the conception of the problem and of context; stresses ecological limits, equity, integration, and holistic perspective

Sources: Alexander et al. (1987), Benveniste (1989), De Bono (1992), Dearden and Mitchell (1998), Friend and Hickling (1997), Gibson (1992), Gordan (1961), Hodgson and White (2001), Hoftstadter (1985), Homer-Dixon (2000), Koestler (1964), Miller (1993), Mulvihill and Keith (1989), PCCRARM (1997b), Porritt (2000), Reckhow (1994), Rowe (1991), Wilson 1998).

creative use of analogies, analogs, and models. It progressively explores, evaluates, refines, and embellishes. It structures incremental adjustments that build toward a coherent whole, effectively fitted within larger systems. Ingenuity is concerned with innovative (both new ideas and novel applications of known ideas) and practical technical and social solutions to difficult problems in complex environments. Social ingenuity (in both an institutional reform and a policy sense) is generally a prerequisite to technical ingenuity. Creativity is often mentioned but rarely understood or systematically applied as a means of coping with uncertainty. Sufficient advances have been made to provide a good general sense of the creative process and to offer numerous practical individual and group techniques for fostering creativity. Strategic choice offers a well-tested mix of frameworks, methods, and procedures for exploring and managing multiple uncertainties in high-pressure planning situations. It employs an array of useful concepts pertaining to problems, comparisons, decisions, and interactions, all within a highly flexible and iterative group decision-making process. Consilience demonstrates how science has and can interlock facts and fact-based theory across disciplines and branches of learning. It offers a science-based explanatory model of the convergence and unity of knowledge, a model potentially capable of coping with complex system uncertainties. Holistic science provides an alternative perspective and framework for spanning boundaries, for escaping the constraints of analytical thought and for understanding the interdependencies and interconnectedness of complex living and nonliving systems.

10.4.5   Risk Assessment and Management

Risk is a combination of a frequency (in the past) or probability (in the future) and a usually harmful consequence for the human or natural environment (Eccelston, 1999b; Erickson, 1994; Whyte and Burton, 1980). Decision makers tend to know the alternatives, but each alternative has several possible outcomes (i.e., outcomes are not certain) (US ACE, 1992). Adverse human outcomes or harm can include injury, disease (morbidity), death (mortality) impaired quality of life, financial loss, property damage, or delay (Wiener and Rogers, 2002). Ecological harm can include damage to individual plants or animals, to species, to ecosystems, and to ecological diversity. There can be economic, health, and environmental risks (Grima et al., 1986). Risks can result from natural (e.g., natural disasters) and from human (e.g., human actions that result in exposures to chemical, microorganisms, radiation) sources. There can be high- or low-probability risks. There are best-risk estimates and high-value-risk estimates (e.g., worst case, varying safety margins) (Kamrin, 1993). There are chronic (e.g., diseases resulting from persistent or repeated exposure) and acute (e.g., from abnormal events) risks. Levels of uncertainty and the magnitude of consequences can vary among risks. There are risks to the overall population and to sensitive or susceptible populations (e.g., asthmatics, fetuses, infants, young children, elderly). There can be objective (quantitative) or subjective (estimated or perceived) risks (US ACE, 1992). All risk estimates or calculations have an element of subjectivity. There are deterministic (exposure quantified as a point estimate) and probabilistic (probability distribution incorporated for each variable) risk estimates. Risk estimates can be based on scientific evidence (empirical), predictive models, or heuristic judgment and qualitative reasoning.

Table 10.5 outlines some major risk concepts. Risk assessment is concerned with how risks are characterized, described, and estimated. Perceived risks are subjective risk interpretations by individuals and groups. Risk communications involves the exchange of risk-related information and opinion between specialists and the public. Comparative risk assessment compares and ranks risk types. Risk evaluation determines the tolerance for, acceptability of, and desirability of options and proposals. *Risk management* is an umbrella term for all activities concerned with identifying, assessing, interpreting, communicating, and evaluating risks. It also includes measures to prevent and reduce risks, measures to take advantage of risk-related opportunities, decision making, implementation, and monitoring. Integrated risk management includes organizational objectives and procedures. A hazard is an intrinsic property that can, under some circumstances, be harmful. A

Table 10.5   Examples of Risk Concepts

Risk Estimation, Analysis, and Assessment

Risk analysis: the systematic, scientific characterization of potential adverse effects of human or ecological exposures to hazardous agents or activities; performed by considering the types of hazards, the extent of exposure to the hazards, and information about the relationship between exposures and responses, including variation in susceptibility

Adverse effects or responses could result from exposures to chemical, microorganisms, radiation, or natural events

Distinctions: between chronic (diseases occurring as a result of repeated or persistent exposures) and acute (abnormal events) sources and effects; between human health and ecological risks

Distinction between deterministic (quantifies exposures as point estimates) and probabilistic (incorporates probability distributions for each variable) risk assessment

Distinctions among empirical risk assessment (based on scientific evidence), model-based assessment (uses predictive models in place of empirical evidence) and qualitative risk assessment (draws on heuristic judgment and qualitative reasoning)

Need to consider risks to highly exposed populations (e.g., asthmatics, fetuses, infants, and young children, socioeconomic groups, elderly)

Methods: relative risk indices, event trees and decision networks, environmental transport and fate models, dose–response models

Perceived Risks

The subjective perception of risk by members of society both individually and collectively; varies from person to person and group to group for the same risk and from one risk to another

Recognizes that lay public knows something that the experts do not and have good reason not to be convinced of all expert evidence

Public has deep emotional investment in beliefs (e.g., anomie, resentment, distrust, sabotage, stress)

Perceived risk affected by person-related (age, gender, personality type, personal stake, sensitive populations), situation-related (e.g., beyond control of individual, involuntary, children at risk, scientific controversy, high media attention, victim identity), and risk-related characteristics (e.g., origin, immediate threat, consequences for health, dread hazard, catastrophic consequences, unfamiliar hazard, uncertainty, controllability, effects on future generations, reversibility, accident history)

Also affected by public trust in institutions, fairness, media attention, benefits, and evidence

Need to consider factors that contribute to outrage (e.g., involuntary exposures, lack of previous knowledge, dread of effects, severe consequences, inadequate or unclear benefits, outside personal control, artificial rather than natural risk, insidious dangers, unknown duration, associated with memorable events, unethical or unfair distribution of risk burden, managed by untrustworthy information sources, effects on children)

Should be serious consideration in determining risk acceptability

Risk Communications

An interactive process involving the exchange among individuals, groups, and institutions of information and expert opinion about the nature, severity, and acceptability of risks and the decisions taken to combat them

(Continued)

Table 10.5 (*Continued*)

Involves providing citizens with scientific information about risk, making risk information genuinely meaningful, facilitating public involvement in processes where risk analyzed and managed, and obtaining public conceptions of risk (risk perceptions); also serves to encourage risk reduction measures, to increase mutual trust and credibility and to resolve conflicts and controversy

Need for two-way interaction: learn about patterns of exposure, peoples' perceptions of risk acceptability, and peoples' concerns, values, and knowledge; should describe risks and uncertainties openly and understandably; must begin before important decisions made

Risk communications strategies and methods adopted can either ameliorate or reinforce perceived risk concerns

Distinction between informational (provides information necessary to understand characteristics and magnitudes of risk faced and methods for ameliorating) and persuasive (goal of changing people's behavior with respect to a particular risk)

Need to explicitly consider uncertainty and public issues and to communicate with diverse ethnic and socioeconomic groups

Elements of risk communications: objectives (why undertaken?), content (what is being conveyed?), form of communications (how should transmit?), feedback from audience (what is being received?)

Comparative Risk Assessment

The process of comparing and ranking various types of risks to identify priorities and to influence resource allocations

Analyzes several different hazards or sources of harm to the same person, or valuable ecosystem site, in terms of relative risk

Relative differences in risks are significant and can assist in settling priorities among alternative environmental programs so as to get the most risk reduction per unit expenditure; CRA can lead to risk-based strategic planning

Examples (progressively less acceptable): first-class risk comparison (same risk/different occasions, risks against existing standards, different estimates of the same risk), second class (with and without activity, risks of different alternatives, same risks in other sites), third class (average risks against most serious risks, risk by source against risks by all sources producing same effects), fourth class (risk/cost ratios, risks vs. benefits, risks vs. risks from same source, risks vs. other causes of same illness or trauma), fifth class (unrelated risks)

Risk Evaluation and Acceptability

Concerned with the desirability of options or proposals; value-full; expertise dispersed throughout society

Need to consider all feasible options (modify wants, modify technology, prevent initiating event, prevent release, prevent exposure, prevent consequences, mitigate consequences)

Risk evaluation: the determination of the importance of risks; risk is context dependent; need to consider cultural, social, and psychological factors; evolves; represents a subjective/ political, value-full decision requiring the involvement of all sectors of society and often necessitating alternative dispute resolution

Need to consider all major consequences (e.g., economics, environment, societal resilience, equity); need to compare against background, alternative actions, other familiar risks and benefits of continuing the project and taking the risk

Takes into account such considerations as predicted effects, public perceptions, risk–benefits, background, and comparative risks

Table 10.5 *(Continued)*

Risk evaluation methods (e.g., professional judgment, costs–benefit analysis, cost-effective-ness analysis, weight scoring, decision analysis such as event tree)

Acceptable risk is a risk whose probability of occurrence is so small, whose consequences are so slight, or whose benefits (real or perceived) are so great that a person, group, or society is willing to take that risk; risk tolerance would be a more accurate characterization of the concept

Risk acceptability of technology dependent on information people exposed to, information choose to believe, values held, social experience, dynamics of stakeholder groups, political process, and historic moment

Risk Management

Risk management is a systematic approach to setting the best course of action by identifying, assessing, understanding, acting on, and communicating risk issues; integrated risk management—process for building into organizational objectives and procedures

The process of identifying, evaluating, selecting, and implementing actions to reduce risk to human health and to ecosystems; answers the question—what shall we do about it?

Goal is scientifically sound, cost-effective, integrated actions that reduce or prevent risks while taking into account social, cultural, scientific, technological, economic, ethical, political, and legal considerations

Risk management: an umbrella term that encompasses risk analysis or assessment, risk evaluation (the determination of the importance of risk), risk mitigation, and monitoring

Some argue should aggressively seek alternatives to command and control (e.g., environ-mental accounting, education, market-based, incentives, consensual decision-making approaches)

Stages: define the problem and put it in context, analyze the risks associated with the problem in context, examine options for addressing the risks, make decisions about which options to implement, take actions to implement the decisions, conduct an evaluation of the actions; conducted in collaboration with stakeholders; uses iterations if new information is developed that changes the need for or nature of risk management

Examples of methods: education/information, incentives, substitution, regulation/prohibition, monitoring, surveillance, research and risk compensation (for the anxiety created as a result of the hazard potential and for the consequences when a hazardous event occurs)

Disasters and Hazards

Hazard is an intrinsic property of a substance, which is activated upon an event; a factor, or circumstance that may under some circumstances be harmful or injurious; can produce a particular type of adverse health or environmental effect

A hazard is a perceived event or source of danger that threatens life or property or both; a disaster is the realization of a hazard

Hazard assessment seeks to recognize things that give rise to concern

Hazard identification: addresses what can go wrong

Examples of hazard identification methods (e.g., literature review, plant visits, brainstorming, hazard and operability studies, failure modes, effects and criticality analysis, safety audit)

Hazard accounting or analysis: establishes boundaries of analysis and determines likelihood of events

Examples of failure/risk assessment methods: preliminary hazard analysis (identifies hazards as early as possible), event tree analysis, fault-tree analysis, failure modes, and effects analysis (attempts every possible way each component or interface among components,

(Continued)

Table 10.5   (*Continued*)

could fail, then considers effect of failure on system), human reliability analysis (identifies how people interacting might cause to fail)

Human Health Risk Assessment

Assesses risk of cancer and from noncancerous (e.g., reproductive, neurotoxic, developmental, immunologic) health effects (alone and in combination); also from abnormal events (acute)

Carcinogen risk assessment includes hazard assessment (whether agent poses carcinogenic hazard to humans and how might be expressed), dose–response assessment (evaluates potential risks to humans at exposure levels of interest), exposure assessment (the qualitative and quantitative determination of magnitude, frequency, and duration of exposure), risk characterization (integrates risk assessment results in nontechnical discussion)

Examples of issues (e.g., animal testing of potential carcinogens, modeling of carcinogenesis, overly conservative exposure assumptions, risk communications, perceptions, and acceptability)

Ecological Risk Assessment

A process used to estimate the likelihood of adverse effects on plants and/or animals from exposure to stressors

Evaluates the probability and resulting adverse effects from one or more environmental hazards or stressors (nonendemic events or chemicals), which when introduced has the potential to accumulate, biomagnify, and genetically mutate species, poison, or in any other way impact a species or ecological system in an area

Examines the extent of damage from a stressor (e.g., defined toxic agents and pollutants) or possible effects to a system or species as a result of a stressor; can be used to predict the likelihood of future adverse effects (prospective) or evaluate the likelihood that effects are caused by past exposure to stressors (retrospective)

Numerous methodological issues (e.g., ranking environmental problems and ecosystem sites, defining endpoints, selecting indicator species, determining scale, managing and quantifying uncertainties, extrapolations across scales, validating predictive tools, valuation, elements of a uniform approach), and areas requiring additional research (e.g., effects of multiple chemical, physical, and biological stressors)

Extensive debate surrounding appropriate ecological risk assessment paradigm and whether same decision process should be used for human health and ecological risk assessment (e.g., no equivalent to lifetime cancer risk estimate)

Example elements-problem formulation, receptor identification (partitioning assessment, biological characterization, system organization), hazard identification, endpoint identification (the target species or system that is subject to an environmental hazard), relationship, exposure characterization, ecological effects characterization, risk characterization, and uncertainty analysis

Sources: Arquiaga et.al. (1992), Barrow (1997), Canter (1993), Carpenter (1997), Covello et al. (1988), CRAM (1993), Dooley (1985), EnHealth Council (2001a), Fischhoff et al. (1981, 1982), Grima et al. (1986), Health Canada (2000a), Hood and Nicholl (2002), Kamrin (1993), Lein (1992), PCCRARM (1997a,b), Powell (1984), Power and Adams (1997), Rahm-Crites (1998), Sandman (1992), Slovic (1987), Stackelberg and Burmaster (1994), Treasury Board (2000), US EPA (1998c), USNRC (1983, 1997), Wiesner (1995), WHOROE (2001c), Whyte and Burton (1980), Yoe (1996).

disaster is the realization of the hazard. Hazard identification determines what can go wrong. Hazard assessment bounds the analysis and determines the likelihood of events. Human health risk assessment estimates individual or cumulative risks to people from abnormal events, from cancer, and from noncancerous health effects. Ecological risk assessment estimates the likelihood of adverse effects on plants or animals from exposure to one or more environmental hazards or stressors. There are numerous methods and methodological issues associated with each risk concept.

Figure 10.2 presents an example of a risk assessment/management process. The process begins by identifying and characterizing the problem and the proposed action. The problem is defined within the context of government requirements, policies, and guidelines and ecological, societal, and political systems (CRAM, 1993). A conceptual model is formulated to provide a framework for generating and evaluating preliminary hypotheses about how and why risk-related effects have or are likely to occur. An analysis plan is prepared describing risk management objectives, options to consider, the scope and focus of analysis, methods, and resource allocation (PCCRARM, 1997b; US EPA, 1998c). The proposed action is scrutinized to identify potential hazards. Hazard identification determines possible sources of harm (usually, based by experience with similar technologies, materials, or conditions), explores causal links, identifies the potential adverse effects, and decides whether the effects warrant further study or management action (Canter, 1993b; Carpenter, 1995; CRAM, 1993; Stackelberg and Burmaster, 1994; Yoe, 1996). Contaminant sources, pathways from contaminant sources to resources, receptors, and endpoints, and potentially affected resources, receptors, and endpoints are identified and characterized. A hazards analysis is undertaken to ascertain the probability of adverse events.

The exposure assessment quantifies (e.g., intensity, frequency, duration) the concentrations of contaminants in the environmental media at the point of human or ecological endpoint contact (Canter, 1993b; Carpenter, 1995; Stackelberg and Burmaster, 1994; Yoe, 1996). With ecological risk assessment it describes the sources of stressors, their distribution in the environment, and their contact or co-occurrence with ecological receptors (US EPA, 1998c). The exposure response assessment determines the relationship of the magnitude of exposure and the probability of effects (Canter, 1993b; CRAM, 1993; Stackelberg and Burmaster, 1994). In the case of human health effects, it involves evaluating how strongly contaminants elicit health response at various doses (Stackelberg and Burmaster, 1994). Ecological risk assessment evaluates stressor–response relationships or evidence that exposure to stressors causes an observed response (US EPA, 1998c). Exposure assessments and exposure response assessments are conducted in parallel and are highly interrelated (CRAM, 1993). Human health and ecological risks, including attendant uncertainties, are presented in a form suitable for public and decision-maker review (CRAM, 1993; Stern and Fineberg, 1996). Risk characterization summarizes the risk analyses, describes the available choices, addresses the implications of uncertainties, and integrates the perspectives and knowledge of interested and affected parties (PCCRARM, 1997b; Stern and Fineberg, 1996; US EPA, 1998c).

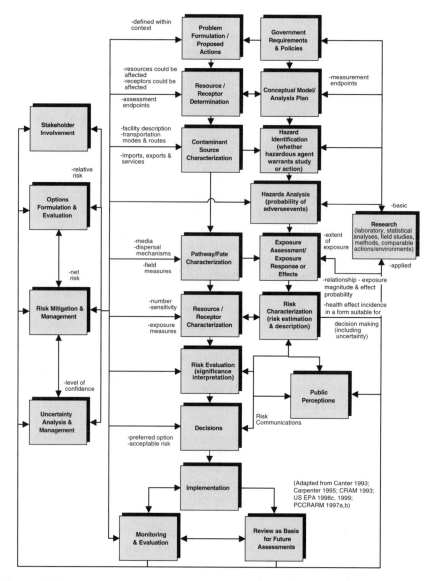

Figure 10.2 Example of a risk assessment management process. (Adapted from Canter, 1993; Carpenter, 1995; CRAM, 1993; PCCRARM, 1997a,b; US EPA, 1998c, 1999.)

Risk evaluation draws on the risk characterization and public perceptions and is aided by risk communications. It interprets the significance of estimated and perceived risks. Risk acceptability or tolerance and option preference decisions are reached. The decisions are implemented (if approved, often with conditions) and monitored. Monitoring tests the validity of predictions, identifies additional research requirements, contributes to methodological advancements, and identifies the need for management actions (CRAM, 1993). Reviews of the process and of monitoring results are instructive for future assessments (US EPA, 1998c). The process is supported by basic and applied laboratory analyses, statistical analyses, field studies, and comparable actions/environments reviews (CRAM, 1993). Stakeholders are involved in each process activity. Options (e.g., regulatory, nonregulatory) and mitigation measures are formulated and evaluated in an ongoing effort to prevent and reduce risks and uncertainties to acceptable or tolerable levels. Implementation and monitoring are structured and guided by a risk management framework (CCRARM, 1997b). The analysis and interpretation of qualitative and quantitative uncertainties is consolidated in an uncertainty analysis. The process is open and iterative (CCRARM, 1997b). Good practice risk assessment and management principles, performance standards, and protocols are formulated, applied, and refined (Canter, 1993b; CCRARM, 1997b; Steinemann, 2000).

Risk assessment application in EIA practice tends to be confined to large, controversial (high levels of perceived risk) undertakings, usually involving nuclear materials or hazardous chemicals or wastes (Carpenter, 1995). Risk assessment and management has much more to offer. It can supplement regulatory standards and guidelines, which often address risks only partially, indirectly and qualitatively. It recognizes the limits of deterministic knowledge and the value of probability analysis (Stackelberg and Burmaster, 1994). It provides a systematic, quantitative set of procedures for analyzing, interpreting, and comparing human health and ecological risks and uncertainties (Arquiaga et al., 1992; Suter, 1993). It systematically explores interrelationships that create exposure and effects (Canter, 1993b). It offers an effective bridge to scientific research and to the needs of regulators (Power and Adams, 1997). It appreciates the uncertainties associated with self-organizing and nondeterministic social and ecological systems (Carpenter, 1995). It provides a host of potentially relevant concepts, principles, distinctions, and methods (Canter, 1993b; Erickson, 1994; Grima et al., 1986; Hunsaker and Lee, 1985).

Risk assessment and EIA are generally mutually supportive concepts (Erickson, 1994; Grima et al., 1986; Westman, 1985). Risk assessment inputs to EIA and EIA can aid risk assessment (Barrow, 1997; Ratanachai 1991). Principles have been formulated for selectively linking and integrating the two fields (Canter, 1993b). They share a common concern with human health and ecological risks. Both grapple with uncertainty, the role of public perceptions, and the interconnections among science, regulatory requirements, environmental management, and public involvement. Risk assessment and cumulative effects assessment both explore interrelationships systematically. EIA and risk assessment processes share many common elements (e.g., problem definition, baseline analysis, impact prediction, mitigation, monitoring) (Dooley, 1985). There are, however, also differences. Risk assessment and management deal only with probabilistic risks. EIA addresses risks and impacts and considers certain, uncertain, and probabilistic effects (Dooley, 1985). Risk assessment tends to place less emphasis on alternatives (Barrow, 1997). It is more oriented toward internal management. It is less prone to consider opportunities as well as threats. It is more often applied to regulate industrial and other activities (Barrow, 1997). Enforcement with EIA tests for compliance. Risk management determines whether the event probabilities are greater than those agreed to (Dooley, 1985).

Before strengthening the links between EIA and risk assessment and management the many criticisms of the latter must first be considered. Some question whether available health and environmental risk data can support the assumptions, models, probability distributions, interpretations, and conclusions (Heinman, 1997; Power and Adams, 1997; SEHN, undated). Concern is raised about the adequacy of risk assessment methods to properly address complex environmental conditions, new technologies, synergistic relationships, latent, indirect, and cumulative effects, exogenous events, vulnerable populations, interconnections across disciplines, processes with large geographical and temporal reaches, and carrying and assimilative capacity (Banken, 1998; Davies and Sadler, 1997; Power and Adams, 1997; Tickner and Raffensperger, 1998). The field is criticized for insufficient consideration of public concerns, values, perspectives, and perceptions, cultural differences, the social context, nontechnological options, and ecological and biospheric limits (Davies and Sadler, 1997; Fischer, 1996; Kamrin, 1993; Raffensperger and deFur, 1997). It is portrayed as relying too heavily on inadequately supported technical and scientific interpretations and opinions (Hardstaff, 2000). It is described as biased in favor of quantitative methods, rational and centralized decision making, technological "solutions," short-term and local effects, and the analysis of individual environmental components rather than entire systems (Fischer, 1996; Heinman, 1997; Raffensperger and deFur, 1997). It is prone to jargon, an unsupportable "aura" of objectivity, and a reliance on the current distribution of power and resources (Heinman, 1997). Where valid, these shortcomings could inhibit democratic debate, heighten public fear and mistrust, exacerbate conflict, reinforce power inequities, undermine political legitimacy, and divert attention and resources away from fundamental social and ethical questions such as acceptable levels of risk, uncertainty, and environmental disruption (Fischer, 1996; Power and Adams, 1997; Raffensperger and deFur, 1997; SEHN, undated; Tickner and Raffensperger, 1998). To respond to these concerns, EIA practice could tightly circumscribe the application of risk assessment and management to proposals, settings, and effects where technological and environmental databases are adequate. Risk assessment and management can be supplemented or combined with other approaches and methods (e.g., consequence analysis, semiquantitative hazards analysis, performance standards, the precautionary principle) (US NRC, 1994). Governments and regulators could provide more good practice requirements and guidance (Hood and Nicholl, 2002). More emphasis could be placed on defining and resolving the problem rather than on adapting a predefined set of methods (US NRC, 1994). More stress could be placed on integrating EIA-related risk assessment and management efforts with organizational risk reporting, assessment, and management procedures and practices (Hood and Nicholl, 2002). Modifications and refinements could be made to minimize potential deficiencies.

10.4.6   Human Health Impact Assessment

A second major uncertainty approach is human health impact assessment (HIA). HIA considers human health effects resulting from certain, probabilistic, and uncertain risks and impacts (BMA, 1998). Health is defined as "a complete state of physical, mental and social well being and not merely the absence of disease" (World Health Organization, 1967). HIA minimizes the negative and accentuates the positive impacts on the health and well-being of a specified population from a proposed action (McIntyre and Petticrew, 1999; NYPHO, 2001). It can be applied to a project, a policy, a program, or a plan. It can be a component of EIA or a stand-alone evaluation for an action subject or not subject to EIA requirements. It is generally instigated when there is uncertainty or concern about possible health risks of a proposal or possible opportunities to increase health gain (Scottish Needs Assessment Programme, 2000). HIA overlaps with and is closely connected with EIA, SIA, risk assessment, and management, and health planning, management, and services (EnHealth Council, 2001a). It integrates knowledge and methods from psychology, sociology, economics, toxicology, and epidemiology (Erickson, 1994). It incorporates personal, social, cultural, economic, and environmental factors and considers the opinions, experience, and expectations of potentially affected parties (Davies and Sadler, 1997; EnHealth Council, 2001; Lehto and Ritsatakis, 1999). It recognizes that human health and environmental integrity are interdependent and essential for sustainability (Davies and Sadler, 1997).

HIA can take the form of a quick screening or audit (to determine if analysis is warranted), a rapid appraisal or mini-impact assessment (available data, minimal quantification, single meeting), an intermediate or standard HIA (standard practice, limited literature review, largely reliant on routine data, impacts quantified, stakeholder participation, nonrigorous, sampling methods), or a comprehensive or maxi-HIA (extensive literature search, primary and secondary data, rigorous with controlled populations where possible, extensive quantification, sampling, and stakeholder participation) (Lehto and Ritsatakis, 1999; Parry and Stevens, 2001; Scott-Samuel et al., 1998). HIA can be prospective (potential health impacts), retrospective (impacts after implementation) or concurrent (assessed during implementation) (NYPHO, 2001). It can adopt a broad (holistic view of health, sociological roots, democratic, general quantification, evidence from key informants and popular concerns, low precision) or tightly defined (defined and observable aspects, epidemiology and toxicology roots, quantification toward measurement, measurement evidence, high precision) perspective (EnHealth Council, 2001b). It can be part of policy preparation, an EIA component, or an element of health advocacy (Lehto and Ritsatakis, 1999).

Figure 10.3 is an example of a comprehensive HIA process. Screening determines which actions require further review or more detailed health-related analysis, which actions have clearly negligible impacts or produce well understood and easily controllable health effects, and which actions require more information (EnHealth Council, 2001b; IPHI, 2001; Lehto and Ritsatskis, 1999). Scoping is based on an overview analysis and extensive stakeholder discussions. It confirms and refines need, identifies issues, specifies potential health concerns and hazards, and determines the type of HIA (EnHealth Council, 2001). It also bounds the analysis, determines the level of detail, establishes the schedule, selects the study team,

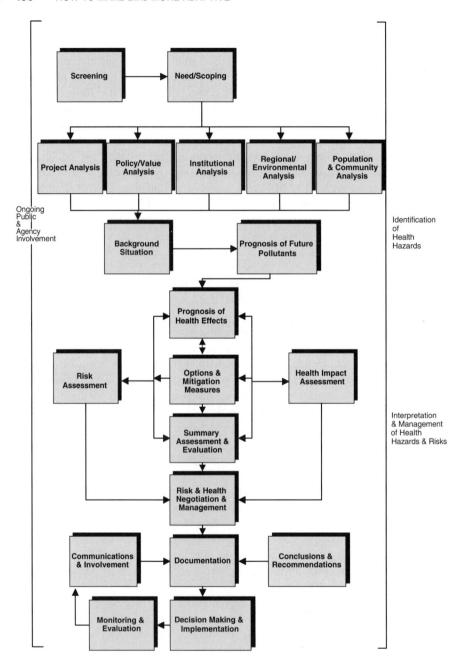

Figure 10.3 Example of a human health impact assessment process. (Adapted from Arquiaga et al., 1994; IPHI, 2001; Lehto and Riksatakis, 1999; NYPHO, 2001; Winters and Scott Samuel, 1997; WHOROE, 2001a,b.)

designs the approach, decides on consultation procedures, determines documentation requirements, and allocates resources (IPHI, 2001; WHOROE, 2001a).

Several analyses establish the basis for assessing health effects. Proposal characteristics (e.g., emissions, effluents) that could induce health effects in target populations (e.g., workers, nearby residents) are identified. Relevant policies are determined. Goals and objectives to guide the process are set. The capacity and capability of health protection agencies to prevent and ameliorate acute and chronic health concerns are determined (Arquiaga et al., 1994; Lehto and Ritsatskis, 1999). The physical, natural, resource, built environmental, and land-use conditions likely to affect the incidence, dispersion, severity, and management of health effects are identified (Lehto and Ritsatskis, 1999). Population (e.g., levels, geographic distribution, food sources and eating habits, age distribution, socioeconomic status, health status, educational levels, genetic endowment) and community (e.g., social support networks, lifestyle and behaviors, community structure, working conditions) characteristics likely to influence the incidence and severity of human health effects are determined (Davies and Sadler, 1997; EnHealth Council, 2001b; Health Canada, 2000a; Lehto and Ritsatskis, 1999). Preexisting health hazard sources (e.g., surface and groundwater water pollution, air pollution, soil and crop contamination, noise, odors, radiation) and health hazards (e.g., communicable diseases, noncommunicable diseases, inappropriate nutrition, injuries, mental disorder) are determined.

The prognosis of future conditions establishes how health risks and hazards could change through the duration of the proposed action. It considers projected population levels and characteristics, planned and anticipated land and resource uses, expectations regarding the dispersion and dilution of pollutants, and projections of the exposure of the target populations to health effects from expected future background health hazards. The estimation of potential health effects from the proposed action involves a risk assessment (where probability distributions for health effects can be predicted) and a health impact assessment (where potential health hazards and benefits are described qualitatively and semiquantitatively). Both identify and assess options and mitigation measures as means for avoiding and reducing adverse health effects and for enhancing benefits. Both also integrate stakeholder perceptions, knowledge, experience, and perspectives (IPHI, 2001; WHOROE, 2001a). The risk assessment predicts the probability of acute and chronic (cancer- and non-cancer-related) health effects. The health impact assessment identifies and predicts direct and indirect health effects on exposed segments of the target populations (Davies and Sadler, 1997). Potential chemical, radiological, biological, physical, and psychological health effects are considered (Arquiaga et al., 1994). Links to physical, natural, social, economic, and service impacts are taken into account (EnHealth Council, 2001b). Hazard agents, exposure conditions, physical health effects, beneficial health effects, effects on health care services, social well-being, social and community health, and psychological well-being are all considered (Davies and Sadler, 1997; Health Canada, 2000a).

The risk assessment and health impact assessment are combined in a summary assessment. Health risk acceptability and impact significance are evaluated, recognizing mitigation and enhancement potential (Davies and Sadler, 1997).

Uncertainties and methods for coping with uncertainties are specified. A health impact management strategy is devised encompassing such matters as objectives, policies, tactics, priorities, roles and responsibilities, contingency and emergency response procedures, mitigation commitments, compensation criteria and procedures, research and information needs, resources for post-project management, and monitoring requirements (EnHealth Council, 2001b; Winters and Scott-Samuel, 1997). Documents detail and summarize all aspects of the process, including conclusions and recommendations (WHOROE, 2001b). Decisions are made based on the documentation and on the consultation activities. The proposed action is implemented (if approved) with environmental and health control conditions (EnHealth Council, 2001b; IPHI, 2001). Health indicators are identified. Key health conditions, hazards, consequences, and compliance with conditions are monitored for a predetermined period (WHOROE, 2001b). The HIA process, methods, consultation, and communications procedures and databases are evaluated. Adjustments are made to the proposed action and to post-approval procedures based on monitoring results (EnHealth Council, 2001b). The process evaluation is widely distributed to assist other HIA processes. Public and agency involvement occurs in all HIA process activities. Agency involvement occurs through an agency steering committee and by means of contacts with individual agencies. Public involvement measures include stakeholder and key informant interviews, surveys of potentially affected populations, a public liaison committee, and periodic consultation events (e.g., open houses).

The typical EIA can be broadened and reoriented to address health impacts systematically. This approach treats HIA as a subset of EIA. HIA could, in turn, be either a subset of SIA or a field that partially overlaps with SIA (Lehto and Ritsatakis, 1999). Alternatively, the HIA process or process activities can be partially integrated, fully integrated, or simply linked with parallel EIA activities (NYPHO, 2001) This approach treats HIA and EIA as separate fields that have the potential for linkage, partial integration, or full integration, depending on the circumstances. Or a more targeted approach can be adopted where selective HIA activities or effects are integrated at key points in the EIA process. This approach views HIA and EIA, selectively and periodically, as partially overlapping fields. The final approach choice is to address health concerns through risk assessment and management (Arquiaga et al., 1994). The selected approach should suit the circumstances.

HIA has many potential benefits for EIA. It ensures greater prominence for human health concerns—an identified EIA practice deficiency and a major public concern (Erickson, 1994; NYPHO, 2001). An enhanced decision-making weight for health is valuable intrinsically and because it contributes to broader social, equity, and sustainability objectives (Davies and Sadler, 1997; NYPHO, 2001). HIA addresses more health effects and uncertainties than can be considered in risk assessment and management. The stress on health benefits counterbalances the EIA and risk assessment preoccupation with minimizing the negative (Davies and Sadler, 1997; NYPHO, 2001). HIA provides a framework for integrating quantitative and qualitative health concerns. It helps bridge EIA and risk assessment and management, EIA and health care planning and services, and EIA and SEA

(NYPHO, 2001; WHOROE, 2001b). HIA provides a means of involving health professionals in EIA practice. It contributes to health impact assessment methodology. It fosters EIA institutional capacity building for addressing health concerns. EIA provides an established set of institutional arrangements for implementing HIA. HIA is an additional evaluation tool for the health care community and for public policy development. HIA (in common with EIA) also contributes to more open, participatory, transparent, systematic, and substantiated planning and policy making.

It is too early to have a clear sense of HIA pitfalls and shortcomings. There are already numerous demands on hard-pressed health planning and management budgets. A desire to assess health impacts more systematically is of little value if necessary expertise and financial resources are not available. HIA must therefore be focused, practical, and realistic. Overlaps, duplication, coordination, and integration with related fields such as EIA, SEA, SIA, and risk assessment and management need to be addressed systematically, without diminishing the genuine need to address human health concerns more effectively (Davies and Sadler, 1997; IPHI, 2001). Capacity building, networking, methodological development, applied research, quality assurance, the clarification of terminology and roles, measures to enhance awareness, and the creative use of limited available resources are all required (Davies and Sadler, 1997; IPHI, 2001; WHOROE, 2001a,b). The many uncertainties associated with HIA needed to be acknowledged and considered. Insight from related fields in uncertainty management will be essential.

10.4.7   The Precautionary Principle

Decision makers face a dilemma. Scientific knowledge of complex environmental and social systems is far from definitive. There will always be scientific uncertainties and varying interpretations of what represents adequate evidence to support a scientific conclusion. Scientists are understandably cautious in coming to firm conclusions. But serious, potentially catastrophic environmental and health consequences can occur as a result of individual and cumulative human actions. It may be too late to avoid such consequences if no actions are taken until scientific standards of proof are satisfied. In the meantime, decisions must be made on a host of proposed activities, which have the potential for environmental and health harm. It is not sufficient simply to approve all activities except those where scientific evidence demonstrates the likelihood of serious or irreversible harm. Nor is it appropriate to reject all proposed actions automatically where there are uncertainties about harm potential and severity. Alternative or supplementary standards of evidence and decision rules are needed to provide a sound and consistent decision-making basis (CEC, 2000). The precautionary principle (PP) is one way to meet this need.

There is no commonly accepted definition of the PP or precautionary approach. Most definitions begin with the *threat* or risk of harm from a proposed activity to the environment or human health. The threat is based on preliminary scientific evaluations that provide reasonable grounds for concern about the potential for dangerous effects on the environment or on human, animal, or plant health (CEC, 2000). Although possible harm is known, the probability of the harm is not known (WHOROE, 2000c). There may be *shortcomings* (e.g., lack of, inconclusive, or insufficient evidence), *uncertainties* (e.g., lack of certainty, some cause-and-effect relationships not fully understood), or *divisions* (e.g., lack of consensus) in the scientific knowledge base (CEC, 2000; Hardstaff, 2000; Wingspread Statement on the Precautionary Principle). The conclusion is drawn that scientific knowledge limitations should not preclude or postpone *actions* to prevent the harm. The PP is not relevant in cases of ignorance (impacts and probabilities are unknown) or when causal relationships are established (certain and preventable, or probabilities can be estimated) (WHOROE, 2000c).

There are multiple interpretations regarding the harm that should trigger the PP (e.g., harm alone, serious or irreversible harm from proposed action, serious or irreversible harm from cumulative actions, varying interpretations of serious) (Tickner and Raffensperger, 1998). Opinions vary concerning scientific evidence standards. Action has been interpreted variously as (1) deciding that inaction to ameliorate harm is not justified by scientific uncertainty (i.e., action generally proceeds but with mitigation to reduce the threat of harm), (2) deciding that the proposed activity is unacceptable because the scientific evidence is inadequate or because the scientific evidence warrants rejection, (3) proceeding with the proposed activity only if it is proven safe scientifically (i.e., reversing the burden of proof and requiring a level of certainty), (4) proceeding only if a reasonably convincing case can be made that the action is safe (i.e., reversing the burden of proof, acknowledging uncertainties, requiring a weight of evidence argument), and (5) proceeding very carefully (i.e., balancing the burden of proof by adopting prudent decision-making criteria, such as safety factors, no or least regrets, best available technology, stringent monitoring) (Gullett, 1997, 1998; Hardstaff, 2000; Wiener and Rogers, 2002). The standards of proof are generally greater if the proposed action is a priori hazardous or new, as in a new technology. Qualifications can be added when applying any interpretation (e.g., proportionality, relative to alternatives, consideration of benefits, additional measures to cope with uncertainties) (Wiener and Rogers, 2002).

These varying interpretations imply thresholds or criteria for threat or risk of harm (which infers a combination of likelihood and severity), thresholds or criteria for deficiencies in scientific knowledge and rules, and principles and procedures for applying the thresholds or criteria. The PP needs to be supported by regulatory authority levels of protection and evidence standards of unacceptable harm (CEC, 2000; Gullett, 1997). Terms such as *threat, harm, serious* or *irreversible, definitive, fully, lack, environment, health*, and *burden of proof* require definition and interpretation, overall, for classes of situations or on a case-by-case basis. A mechanism for determining whether the PP is to be applied is required. Criteria, procedures, decision rules, and institutional arrangements for applying the PP are needed (CEC, 2000). The relationship of PP requirements to risk regulation (e.g., an overarching principle, a risk acceptability criterion applied after risk assessment) and to EIA requirements needs to be addressed (Wiener and Rogers, 2002). Some argue that the PP also necessitates a reversed burden of proof from victims to proponents, an open, transparent, and democratic decision-making process, a systematic analysis of all alternatives for reducing (to acceptable levels) or eliminating the harm, a greater weight to "ignorance" in decision making, preventive anticipation or risk avoidance as decision norms, and a proactive effort to safeguard ecological space, to minimize serious or irreversible environmental damage, to avoid social deprivation, to operate within ecological and biosphere limits, to pay past ecological debts, and to protect the interests of future generations (O'Riordan and Cameron, 1994; Porritt, 2000; Raffensperger and deFur, 1997; SEHN, undated; Tickner and Raffensperger, 1998; WHOROE, 2001b,c).

Figure 10.4 is an example of a precautionary EIA process. A decision is first made regarding whether the precautionary trigger applies to the action proposed. The PP is commonly triggered when there is a potential for serious or irreversible environmental or human health harm, a scientific evaluation, and scientific uncertainty (CEC, 2000). The strength of the connection between harm and evidence ranges from significant risk, through likelihood of damage, to reasonable grounds for concern that harm may be caused, to potential for damage and no proof of harmlessness (Gullett, 1997). Very general PP requirements and guidelines maximize the ability to make commonsense adjustments to individual circumstances but increase the potential for arbitrary, biased, and inconsistent interpretations and judgments. A scientific evaluation identifies the potential threat, characterizes the problem, and assesses knowledge and uncertainty levels (CEC, 2000). Qualifications to the principle are added where appropriate (CEC, 2000; EnHealth Council, 2001b; Government of Canada, 2001). Key terms are defined. Links to EIA, to risk management (a framework for or a tool within), and to other environmental management requirements are identified (CEC, 2000; Government of Canada, 2001). Relevant implications are noted. A clear rationale is provided for each interpretation. The input requirements to apply the PP are specified. The overall precautionary approach is consolidated. Precautionary goals are set (Tickner and Raffensperger, 1998). The precautionary elements of the study design are prepared. A precautionary perspective is applied to the project purpose, to the assessment of need, and to the identification of alternatives (Gullett, 2000).

What is known (certainties) and what is not known (types and sources of uncertainties) are determined (Tickner and Raffensperger, 1998). Harm and scientific evidence thresholds and criteria are established. The harm criteria include such considerations as magnitude, temporal and spatial scale, reversibility, degree of complexity and connectivity, vulnerable environments and populations, error friendliness, catastrophic potential and availability of alternatives to reduce or eliminate harm (Tickner, 1998). The scientific or causal inference criteria pertain to such matters as amount, strength, and consistency of evidence across a wide range of circumstances, knowledge coherence, plausibility of effect, consideration of all evidence and plausible hypotheses, study power to detect effect, statistically significant evidence, public health significance, and causal relatedness based on previous experience (Tickner, 1998). Precautionary decision rules, thresholds and criteria, and application principles and procedures are formulated. The decision rules determine what, for example, represents a basis for action rejection, action

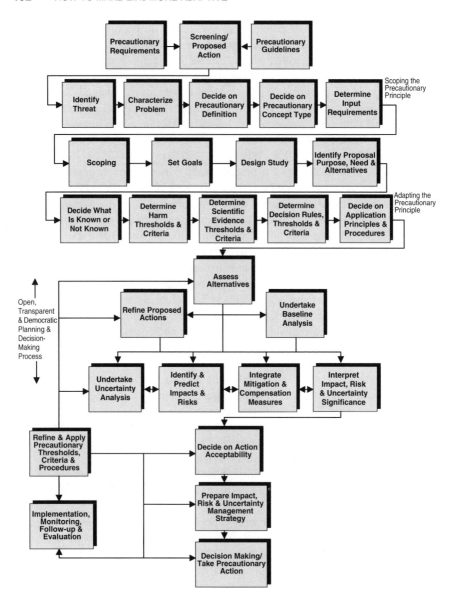

Figure 10.4   Example of a precautionary EIA process.

deferral, additional study (e.g., a risk assessment), and specific approval conditions. Further refinements to the precautionary approach occur through the balance of the EIA process.

The customary EIA process activities are undertaken. The PP contributes to assessing which alternatives are acceptable, which elements of the proposed action could pose an unacceptable harm, how uncertainties are to shape the application of the principle, which predicted risks and impacts could represent an unacceptable harm, whether mitigation and management measures are likely to reduce the harm to acceptable levels, whether the anticipated risks, impacts, and uncertainties are significant and whether the proposed action is acceptable (Gullett, 1997, 1998, 2000). Precautionary measures to manage anticipated impacts, risks, and uncertainties are integrated into management strategies (Gullett, 1998). The PP affects decision making (e.g., the taking of precautionary action, the weight of uncertainties in final decisions) (Gullett, 1999, 2000). It influences implementation, monitoring, follow-up, and evaluation (i.e., precautionary measures to be followed unless compelling reason for not doing so) (Gullett, 1998). Care is taken to ensure that the action stays within the precautionary acceptability levels. The PP application is evaluated, both to facilitate post-approval adaptations and to assist in future applications. The EIA process is open, transparent, and democratic (Gullett, 2000; Tickner and Raffensperger, 1998). The public and government agencies are involved in scoping, adapting, and applying the PP. The precautionary analyses draw upon multiple disciplines, sources of information, forms of expertise (including local, lay, and traditional knowledge), values, goals, and ways of reasoning (Harremoës et al., 2002). They also contribute to reducing and coping with uncertainties and errors (especially type II errors) (Gullett, 2000).

The PP can reduce the incidence and severity of serious, irreversible, and catastrophic environmental harm—a response to the "tragedy of the commons" dilemma (Gullett, 1997). It provides a decision-making tool for consistently addressing uncertainties (Gullett, 1997). It underscores the need to consider systematically all available options for avoiding harm (Gullett, 2000; Tickner and Raffensperger, 1998). It reshapes decision making by shifting the burden of proof from the public to the proponent, by seriously assessing proposal acceptability, by broadening decision making beyond science, and by contributing to more environmentally prudent decision making (Tickner and Raffensperger, 1998). It reinforces democratic and substantive environmental and social values and imperatives (Government of Canada, 2001). It is sufficiently flexible to adapt to varying contexts and to proposal and action-specific circumstances.

Depending on how it is interpreted and applied, the PP can have serious drawbacks. There will always be risks and uncertainties when seeking to predict and manage environmental change. Few, if any, EIA proposed actions would be acceptable if the lack of proof of safety or acceptable levels of environmental impact is sufficient grounds for rejecting or deferring a proposed action (Bailey, 1997a; Holm and Harris, 1999). EIA practice and the PP both operate somewhere between the two extremes of scientific certainty of no harm and scientific certainty of harm. Presumably, the goal of the PP is to place more weight on scientific uncertainty about harm potential and less on scientific uncertainty about no harm. Insisting on more decision-making weight for uncertainty, however, could oversimplify and distort an evaluation (Holm and Harris, 1999). Other, perhaps equally or more valid and compelling perspectives, values, and positions could receive no, minimal, or insufficient consideration (Bailey, 1997). Decision making might be reduced to a hedging mechanism (Wildavsky, 1995).

The breadth of PP interpretations could be used to justify everything from minimal changes in conventional approaches (e.g., identify uncertainties and be careful) to rejecting almost any proposed action, many with significant environmental, social, and economic benefits (Appell, 2001; EnHealth Council, 2001b; Whelen, 1996). Potential negative outcomes could include arbitrary, inconsistent, and distorted decisions, the stifling of innovation, application abuses (e.g., trade protectionism), the advancement of agendas of dubious validity and with limited public support, unwarranted public and private costs and delays, the neglect of legitimate risks, the exacerbation of unwarranted fears, the rejection of scientific knowledge, and the misuse of scarce environmental management resources (Appell, 2001; Bailey, 1997b; Foster et al., 2000; Government of Canada, 2001; Hardstaff, 2000; Holm and Harris, 1999; Whelen, 1996). These potential shortcomings point to the need to recognize the valid concerns underlying the principle while avoiding more extreme interpretations and being wary of overly vague and discretionary requirements and guidelines. A prudent, open, and democratic process for both formulating and applying the PP is essential. Vigilance is required to prevent and minimize abuses, distortions, and negative propensities (Wiener and Rogers, 2002). The risks and uncertainties of action and inaction both require consideration (Wiener and Rogers, 2002). Specific requirements and guidelines are needed concerning how to apply the principle within the EIA process. The limited experience with the PP underscores the need to compile a good practice knowledge base (Gullett, 1997).

10.4.8   Adaptive Environmental Assessment and Management

Adaptive environmental assessment and management (AEAM) treats environmental management as a quasiexperiment (i.e., probing ecosystem responses to human actions) (Johnson, 1999; Lee, 1999). Managers learn while doing. Subsequent decisions are adjusted and enhanced from feedback (Reinke and Swartz, 1999; Wieringa and Morton, 1996). The AEAM process is an iterative cycle of planning, implementation, monitoring, research, and reexamination (IEMTF, 1995). Each cycle facilitates the selection of more appropriate management actions, helps change stakeholder behavior, and provides a learning opportunity (Lal et al., 2001).

An example of an AEAM process is depicted in Figure 10.5. The process begins by identifying relevant environmental conditions, stakeholders, institutions, resources, and values (Haney and Power, 1996; Iles, 1996). This ensures that the process suits the context. Stakeholders agree on an initial agenda of questions (Lee, 1999). The overall process is then designed and planned. A manager, a core group, and specialist support staff are selected (Holling, 1978). The core group provides continuity by keeping the process focused and on track (Hegmann and Yarranton, 1995). The specialists are organized into technical working groups. A preliminary problem description is prepared. Key problem features are highlighted. Participants for the first workshop are identified. The workshops involve policy people, managers, scientists and a diversity of stakeholder representatives. Each workshop draws upon specialist advisors (Holling, 1978). The choice of participants depends on the workshop purpose (Holling, 1978).

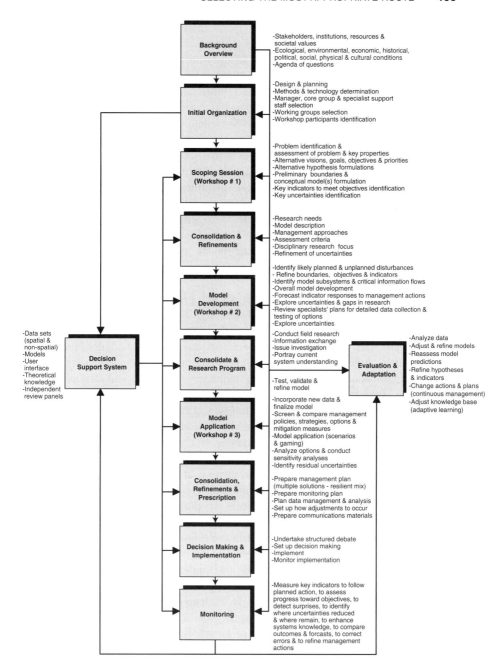

Figure 10.5 Example of an adaptive environmental assessment and management process. (Adapted from: GBC, 1999; Haney and Power, 1996; Holling, 1978; Hyman et al., 1988; IEMTF, 1995; Johnson, 1999; Iles, 1996; Lal et al., 2001; Morgan, 1998.)

The initial workshop helps key participants understand the problem and share a common sense of identity and purpose (Hyman et al., 1988). The first workshop (the equivalent of scoping) begins by identifying and assessing the management problem (GBC, 1999; Jones and Greig, 1988). Alternative visions, goals, priorities, and measurable objectives are formulated. Initial alternative hypotheses are identified (Walters, 1986). Preliminary temporal and spatial boundaries and scales are established. An initial effort is made to formulate a conceptual model or models. The model illustrates critical interrelationships among key environmental components potentially susceptible to possible human activities. Key indicators (social, economic, resource, and environmental) are identified (Jones and Greig, 1988; Walters, 1986). Major uncertainties are identified and assessed (GBC, 1999). Following the first workshop research needs are determined. The model and uncertainty descriptions are refined. Possible management approaches are identified. Criteria for assessing options and activities are formulated. Individual disciplinary research is undertaken to elaborate on the model, to narrow and explore uncertainties, and to describe possible human activities more precisely. Preparations are made for the second workshop.

The second workshop defines, with greater precision, probable planned and unplanned environmental disturbances (Holling, 1978). Study boundaries, objectives, and indicators are refined. The model, composed of working hypotheses, is formulated to clarify the problem, to represent current understanding of the system, to assess the significance of data gaps and uncertainties, to provide options, to facilitate communications among scientists, managers, and other stakeholders, and to predict the system effects of alternative management actions (Haney and Power, 1996; Hegman and Yarranton, 1995; Johnson, 1999; Walters, 1986). The model subsystems are identified and critical internal (among subsystems) and external linkages are determined. The model, depending on the situation, can be quantitative, qualitative, or a combination. Multiple linked models are applied as appropriate. Other techniques, such as scenario formulation are employed to supplement the modeling work, where warranted. Preliminary forecasts of how indicators might respond to potential management actions are prepared, bearing uncertainties in mind (Noble, 2000b). Uncertainties are explored and research gaps are noted. Specialist research plans for detailed data collection to test options and to address uncertainties are consolidated and reviewed by the full team and by decision makers (Hyman et al., 1988). Methods and apparatus are designed to take the appropriate measurements (Lee, 1999). The specialists undertake the field research following the workshop. Information is exchanged among the specialists and between the specialists and the managers. Issues emerging from the workshop are explored. A summary compilation of knowledge about the system is prepared (Lee, 1999). The model is tested, checked for validity, and refined. The third workshop is organized.

The third workshop incorporates the research program data and finalizes the model. The model is used to screen and compare management policies and strategies, to test options and mitigation/enhancement measures, and to explore uncertainties (GBC, 1999), Sensitivity analyses assess the implications of varying the model's time and space boundaries, basic assumptions, and uncertainty ranges (GBC, 1999; Hyman et al., 1988). Sometimes, gaming is employed to open up discussion and to search for creative solutions to problems. Innovative approaches and win–win situations are sought actively. Residual uncertainties are noted. Following the workshop a management plan is prepared. The management plan includes a resilient mix of strategies and a range of acceptable outcomes, while avoiding catastrophes and irreversible negative effects (Holling, 1978; Johnson, 1999). An experimental design for implementation and a monitoring plan are prepared. The monitoring plan specifies how monitoring data are to be managed and analyzed. Possible responses to potential monitoring results are considered (GBC, 1999). Communications materials are prepared to facilitate stakeholder discussions about process results and proposed management and monitoring actions. Process outputs are consolidated and documented in a form suitable for decision making.

The key indicators are monitored through implementation to follow the planned actions, to indicate deviations from the plan, to identify changing circumstances, events, and decisions (i.e., surprises), to assess progress toward objectives (effectiveness), to identify where uncertainties are reduced and where they remain, to enhance systems knowledge, to compare actual outcomes to forecasts, to correct errors, and to refine management actions (GBC, 1999; Noble, 2000b). The data obtained during monitoring are analyzed, documented, and fed back into each process activity (Haney and Power, 1996). The process is supported by theoretical knowledge, by spatial and nonspatial data sets, by models, and by procedures for users to interface with the decision support system (Lal et al., 2001). Independent review panels provide additional insight and advice (Wieringa and Morton, 1996). Interested parties are involved both through the workshops and in broader consultation opportunities at key decision points (Johnson, 1999).

AEAM has much to offer EIA, especially in actively adapting to and managing the uncertainties associated with complex problems and/or complex environments (Hyman et al., 1988; Johnson, 1999; Noble, 2000b). It recognizes the value of science in identifying and diagnosing surprise but acknowledges the limits of scientific methods, prediction, and control (Lee, 1999). It appreciates that the uncertain, the unexpected, and the unknown are normal facets of planning and management (Dearden and Mitchell, 1998). It facilitates rapid knowledge acquisition rates and rapid detection of changes (McLain and Lee, 1996; Smith, 1993). It recognizes that more information is not always desirable and can hinder decision making (Hyman et al., 1988). It permits learning by doing and underscores the value of monitoring (Morgan, 1998). It is able to handle indirect effects and cumulative effects (Hyman et al., 1988). It provides an integrative systems approach, which links and transcends disciplines and perspectives and helps alleviate the problems associated with fragmented research and coordination (Holling, 1978; Noble, 2000b). The workshop format, built around model building and testing, provides a potentially constructive and nonconfrontation approach for stakeholders to build a common understanding of the problem, to synthesize existing knowledge, to highlight key uncertainties, to clarify assumptions, to stimulate creativity, and to generate innovative options (GBC, 1999; McLain and Lee, 1996; Noble, 2000b; Smith, 1993).

The EIA and AEAM processes are very similar, especially at the strategic planning level. The project level is potentially more problematic. Projects, particularly those involving large up-front costs, allow for contingencies but are often not amenable to midcourse corrections, which radically depart from original project objectives (Carpenter, 1997). Project-level EIA is largely oriented to obtaining data and to making specific predictions for decision-making purposes. AEAM is more focused on reaching a policy, resource, or environmental management strategy consensus (Morgan, 1998). Organizational resistance to AEAM will often occur because of a reluctance to admit uncertainty, to make mistakes, or to try new solutions, a lack of interest in developing an organizational learning capacity, a perception that it will challenge bureaucratic self-interests, short-term perspectives, and an expectation that scientific research will be costly and of little administrative and political value (GBC, 1999; Gunderson, 1999; Lee, 1999; Walters, 1997). AEAM principles may be less applicable in unique situations where lessons are not transferable (i.e., no spatial replication), where impacts are curable rather than chronic, where uncertainties are limited and manageable, where reasonably accurate impact predictions can be formulated, where continuing surveillance of environmental systems is unwarranted, where ecological components and systems are not resilient, where stakeholders are inflexible (i.e., major value conflicts), where multiple systems are involved, and where there are potentially significant irreversible risks and impacts associated with experimentation (Gunderson, 1999; Johnson, 1999; Morgan, 1998; Noble, 2000b; Walters, 1997). The full application of AEAM modeling and field experimentation can be costly (Walters, 1997). AEAM is information dependent. The information needed to support AEAM may not be available, and there may be no means to develop the information (IEMTF, 1995).

AEAM tries to cover multiple objectives by including representatives of various disciplinary backgrounds in the study team (Hyman et al., 1988). Since the process does not specify a systematic way of dealing with multiple objectives, the results are very sensitive to study team composition (Hyman et al., 1988). AEAM is based largely on applying ecological, often linear systems models (McLain and Lee, 1996). Such models can have difficulty addressing cross-scale linkages (e.g., between physical–chemical and ecological processes), the nonadditivity of parameters and effects, and difficult and emergent processes (Walters, 1997). Nonscientific and qualitative information, knowledge, and experiences are sometimes discounted (McLain and Lee, 1996). Such models may be unable to accommodate fundamental conflicts among scientists regarding facts and assumptions and among policy makers concerning community preferences (McLain and Lee, 1996). Most AEAM literature focuses on procedural elements (Smith, 1993). More attention could be devoted to social, cultural, and economic concerns and to substantial contributions to sustainability (Smith, 1993; UNEP, 1997). Methods of obtaining institutional support, the institutional structures required for AEAM to work, and the procedures for overcoming data inadequacies, model inadequacies, and misunderstandings about AEAM concepts and methods all require additional attention (Jones and Greig, 1988; McLain and Lee, 1996).

10.5 INSTITUTING AN ADAPTIVE EIA PROCESS

10.5.1 Management at the Regulatory Level

The four jurisdictions (the United States, Canada, the European Union, and Australia) all address uncertainty in EIA and in related fields. EIA requirements and guidelines in the four jurisdictions include numerous general and specific references to uncertainty and adaptation. More extensive and detailed references are made (a different combination in each jurisdiction) to risk assessment and management, to the precautionary principle or approach, to human health impact assessment (HIA), and to adaptive environmental assessment and management (AEAM). The only difficult problems referred to specifically are transboundary effects and catastrophes. Minimal references are made to chaos and complexity.

Uncertainty The U.S. National Environmental Policy Act (NEPA) refers to "unintended consequences." CEQ regulations identify the "degree to which the possible effects on the human environment are highly uncertain or involve unique or unknown risks" as an impact significance factor. The regulations provide general advice concerning impact significance interpretations when there is incomplete or unavailable information. Reference is made to such matters as disclosure, availability, and ease of obtaining information, information relevance, the probability of occurrence, the potential for catastrophic effects, and the credibility of the evidence and methods (e.g., not based on pure conjecture, within the rule of reason, reasonably foreseeable). Social and cumulative effects guidelines refer to managing uncertainty by monitoring, mitigation, and adaptive management and to the importance of communicating uncertainties. Uncertainty management also is addressed in individual agency guidelines (e.g., U.S. Army Corp of Engineers), usually in conjunction with risk management (US ACE, 1992; Yoe, 1996; Yoe and Skaggs, 1997).

The Canadian Environmental Assessment Act identifies uncertainty as a screening consideration and as a factor in determining whether to refer a comprehensive study back to the responsible authority. EIA guidelines identify uncertainty as an impact significance factor, as the critical question for implementing a follow-up program and as a cumulative effects consideration. Mention is also made of type 1 and type 2 statistical errors, the differences among certain, reasonably foreseeable, and hypothetical future actions, uncertainty sources, and methods for addressing uncertainty. The five-year review of the act identifies federal process application inconsistencies and uncertainties and inconsistent assessment quality as major concerns. Measures to make the process more certain, predictable, and timely are proposed.

The European EIA and SEA directives identify monitoring as a means of identifying unforeseen adverse effects and of undertaking appropriate remedial actions. Both require that information compilation difficulties, such as technical deficiencies or lack of know-how, be identified. General EIA guidelines identify uncertainty as a significance determination factor. They also refer to identifying uncertainties associated with existing environmental conditions, with impact prediction methods, with the proposed project, with data compilation, and with mitigation effectiveness. Indirect and cumulative effects guidelines detail the uncertainty types encountered in determining boundaries, in establishing baseline conditions, and in understanding interactions and pathways. The guidelines stress the need for flexibility, for understanding and documenting assumptions, and for justifying uncertainty interpretations.

The Australian EIA legislation and regulations require that each EIA document type describe information timeliness, reliability, and uncertainties. The administrative guidelines on significance identify degree of confidence with which impacts are known or understood as a significance factor. The ANZECC EIA guidelines suggest that a formal EIA process may be justified if there is a high level of uncertainty or a large number of unknowns. The guidelines also refer to knowledge limits and uncertainties regarding ecosystem resilience, project design and technology, human environmental conditions, the ability to predict, manage and monitor impacts, changing community values, data adequacy and accuracy, associated programs, and links to monitoring and management.

The four jurisdictions collectively address many aspects of uncertainty and uncertainty management. A more systematic effort could be make to identify relevant uncertainty forms and sources, to describe key uncertainty concepts, to identify where and how uncertainties arise in the EIA process, and to provide examples of uncertainty management methods. Further direction and advice could be provided regarding documenting uncertainties and concerning the role of uncertainty in EIA-related decision making (e.g., screening, scoping, significance determination, option rejection and comparison, the triggering of mitigation, and monitoring requirements). Each jurisdiction could consider the specific sources of uncertainties in EIA requirements and guidelines (from the perspective of stakeholders) that have led to unwarranted inconsistencies.

Risk Assessment and Management   NEPA refers to risk to health or safety. CEQ regulations identify unique or unknown risks, from both natural hazards and accidents, as significance determination factors. The regulations also refer to impacts with catastrophic consequences and to a requirement for a reasonably foreseeable analysis. An executive order (EO 13045) addresses the protection of children from environmental health and safety risk. EIA guidelines refer to accident effects. Federal agencies, including the Environmental Protection Agency, the Department of Energy and the U.S. Army Corps of Engineers, have issued a host of guidelines, research reports, case studies, and forum reports on human health and ecological risk assessment. Additional risk assessment and risk management guidance is provided by reports prepared by or for the National Research Council

(NRC) and various commissions, committees, and boards (e.g., the Presidential/ Congressional Commission on Risk Assessment and Management, the Committee on Environmental and Natural Resources, the Committee on Risk Assessment and Hazardous Air Pollutants, the Committee on Risk Assessment Methodology).

Canadian EIA requirements refer to the environmental effects of malfunctions and accidents. The probability or predictability of an impact occurring is a significance determination criterion. Risks associated with effects, such as exposure of humans to contaminants or pollution or from accidents, are identified as factors for determining the level of effort for assessing policy, plan, or program proposals. EIA guidelines refer to risks from environmental hazards and possible malfunctions and accidents. Quantitative risk assessment is cited as a method for determining significance. The relationships between risk management and EIA and between risk assessment and management and health impact assessment have been considered (Grima et al., 1986; Health Canada, 2000a). An interdepartmental risk management framework has recently been released. Health Canada and Environment Canada, respectively, have formulated requirements and guidelines concerning health risk determination and environmental risk assessment.

The European EIA and SEA directives refer to risks to human health or the environment from accidents. EIA guidance documents refer to projects involving unique or unknown risks, to environmental damage and risks from natural disasters, to actual or perceived human health risks, to risks of accidents and abnormal events, to the occurrence of disease or disease vectors, and to especially vulnerable groups of people. Human health and ecological risk assessment technical requirements and guidelines (pertaining largely to chemical regulation) have been prepared.

Australian EIA requirements and guidelines include general references to risk (notably, regarding accidental events) and significance determination. Several general health and environmental risk assessment and management guidelines have been prepared. Risk assessment and management has been identified as part of the health impact assessment process (EnHealth Council, 2001a,b). A major survey has recently been completed of Australian health risk perceptions (Starr et al., 2000).

The four jurisdictions all refer to health risks in EIA requirements. More attention is devoted to human health risks and to risks from accidents and natural disasters than to chronic health risks, ecological risks, and perceived risks. The United States provides detailed human health and ecological risk assessment and management requirements and guidance. The other jurisdictions concentrate more on risk management. All jurisdictions could devote more attention to potential risk assessment and management roles in the EIA process. Risk assessment and management strengths and limitations, the measures introduced to ameliorate limitations, and similarities and interconnections between EIA and risk assessment and management should be addressed.

Human Health Impact Assessment   Stimulating the health and welfare of humans is one the purposes of the U.S. National Environmental Policy Act

(NEPA). The human environment includes the relationship of people with the environment. The degree to which the proposed action affects public health and safety is an impact significance criterion. EIA guidelines and checklists refer to radiological impacts under normal operating conditions, to accident scenarios and accident conditions, and to toxic and carcinogenic health effects from exposure to hazardous chemical. The U.S. Environmental Protection Agency (EPA) and the U.S. Office of Research and Development have numerous research programs and strategies concerning links between health effects and environmental impacts.

The Canadian Environmental Assessment Act (1992) (CEAA) refers to health effects resulting from environmental effects. General EIA guidelines refer to health effect type, cumulative health effects, health effects significance, and to measures to mitigate significant adverse health effects. Human impact assessment is a research priority. Health Canada has prepared an HIA handbook (Health Canada, 2000a). The handbook addresses such matters as health and EIA basics, HIA decision making, health practitioner roles, HIA-related concepts, methods and procedures, and links between health and traditional knowledge, sustainable development, risk management, SIA, economics, and public health.

The European EIA and SEA directives refer to protecting human health and to human health effects. EIA guidelines elaborate on types of potential health effects. Ensuring a high level of health protection is prominently featured in European legislation (e.g., Maastricht Treaty) and resolutions. The World Health Organization (WHO) Regional Office for Europe has assumed a lead role in developing and promoting HIA as a policy measure to facilitate health protection. The WHO has prepared guidance documents that describe HIA concepts, approaches, frameworks, procedures, and methods. The European Public Health Strategy also identifies HIA as a means to promote health protection. Several European countries (e.g., Netherlands, Ireland, Britain, Sweden) have prepared HIA guidelines.

The Australian EIA legislation addresses health effects as part of effects on people. Reference is made to species posing a threat to human health. The ANZECC guidelines (1996) identify the possibility of health impacts or unsafe conditions as a significance determination factor. HIA guidelines have been prepared (EnHealth Council, 2001b). The guidelines address such matters as definition and scope, principles, the HIA process, roles, and preparing a health impact statement. Health impact assessment has been a requirement in Tasmania since 1996, where HIA is integrated with EIA requirements.

Health effects are mentioned in the EIA requirements of all four jurisdictions. But the treatment of health effects is general and fragmentary. The numerous recent HIA initiatives and guidelines are correcting this deficiency. More attention needs to be devoted to the interrelationships between EIA and HIA and between HIA and risk assessment and management. The role of the health community in EIA and HIA also requires additional consideration. The experience base for HIA is very limited. The effectiveness of HIA requirements and guidelines needs to be monitored and evaluated. The many uncertainties associated with identifying, predicting, and managing human health effects should receive particular consideration.

The Precautionary Principle The U.S. government does not recognize an overriding precautionary principle. It has selectively adopted precautionary approaches in EIA and in risk management. NEPA refers to the relationship between the short-term uses of the environment and the maintenance and enhancement of long-term productivity and to any irreversible and irretrievable commitment of resources. EIA regulations and guidelines point to using conservative assumptions and safety factors and to promoting pollution prevention. The President's Council on Sustainable Development identifies the precautionary principle as a core belief (Tickner and Raffensperger, 1998). The U.S. government's position is more tightly circumscribed. It supports precautionary approaches to risk management (e.g., inquiring about the degree of precaution embedded in a risk assessment) but stops short of recognizing any universal precautionary principle (Graham, 2002). It maintains that EIA and other regulatory review requirements address the concerns represented by the principle without opening the door to the abuses and adverse consequences potentially associated with its universal application (Wiener and Rogers, 2002). This position appears consistent with a U.S. Supreme Court ruling that there must be a demonstration of significant risk before regulating.

Changes to the purposes of the Canadian EIA act refer to the precautionary principle. The act and the SEA cabinet directive both refer to irreversibility. EIA significance guidelines identify reversibility as a significance determination factor. Two recent environmental assessment panel decisions include precautionary requirements (Gibson, 2000). Two federal statutes, two provincial statutes, and several proposed laws mention the precautionary principle. A Supreme Court of Canada decision on the use of pesticides applies the precautionary principle in support of the majority interpretation of a municipal bylaw (Government of Canada, 2001). The government of Canada, in an effort to ensure a greater level of consistency, has issued a report entitled *A Canadian Perspective on the Precautionary Approach/Principles* (Government of Canada, 2001). The report explains the need for a federal precautionary approach framework. It also describes overarching considerations, guiding principles, and principles for precautionary measures.

The European SEA directive (June 2001) refers to the prudent and rational utilization of natural resources, based on the precautionary principle. Both the EIA and SEA directives mention reversibility of effects. The precautionary principle is endorsed in the 1992 Maastricht Treaty of the European Union. The European Court of Justice has defined the conditions for applying precautionary measures in community law and has explained the rationale for precautionary measures (Coleman, 2002). The Commission of the European Communities has issued a *Communications on the Precautionary Principle* (2000). The *Communications* contains application guidelines. The European Commission indicates that uncertainty does not justify inaction but adds several qualifications (Wiener and Rogers, 2002). The European Environment Agency has prepared a report detailing 14 case studies concerned with applying precaution in policy making (Harreemoës et al., 2002).

The Australian EIA legislation includes a definition of the precautionary principle. The act specifies the decisions for which the minister must take account of the precautionary principle. The decisions include whether an action is a controlling action and whether or not to approve the taking of an action. *The Administrative Guidelines on Significance* (2000) note that the minister must take the precautionary principle into account when deciding whether an action is likely to have a significant impact on a matter of national environmental significance. The Health Impact Assessment Guidelines provide an overview of the precautionary approach as part of the HIA process (EnHealth Council, 2001b). The precautionary principle also is referred to in Australia's National Strategy for Ecologically Sustainable Development (1992), in the Intergovernmental Agreement on the Environment, and in the National Framework for Environmental and Health Impact Assessment (1994).

There is considerable variation among the jurisdictions as to if and how the precautionary approach or principle is addressed in environmental requirements. When the precautionary principle is inappropriate, EIA regulations and guidelines should explain how other mechanisms are to address the relationship of uncertainty and potentially severe consequences. If the principle could be applied, EIA requirements should define the principle and specify which harmful effects, uncertainties, and actions trigger its application. Conditions for applying the principle and the decisions to which it applies should be indicated. Guidelines can provide more specific advice regarding roles within the EIA process, possible criteria, thresholds, and decision rules and links to risk assessment and management, human health impact assessment, and uncertainty analyses. The principle's strengths, potential drawbacks, and means of avoiding and reducing potential shortcomings should be assessed. Additional applied research would be helpful.

Adaptive Environmental Assessment and Management (AEAM) U.S.

EIA requirements address AEAM elements indirectly with references to harmonious relationships with the environment, a systematic approach, the relationship between the short-term use of the environment and long-term productivity, and the limits of predictive capability (Carpenter, 1997). The CEQ regulations provide for monitoring when applicable for mitigation and to ensure that decisions are implemented. Cumulative effects guidelines suggest a role for monitoring and adaptive management in addressing uncertainty (US CEQ, 1997a). Adaptive management advantages are described briefly. The NEPA effectiveness study (US CEQ, 1997b) advocates a science-based and flexible management paradigm: predict, mitigate, monitor, and adapt. Monitoring and adaptation are described as helping determine prediction accuracy, facilitate midcourse corrections, and move iteratively toward goals, all in the face of uncertainty. Adaptive environmental management is identified as a means of attaining both NEPA's goals and an agency's mission (US CEQ, 1997b). A federal agency task force concerned with the ecosystem approach identified an implementation role for adaptive management (IEMTF, 1995). The task force recommended that agencies develop regional ecosystem plans, coordinated under NEPA. Four land management federal agencies (the U.S. Forest Office, the Department of the Interior's Bureau of Land Management, the Parks Service, and the Fish and Wildlife Service) have adopted or are in the process of adopting an adaptive management approach (Carpenter, 1997). The

NEPA task force was established by the Council on Environmental Quality in May 2002 and involves a variety of federal agencies. The task force is addressing, among other things, the role of adaptive management in NEPA implementation practices and procedures.

Adaptive environmental management is partially applied in the other jurisdictions. EIA requirements and guidelines include few references to AEAM. A review of AEAM was undertaken in Canada (ESSA, 1982). An amendment to the Canadian EIA legislation mentions using follow-up programs to implement adaptive management measures and to improve environmental assessment quality. The European SEA directive includes monitoring provisions. Reference also is made to impact reversibility. The Australian EIA requirements identify effective and adaptive management as a reserve management principle. Impact significance guidelines refer to the resilience of the environment to cope with change (ANZECC, 1996).

EIA requirements could provide more explicitly for AEAM, appreciating the differences between EIA and AEAM. EIA guidelines could address potential AEAM and EIA interrelationships. The potential role of AEAM in identifying and coping with uncertainties should receive particular attention. More consideration could be given to AEAM strengths, limitations, and measures to reduce limitations. Applied research could explore AEAM adaptations for addressing social and economic concerns, AEAM roles in assessing nonresource management proposals and steps to ameliorate organizational resistance and inflexibility.

Difficult Problems, Chaos, and Complexity   The U.S. Council on Environmental Quality (CEQ) provides transboundary impact analysis guidance. Canadian EIA requirements mention transboundary and international effects. Transboundary effects between provinces and territories are considered in EIA harmonization agreements and accords. Canada is a party to the United Nations Economic Commission for Europe (UNECE) Convention on EIA in a Transboundary Context (1991). The Commission for Environmental Cooperation (CEC) [a body created under the North American Free Trade Agreement (NAFTA)] considers significant adverse transboundary impacts between Canada and the United States and between the United States and Mexico. Transboundary effects in Europe are addressed through the UNECE Conventions on EIA in a Transboundary Context, the Protection and Use of Transboundary Watercourses and International Lakes and Transboundary Effects of Industrial Accidents. The EIA and SEA directives refer to transboundary context, transboundary consultations, and transboundary effects.

Aside from catastrophic effects, EIA requirements and guidelines in the four jurisdictions do not refer to other types of difficult problems or to chaos and complexity. There are scattered references to unusual and complex interactions and effects. Such conditions can affect impact significance interpretations. EIA guidelines could devote more attention to other types of difficult problems (e.g., transscientific, latent time bombs) in EIA practice. Guidelines and applied research could also identify insights and implications from chaos and complexity theories for EIA practice.

10.5.2 Management at the Applied Level

This chapter presents three sets of concepts (related to problem types, environment types, and uncertainties) and five processes dealing with uncertainties (general adaptation strategies and tactics, risk assessment and management, health impact assessment, the precautionary principle, adaptive environmental assessment and management). An adaptive EIA process could, as illustrated in Figure 10.6, combine elements from some or all of these concepts and processes.

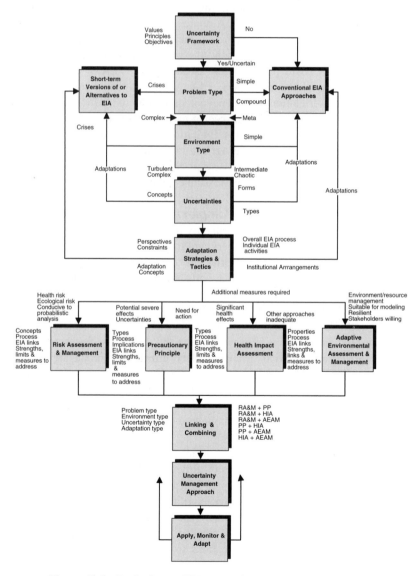

Figure 10.6 Designing an EIA uncertainty management approach.

The process first determines whether a risk and uncertainty management framework is warranted. An analysis of the problem is undertaken next. A conventional EIA approach is applied if the problem and environmental conditions appear simple and manageable. If the problem and/or environment are determined to be crises, then abbreviated versions of or alternatives to EIA are instigated. Uncertainties are considered regardless of the approach adopted. Uncertainty is a central feature of the process if a complex or metaproblem are involved and/or if environmental conditions are complex or chaotic. Pertinent uncertainty forms and sources are identified and analyzed. Uncertainty concepts are applied where appropriate. An adaptive EIA process is designed. Uncertainties are considered for every EIA activity and for every interconnection among activities. The process is iterative, flexible, heuristic, open, continuous, cyclical, interactive, and boundary spanning. Institutions are modified and reformed to facilitate and accommodate adaptive EIA processes.

Consideration is given to whether one or more of (1) risk assessment and management (RAM), (2) the precautionary principle (PP), (3) health impact assessment (HIA), and (4) adaptive environmental assessment and management (AEAM) should also be applied. RAM is more suited to situations where human health and/or ecological risks are major concerns and uncertainties are conducive to probabilistic analyses. The PP is usually instigated when there are potentially severe or irreversible adverse environmental impacts, a need for action, and significant scientific uncertainties. HIA is undertaken when significant positive and/or negative health effects are likely, adequate health-related resources are available, and other approaches are unlikely to address health concerns adequately. AEAM is applied more commonly in environmental and resource management situations involving complex but not unique situations, where ecological systems are resilient and conducive to modeling and where stakeholders are willing to engage in workshops.

The characteristics, variations, procedures, methods, EIA links, strengths, limitations, and measures to reduce limitations of each approach are considered before it is decided if it should be applied. A clear rationale is prepared for whether the approach is to be applied, and if so, how it is to be linked to or combined with EIA. A rationale also is provided for how uncertainty management approaches are combined when more than one approach is used and how the approach(es) are matched to the relevant problem, environment, and uncertainty types. The overall uncertainty management approach integrates all elements into a coherent whole. The approach is applied, monitored, and adapted as needed. Several iterations are required before the approach is finalized. Provision is made for early and ongoing stakeholder involvement in the host of uncertainty management interpretations and judgments.

10.6 ASSESSING PROCESS EFFECTIVENESS

Table 3.5 (in Chapter 3) presents criteria and indicators for assessing the positive and negative tendencies of the EIA processes described in Chapters 3 to 10. In this section we summarize the positive and negative tendencies of adaptive EIA processes.

Risk assessment and adaptive environmental assessment and management methods are consistent with scientific *rigor*. Chaos and complexity theory are derived from science. Adaptive EIA processes integrate scientific knowledge but recognize the limits of science. They generally make assumptions, findings, interpretations, conclusions, and recommendations explicit. Some PP versions define the role of scientific knowledge narrowly. Contributions to applied problem solving are usually a higher priority than scientific knowledge contributions.

Adaptive EIA processes are *comprehensive* in that they define problems and opportunities broadly and generally adopt a holistic perspective. They usually consider a diversity of relevant physical, biological, social, cultural and economic effects, but they focus on uncertainties. RAM and HIA address the often-neglected concern—human health effects. RAM and sometimes AEAM can underemphasize qualitative impacts. The PP and HIA help offset this potential bias. Social, cultural, and economic effects can be secondary considerations in RAM (except for perceived risks), in HIA (unless health is defined broadly), and in AEAM. RAM, HIA, and AEAM all trace through webs of interrelationships.

Adaptive EIA processes are *systematic*. They correct the tendency of conventional approaches not to address uncertainties systematically. Their broad definitions of problems and environments open up the process to a wide-ranging analysis of options and effects. The stress on monitoring and iterative processes contributes to more grounded decision making, both prior and subsequent to approvals. The focus on identifying and managing uncertainties can result in the neglect of other EIA objectives, methods, and decision-making considerations.

The PP (severe environmental harm), RAM (human and ecological risk), HIA (human health effects), and AEAM (ecological integrity and resilience) are all environmentally *substantive*. Each seeks to avoid, prevent, and minimize human and ecological risks and hazards. General uncertainty management approaches also seek to "hedge away" from severe adverse environmental consequences. RAM, HIA, and AEAM (to a lesser extent) sometimes stress expert over nonexpert knowledge. Some PP versions can create the opposite imbalance. HIA is the only adaptive approach that encompasses environmental benefits and positive actions. Overly conservative estimates and risk and harm interpretations can result in the rejection of proposals with major net benefits and only limited uncertainties. Environmental quality is generally maintained and protected with adaptive EIA processes. Sustainability is generally addressed only indirectly, although arguably the PP is a prerequisite to sustainability.

Adaptive EIA processes are *practical* when they focus on critical environmental components, interactions, effects, and uncertainties. They are realistic about knowledge limits. RAM methods are well established, but the field remains highly controversial. Experience with the PP, HIA, and AEAM is limited and mixed. The more comprehensive forms of RAM, HIA, and AEAM can be costly and time consuming. The effective management of uncertainties often leads to major cost savings. The priorities of adaptive EIA processes tend to correspond with those of regulators and the public. Organizations often resist adaptation and are sometimes reluctant to engage in processes that admit uncertainties, treat errors as opportunities, and revisit decisions repeatedly. The limited experience with applying the PP and the fear of its abuse has led to both opposition and a propensity to add multiple conditions.

Adaptive EIA processes both facilitate and sometimes inhibit *democratic local influence*. Stakeholders, consistent with adaptive processes, are often skeptical of claims of certainty, predictive accuracy, and mitigation effectiveness. Human health and ecological risks are commonly major public concerns. The public tends to define problems broadly and is highly sensitive to the possibility of human and natural disasters. The PP and HIA (if it is defined broadly) can accommodate traditional perspectives and knowledge. The emphasis on perceived risks, two-way risk communications, and multiple risk management perspectives can further local influence. AEAM provides a forum for stakeholder involvement. Adaptive EIA processes do not generally advance local influence proactively. Political implications are rarely a priority. The PP comes closest to raising local concerns to a level where they become a major factor in determining the fate of a proposed action. The PP is more likely to be conducive to stakeholder acceptance (i.e., taking the public's fears and concerns seriously). The other uncertainty management approaches can, on occasion, inhibit local influence by relying largely on expert knowledge and opinion or by limiting public influence to access and involvement.

Adaptive EIA processes can be *collaborative*. They stress, consistent with many stakeholders, anticipating and managing uncertainties and risks, broadly defining the problem, and an EIA process that is open, responsive, and flexible. Adaptive EIA processes share the public's skepticism of the ability of proponents and governments to predict and manage environmental change. Collaboration can be furthered by joint planning forums (as occurs with AEAM), with serious consideration of public risk perceptions, risk communications, and multiple risk management perspectives (as can occur with RAM), with a shift in the burden of proof toward proponents and regulators (as can occur with PP) and with a broadly defined and comprehensive treatment of health concerns (as can occur with HIA). Collaboration can be inhibited if expert knowledge is overemphasized. Consensus building and conflict resolution techniques are not usually well developed with adaptive EIA processes.

Procedural and distributional fairness are rarely adaptive EIA process priorities. Fairness and equity can be considered indirectly because avoiding catastrophic and severe environmental and health consequences are usually public priorities. By shifting the burden of proof, the PP helps ameliorate or correct (depending on the PP interpretation adopted) resource inequities, which currently favor proponents and regulators. A focus on risk perceptions can help offset an expert interpretation bias. Adaptive EIA processes, however, are seldom guided and shaped by *ethical* imperatives or standards. Rights and responsibilities usually receive limited consideration. Ethical issues, implications, trade-offs, and dilemmas are rarely addressed. Social and environmental fairness, equity, and justice concerns are not considered from multiple perspectives.

Adaptive EIA processes are, by definition, *adaptable*. They anticipate and respond rapidly to changing circumstances. Problems and opportunities are broadly defined. Design, ingenuity, and creativity help escape the boundaries of conventional thinking and problem solving. Adaptive processes test multiple assumptions, scan ahead, and reconsider past decisions. They are highly sensitive to changing and evolving environmental conditions. They consider risks and uncertainties systematically. More comprehensive RAM and HIA forms can be cumbersome and less adaptable. The focus on uncertainties can inhibit creative approaches to managing more certain impacts.

Adaptive EIA processes focus on difficult problems in complex environments. They accept knowledge limits. Consequently, they foster the *integration* of diverse values, forms of knowledge, perspectives, and ideals. AEAM stakeholder forums can reinforce this tendency. Too much stress on disciplinary expertise, as can occur with RAM and with AEAM, can inhibit integration. Concepts such as strategic choice, consilience, and holistic science emphasize links to related decisions and the need to span boundaries among disciplines, professions, and EIA types. RAM, AEAM, the PP, and HIA all explore patterns of interconnections that transcend conventional categories and operate in related forms of environmental management. The social side of adaptive EIA processes is not as well developed. The four adaptive process approaches (RAM, PP, AEAM, and HIA) all receive more attention outside than within EIA practice. Links between these approaches and the EIA process should be considered further. Integration with more certain EIA aspects also requires more attention.

10.7 SUMMING UP

In this chapter we address how EIA processes anticipate and respond adaptively to the uncertainties associated with difficult problems in chaotic and complex environments. The two stories illustrate two aspects of uncertainty management. In the first story a collaborative approach helped manage the uncertainties stemming from both process flaws and circumstances that could not be anticipated. In the second story, uncertainties regarding jurisdiction necessitated both adaptive strategies within defined structures and procedures and a redefinition of formal and informal institutional arrangements to respond to unique project-related circumstances. EIA uncertainty management involves many more considerations than are portrayed in the stories.

The problem is a combination of confusion regarding the nature of uncertainty, risk, and health effects and ambivalence concerning the most appropriate approach, or combination of approaches, for managing EIA process uncertainties. The direction involves an enhanced understanding of uncertainty, difficult problems, chaotic and complex environments, and adaptation coupled with a selective blending of general adaptation strategies and tactics, risk assessment and management, the precautionary principle, human health impact assessment, and adaptive environmental assessment and management. Table 10.6 is a checklist for formulating, applying, and assessing an adaptive EIA process.

Table 10.6   Checklist: An Adaptive EIA Process

*The response to each question can be yes, partially, no, or uncertain. If yes, the adequacy of
the process taken should be considered. If partially, the adequacy of the process and the
implications of the areas not covered should be considered. If no, the implications of not
addressing the issue raised by the question should be considered. If uncertain, the reasons
why it is uncertain and any associated implications should be considered.*

Uncertainty

1. Is uncertainty defined clearly and applied consistently?
2. Are the different forms of uncertainty identified?
3. Are the sources of uncertainty identified?
4. Are relevant uncertainty concepts applied?

Difficult Problems

1. Are the characteristics of the problem specified?
2. Is the EIA process designed to match the problem characteristics?
3. Does the process take into account, where relevant, the special characteristics of difficult
   problems?

Chaotic and Complex Environments

1. Are the characteristics of the environment specified?
2. Is the EIA process designed to match the environmental characteristics?
3. Does the process take into account, where relevant, the special characteristics of chaotic
   and complex environments?

Adaptability in the EIA Process

1. Is uncertainty viewed as a fundamental attribute of the EIA process?
2. Are uncertainties recognized and addressed in:
   a. Problem definition?
   b. Scoping?
   c. Proposal(s) characteristics?
   d. Option identification?
   e. Option evaluation?
   f. Baseline analysis?
   g. Impact analysis?
   h. Impact management?
   i. Uncertainty management?
   j. Public and agency consultation?
   k. Study team management?
   l. Decision making?
   m. EIA institutional arrangements?
3. Are general adaptation concepts and methods incorporated into the EIA process, where
   relevant?

Risk Assessment and Management

1. Is risk defined and applied consistently?
2. Are the different types of risk taken into account?

(Continued)

Table 10.6   (*Continued*)

3. Is the situation appropriate for the application of RAM?
4. Are risk concepts incorporated into the EIA process, where relevant?
5. Is a RAM process described?
6. Are RAM and EIA systematically linked and, where appropriate, integrated, taking into account the similarities and differences between the two fields?
7. Are the strengths and limitations of RAM taken into account?
8. Are effective steps taken to prevent and reduce the limitations of RAM?

Human Health Impact Assessment

1. Is human health defined and applied consistently?
2. Are the objectives of HIA recognized?
3. Is the situation appropriate for the application of HIA?
4. Is the appropriate form of HIA applied?
5. Is an HIA process described?
6. Are HIA and EIA linked systematically and, where appropriate, integrated, taking into account the similarities and differences between the two fields?
7. Are the strengths and limitations of HIA taken into account?
8. Are effective steps taken to prevent and reduce the limitations of HIA?

The Precautionary Principle

1. Is the PP defined and applied consistently?
2. Are the varying interpretations of the PP taken into account?
3. Is a clear rationale provided for the interpretation selected?
4. Are the implications of the PP considered?
5. Is the situation appropriate for application of the PP?
6. Is a clear rationale for applying or not applying the PP in the EIA process presented?
7. Is a precautionary EIA process presented, where appropriate?
8. Are the strengths and limitations of the PP taken into account?
9. Are effective steps taken to prevent and reduce the limitations of the PP?

Adaptive Environmental Assessment and Management

1. Are the properties of AEAM identified and taken into account?
2. Is the situation appropriate for the application of AEAM?
3. Is an AEAM process described?
4. Are AEAM and EIA systematically linked and, where appropriate, integrated, taking into account the similarities and differences between the two fields?
5. Are the strengths and limitations of AEAM taken into account?
6. Are effective steps taken to prevent and reduce the limitations of AEAM?

Management at the Regulatory Level

1. Do EIA requirements and guidelines facilitate the systematic identification and integration of uncertainty concepts, methods, and procedures into the EIA process?
2. Do the EIA requirements and guidelines facilitate the systematic consideration of human health and ecological risks, including the potential application of risk assessment and management, where appropriate?

Table 10.6 (*Continued*)

3. Do the EIA requirements and guidelines facilitate the systematic consideration of positive and negative health effects, including the potential application of HIA, where appropriate?
4. Do the EIA requirements and guidelines facilitate the systematic application of the precautionary principle, where appropriate?
5. Do the EIA requirements and guidelines facilitate the systematic application of AEAM, where appropriate?
6. Do the EIA requirements and guidelines facilitate the systematic identification and management of difficult problems?
7. Do the EIA requirements and guidelines facilitate undertaking EIA in complex and chaotic environments?

Management at the Applied Level

1. Is the EIA process established within the context of an uncertainty framework?
2. Is a clear rationale provided for whether a conventional EIA approach, a crisis-driven EIA approach, or an uncertainty-based EIA approach is applied?
3. Is the applicable problem type identified and addressed systematically through the EIA process?
4. Is the applicable environment type identified and addressed systematically through the EIA process?
5. Are applicable uncertainty forms, types, and concepts identified and addressed systematically through the EIA process?
6. Are applicable general adaptation strategies and tactics identified and applied systematically through the EIA process?
7. Is RAM applied in the EIA process, where applicable, and in an appropriate manner?
8. Is the PP applied in the EIA process, where applicable, and in an appropriate manner?
9. Is HIA applied in the EIA process, where applicable, and in an appropriate manner?
10. Is AEAM applied in the EIA process, where applicable, and in an appropriate manner?
11. Are the various uncertainty management concepts and methods linked effectively and, where appropriate, combined?
12. Is an overall uncertainty management approach formulated?
13. Is the uncertainty management approach, applied, monitored, and adapted, as needed?

Process Effectiveness

1. Does the process balance adaptability with an adequate level of applied scientific performance?
2. Is the process broadened to encompass risk, uncertainty, and health concerns without neglecting other environmental components and considerations?
3. Does the process systematically address uncertainty-related concerns without neglecting other EIA considerations?
4. Does the process fully address substantive human health and ecological effects, risks, and uncertainties while adequately addressing other substantive environmental concerns?
5. Is the process practical in its treatment of uncertainties, risks, and health effects, without neglecting other practical concerns?
6. Does the process facilitate local influence in identifying and addressing uncertainties, risks, and health effects without inhibiting local influence in other EIA matters?
7. Is the process collaborative in its treatment of uncertainties, risks, and health effects, while allowing for collaboration on other relevant matters?

(*Continued*)

Table 10.6   (*Continued*)

---

8. Does the process integrate ethical concerns into the treatment of uncertainties, risks, and health effects and still address ethics in other EIA-related matters?

9. Does the process creatively and promptly anticipate, and adapt to the problem, the environment, and changing circumstances?

10. Is the process effectively linked to proposal planning, related decisions, and related management forms?

11. Does the process effectively integrate diverse values, forms of knowledge, perspectives, and ideals?

12. Does the process effectively span and transcend disciplines and EIA types?

Uncertainty, broadly defined, is any situation where we are not absolutely sure. There are many uncertainty forms (e.g., quantitative, qualitative, objective, subjective, methodological, perceived, knowledge, values, past, present, or future conditions). Uncertainties can pertain to any EIA process activity. There are many possible uncertainty sources (e.g., data or knowledge deficiencies, theoretical or methodological deficiencies, resource limits, poor communications, natural variations, novel situations). Several uncertainty-related concepts are potentially relevant (e.g., ignorance, errors, indeterminism, vagueness, ambiguity, doubt, confusion, surprise).

Problems are triggered by a question or a situation, are negative, and need to be addressed. Problem-solving processes identify, define, bound, and state the problem. The problem is then refined and addressed progressively. There are simple or tame problems, compound or semistructured problems, complex or ill-structured problems, and crises or metaproblems. Simple and compound problems can be addressed by routine and conventional EIA procedures, respectively. Complex and metaproblems are more difficult. They are real, complex, messy, transcend boundaries and disciplines, are prone to dilemmas, impossibilities, and crises, and require ingenuity. An adaptive EIA process is needed to cope properly with difficult problems.

EIA processes should suit the *environment* or context. There are many environmental components or systems (e.g., ecological, social, economic, institutional, technological). There are simple, moderately complex, and highly chaotic and/or complex environmental systems. Command and control and conventional EIA processes, respectively, operate effectively in simple and moderately complex environments. Chaotic or complex environmental systems are more problematic. They exhibit such properties as self-organizing, emergent, turbulent, nonlinear, irreducible, random, incoherent, unpredictable, interdependent, resilient, and unstable. Adaptive EIA processes and organizations can operate more effectively in chaotic or complex environments.

There are many ways of addressing *uncertainty in the EIA process*. A perspective change is first required (e.g., uncertainty as a fundamental attribute of the process). Measures can be introduced into each EIA activity to anticipate, cope with, learn from, and manage uncertainties. Uncertainty management measures can be integrated into problem definition, scoping, proposal characteristic determination, option identification and evaluation, individual and cumulative impact identification, prediction and interpretation, mitigation and compensation, impact and uncertainty management, public and agency consultation, study team management, decision making, monitoring, and EIA institutional arrangement reform. Uncertainty management is facilitated by insights and lessons from design, ingenuity, creativity, strategic choice, consilience, and holistic science.

Risk combines frequency or probability with a harmful environmental consequence. There are many risk types (e.g., economic, health, environmental, from natural or human sources, chronic, acute, for overall and for sensitive populations, deterministic, and probabilistic). Potentially relevant risk-related concepts include risk assessment, perceived risks, risk communications, comparative risk assessment, risk acceptability or tolerance, risk management, disasters and hazards, human health risk assessment, and ecological risk assessment. Risk assessment processes include, for example, problem and analysis plan formulation, receptor determination, pathway and receptor characterization, hazard identification and analysis, exposure assessment and response, risk characterization, risk evaluation, decision making, implementation, and monitoring. Risk assessment processes integrate research, public perceptions, stakeholder concerns and preferences, option analyses, mitigation measures, and uncertainty analyses. There are many similarities but also important differences between EIA and risk assessment and management. Similarities, differences, and risk assessment and management strengths, deficiencies, and measures to address deficiencies should all be considered when linking and integrating EIA and risk assessment and management.

Human health impact assessment (HIA) is concerned with positive and negative, certain and uncertain human health effects. HIA is closely connected with other types of EIA, draws on an interdisciplinary knowledge base, and can contribute to sustainability. It assumes many forms (e.g., quick screening, rapid appraisal, standard HIA, comprehensive HIA). It can be prospective, retrospective, or concurrent. It can be broadly or narrowly defined. HIA processes tend to begin with screening, scoping, a background analysis, and a prognosis of future health-related environmental conditions. Health effects associated with options and before and after mitigation are predicted, summarized, and evaluated. HIA is supported by quantitative (e.g., risk assessment) and qualitative (e.g., health impact assessment) procedures. The health and risk analyses provide the basis for management measures, documentation, conclusions, recommendations, and decision making. Results are monitored and evaluated. Agencies and the public are heavily involved in the process. Health effects are addressed by broadening EIA, merging EIA and HIA, selectively integrating EIA and HIA, or broadening risk assessment and management. HIA is a newly emerging field of EIA practice. It has many attributes, strengths, and limitations that should be considered carefully, especially its uncertainty management procedures.

The *precautionary principle* responds to the dilemma of what to do when there is a need to take action because of potentially severe environmental consequences but shortcomings in the scientific knowledge base. There are multiple interpretations of what represents severe harm potential, inadequate scientific evidence, and the basis for action (e.g., inaction not justified, rejection of proposal, only proceed if proven safe, proceed if reasonable case can be made, proceed with caution). Applying the precautionary principle requires thresholds, criteria, decision rules, definitions for key terms, and institutional arrangements. Some argue that the principle also requires a reversed burden of proof; open, transparent, and democratic decision making; systematic alternatives analyses; and greater decision-making weight on prevention, risk avoidance, ignorance, and environmental values. A precautionary EIA process involves screening (whether the principle is to be applied), scoping, goal setting, study design, an analysis of need and alternatives, adaptations to the principle to suit the situation, refining and applying precautionary thresholds, criteria and procedures, precautionary decision making, the taking of precautionary action, implementation, monitoring, follow-up, and evaluation, all within an open, transparent, and democratic EIA process. The precautionary principle is highly controversial but addresses a valid concern. Ascribed strengths and drawbacks need to be considered carefully. A clear rationale should be presented for if and how the principle (or an alternative approach) is applied in EIA practice.

Adaptive environmental assessment and management (AEAM) treats environmental management as a quasiexperiment (i.e., probing ecosystem responses to human activities). The AEAM process is an iterative cycle of planning, implementation, monitoring, research, and reexamination. AEAM processes are typically built around a series of workshops. The workshops construct and apply a model that characterizes critical environmental conditions and interactions and tests possible management actions and alternative assumptions. The periods between workshops are devoted to consolidation and refinement. The process is guided by a core group and by specialist support staff. Workshops involve policy people, managers, and a diversity of stakeholders. Key indicators are monitoring throughout implementation. Data obtained during monitoring are analyzed, documented, and fed into each process activity. The process is open, continuous, cyclical, evolving, and highly iterative. AEAM has much to offer EIA, but there also are important differences between the two fields. These differences and AEAM strengths and shortcomings should be considered carefully when connecting or integrating EIA and AEAM.

The *four jurisdictions* (the United States, Canada, the European Union, Australia) address many aspects of uncertainty. More guidance could be provided concerning uncertainty forms and sources, uncertainty concepts, the role of uncertainty in the EIA process, and uncertainty management methods. All four jurisdictions mention health risks in EIA requirements. More attention could be devoted to risk in the EIA process, especially chronic health risks, ecological risks, and perceived risks. The relationship between risk assessment and management and EIA also could receive further consideration. Health effects are a concern but receive general and fragmentary treatment in the four jurisdictions. HIA initiatives should help ameliorate this deficiency. Links between HIA and EIA and the effectiveness of HIA requirements and guidelines should receive more attention. The jurisdictions vary greatly in if and how the precautionary approach or principle is addressed in environmental requirements. More specific requirements and guidelines are needed regarding if and how the principle could be applied in EIA practice. EIA requirements could include more explicit provisions for applying AEAM, appreciating the differences between the two fields. EIA guidelines could address potential interrelationships between EIA and AEAM. The four jurisdictions could devote more attention to relationships between EIA and difficult problems and complex and chaotic environments.

Uncertainty management at the applied level involves selectively combining concepts (related to problem types, environment types, and uncertainties) and approaches (general adaptation strategies and tactics, risk assessment and management, health impact assessment, the precautionary principle, adaptive environmental assessment and management), within an adaptive EIA process. Designing an EIA uncertainty management approach entails formulating an uncertainty framework (to identify relevant values, principles, and objectives), identifying the applicable problem and environment type (to determine the appropriate EIA approach), characterizing uncertainties, formulating and applying general adaptation strategies and tactics, determining whether and how risk assessment and management, the precautionary principle, health impact assessment, and adaptive environmental assessment and management could be used to manage risks, uncertainties, and health effects, linking and combining the concepts and approaches, formulating an overall uncertainty management approach, and applying, monitoring, and adapting the approach.

Adaptive EIA processes integrate most forms of scientific knowledge and methods. They recognize science limits but are more concerned with problem solving than with scientific knowledge contributions. They broadly define problems and opportunities, address the valid issues of health effects and human and ecological risks and uncertainties, and consider interrelationships systematically. Other types of effects and more certain consequences receive less attention. Adaptive EIA processes focus on practical concerns relevant to avoiding and reducing severe environmental consequences and related effects. Adaptive methods are sometimes cumbersome, have a mixed record, and are often resisted by organizations. Stakeholders frequently share adaptive EIA process perspectives. Adaptive approaches, although generally supportive of collaboration and local influence, provide little specific consensus building and conflict resolution direction. Fairness, equity, and other ethical concerns are considered only selectively and indirectly. Adaptive EIA processes are generally flexible and often creative, especially in managing uncertainties. They focus on interrelationships and span boundaries among disciplines and between EIA and other environmental management forms. Adaptive approaches are more fully developed outside than within EIA practice.

CHAPTER 11

HOW TO CONNECT AND COMBINE EIA PROCESSES

11.1 HIGHLIGHTS

In this chapter we explore how the individual EIA processes, presented in Chapters 2 to 10, might be connected and combined, at both the regulatory and applied levels. It also identifies residual challenges and future action priorities.

- The analysis begins in Section 11.2 with two applied anecdotes. These stories describe applied experiences associated with efforts to balance multiple perspectives and demands.
- The analysis in Section 11.3 then defines the problem, which is how to adapt, connect, and combine EIA processes to suit the situation. The direction is frameworks and procedures for matching EIA processes and contexts and for connecting and integrating EIA processes at the regulatory and applied levels.
- In Section 11.4 we explore how EIA legislation, regulations, and guidelines could be reformed and refined to better address the regulatory deficiencies and opportunities described in Chapters 2 to 10.
- In Section 11.5 we identify EIA process attributes likely to affect the match between process and context. It describes ways of facilitating the fit between process and context.

Environmental Impact Assessment: Practical Solutions to Recurrent Problems, By David P. Lawrence
ISBN 0-471-45722-1   Copyright © 2003 John Wiley & Sons, Inc.

- In Section 11.6 we identify links, overlaps, and middle-ground concepts between pairs of EIA process types. It describes ways of considering interconnections between EIA process types.

- In Section 11.7 we present examples of how composite EIA processes could be formulated and applied.

- In Section 11.8 we identify residual challenges that require further attention and could inhibit efforts to enhance EIA process management. Priorities for future action are also identified.

- In Section 11.9 we revisit the not-so-hypothetical scenario presented in Chapter 1. We consider whether the analyses presented in Chapter 2 to 10 could help prevent and ameliorate the types of problems that arose in the scenario.

- In Section 11.10 we provide an overview of the major insights and lessons derived from the analysis. A summary checklist is provided.

11.2 INSIGHTS FROM PRACTICE

11.2.1 Power, Sustainability, and Adaptation Within the Context of Environmental Conflict Resolution*

The Voisey's Bay case, in the Canadian province of Newfoundland and Labrador, concerns a proposed mine and mill project that would exploit an exceptionally rich nickel–copper–cobalt ore body and is located conveniently near tidewater. The project proponent is Voisey's Bay Nickel Company, a subsidiary of Inco Ltd. The government of Newfoundland and Labrador is the main government player, since provincial constitutional authority covers crown lands and most associated resources. The government of Canada carries federal constitutional authority for harbors, coastal waters, fisheries, trade, and commerce. It has tax and subsidy powers that can have a major influence on project decision making, and it has fiduciary responsibility for native people. Both the Inuit (traditionally, coastal people) and the Innu (hunters of the interior) have used and occupied the Voisey's Bay area. It remains Innu and Inuit land, at least inasmuch as aboriginal title is recognized, since neither people ever signed away title to these lands.

The Voisey's Bay ore body was discovered in 1993. The EA planning and decision-making process extended over five years. It began in January 1997 with a Canada, Newfoundland/Labrador, Innu Nation, and Labrador Inuit Association Memorandum of Understanding for a joint environmental assessment and review under the direction of an appointed independent panel, featuring public hearings

---

*A more detailed case study paper concerning this project was presented at the conference *Towards Adaptive Conflict Resolution: Lessons from Canada and Chile*, Liu Centre for the Study of Global Issues, University of British Columbia, Vancouver, September 25, 2002.

with intervener funding. The panel guidelines introduced a contribution to sustainability test, requiring the proponent not just to avoid and minimize adverse environmental effects but also to make a positive contribution to local and regional ecological and community sustainability. In March 1999, the panel recommended acceptance of the project but included 107 recommendations. The most significant of these recommendations concern extending the lifespan of the project (to establish lasting benefits), to complete land claims negotiations first, and to ensure that specific agreements were reached with aboriginal groups on project impacts and benefits and on co-managing environmental reviews during project implementation.

Subsequent negotiations and court challenges extended over an additional three years, until June 2002, when the main agreements were signed and ratified. A key agreement between the provincial government and the proponent concerning facility characteristics and operations extended the predicted life span of the agreement to greater than 30 years, an adjustment intended to facilitate community capacity building and continued viability after mining ended. Also important were impacts and benefits agreements between the company and the Innu Nation and the Labrador Inuit Association pertaining to revenue sharing, local employment and contracting, training programs, and community roles in the review of project implementation. Associated environmental co-management agreements established a joint body (two representatives from each party) to monitor project effects, to review new and to-be-specified project-related actions including tailings and waste rock disposal options, and to recommend necessary adjustments. Together, these agreements address a diversity of sustainability-related concerns and should ensure a continuing flow of benefits. Initial project construction is now under way.

The Voisey's Bay assessment and decision-making process was successful for a combination of reasons. The Innu and Inuit made effective use of multiple strategies and venues (e.g., negotiations, media events, court actions, site occupations) to become powerful players in the project planning and decision making. The proponent employed an iterative project design that integrated corporate planning, EIA-related deliberations, biparty and multiparty negotiations and wider deliberations in the media, the financial world, government bodies, and the courts. Especially because of its sustainability-centered evaluation criteria, the panel was very successful in addressing the Innu and Inuit concerns in ways that the federal and provincial governments and the proponent could accept. It was capable, independent, and operated under a mandate broad enough to encompass the full suite of issues, especially the key community concerns about the durability of benefits from exploitation of their traditional and future lands. The process also benefited from the panel's careful and explicit attention to uncertainties and risk avoidance; its evident attention to the implications of aboriginal title and rights; its use of and respect for local knowledge; the provision of funding for the Innu, Inuit, and other community and public interest participants in the hearings; and the extensive set of recommendations specifying appropriate project approval conditions addressing communities' concerns.

The Voisey Bay case underscores the importance of ensuring that Indigenous people (and other interests whose voices are relevant but traditionally underrecognized) have substantial negotiating power in the planning and decision making. It demonstrates the advantages of broadly scoped planning and assessment that require integrated attention to social, cultural, economic, and ecological aspects covering the project's full life cycle and influence on subsequent conditions. The case also points to the value of processes for continued iterative deliberation and review throughout the project life. The Voisey EIA process combines elements of each of eight themes. The process was collaborative but also entailed a shift in political power. It addressed and advanced substantive economic, social, and ecological concerns but within the constraints of what was possible and practical for available planning and decision-making structures and procedures. The process was supported by scientific, technical, and rational analyses, including local knowledge, but also was sufficiently flexible to adapt to changing and emerging needs, concerns, and priorities. The process addressed both procedural equity (e.g., through participant funding) and substantive equity (e.g., through the impacts and benefits agreements) concerns.

ROBERT B. GIBSON
Environment and Resources Studies, University of Waterloo

11.2.2   The Effective Marriage of Process and Substance*

Ski fields are confined to the southeastern corner of Australia. Within the state of Victoria there are six designated alpine resorts. One of these, Mt. Stirling, lies about 240 kilometers north of Melbourne (the capital of Victoria, with a population of about 4 million), and is adjacent to a highly developed downhill ski resort at Mt. Buller. Although Mt Stirling, with an elevation of 1746 meters, is used in the winter for cross-country (Nordic) skiing and in the summer for a variety of nature-based and educational pursuits, it is relatively undeveloped. For many years it was government policy that Mt. Stirling would be developed for downhill skiing in response to demand pressures.

In March 1994, the then Victorian Minister for Natural Resources entered into a preliminary agreement with Buller Ski Lifts Ltd. for the construction of a cable car linking Mt. Buller to Mt. Stirling and the development of downhill skiing on the southern slopes of Mt. Stirling. The wide public controversy over this action came to a head when a lessee served a supreme court injunction on the government (the Alpine Resorts Act required consultation with lessees, which had not been undertaken). In response, the government committed to an environment effects statement (EES) on the future development of Mt. Stirling, to be prepared by the Environment Assessment Branch within the Department of Planning.

---

*A fuller treatment of the Mt. Stirling case study can be found in the paper by Robin Saunders and Ashley Stephens, "Mount Stirling: Political and Environmental Convergence for Sustainable Development," *Environmental Impact Assessment Review*, Volume 19, No. 3, May 1999.

A consultative committee comprising representatives from all principal stakeholders and interest groups (government agencies, the local council, tourist operators, and community groups representing downhill skiing, Nordic skiing, the community, and environmental interests) was established to guide the preparation of the EES. Management of the consultative committee, with members holding widely divergent views, required considerable conflict management skills. A further series of community groups, interested in such activities as bush walking, land care, four-wheel-drive touring, and other forms of nature-based tourism, received all committee reports, minutes, and papers and were invited to observe meetings, make presentations, and ask questions. The Environment Assessment Branch prepared a detailed study brief, which set out the scope of the EES and called for the examination of a range of alternatives. The consultative committee worked through the brief in detail until consensus was gained. Several members of the consultative committee were involved in the consultant selection process. Throughout the study, there was an emphasis on extensive public consultation using a diversity of techniques.

A framework for evaluating alternatives, based on sustainable development principles, was developed. The evaluation of the six alternatives (three involving downhill skiing, the others having more emphasis on Nordic skiing and nature-based activities) involved the detailed study of landform stability, water resources, future global warming, ecological values, visual impacts, aboriginal and European heritage values, existing and potential future uses, social impacts on the local community, and economic viability. The alternatives were evaluated systematically against objectives framed for each topic.

After public exhibition and submissions, an independent panel of inquiry appointed by the Minister for Planning held public hearings. Over 600 submissions were received and over 60 groups and individuals made personal presentations to the panel. The panel concluded that there was no demonstrated need for downhill skiing at Mt. Stirling, that downhill skiing would prejudice water quality and ecosystem quality and diversity while requiring considerable public-sector investment, and that nature–based tourism development was a more appropriate form of development. The government accepted the inquiry recommendations, which were welcomed and supported publicly by all interest groups.

The EES process resolved the controversy about the future of Mt. Stirling through the use of a democratic and collaborative process, by examining independently all aspects of the environment (including social and economic factors) and by systematically evaluating alternatives against broad objectives. It resulted in the protection of Mt. Stirling from the environmental degradation that would have resulted from large-scale downhill ski development. This story describes an EIA process that integrates elements of all eight EIA process types. It was structured around a form of collaborative planning, with a high degree of community influence. It was rigorous and independent. It was supported by the systematic and rational analysis of a broad array of alternatives. It operated within practical decision-making constraints and procedures. It consciously sought to identify and advance environmental and sustainability values. It fostered procedural equity and sought to maintain and enhance both distributive (decision-making outcomes) and relational (maintaining social relations) justice. It represents a potentially useful model for applying SEA principles in difficult situations where government must decide about the future of a sensitive environmental resource, there are strong conflicting interests and there is public mistrust about decision-making integrity.

ROBIN SAUNDERS
Robin Saunders Environmental Solutions Pty Ltd.

11.3   DEFINING THE PROBLEM AND DECIDING ON A DIRECTION

The two stories illustrate that it is possible in practice to effectively integrate aspects of all eight EIA processes into a successful EIA process. Success is defined in terms of a satisfactory (to all parties) planning and decision-making process and net environmental benefits, not in terms of proposal approval. Although it appears that both stories offer insights that could have broader application, there is always the possibility that the satisfactory outcomes are largely the result of unique local circumstances. Before it is possible to assert that either, or the two in combination, represent a model for EIA process management, it is necessary to scrutinize more carefully the various ways in which the EIA process types can be linked, integrated, and transcended.

This book identifies (in Chapter 2) numerous choices for managing conventional EIA processes. It then presents (in Chapters 3 to 10) eight different EIA processes, each responding to a different constellation of recurrent problems often encountered in EIA practice. Chapters 2 to 10 provide an array of potentially valuable EIA process management tools. But they do not address overall process management regulatory roles, when to apply the tools (i.e., matching process to context), how to deal with overlaps among EIA process types (i.e., process interconnections), and how to respond when there are multiple problems (i.e., formulating composite EIA processes).

More specifically, it is necessary to consider (1) how far to go, at the regulatory level, in directing and guiding EIA process management (given the range of processes and contexts); (2) how to match process and contextual characteristics; (3) how to identify and deal with interconnections between EIA process types; (4) how to design and manage composite EIA processes; and (5) how to cope with remaining challenges and to establish priorities.

11.4   COMPOSITE REGULATORY FRAMEWORKS

EIA regulatory practitioners are engaged in a delicate balancing act. They define, through legislation and regulations, minimum levels of adequate EIA practice. EIA legislation and regulations also contain broad goals and principles to provide a rationale for requirements and a direction for current and future regulatory and applied practice. Both minimum standards and ideal characteristics change and evolve. EIA guidelines and applied research help ensure that requirements are achieved. Because of their greater flexibility, they also contribute to practice levels that often exceed the minimum, thereby narrowing the gap between the adequate and the ideal. EIA requirements and guidelines need to be neither too general (which could result in a low and inconsistent level of practice) nor too precise (which could unduly restrict and limit adaptation and innovation). EIA requirements operate within the context of a complex set of related environmental requirements, policies, and objectives, a rapidly evolving field of theory and practice, and a multiplicity of ecological, social, economic, cultural, and political conditions, events, patterns, trends, and uncertainties.

How then should the EIA process be addressed as part of this balancing act? The EIA process should not be ignored by concentrating exclusively on document content and on administrative procedures. The EIA process provides the framework for conducting all EIA activities and for applying all EIA methods. A poorly designed and executed EIA process (a circumstance that occurs all too frequently in practice) can readily undermine individual EIA activities and methods. EIA documents should be outputs from the EIA process rather than ends in themselves. In Chapter 2 we demonstrate that a single standardized EIA process is a dangerous myth. This does not preclude identifying core EIA process attributes, if considerable discretion is left to EIA process participants in choosing among process management choices. Guidelines could provide participants with a sense of the range of potentially appropriate EIA process choices.

Chapter 2 offers an initial sense of core EIA process attributes that could be incorporated into EIA requirements and guidelines. In it we describe good regulatory practice for undertaking screening (proponent-driven, action-driven, environment-driven, combinations of proponent-, environment-, and action-driven, significance determination), for conducting individual EIA activities (general, scoping, proposal characteristics, baseline analysis, proposal characteristics, impact analysis and synthesis, alternatives analysis, mitigation and enhancement, methods, documents, management, auditing, participation, review, and decision making), and for addressing interrelationships (among EIA activities, with international EIA activities, among government levels, with related governmental requirements and action, with the EIA knowledge base). In Chapter 2 we also provide an overview of available EIA process management choices. EIA requirements could specify the core EIA process elements that must be addressed in EIA documents. EIA guidelines could elaborate on good-practice EIA process management and could provide examples of available management choices.

As described above, regulatory measures to address the EIA process, although necessary, are unlikely to respond adequately to the recurrent shortcomings. In Chapters 3 to 10 we describe in detail the many measures introduced by the four jurisdictions to avoid and minimize the recurrent shortcomings. It is unclear how effective these measures have been in reducing the likelihood and severity of the shortcomings. References continue to be made to the shortcomings. Each of Chapters 3 to 10 identifies specific ways in which current regulatory approaches could be enhanced to prevent and ameliorate the recurrent problems. Numerous good practices and potential pitfalls are identified. Several relevant practice-based anecdotes are presented. A variety of approaches are described for addressing each problem at the regulatory and applied levels. EIA requirements and guidelines should be broad enough to allow for a diversity of approaches for ameliorating the shortcomings while ensuring consistency with the purpose and objectives of EIA requirements. Approaches applied in other jurisdictions should not be borrowed uncritically. Contextual factors vary greatly among jurisdictions. It is foolhardy to apply a measure simply because it has been applied elsewhere, especially if the effectiveness of the measure has not been assessed. Considerable benefits, however, are likely to accrue from jurisdictions sharing experiences and coordinating applied research efforts.

Regulators should proceed cautiously in controlling and guiding how EIA process types are linked and combined. Several examples of composite processes are described briefly later in this chapter. EIA requirements and guidelines can help ameliorate the recurrent shortcomings with individual measures and by recognizing that it often is necessary to juggle multiple, sometimes conflicting, values and approaches. Over time it could be possible to identify, from applied research, which approach combinations are best suited to which categories of situations. These patterns can be noted in EIA guidelines, while acknowledging the need to make proponent-, proposal-, environment-, and stakeholder-specific adjustments. Some variations in EIA regulatory requirements already occur, depending on proponent, action, and environment types and based on significance determinations. It is unlikely that the evidence from effectiveness reviews will be sufficiently definitive to enshrine the circumstances under which particular EIA process combinations must and must not be applied to prevent and ameliorate various mixes of recurrent problems. It may be possible for individual agencies to establish, as a policy, based on their experiences in dealing with the recurrent problems, the EIA process approaches that they will generally take for different classes of situations.

11.5 MATCHING PROCESS AND CONTEXT

The concept of a plurality of approaches or paradigms, where application choices are contingent on contextual characteristics, is well-trodden ground in planning, in the social sciences and, to some degree, in the natural sciences (Patterson and Williams, 1998; Ritzer, 1996; Rothman and Sudarshan, 1998; Schön and Rein, 1994). Alternative approaches or processes have been advanced in EIA (e.g., traditional scientific vs. AEAM, political vs. technical approaches) but are not as fully developed and are not generally matched with classes of contextual characteristics. The general thrust of the pluralistic/contingent approach is (1) to differentiate key approach characteristics, strengths, and limitations; (2) to identify contingent factors or classes of situational characteristics; and (3) to match compatible approach and situational characteristics.

A variation of the pluralistic/contingent approach is to consider both the similarities and the differences among alternative approaches or processes. The similarities or shared characteristics provide the basis for core elements. The differences provide the basis for matching approach and contextual characteristics. The eight EIA process types are based on a desire to prevent and minimize specific recurrent problems. Matching process and context therefore entails applying a scientific EIA process when scientific rigor is a priority, a rational EIA process when rationality is a priority, and so on. More specifically:

- A *rigorous* EIA process is more appropriate in situations where the environment is amenable to scientific analysis; where causal webs of direct and indirect effects can be identified, measured, predicted, and monitored; where scientific knowledge and methods are likely to make a significant contribution to decision making; and where there are adequate resources and stakeholder support for scientific analyses.

- A *rational* EIA process is well suited to stable systems with well-defined proposals, options, and effects. The systematic screening and comparison of multiple options is a priority. It should be possible to aggregate preferences and effects, either quantitatively or qualitatively. This suggests an open and nonoppressive environment where stakeholders are willing to engage, to communicate, and to be "reasonable." Positions, values, and interests are not polarized. A high premium is placed on scientific and technical knowledge and evidence.

- A *substantive* EIA process is well adapted for comprehensive medium- to long-term efforts to advance environmental or sustainability objectives. Such processes may function more effectively at a strategic (policy, plan, program) level, where the scale of analysis is regional or greater. Major project-level EIAs can apply this process but tend to be more effective if undertaken within the context of an array of larger-scale environmental management frameworks and indicator systems.

- A *practical* EIA process is well matched to short-term, politicized environments, characterized by limited resources, a limited ability to control the environment, and high levels of uncertainty beyond the immediate future. Stakeholders are generally resistant to change but are prepared to bargain. Changes generally take the form of mutually beneficial, incremental adjustments from the status quo. A heavy emphasis is placed on satisfying decision-making requirements, on efficiently working within available resources, and on ensuring implementation.

- A *democratic* EIA process tends to work especially well when clearly identifiable stakeholders wish to control their own destinies. Proponents must be willing and able to delegate, for selective decisions, their decision-making authority. Local stakeholders must be willing and able to select, maintain contact with, and support representatives. Stakeholder representatives must be prepared for and capable of participating in a time-consuming, demanding, complex, and sometimes controversial planning and decision-making process with other parties. It is essential that there be sufficient time and resources to support the process.

- A *collaborative* EIA process seems best suited to situations where major stakeholders can work together collaboratively to achieve mutually agreed upon ends by mutually agreed upon means. It should be possible for stakeholders to select and support representatives. Stakeholder representatives should be willing and able to participate in a time-consuming and often protracted joint planning endeavor. They should be prepared to maintain close contact with their constituents and to accept that other parties retain final decision-making authority. There should be sufficient resources to support the process and sufficient time for the process to proceed at its own pace.

- An *ethical* EIA process is especially appropriate when issues of fairness, equity, and social and/or environmental justice predominate. There should be both procedural and distributional ethical concerns. All major parties should desire that ethical rights and duties be identified and formalized. There should be a willingness to identify and reconcile conflicting ethical procedural and distributional principles, rules, rights, and duties. All parties should be comfortable with ethical concerns taking a lead role in each EIA activity.

- An *adaptive* EIA process is more effective for complex problems in turbulent, unstable, and complex environments. Concerns with risk, uncertainty and human and ecological health should predominate. There should be a general acceptance of the need to anticipate and to manage uncertainties rapidly and effectively. There should be a willingness to recognize knowledge limits, to ensure that uncertainty-related concerns assume a lead decision-making role, and to commit to an iterative, learning, and adaptive EIA process and organizational structure.

Several additional refinements are necessary to reflect the complexities of both the processes and the context.

1. The EIA processes, described in Chapters 3 to 10, include numerous subsets and variations. Part of matching process to context necessitates selecting the process variations that best match the context.
2. The boundaries between core process elements and EIA process types are fuzzy and permeable. Iterative adjustments to both the process and the core elements are necessary to optimize responses to recurrent problems.
3. Each EIA process has strengths and limitations. It is necessary to ameliorate the relevant limitations and to reinforce the relevant strengths. This can entail drawing upon other EIA process types, concepts, and methods, as appropriate.
4. More than one recurrent problem is likely in any given situation. This means that aspects of more than one EIA process will need to be incorporated into most overall EIA processes.

5. EIA process types often overlap and sometimes conflict. Interactions among process types will need to be considered. There should be minimal duplication (unless deliberate), and individual process elements should be mutually supportive or counterbalancing.

6. Matching process to context is not a one-step procedure. There should be an initial matching of process to context type. Proposal- and environment-specific adjustments should then be made. Ongoing adjustments will be necessary as the process and context co-evolve.

Matching process to context, although helpful in establishing priorities, oversimplifies how EIA processes might be applied. Elements of all EIA processes are necessary in any EIA process. It would be foolish not to draw upon relevant *scientific* knowledge and expertise and to apply pertinent scientific methods where practical and appropriate. EIAs invariably have team members with specialized natural and social scientific expertise. Links to government scientists and to other members of the scientific community are essential. Sound decision making can occur only if choices and impacts are systematically, consistently, and *rationally* analyzed and evaluated. Carefully reasoned arguments are essential to informed debate and to deliberate and substantiated documentation and decision making.

EIA is expected to make a *substantive* contribution to minimizing adverse environmental consequences and to enhancing environmental benefits. An EIA, that does not focus directly on the tangible environmental implications of choices and proposals is no more than an empty procedural exercise. An *impractical* EIA process is unfocused, disconnected from reality, weak on implementation, of variable quality, and slow to learn from experience and practice.

EIA processes are not likely to succeed if interested and affected people are powerless in decisions that affect their lives. The appropriate balance of *power* may be a source of debate, but there should be no doubt about whether affected parties should contribute to decision making. Public participation should not be an EIA process management option. One-way communications and education, although necessary, are insufficient. An EIA process necessarily provides for two-way consultation and communications. Public participation is usually more effective when broadened to encompass mutual education, negotiations, and *collaboration.*

Environmental and social fairness, equity, and justice issues are raised, directly or indirectly, in almost any EIA process. It is essential that the EIA process explicitly address procedural and distributional *ethical* concerns and trade-offs. The rights and duties of the various parties in the process also should be explicitly identified and substantiated. EIA is almost never fully certain. There will usually be *uncertainties* associated with every EIA activity and with every interaction among EIA activity. Totally accurate impact predictions and complete control are equally unrealistic. Thus any EIA process must be adaptive, must address uncertainties, and often must consider human health and ecological risks.

If elements of each EIA process type are potentially relevant in any EIA process, matching process and contextual characteristics is more commonly a case of

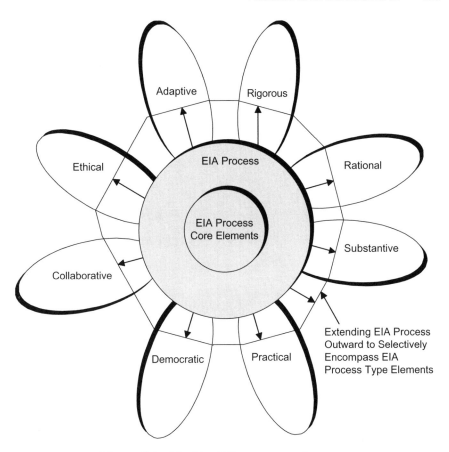

Figure 11.1   Matching EIA processes and context.

determining the roles for each EIA process type than of selecting a single process to apply in a specified class of situations. As Figure 11.1 illustrates, EIA process management necessarily extends from core EIA process elements and applies appropriate elements selectively from each EIA process-type. Decisions regarding EIA process-type roles will depend partly on the context and partly on the potential contributions (both positive and negative) that each EIA process type could make at various points in the EIA process. Defining appropriate roles also involves considering interconnections among EIA process types.

11.6   PROCESS INTERCONNECTIONS

EIA process types can be connected in many ways, as highlighted in Table 11.1. One process could provide inputs to or derive outputs from another process. Both processes could provide elements that fit within a larger framework (e.g.,

Table 11.1 Examples of Process Interconnections, Overlaps, and Middle-Ground Concepts

Rigorous and Rational	Rigorous and Substantive	Rigorous and Practical	Rigorous and Democratic
Complementary methods	New or holistic science	Both grounded in reality and inductive (i.e., experience-based)	Scientific analyses and knowledge inform democratic processes
Reliance on specialist advisors	Emphasis on experimentation and modeling	Balance of rigor and relevance and of explanation and prescription	Both explicit and traceable
Emphasis on comprehensiveness, explicitness, consistency, and transparency	Scientific knowledge informs substantive process	Applied science informs decision making; practicalities bound scientific analysis	Scientific interpretations informed and facilitated
Both stress analysis more than synthesis and seek systematic, traceable process	Substantive goals guide applied science	Middle ground: phenomenological sociology, interpretative social science, pragmatism, empiricism	Middle ground: civic, lay, and holistic science, critical social science
Middle ground: scientific rationality	Middle ground: holistic science, AEAM, ecological theory, positivistic social science, interpretative social science		

Rigorous and Collaborative	Rigorous and Ethical	Rigorous and Adaptive	Rational and Substantive
Scientific analyses inform collaborative processes	Both seek to avoid bias	Both explicitly address risks and uncertainties and stress importance of monitoring	Rational methods inform substantive processes
Scientific interpretations informed and facilitated	Ethical standards for scientific research	Adaptation recognizes scientific limits; precautionary principle combines scientific limits with severe harm potential	Substantive framework for rational methods
Scientific analysis and collaborative synthesis complementary	Complementary interpretations	Shared interest in complex and chaotic systems	Rational analysis and substantive synthesis complementary
Middle ground: AEAM, civic science, peer review, scientific community, SIA (collaborative)	Middle ground: research ethics, holistic science	Shared methods (e.g., modeling)	Middle ground: ecological and social rationality, rational systems, substantive/purposive rationality, ecological impact assessment, SIA (technical), sustainability assessment
		Middle ground: AEAM, new science, social learning, uncertainty analysis, uncertainty principle	

Rational and Practical	Rational and Democratic	Rational and Collaborative	Rational and Ethical
Rationality informed by experience Practical bounds to rationality Middle ground: mixed scanning, strategic planning, purposive incrementalism, bounded rationality, effective planning, practical rationality, legal rationality, knowledge and reflection in action, prescriptive pragmatism	Rational analyses inform democratic processes Democratic offsets autocratic tendencies of rational methods; basis for interpretations and conclusions Middle ground: political rationality, advocacy planning, critical, social, political, structural, and reform rationalism	Rational analyses inform collaborative processes Rational argumentation in collaboration Collaborative inputs to rational conclusions Middle ground: communicative rationality, rational argumentation, procedural rationality, rational discourse, communicative planning, joint fact-finding, collaborative planning	Ethical conscience to rationality Rational analysis of distributional consequences Rational testing of ethical assumptions and assertions Ethical goals and principles to guide rational process Middle ground: professional ethics, value or normative rationality, equity, or progressive planning

Rational and Adaptive	Substantive and Practical	Substantive and Democratic	Substantive and Collaborative
Integration of risk and uncertainty into rational analysis (e.g., risk assessment) Uncertainties about rational processes Middle ground: strategic choice, strategic planning and management, risk assessment and management	Both real, collective, and experience-based Short-term steps toward long-term needs Practical bounds on substantive aspirations Middle ground: purposive incrementalism, environmental management, integrated environmental, and resource management Shared methods (e.g., rapid rural appraisal)	Substantive analyses inform democratic processes Substance guides and frames procedures Both stress value of local knowledge Middle ground: co-management, bio-regionalism, environmental and social movements and advocacy, traditional knowledge, political SIA, deep ecology, ecological politics, eco-feminism	Substantive analyses inform collaborative processes Substance guides and frames procedural Both stress dialogue, local knowledge, and consensus building Both elements of sustainability Shared methods (e.g., scenario writing, storytelling) Middle ground: co-management, shared vision planning

(Continued)

Table 11.1 *(Continued)*

Substantive and Ethical	Substantive and Adaptive	Practical and Democratic	Practical and Collaborative
Both stress procedural equity. Both concerned with intergenerational equity and needs of disadvantaged. Distributional equity conducive to sustainability. Middle ground: social and environmental justice, traditional knowledge, eco-feminism, teleological or consequentialist ethics, distributional or outcome equity or fairness, environmental ethics, sustainability ethics, normative ethics	Both seek to avoid crises and recognize need for adaptive processes and organizations. Substantive uncertainties. Ecological, social, and health all part of sustainability. Middle ground: AEAM, ecosystem approach, social learning, latent time bombs, human health and ecological risk assessment, disaster and hazard analysis. Shared methods (e.g., modeling, visioning, life-cycle analysis)	Both are procedural, assume conflicting interests, are overtly political and emphasize dialogue, bargaining, and experience-based knowledge. Practical bounds to democratic processes. Democratic principles to guide and limit practical processes. Middle ground: critical pragmatism, capacity building, participant and intervener funding	Both are procedural and involve bargaining and dialogue. Practical bounds to collaborative processes. Collaborative principles to guide and limit practical processes. Middle ground: dialogical incrementalism, communicative action, practical communications, practical deliberative learning. Shared methods (e.g., participatory rural appraisal)

Practical and Ethical	Practical and Adaptive	Democratic and Collaborative	Democratic and Ethical
Ethics a conscience for practical processes. Practicality ensures that ethical standards and decision rules can be implemented. Middle ground: critical pragmatism, purposive incrementalism, applied or practical ethics	Both recognize major uncertainties and control limits, stress flexibility, and hedge away from disasters. Effectiveness reviews to anticipate and adapt to uncertainties. Practical bounds to adaptation. Middle ground: practical, deliberative learning, real problems, ingenuity gaps	Both procedural and participatory. Both value local and traditional knowledge and involve consultation, communications, mutual education, negotiations, and collaboration. Collaboration as a means for delegation. Middle ground: critical dialogue, coalition building and networking, active mediation, co-management, constructive engagement	Procedural equity facilitates democratic processes. Both concerned with rectifying power inequities. Ethical analyses inform democratic processes. Ethical analysis informed by democratic perspectives. Middle ground: communitarian principles, traditional knowledge, community development and empowerment, critical, egalitarian, and feminist ethics

Democratic and Adaptive	Collaborative and Ethical	Collaborative and Adaptive	Ethical and Adaptive
Both iterative and interactive Risk and uncertainty analyses inform democratic processes Local and traditional knowledge consistent with a learning process Uncertainties about procedures Middle ground: critical and transformative learning, critical EIA education, traditional ecological and indigenous knowledge	Procedural equity facilitates collaboration Ethical analyses inform collaboration Open collaborative process consistent with multiple ethical perspectives Negotiations about procedural and distributional rights and duties Middle ground: communicative or discourse ethics, procedural justice, communitarian ethics	Both iterative, interactive and stress mutual education Risk and uncertainty analyses inform collaborative processes Collaboration conducive to addressing risk and uncertainty perceptions, evaluation and management Uncertainties about collaborative procedures and outcomes Middle ground: collaborative learning, linguistic imprecision and confusion, risk communications, perceived risk	Uncertainties about values and ethical interpretations Ethical perspectives inform adaptive processes Ethical processes can be informed by risk and uncertainty analyses Middle ground: precautionary principle, risk assessment professional standards, human and ecological health standards

503

sustainability). Combining two processes can offset the negatives or reinforce the positives associated with individual processes. Where two processes are combined, areas of duplication can be eliminated (to facilitate efficiency) or retained (to provide multiple perspectives). Processes can be applied simultaneously (separately, partially integrated, or fully integrated) or sequentially (perhaps applying to different EIA activities).

Complementary process subsets, variations, methods, procedures, concepts, frameworks, and institutional arrangements can be integrated partially or completely. Middle-ground concepts provide an opportunity to build outward from the overlaps to take in appropriate EIA process-type elements. Conflicts can sometimes be reduced through role definition and by applying potentially conflicting process elements to different EIA activities. Or it may be helpful to retain conflicting perspectives in a dramatic tension, consistent, for example, with societal divisions.

Process integration can involve tracing how each process evolved; identifying the needs met by the processes individually and collectively; determining the advantages and disadvantages of partial or complete integration; focusing on major process characteristics, similarities, differences, and interactions; identifying potential forms of synthesis; strengthening mutually complementary relationships; offsetting individual and mutual weaknesses; and retaining diversity where supportive of individual or joint process needs (Boothroyd, 1995; Nooteboom and Wieringa, 1999; Yiftachel, 1989). The end result could be separated or partially or completely integrated composite EIA processes.

11.7   COMPOSITE EIA PROCESSES

A composite EIA processes could be formulated in any one of many ways. Elements from other EIA process types could then be added selectively if they are mutually supportive and appropriate to the situation. This approach, although attractive in its simplicity, will probably result in only modest improvements to current EIA good practice. It is not as if current good practice does not address the recurrent shortcomings. But the frequency and intensity with which the shortcomings continue to be referenced suggest that more fundamental EIA process management reforms are required.

Another, relatively simple approach can take place when the major issues are consistent with the conditions most suited to one of the eight EIA process types. If, for example, the issues are largely ethical, the point of departure would be an ethical EIA process. The ethical process would be supplemented by core process choices (as described in Chapter 2) and other EIA process elements, as warranted and appropriate. The EIA process would focus on identifying and managing ethical issues and trade-offs. The same core–supplemental pattern would be followed if the issues revolved primarily around risk and uncertainty management, the rational evaluation of available choices, contributions to substantive environmental objectives, the practical resolution of approval and implementation-related issues, and so on. Situations where issues are so highly focused are likely to be rare. A variation of this approach would be to undertake two or more processes simultaneously or successively to provide varying perspectives at each decision.

A more realistic combination would be to build the process around clearly complementary (e.g., rational and scientific or collaborative and democratic) or counterbalancing (e.g., substantive and practical or rational and adaptive) EIA process types. Other EIA process types would assume a support role. Figure 11.2 portrays a composite EIA process with a collaborative–democratic core. Some EIA decisions would be delegated. Some would be shared. Proponents and regulators would retain final decision-making authority for the remaining decisions, although there would be extensive consultation. This type of composite process is best suited to situations where issues are highly clustered. There should be broad agreement concerning issue clusters and the rationale for distinguishing between core and support roles.

A more complex and perhaps more realistic composite EIA process is illustrated in Figure 11.3. The process begins from core EIA process elements. The procedural elements are structured around a combination of collaborative and democratic decision making. Practical perspectives, strategies, and constraints are counterbalanced against substantive visions, goals, and indicators. The process is supported and informed by rational analyses; scientific analyses; adaptive, risk, and uncertainty perspective and analyses; and ethical perspectives, principles, and analyses. It is structured within public participation, ethical, risk and uncertainty, and environment and sustainability frameworks. It is adapted to an array of contexts. It draws upon a variety of tools, fully involves all interested and affected participants, and produces both direct (e.g., documents, conclusions) and indirect (e.g., environmental quality changes, institutional changes, mutual education, contributions to the state of EIA practice) products.

Another approach, illustrated in Figure 11.4, structures the process around temporal and spatial distinctions. Practical considerations predominate in the short term. The focus during this period is at the micro level (i.e., incremental adjustments from current practices). Rigorous, rational, adaptive, ethical, and substantive perspectives, methods, and analyses all contribute to the analysis but are filtered through a practical perspective. Both micro and macro analyses occur in the medium term. Rigorous, rational, adaptive, and ethical EIA processes all contribute directly to medium-term analyses. Practical and substantive EIA processes make indirect contributions (i.e., filtered through the directly contributing approaches) during the medium term. Substantive methods, insights, and ideals take the lead in the long term—macro analysis. Ethical and adaptive perspectives, methods, and analyses assume a support role. The practical, rigorous, and rational EIA processes make a minor contribution. A blending of democratic and collaborative decision making occurs for all time horizons and for all scales of analysis. Review and approval steps occur in association with each time horizon.

Constructive dialogue and negotiations procedures blend rational argumentation, practical considerations, ethical principles, substantive imperatives, the delegation and sharing of power, mutual education and effective communications, consultation, and collaboration (Forester, 1999; Healey, 1997; Sager, 1994). Dialogue-centered processes can be aided by substantive studies and by rational, risk,

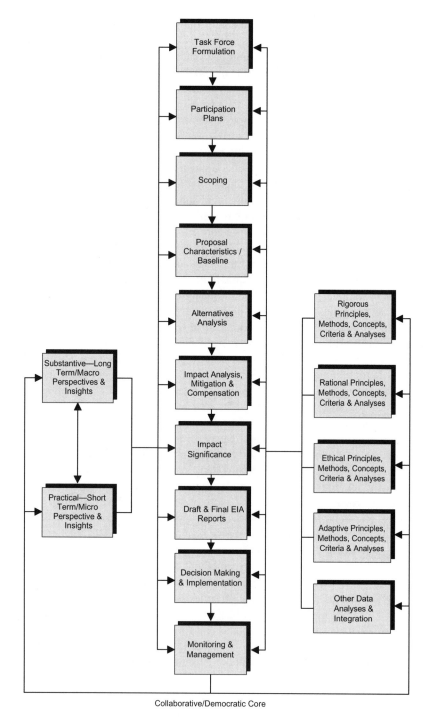

Collaborative/Democratic Core

Figure 11.2 Example composite EIA process: collaborative democratic core.

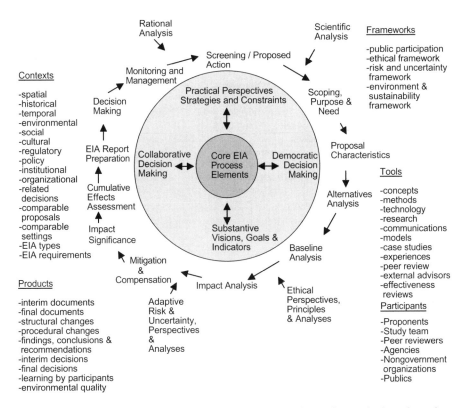

Figure 11.3 Example composite EIA process: merging of practical, substantive, democratic, and collaborative.

uncertainty, ethical, and scientific analyses. The focus of this composite EIA process is on creating and maintaining the conditions required for constructive and creative stakeholder dialogue. A related approach is to build the process around converging interests or perspectives (e.g., empowerment, participation, sustainability, fairness and equity, education, communications) (Hoffman and Davidson, 1997; Smith et.al., 1997).

A further approach concentrates on areas of divergence. EIA processes often deal with dichotomies. Examples include rigor vs. relevance, certain vs. uncertain, subjective vs. objective, conflict vs. consensus, procedural vs. substantive, top-down vs. bottom-up control, ends vs. means, apolitical vs. political, human limits vs. human potential, predict and control vs. adapt and manage, fundamental change vs. incremental change, simple vs. complex problems and environments, technical/scientific knowledge vs. community and traditional knowledge, closed vs. open, and short term vs. long term. The EIA process could focus on determining how best to address each dichotomy. It could be appropriate in some cases to emphasize one

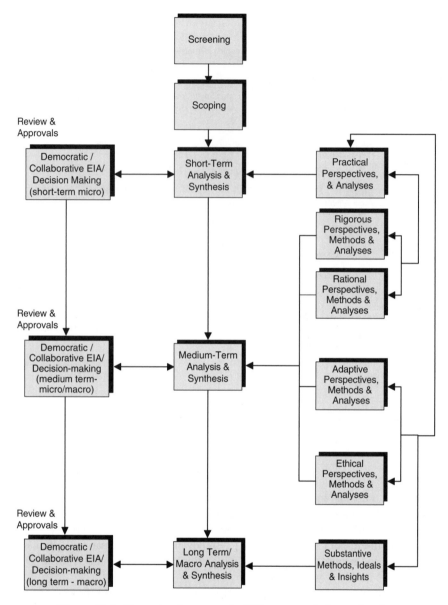

Figure 11.4   Example of a composite EIA process: timing and scale.

side of the dichotomy over the other. In others a middle ground may be possible. Perhaps it is possible to achieve both (i.e., a false dichotomy). Or the two ends sides of the dichotomy could be treated as valid perspectives to be maintained in a dynamic balance. The EIA process types would be knowledge sources to draw upon as the EIA process is structured around the dichotomies.

Existing integrative frameworks are an additional way to formulate a composite EIA process. An obvious candidate is sustainability assessment. As illustrated in Figure 11.5, sustainability assessment combines the substantive (broadly defined but focused on critical needs and aspirations) with the ethical (e.g., intergenerational equity, resource equity, equity in the distribution of power). It fosters planning and decision making that is rational, adaptive, practical, collaborative, and democratic. Both the substantive and the procedural operate within a sustainability framework. Sustainability assessment recognizes the need for institutional reform and for mutually supportive links to other forms of sustainability management, to sustainability indicators, and to sustainability limits and carrying capacity. A sustainability-oriented composite EIA process would selectively draw upon insights, methods, and perspectives from the EIA process types. The resulting modifications and refinements would reinforce the strengths and help offset the limitations (e.g., politically naïve, limited consideration of uncertainties, weak on consensus building, and conflict resolution methods) of sustainability assessment. It also would facilitate a more effective fit between process and context.

Another integration possibility would be to blend adaptive environmental assessment and management (AEAM) and the strategic choice method. AEAM combines

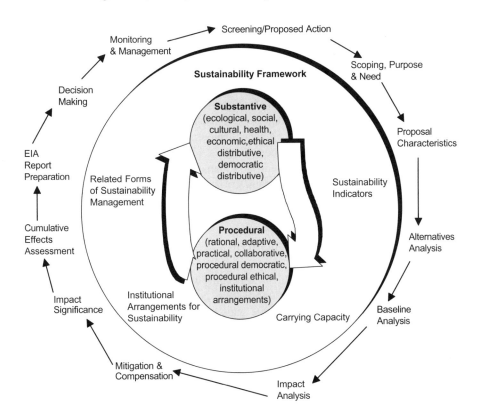

Figure 11.5   Example of a composite EIA process: sustainability shaped.

elements of scientific rigor, adaptation and collaborative decision making with a focus on substantive environmental improvements. The strategic choice method provides a range of systematic, rational, and collaborative uncertainty management methods. Additional attention could be directed to political power, to ethical and practical concerns, and to social, health, and sustainability effects. Selective reference could be made to core EIA process elements and to other EIA process types.

An additional candidate would be the ecosystem approach. The ecosystem approach combines substantive environmental visions, principles, and objectives with collaborative decision making, practical political realities, adaptive planning, and a sound foundation of scientific and technical analysis. More consideration would be given to social, cultural, and economic concerns. A greater effort would be made to identify and resolve ethical concerns and trade-offs, to offset power inequities, and to ensure that choices are screened and compared systematically. Particular attention would be devoted to the differences between the ecosystem approach and EIA. Two other existing framework possibilities are traditional knowledge (a blending of the ecological, scientific, social, spiritual, cultural, ethical, historic, collaborative, and local control) and integrated environmental assessment (starts from a combination of environment and EIA types). Building from existing frameworks is helpful because the integration process has already begun. However, it is still necessary to add missing elements, supplement elements that have received insufficient attention, ameliorate shortcomings, and explore the implications of differences from EIA.

The EIA process characterizations, presented in Chapters 2 to 10, include a host of methods, concepts, and frameworks, which could enhance a composite EIA process. A few examples include forms of rationality, life-cycle assessment, ecological footprint analysis, social and transformative learning, deliberative practice, effective planning, empowerment, lay science, integrated environmental and resource management, scenario writing, discourse ethics, bioregionalism, fuzziness, the precautionary principle, risk assessment, and management and ingenuity. Potentially relevant concepts, methods, and frameworks should be reviewed to determine possible roles in a composite EIA process.

11.8   CHALLENGES AND PRIORITIES

Enhancing EIA process management, especially with reference to recurrent shortcomings, is a daunting task. Although there is considerable potential for improvement, there also are major knowledge gaps regarding the processes and process elements most and least effective in various classes of situations in an EIA process. The process-related knowledge base is incomplete and scattered, both within and outside EIA literature. Most EIAs involve multiple issues, a diversity of stakeholders, and a complex and evolving environment. EIA institutional arrangements are highly interrelated with other regulatory instruments. It is difficult to separate the influence of process characteristics from other variables that affect outcomes. Definitions of process success and failure vary greatly. There are generally multiple perceptions and perspectives of what occurred and should have occurred. EIA usually occupies a peripheral position relative to project planning, policy making, and decision making. EIA is, at best, a secondary mission for most proponents and regulators. EIA process management problems exacerbate the organizational tendency to bypass, inhibit, or starve for funds agency EIA functions.

Not surprisingly, process participants are inclined to and frequently have a vested interest in describing and interpreting the process to their own best advantage. Memories are selective. Documentation is rarely complete. It is never easy to separate the unique characteristics of an EIA process from those process attributes with potentially broader implications, either as pitfalls to be avoided or as models to be followed. Effectiveness reviews of EIA process experience at both the regulatory and applied levels are limited, often anecdotal, and generally are presented from only a single perspective. Poorly designed and executed EIA processes contribute to a legacy of mistrust and conflict, which, in turn, widens the gulf between EIA practitioners and other stakeholders. Breaking out of this cycle requires a major effort.

These constraints are compounded by a persistent belief in the myth of a one-size-fits-all EIA process. The desire for or assumption of a blueprint EIA process (based on a misplaced wish for certainty in an uncertain world) contributes to a common expectation that EIA process management is no more than the scheduling of a standard set of EIA activities, in combination with agency consultation, public consultation, study team management, and EIA document preparation. The assumed simplicity of the EIA process reinforces the belief that EIA process management requires no specialized knowledge beyond normal project management skills and a working knowledge of regulatory requirements. Without a perception of either the EIA process knowledge base or of recurrent EIA process management shortcomings, there is minimal incentive for improvement. This reluctance to alter prevailing practice dovetails with the organizational propensity to resist change. When the recurrent problems arise, the tendency is to revert to explanations that focus on unique circumstances, beyond the ability of the EIA process manager to either anticipate or control. Flawed practice is repeated. When outcomes continue to disappoint, there are always new reasons for the new problems.

Notwithstanding these challenges, there is considerable potential for improvement if initiatives are focused, efficient, and demonstrably effective. The first priority should be demonstrating the negative consequences associated with failing to address the recurrent shortcomings adequately. If an effective case can be made for reforming current practice, based on the experiences and perceptions of critical stakeholders, an incentive for action is created. It is next necessary to demonstrate that practical improvements are possible, are not likely to be costly, can be readily implemented, and can produce short-term benefits to the major parties.

The knowledge base for enhanced EIA process management, especially concerning recurrent problems, is considerable, as demonstrated in the preceding chapters. Possibly, this book and related references can spark a dialogue about how to facilitate better EIA process management and how to move beyond the recurrent problems. Relevant models, concepts, and frameworks can be refined and tested.

Process-related regulatory requirements and guidelines can be compared and evaluated. More documentation of EIA process management experiences could occur. Further consideration could be given to the relationship between EIA processes and contextual characteristics.

There is a particular need for comparative evaluations of process management experiences, EIA process case studies from multiple perspectives, and more effective links to process management efforts in related fields. The net result of such efforts could be an enhanced repertoire of process management tools and skills and a better definition of good and bad process management practice. However, before significant steps forward can be taken, a serious effort must be made to prevent and ameliorate recurrent EIA process management problems. Hopefully, this book will facilitate such efforts.

11.9 SCENARIO POSTSCRIPT

This book began with a not-so-hypothetical scenario. The scenario describes how a well-intentional EIA process came apart at the seams. The process broke down because of a failure to anticipate, acknowledge, and respond adequately to a series of problems that emerged through the process. The problems arose from inadequately addressed stakeholder demands. The demands (i.e., make the process more rigorous, rational, substantive, practical, democratic, collaborative, ethical, and adaptive) reflect common stakeholder perspectives. They also represent valid and recurrent perspectives on how the EIA process should be conducted.

As the book demonstrates, the problems described in the scenario are far from hypothetical. They frequently arise both in EIA practice and in the practice of related environmental management fields. The potential for preventing and ameliorating the problems is considerable, as detailed in Chapters 3 to 10. Multiple processes, concepts, frameworks, and methods are available to EIA practitioners for addressing the problems portrayed in the scenario. But a suite of tools does not guarantee that the problems will be solved or even ameliorated. There will always be unique aspects to any EIA process. The match between process and context may not be effective. The appropriate tools may or may not be selected. The methods and procedures may or may not be applied effectively. Especially if there is a legacy of mistrust, stakeholders could still resist all initiatives. The procedures and methods presented in the preceding chapters are, at best, conducive to avoiding and ameliorating the recurrent problems. Failing to acknowledge and respond to the problems, however, will ensure that the problems persist and almost certainly will make the situation worse.

11.10 SUMMING UP

This chapter begins with two EIA process success stories, both of which describe processes that effectively integrated elements of all EIA process types. Success is defined as a process satisfactory to all parties and net environmental benefits, not as proposal approvals. Unresolved is the extent to which the apparently satisfactory processes and outcomes were the result of unique local circumstances or could be a model with potential for broader application. A closer and more systematic scrutiny of ways in which the EIA process types can be linked, integrated, and transcended is required. In this chapter we therefore go on to describe how the individual EIA processes presented in Chapters 2 to 10 might be connected and combined at both the regulatory and applied levels. We also identify residual challenges and priorities for future action. Table 11.2 is a checklist for formulating, applying, and assessing a composite EIA process.

Table 11.2   Checklist: A Composite EIA Process

The response to each question can be yes, partially, no, or uncertain. If yes, the adequacy of the process taken should be considered. If partially, the adequacy of the process and the implications of the areas not covered should be considered. If no, the implications of not addressing the issue raised by the question should be considered. If uncertain, the reasons why it is uncertain and any associated implications should be considered.

Composite Regulatory Frameworks

1. Do EIA requirements establish a minimum EIA process standard?
2. Do EIA requirements convey a sense of ideal EIA process characteristics?
3. Do EIA requirements and guidelines provide a sense of core EIA process attributes?
4. Do EIA requirements and guidelines provide a sense of available EIA process management choices?
5. Do EIA guidelines provide examples of good practice EIA process management?
6. Do EIA requirements and guidelines acknowledge the recurrent problems?
7. Do EIA requirements and guidelines provide examples of ways in which the recurrent problems can be avoided and minimized?
8. Do EIA requirements and guidelines acknowledge the need to link and combine EIA process types?
9. Do EIA requirements and guidelines recognize the need to adjust the EIA process for different classes of situations and for proposal and environment-specific circumstances?
10. Are EIA requirements and guidelines sufficiently flexible not to unduly constrain EIA process management?

Matching Process and Context

1. Is the need for the process to match the context recognized?
2. Are the process elements that correspond to the EIA process types appropriate for the conditions where they are applied?
3. Are pertinent subsets and variations of EIA process types recognized and considered?
4. Are links between core process elements and EIA process types considered?
5. Are the relevant strengths and limitations of the applicable EIA process types taken into account?
6. Is the likelihood of more than one recurrent problem acknowledged?
7. Are steps taken to ensure that EIA process elements are not redundant and are mutually supportive?

(Continued)

Table 11.2   (*Continued*)

8. Are iterative adjustments made to the EIA process to maintain an ongoing fit between process and context?
9. Are relevant elements of each EIA process type incorporated into the process?

Process Interconnections

1. Does the EIA process take into account interconnections between EIA process types?
2. Are overlaps between EIA process types considered?
3. Are relevant middle ground concepts between EIA process types considered?
4. Are alternative ways of linking EIA process types explored?
5. Are potential conflicts between process elements taken into account?
6. Are steps taken to minimize or manage potential conflicts between EIA process elements?

Composite EIA Processes

1. Is the possibility of combining a conventional EIA process with selective elements from other process types considered?
2. Is the possibility considered of applying one of the EIA process types, taking into account major issues and integrating elements from other EIA process types where appropriate?
3. Is the possibility of applying two or more EIA process types, either simultaneously or sequentially, considered?
4. Is the possibility considered of combining two or more complementary or counterbalancing EIA process types as the core of the process, with elements of other EIA process types assuming a support role?
5. Is the possibility considered of building a composite process around constructive dialogue and negotiations?
6. Is the possibility considered of structuring a composite process around areas of divergence?
7. Is the possibility considered of building a composite process around existing integrative frameworks (e.g., sustainability assessment, AEAM, the ecosystem approach, strategic choice, traditional knowledge)?
8. Is consideration given to selectively integrating into the composite EIA process pertinent and applicable methods, concepts, and frameworks (as presented in Chapters 2 to 10)?

Challenges and Priorities

1. Are the constraints to reforming and adapting prevailing approaches to EIA process management recognized?
2. Are the EIA process management recurrent shortcomings acknowledged?
3. Are the dangers associated with a blueprint approach to EIA process management appreciated?
4. Is the need for specialized EIA process management knowledge recognized?
5. Is the need for reforming EIA process management accepted?
6. Is effective use made of process management literature from EIA and from related fields?
7. Is effective use made of EIA process management experience?
8. Is a concerted effort made to enhance EIA process management practice?
9. Are lessons and insights derived from EIA process management experiences shared with others?

The problem is when and how to apply the EIA process types, both individually and collectively. The direction is toward approaches for regulating and guiding EIA process management, for matching process and context, for linking EIA process types, for formulating composite EIA processes, for determining challenges, and for establishing priorities.

EIA requirements and guidelines should define minimum levels of EIA process-related practice and establish process-related goals and principles. EIA guidelines should narrow the gap between the minimum and the ideal. Core EIA process attributes should be identified. A sense should be provided of the available EIA process management choices. EIA requirements and guidelines should acknowledge recurrent process-related problems and facilitate efforts to avoid and reduce the problems. The need to adapt the EIA process to contextual factors and to link and integrate EIA process types should be recognized. EIA requirements and guidelines should draw upon process-related experiences, applied research, and effectiveness reviews.

The EIA process should be appropriate to the situation. The contextual characteristics best suited to each EIA process type are identified. Matching process and context also involves considering process subsets and variations, links between core process elements, process-type strengths and limitations, the possibility of multiple problems, process overlaps and conflicts, and process adjustments to match an evolving context. Elements of each EIA process type need to be integrated into any EIA process.

The EIA process types are highly interconnected. There are multiple links, overlaps, and middle-ground concepts. EIA process elements can be arranged sequentially or simultaneously. They can be complementary, offsetting, or conflicting. They can be used to apply multiple perspectives or to explore conflicting perspectives. Integrating EIA processes necessitates determining the need for integration, focusing on major characteristics, similarities, differences, and interactions; identifying possible synthesis forms; strengthening complementary relationships; offsetting individual and mutual weaknesses; and retaining diversity.

Composite EIA processes can assume many forms. Conventional EIA processes can be adapted to incorporate EIA process-type elements. A single EIA process type can be adapted to incorporate, as supplementary considerations, elements from other EIA process types. A composite process can use elements from two or more EIA process types as the process core, in either a complementary or counterbalancing relationship. Elements from other EIA process types can be added as supplementary concerns. Composite EIA processes can be designed around temporal (e.g., short-term, medium term, long-term) and spatial (e.g., micro, macro) distinctions. They can be centered on facilitating constructive dialogue and negotiations. They can be structured around areas of convergence or divergence. They can start from existing integrative frameworks such as sustainability assessment, adaptive environmental assessment and management, strategic choice, the ecosystem approach, traditional knowledge, and integrated environmental assessment. The differences between EIA and other fields of practice should be considered when integrative frameworks, derived from other forms of environmental management, are employed. Numerous additional methods, concepts, and frameworks can be incorporated into a composite EIA process.

There are major gaps in the EIA process management knowledge base. It is very difficult to separate out and interpret the effects of EIA process management choices. EIA in general and EIA process management in particular are rarely an agency priority. Process management limitations and negative repercussions are seldom acknowledged or documented. There continues to be a widespread belief in a single, infinitely adaptable EIA process. There is little recognition of the need for specialized EIA process management skills and knowledge. The EIA process knowledge base is poorly understood. Recurrent EIA process management shortcomings are not widely acknowledged. Notwithstanding these challenges, there remains considerable potential for improvement. The first priority should be demonstrating the negative consequences of failing to address recurrent shortcomings adequately. The EIA process knowledge base should then be supplemented with applied examples, methodological refinements, effectiveness assessments, information and perspective exchanges, and better links to related fields.

The types of problems identified in the scenario (presented in Chapter 1) can often be avoided and ameliorated. Chapters 2 to 10 provide a host of procedures and concepts conducive to enhanced EIA process management. Definitive prescriptions cannot be provided because of the complexity of the field and situation-specific circumstances. But a sufficient knowledge base has been established to suggest numerous improvement possibilities. Enhancing EIA process management necessarily begins with an open mind, a willingness to consider the possibility of recurrent, avoidable problems, and a commitment to explore how best to prevent, ameliorate, and move beyond those problems.

REFERENCES

Alberti, M., and J. D. Parker, "Indices of Environmental Quality: The Search for Credible Measures," *Environmental Impact Assessment Review* **11**, 95–101 (1991).

Albrecht, S. L., "Equity and Justice in Environmental Decision-making: A Proposed Research Agenda," *Society and Natural Resources* **8**, 67–72 (1995).

Alexander, C., H. Neis, A. Anninou, and I. King, *A New Theory of Urban Design*, Oxford University Press, New York, 1987.

Alexander, E., *Approaches to Planning*, Gordon and Breach, New York, 1986.

Alexander, E., "Rational Revisited: Planning Paradigms in a Post-modern Perspective," *Journal of Planning Education and Research* **19**, 242–256 (2000).

Al-Kodmany, K., "Visualization Tools and Methods in Community Planning: From Freehand Sketches to Virtual Reality," *Journal of Planning Literature* **17**, 189–211 (2002).

Allen, G. M., and E. M. Gould, "Complexity, Wickedness and Public Forests," *Journal of Forestry* **84**, 20–23 (1986).

Amy, D. J., *The Politics of Environmental Mediation*, Columbia University Press, New York, 1987.

Anderson, D., *Strengthening Environmental Assessment for Canadians*, Report of the Minister of the Environment to the Parliament of Canada on the Review of the Canadian Environmental Assessment Act, Canadian Environmental Assessment Agency, Hull, Québec, Canada, 2001.

Andrews, R. N. L., in R. Clark and L. Canter, eds., "The Unfinished Business of National Environmental Policy," *Environmental Policy and NEPA: Past, Present and Future*, St. Lucie Press, Boca Raton, Florida, 1997.

Environmental Impact Assessment: Practical Solutions to Recurrent Problems, By David P. Lawrence
ISBN 0-471-45722-1   Copyright © 2003 John Wiley & Sons, Inc.

Antunes, P., P. Santos, and L. Jordão, "The Application of Geographical Information Systems to Determine Impact Significance," *Environmental Impact Assessment Review* **21**, 511–535 (2001).

Appell, D., "The Uncertainty Principle," *Scientific American*, January 2001.

Armitage, D., "An Integrative Methodological Framework for Sustainability Environmental Planning and Management," *Environmental Management* **19**, 469–479 (1995).

Armour, A., "Integrating Impact Assessment in the Planning Process: From Rhetoric to Reality," *Impact Assessment Bulletin* **8**, 3–13 (1990a).

Armour, A., *Socially Responsible Facility Siting*, doctoral dissertation, University of Waterloo, Waterloo, Ontario, Canada, 1990b.

Armour, A., and B. Sadler, "Synopsis of Workshop Discussions," *The Place of Negotiations in Environmental Assessment*, Canadian Environmental Assessment Research Council, Hull, Québec, 1990.

Arnstein, S. R., "A Ladder of Citizen Participation," *Journal of the American Institute of Planners* **35**, 216–224 (1969).

Arquiaga, M. C., L. W. Canter, and D. I. Nelson, "Risk Assessment Principles in Environmental Impact Studies," *Environmental Professional* **14**, 204–219 (1992).

Arquiaga, M. C., L. W. Canter, and D. I. Nelson, "Integration of Health Impact Considerations in Environmental Impact Studies," *Impact Assessment* **12**, 175–198 (1994).

Auditor-General, *Referrals, Assessment and Approvals Under the Environmental Protection and Biodiversity Conservation Act*, Canberra, Commonwealth of Australia, Audit Report, No. 38, 2002–2003.

Australian and New Zealand Environmental Conservation Council (ANZECC), *Guidelines and Criteria for Determining the Need for and Level of EIA in Australia*, Canberra, Australia, 1996.

Axelrod, R., and M. D. Cohen, *Harnessing Complexity*, Free Press, New York, 1999.

Bagri, A., J. McNeely, and F. Vorhies, *Biodiversity and Impact Assessment*, paper presented at a Workshop on Biodiversity and Impact Assessment, Christchurch, New Zealand, 1998.

Bailey, J., "Environmental Impact Assessment and Management: An Unexplored Relationship," *Environmental Management* **21**, 317–327 (1997a).

Bailey, P. D., "EIA: A New Methodology for Environmental Policy," *Environmental Impact Assessment Review* **10**, 221–227 (1997b).

Banfield, E. C., in M. Meyerson and E. C. Banfield, eds., "Notes on a Conceptual Scheme," *Politics, Planning and the Public Interest*, Free Press, New York, 1955.

Banken, R., in A. L. Porter and J. J. Fittipaldi, eds., "Public Health in Environmental Assessments," *Environmental Methods Review: Retooling Impact Assessment for the New Century*, Army Environmental Policy Institute and International Association for Impact Assessment, Fargo, North Dakota, 1998.

Bardwell, L. V., "Problem-Framing: A Perspective on Environmental Problem-Solving," *Environmental Management* **15**, 603–612 (1991).

Barker, A., and C. Wood, "An Evaluation of EIA Performance in Eight EU Countries," *Environmental Impact Assessment Review* **19**, 387–404 (1999).

Barrow, C. J., *Environmental and Social Impact Assessment: An Introduction*, Arnold, London, 1997.

Barrow, J. D., *Impossibility: The Limits of Science and the Science of Limits*, Oxford University Press, New York, 1998.

Bartlett, R. V., *Policy Through Impact Assessment*, Greenwood Press, New York, 1989.

Bartlett, R. V., in R. Clark and L. Canter, eds., "The Rationality and Logic of NEPA Revisited," *Environmental Policy and NEPA: Past, Present and Future*, St. Lucie Press, Boca Raton, Florida, 1997.

Bass, R., "Evaluating Environmental Justice Under the National Environmental Policy Act," *Environmental Impact Assessment Review* **18**, 83–92 (1998).

Bass, R. E., and A. I. Herson, *Mastering NEPA: A Step-by-Step Approach*, Solano Press, Point Arena, California, 1993.

Bauer, M. R., and J. Randolph, "Characteristics of Collaborative Planning and Decision-making Processes," *Environmental Practice* **2**, 156–165 (2000).

Baxter, W., W. A. Ross, and H. Spaling, "Improving the Practice of Cumulative Effects Assessment in Canada," *Impact Assessment and Project Appraisal* **19**, 253–262 (2001).

Beanlands, G. E., and P. N. Duinker, *An Ecological Framework for Environmental Impact Assessment in Canada*, Institute for Resource and Environmental Studies and Federal Environmental Assessment Office, Hull, Québec, Canada, 1983.

Beatley, T., "Environmental Ethics and Planning Theory," *Journal of Planning Literature* **4**, 1–32 (1989).

Beatley, T., "Planning and Sustainability: The Elements of a New (Improved?) Paradigm," *Journal of Planning Literature* **9**, 383–394 (1995).

Beatley, T., "Preserving Biodiversity," *American Planning Association Journal* **66**, 5–20 (2000).

Beder, S., "Bias and Credibility in Environmental Impact Assessment," *Chain Reaction* **6** (1993).

Behn, R. D., "Managing by Groping Along," *Journal of Policy Analysis and Management* **7**, 643–663 (1988).

Bendix, S., in L. L. Hart, G. A. Enk, and W. F. Hornick, eds., "How to Write a Socially Useful EIS," *Improving Impact Assessment*, Westview Press, Boulder, Colorado, 1984.

Benveniste, G., *Mastering the Politics of Planning*, Jossey-Bass, San Francisco, 1989.

Berke, P. R., "Does Sustainability Development Offer a New Direction for Planning? Challenges for the Twenty-first Century," *Journal of Planning Literature* **17**, 21–36, 2002.

Berkes, F., in J. T. Inglis, ed., "Traditional Ecological Knowledge in Perspective," *Traditional Ecological Knowledge: Concepts and Cases*, International Program on Traditional Ecological Knowledge and International Development Research Centre, Ottawa, Canada, 1993.

Beauregard, R. A., "The Object of Planning," *Urban Geography* **8**, 367–373 (1987).

Bingham, G., *Resolving Environmental Disputes: A Decade of Experience*, Conservation Foundation, Washington, D.C., 1986.

Bingham, G., *What Is Consensus-Building and Why Is It Important for Resource Management?* Resolve: Center for Environmental and Policy Dispute Resolution, Washington, D.C., 2001.

Bingham, G., and L. M. Langstaff, in R. Clark and L. Canter, eds., "Alternative Dispute Resolution in the NEPA Process," *Environmental Policy and NEPA: Past, Present and Future*, St. Lucie Press, Boca Raton, Florida, 1997.

Bird, A., *Philosophy of Science*, McGill–Queen's University Press, Montreal and Kingston, Canada, 1998.

Birkeland, J., "The Relevance of Ecofeminism to the Environmental Professions," *The Environmental Professional* **17**, 55–71 (1995).

Bishop, A. B., *Structuring Communications Programs for Public Participation in Water Resource Planning*, IWR Contract Report, 75–2, U.S. Army Engineer Institute for Water Resources, Fort Belvoir, Virginia, 1975.

Bishop, A. B., in J. Creighton, J. D. Priscoli, and C. M. Dunning, eds., "Communications in the Planning Process, *Public Involvement Techniques: A Reader of Ten Years Experience at the Institute for Water Resources*, U.S. Army Corps of Engineers, IWR Research Report, 82-R, Washington, D.C., 1983.

Bisset, R., *Environmental Impact Assessment: Issues, Trends and Practice*, United Nations Environment Program, Scott Wilson Resource Consultants, Nairobi, Kenya, 1996.

Bisset, R., in N. Lee and C. George, eds., "Methods of Consultation and Public Participation," *Environmental Assessment in Developing and Transitional Countries*, John Wiley & Sons, Chichester, UK, 2000.

Blanco, H., *How to Think About Social Problems: American Pragmatism and Idea of Planning*, Greenwood Press, Westport, Connecticut, 1994.

Bolan, R. S., "The Practitioner as Theorist," *Journal of the American Planning Association* **46**, 261–274 (1980).

Bond, A., and G. Stewart, "Environmental Agency Scoping Guidance on the Environmental Impact Assessment of Projects," *Impact Assessment and Project Appraisal* **20**, 135–142 (2002).

Boothroyd, P., in F. Vanclay and D. A. Bronstein, eds., "Policy Assessment," *Environmental and Social Impact Assessment*, John Wiley & Sons, New York, 1995.

Boothroyd, P., and W. Rees, "Impact Assessment from Pseudo Science to Planning Process: An Educational Response," *Impact Assessment Bulletin* **2**, 9–17 (1985).

Bowler, P. J., *The Fontana History of the Environmental Sciences*, Fontana Press (Harper-Collins), London, 1992.

Boyce, D. E., in A. G. Wilson, ed., "The Metropolitan Plan Making Process: Its Theory and Practical Complexities," *Urban and Regional Planning*, Pion, London, 1971.

Boyce, J. J., "Equity and the Environment," *Alternatives* **21**, 12–17 (1995).

Boyer, M. C., *Dreaming the Rational City*, MIT Press, Cambridge, Massachusetts, 1983.

Branch, K., D. A. Hooper, J. Thompson, and J. Creighton, *Guide to Social Assessment: A Framework for Assessing Social Change*, Westview Press, Boulder, Colorado, 1993.

Brascoupé, S., and J. Mann, *A Community Guide to Protecting Indigenous Knowledge*, Research and Analysis Directorate, Department of Indian Affairs and Northern Development, Canada, June 2001.

Braybrooke, D., and C. E. Lindblom, *A Strategy of Decision*, Free Press, New York, 1963.

Briassoulis, H., "Theoretical Orientations in Environmental Planning: An Inquiry Into Alternative Approaches," *Environmental Management* **13**, 381–392 (1989).

Briassoulis, H., "Who Plans Whose Sustainability? Alternative Roles for Planners," *Journal of Environmental Planning and Management* **42**, 889–902 (1999).

British Columbia Environmental Assessment Office (BC EAO), "Part D: Environmental Assessment Aboriginal Issues and Consultation," *Guide to the BC EA Process*, Victoria, B. C., Canada, January 2001.

British Medical Association (BMA), *Health and Environmental Impact Assessment: An Integrated Approach*, Earthscan Publications Ltd., London, 1998.

Bronfman, L. M., "Setting the Social Agenda: An Organizational Perspective," *Environmental Impact Assessment Review* **11**, 69–79 (1991).

Brooke, C., *Biodiversity and Impact Assessment*, paper prepared for the Conference on Impact Assessment in a Developing World, Manchester, UK, 1998.

Brown, A. L., in A. L. Porter and J. Fittipaldi, eds., "Decision-Scoping," *Environmental Methods Review: Retooling Impact Assessment for the New Century*, Army Environmental Policy Institute and International Association for Impact Assessment. Fargo, North Dakota, 1998.

Brown, J. H., "Commentary 1," *Proceedings of the Workshop on Cumulative Environmental Effects: A Bi-national Perspective*, Canadian Environmental Assessment Research Council and the U.S. National Research Council Board of Basic Biology, Ottawa, Ontario, Canada, 1986.

Buckley, R., in A. L. Porter and J. Fittipaldi, eds., "Improving the Quality of Environmental Impact Statements (EISs)," *Environmental Methods Review: Retooling Impact Assessment for the New Century*, Army Environmental Policy Institute and International Association for Impact Assessment, Fargo, North Dakota, 1998.

Burdge, R. J., *A Conceptual Approach to Social Impact Assessment*, Social Ecology Press, Middleton, Wisconsin, 1994.

Burdge, R. J., "Why Is Social Impact Assessment the Orphan of the Assessment Process?" *Impact Assessment and Project Appraisal* **20**, 3–9 (2002).

Burdge, R. J., and R. A. Robertson, "Social Impact Assessment and Public Involvement," *A Conceptual Approach to Social Impact Assessment*, Social Ecology Press, Middleton, Wisconsin, 1994.

Burdge, R. J., in F. Vanclay and D. A. Bronstein, eds., "Social Impact Assessment," *Environmental and Social Impact Assessment*, John Wiley & Sons, New York, 1995.

Caldwell, L. K., *The National Environmental Policy Act: An Agenda for the Future*, Indiana University Press, Bloomington, Indiana, 1988.

Caldwell, L. K., "Analysis–Assessment–Decision: The Anatomy of Rational Policymaking," *Impact Assessment Bulletin* **9**, 81–92 (1991).

Caldwell, L. K., in R. Clark and L. Canter, eds., "Implementing NEPA: A Non-technical Political Task," *Environmental Policy and NEPA: Past, Present and Future*, St. Lucie Press, Boca Raton, Florida, 1997.

Caldwell, L. K., R. A. Minard, and S. Coyle, Jr., *Managing NEPA at the Department of Energy*, A Study for the Department of Energy, Center for the Economy and the Environment, Washington, D.C., June 1998.

Calow, P., and V. E. Forbes, "Science and Subjectivity in the Practice of Ecological Risk Assessment," *Environmental Management* **21**, 803–830 (1997).

Campbell, M. C., and D. W. Floyd, "Thinking Critically About Environmental Mediation," *Journal of Planning Literature* **10**, 235–247 (1996).

Canadian Environmental Assessment Agency (CEAA), *The Canadian Environmental Assessment Act Training Compendium*, Ministry of Supply and Services, Hull, Québec, 1994.

Canadian Environmental Assessment Agency (CEAA), *A Guide on Biodiversity and Environmental Assessment*, Hull, Québec, Canada, 1996.

Canadian Environmental Assessment Agency (CEAA), *A Reference Guide for the Canadian Environmental Assessment Act: Addressing Cumulative Effects Assessment*, Hull, Québec, Canada, 1998a.

Canadian Environmental Assessment Agency (CEAA), *Operational Policy Statement: Need, Purpose, Alternatives to and Alternative Means*, Hull, Québec, Canada, 1998b.

Canadian Standards Association (CSA), *A Guide to Public Involvement*, Canadian Standards Association, Etobicoke, Ontario, Canada, 1996.

Canter, L. W., *Health Risk Impact in Environmental Impact Statements*, Environmental and Ground Water Institute, University of Oklahoma, Norman, Oklahoma, 1990.

Canter, L. W., "The Role of Environmental Monitoring in Responsible Project Management," *Environmental Professional* **15**, 76–87 (1993a).

Canter, L. W., "Pragmatic Suggestions for Incorporating Risk Assessment Principles in EIA Studies," *Environmental Professional* **15**, 125–138 (1993b).

Canter, L. W., *Environmental Impact Assessment*, 2nd ed., McGraw-Hill, New York, 1996.

Cardinall, D., and J. C. Day, "Embracing Value and Uncertainty in Environmental Management and Planning: A Heuristic Model," *Environments* **25**, 110–125 (1998).

Carpenter, R. A., "Risk Assessment," *Impact Assessment* **13**, 153–187 (1995).

Carpenter, S. L., *Managing Public Disputes*, Jossey-Bass, San Francisco, 1991.

Carpenter, S. L., in R. Clark and L. Canter, eds., "The Case for Continuous Monitoring and Adaptive Management under NEPA," *Environmental Policy and NEPA: Past, Present and Future*, St. Lucie Press, Boca Raton, Florida, 1997.

Cartwright, T. J., "Problems, Solutions and Strategies: A Contribution to the Theory and Practice of Planning," *Journal of the American Institute of Planners* **39**, 179–187 (1973).

Cartwright, T. J., "Planning and Chaos Science," *American Planning Association Journal* **57**, 44–56 (1991).

Casti, J., *Complexification*, Harper-Collins, New York, 1994.

Chase, A., "Anthropology and Impact Assessment: Development Pressures and Indigenous Interests in Australia," *Environmental Impact Assessment Review* **10**, 11–23 (1990).

Checkoway, B., "Paul Davidoff and Advocacy Planning in Retrospect," *American Planning Association Journal* **60**, 139–143 (1994).

Clark, B., "Improving Public Participation in Environmental Impact Assessment," *Built Environment* **20**, 294–308 (1994).

Clark, R., in R. Clark and L. Canter, eds., "The Rational Approach to Change," *Environmental Policy and NEPA—Past Present and Future*, St. Lucie Press, Boca Raton, Florida, 1997.

Coates, J. F., "Impacts We Will Be Assessing in the 21st Century," *Impact Assessment Bulletin* **9**, 9–26 (1990).

Cole, S., "Dare to Dream—Bringing Futures into Planning," *American Planning Association Journal* **67**, 372–375 (2001).

Coleman, R. J., *Address on the US, Europe and Precaution: A Comparative Case Study Analysis of the Management of Risk in a Complex World*, Director General, Health and Consumer Protection Directorate, European Commission, Brussels, Belgium, 2002.

Coleman, W. G., "Biodiversity and Industry Ecosystem Management," *Environmental Management* **20**, 815–825 (1996).

Commission of the European Commission (CEC), *Communications from the Commission on the Precautionary Principle*, COM (2000) 1 Final, Brussels, Belgium, 2000.

Commission of the European Commision (CEC), *Communication From the Commission on Impact Assessment*, COM (2002) 276 Final, Brussels, Belgium, 2002.

Committee on Risk Assessment Methodology (CRAM), *Issues in Risk Assessment*, National Research Council, National Academic Press, Washington, D.C., 1993.

Committee on the Challenges of Modern Society (CCMS), *Evaluation of Public Participation in EIA*, Report 207, North Atlantic Treaty Organization, Brussels, Belgium, 1995.

Conçalves, M. E., "Implementation of EIA Directives in Portugal: How Changes in Civic Culture Are Challenging Political and Administrative Practice," *Environmental Impact Assessment Review* **22**, 249–269 (2002).

Constanza, R., "Visions of Alternative (Unpredictable) Futures and Their Use in Policy Analysis," *Conservation Ecology* **4**, 5–24 (2000).

Court, J. D., C. J. Wright, and A. S. Guthie, *Assessment of Cumulative Impacts and Strategic Assessment in Environmental Impact Assessment*, prepared for the Australian Environmental Protection Agency by J.D. Court and Associates Pty Ltd. and Guthie Consulting, Canberra, Australia, May 1994.

Covello, V. T., P. Sandman, and P. Slovic, *Communications, Risk Statistics and Risk Comparisons: A Manual for Plant Managers*, Chemical Manufacturers' Association, Washington, DC, 1988.

Coveney, P., and R. Highfield, *Frontiers of Complexity: The Search for Order in a Chaotic World*, Faber & Faber, London, 1995.

Craib, I., *Modern Social Theory: From Parsons to Habermas*, Wheatsheaf Books Ltd., Sussex, UK, 1984.

Craig, D., "Social Impact Assessment: Politically Oriented Approaches and Applications," *Environmental Impact Assessment Review* **10**, 37–54 (1990).

Creighton, J. L., in J. L. Creighton, C. M. Dunning, J. D. Priscoli and D. B. Ayres, eds., "Comparing Public Involvement and Environmental Mediation," *Public Involvement and Dispute Resolution: A Reader on the Second Decade of Experience at the Institute for Water Resources*, Institute for Water Resources, U.S. Army Corps of Engineers, Alexandria, Virginia, 1999.

Creighton, J. L., C. M. Dunning, J. D. Priscoli, and D. B. Ayres, eds., *Public Involvement and Dispute Resolution: A Reader on the Second Decade of Experience at the Institute for Water Resources*, Institute for Water Resources, U.S. Army Corps of Engineers, Alexandria, Virginia, October 1999.

Creighton, J. L., J. D. Priscoli, and C. M. Dunning, *Public Involvement Techniques: A Reader of Ten Years Experience at the Institute for Water Resources*, Institute for Water Resources, U.S. Army Corps of Engineers, Ft. Belvoir, Virginia, May 1983.

Cronin, C., in J. Habermas, "Translator's Introduction," *Justification and Application: Remarks on Discourse Ethics*, MIT Press, Cambridge, Massachusetts, 1993.

Crossley, J. W., "Managing Ecosystems for Integrity: Theoretical Considerations for Resource and Environmental Managers," *Society and Natural Resources* **9**, 465–482 (1996).

Culhane, P. J., "Post-EIS Auditing: A First Step to Make a Rational Environmental Assessment a Reality," *Environmental Professional* **15**, 66–75 (1993).

Culhane, P. J., H. P. Friesema, and J. A. Beecher, *Forecasts and Environmental Decision-making*, Westview Press, Boulder, Colorado, 1987.

Cumulative Effects Assessment Working Group and AXYS Environmental Consulting Group (CEAWG and AXYS), *Cumulative Effects Assessment Practitioners Guide*, Canadian Environmental Assessment Agency, Hull, Québec, Canada, February 1999.

Curtis, F., and H. Epp, "Role of Deductive Science in Evaluation of Alternatives in Environmental Assessment," *Impact Assessment and Project Appraisal* **17**, 3–8 (1999).

Dalal-Clayton, B., and B. Sadler in A. Donnelly, B. Dalal-Clayton, and R. Hughes, eds., "Strategic Environmental Assessment: A Rapidly Evolving Approach," *A Directory of Impact Assessment Guidelines*, Russell Press, Nottingham, UK, 1998.

Damasio, A. B., *Descartes' Error*, Avon Books, New York, 1994.

Daniels, S. E., and G. B. Walker, "Collaborative Learning: Improving Public Deliberation in Ecosystem-based Management," *Environmental Impact Assessment Review* **16**, 71–102 (1996).

Davidoff, P., "Advocacy and Pluralism in Planning," *Journal of the American Institute of Planners* **3**, 331–338 (1965).

Davidoff, P., in R. W. Burchell and G. Sternlieb, eds., "The Redistributive Function in Planning: Creating Greater Equity Among Citizens of Communities," *Planning Theory in the 1980's*, Center for Urban Policy Research. New Brunswick, New Jersey, 1978.

Davidoff, P., and T. A. Reiner, "A Choice Theory of Planning," *Journal of the American Institute of Planners* **28**, 102–109 (1962).

Davies, K., and B. Sadler, *Environmental Assessment and Human Health: Perspectives, Approaches and Future Directions*, A Background Report for the International Study of the Effectiveness of Environmental Assessment, Health Canada and International Association for Impact Assessment, Ministry of Supply and Services Canada, Ottawa, Canada, May 1997.

Dawkins, R., *Unweaving the Rainbow: Science, Delusion and the Appetite for Wonder*, Houghton Mifflin Company, Boston, 1998.

Day, D., "Citizen Participation in the Planning Process: An Essentially Contested Concept?" *Journal of Planning Literature* **11**, 421–434 (1997).

Deakin, M., S. Curwell, and P. Lombardi, "Sustainable Urban Development: The Framework and Directory of Assessment Methods," *Journal of Environmental Assessment Policy and Management* **4**, 171–197 (2002).

Dearden, P., and B. Mitchell, *Environmental Change and Challenge*, Oxford University Press, Toronto, Canada, 1998.

De Bono, E., *Serious Creativity*, HarperCollins, Toronto, Canada, 1992.

Dennis, N. B., in R. Clark and L. Canter, eds., "Can NEPA Prevent "Ecological Train Wrecks?"" *Environmental Policy and NEPA: Past, Present and Future*, St. Lucie Press, Boca Raton, Florida, 1997.

Denq, F., and J. Altenhofel, "Social Impact Assessments Conducted by Federal Agencies: An Evaluation," *Impact Assessment* **15**, 209–231 (1997).

Department of the Environment, Transport and the Regions (DRTR), *Public Participation in Making Local Environmental Decisions, The Aarhus Convention Newcastle Workshop*, DETR, London, U.K., 2000.

Dery, D., "Coping with 'Latent Time Bombs' in Public Policy," *Environmental Impact Assessment Review* **17**, 413–425 (1997).

DeSario, J., and S. Langton, eds., *Citizen Participation in Public Decision-making*. Greenwood, Westport, Connecticut, 1987.

Devall, B., *Deep Ecology*, Gibbs Smith, Salt Lake City, Utah, 1985.

Devuyst, D., "Sustainability Assessment: The Application of a Methodological Framework," *Journal of Environmental Assessment Policy and Management* 1, 459–488, 1999.

Dickman, M., "Failure of Environmental Impact Assessment to Predict the Impact of Mine Tailings of Canada's Most Northerly Hypersaline Lake," *Environmental Impact Assessment Review* 11, 171–180 (1991).

Diduck, A. P., and A. J. Sinclair in A. J. Sinclair, ed., "Testing the Concept of Critical Environmental Assessment (EA) Education," *Canadian Environmental Assessment in Transition*, Department of Geography, University of Waterloo, Waterloo, Ontario, Canada, 1997.

Diffenderfer, M., and D. Birch, "Bioregionalism: A Comparative Study of the Adirondacks and the Sierra Nevada," *Society and Natural Resources* 10, 3–16 (1997).

Dooley, J. E., in V. W. Maclaren and J. B. Whitney, eds., "Risk and Environmental Management," *New Directions in Environmental Impact Assessment in Canada*, Methuen, Toronto, Canada, 1985.

Dunning C. M., in J. L. Creighton, C. M. Dunning, J. D. Priscoli, and D. B. Ayres, eds., "Collaborative Problem Solving for Installation Planning and Decision Making," *Public Involvement and Dispute Resolution—A Reader on the Second Decade of Experience at the Institute for Water Resources*, Institute for Water Resources, U.S. Army Corps of Engineers, Alexandria, Virginia, 1999.

Eccleston, C. H., *The NEPA Planning Process: A Comprehensive Guide with Emphasis on Efficiency*, John Wiley & Sons, New York, 1999a.

Eccleston, C. H., "Determining If and When Impacts of Potential Accidents Need To Be Evaluated in a NEPA Analysis," *Environmental Practice* 1, 89–96 (1999b).

Economic Commission for Europe (ECE), *The Aarhus Convention: An Implementation Guide*, United Nations, New York and Geneva, 2000.

Elling, B. "Integration of Strategic Environmental Assessment into Spatial Planning," *Impact Assessment and Project Appraisal* 18, 233–244 (2000).

Emond, D. P., "Accommodating Negotiation/Mediation Within Existing Assessment and Approval Processes," *The Place of Negotiation in Environmental Assessment*, Canadian Environmental Assessment Research Council, Hull, Québec, Canada, 1990.

EnHealth Council, *Environmental Health Risk Assessment: Guidelines for Assessing Human Health Risks from Environmental Hazards*, Commonwealth of Australia, Canberra, Australia, 2001a.

EnHealth Council, *Health Impact Assessment Guidelines*, Commonwealth of Australia, Canberra, Australia, September 2001b.

Enk, G. A., and W. F. Hornick in F. A. Rossini and A. L. Porter, eds., "Human Values and Impact Assessment," *Integrated Impact Assessment*, Westview Press, Boulder, Colorado, 1983.

Ensminger, J. T., and R. B. McLean, "Reasons and Strategies for More Effective NEPA Implementation," *Environmental Professional* 15, 46–56 (1993).

Environment Australia, *Report on the Operation of the Environmental Protection and Biodiversity Conservation Act*, Department of the Environment and Heritage, Commonwealth of Australia, Canberra, Australia, December 2001.

Environmental and Social Systems Analysts (ESSA), *Review and Evaluation of Adaptive Environmental Assessment and Management*, Environment Canada, Vancouver, Canada, 1982.

Environmental Impact Screening Committee (EISC), "Benefit Guidelines Associated with Mineral Exploration in the Inuvialuit Settlement Region," *Operating Guidelines and Procedures*, EISAC, Inuvik, Northwest Territories, Canada, February 1999.

Environmental Resources Management (ERM), *Revision of EU Guidance Documents on EIA: First Interim Report*, European Commission DGXI, Edinburgh, Scotland, July 2000.

Environmental Resources Management (ERM), *Guidance on EIA Screening*, European Communities, Edinburgh, Scotland, June 2001a.

Environmental Resources Management (ERM), *Guidance on EIA Scoping*, European Communities, Edinburgh, Scotland, June 2001b.

Environmental Resources Management (ERM), *Guidance on EIA EIS Review*, European Communities, Edinburgh, Scotland, June 2001c.

Erckmann, J., "Commentary II," *Proceedings of the Workshop on Cumulative Environmental Effects: A Binational Perspective*, Canadian Environmental Assessment Research Council and U.S. National Research Council, Board of Basic Biology, Ottawa, Canada, 1986.

Erickson, P. A., *A Practical Guide to Environmental Impact Assessment*, Academic Press, San Diego, California, 1994.

Etzioni, A., "Mixed Scanning: A 'Third' Approach to Decision-making," *Public Administration Review* **27**, 385–392 (1967).

Etzioni, A., "Mixed Scanning Revisited," *Public Administration Review* **46**, 8–14 (1986).

Etzioni, A., *The Spirit of Community*, Simon and Schuster, New York, 1993.

Etzioni, A., *Rights and the Common Good*, St. Martin's Press, New York, 1995.

European Union, *Directive 2003/4/EC of the European Parliament and of the Council of January 2003 on Public Access to Environmental Information and Repealing Council Directive 90/313/EEC*, Official Journal of the European Union, Brussels, Belgium, 2003.

Euston, A. F., in R. Clark and L. Canter, eds., "NEPA's Interdisciplinary Mandate: Redirection for Sustainability," *Environmental Policy and NEPA: Past, Present and Future*, St. Lucie Press, Boca Raton, Florida, 1997.

Fainstein, S. S., and N. I. Fainstein, "Economic Restructuring and the Rise of Urban Social Movements," *Urban Affairs Quarterly* **21**, 187–206 (1985).

Faludi, A., *Critical Rationalism and Planning Methodology*, Pion, London, 1986.

Fell, A., and B. Sadler, "Public Involvement in Environmental Assessment and Management: Preview of IER Guidelines on Good Practice," *EA* June (1999).

Filyk, G., and R. Côté in R. Boardman, ed., "Pressure from Inside: Advisory Groups and the Environmental Policy Community," *Canadian Environmental Policy: Ecosystems, Politics and Process*, Oxford University Press, Toronto, Canada, 1992.

Finsterbusch, K., "In Praise of SIA: A Personal Review of the Field of Social Impact Assessment: Feasibility, Justification, History, Methods, Issues," *Impact Assessment* **13**, 229–252 (1995).

Firth, L. J., "Role of Values in Public Decision-making: Where Is the Fit?" *Impact Assessment and Project Appraisal* **16**, 325–329 (1998).

Fischer, F., in S. Campbell and S. Fainstein, eds., "Risk Assessment and Environmental Crisis: Toward an Integration of Science and Participation," *Readings in Planning Theory*, Blackwell Publishers Inc., Cambridge, Massachusetts, 1996.

Fischer, J. B., "Strategic Environmental Assessment Performance Criteria: The Same Requirements for Every Assessment?" *Journal of Environmental Assessment Policy and Management* **4**, 83–99 (2002).

Fischler, R., "Communicative Planning Theory: A Foucauldian Assessment," *Journal of Planning Education and Research* **19**, 358–368 (2000).

Fischhoff, B., P. Lichtenstein, P. Slovic, S. L. Derby, and R. L. Keeney, *Acceptable Risk*, Cambridge University Press, Cambridge, 1981.

Fischhoff, B. P., P. Slovic, and L. Lichtenstein, "Lay Foibles and Expert Judgments About Risks," *The American Statistician* **36**, 240–255 (1982).

Forester, J., "Bounded Rationality and the Politics of Muddling Through," *Public Administration Review* **44**, 23–31 (1984).

Forester, J., *Planning in the Face of Power*, University of California Press, Berkeley, California, 1989.

Forester, J., *The Deliberative Practitioner*, MIT Press, Cambridge, Massachusetts, 1999.

Foster, K. R., P. Vecchia, and M. H. Rapacholi, "Science and the Precautionary Principle," *Science* May 12, 979–981 (2000).

Freudenburg, W. R., in G. A. Daneke, M. W. Garcia, and J. D. Priscoli, eds., "The Promise and Peril of Public Participation in Social Impact Assessment," *Public Involvement and Social Impact Assessment*, Westview Press, Boulder, Colorado, 1983.

Freudenburg, W. R., "Social Impact Assessment," *Annual Review of Sociology* **12**, 451–478 (1986).

Friedmann, J., *Planning in the Public Domain: From Knowledge to Action*, Princeton University Press, Princeton, New Jersey, 1987.

Friend, J., and A. Hickling, *Planning Under Pressure: The Strategic Choice Approach*, Butterworth-Heinemann, Oxford, 1997.

Gadgil, M., F. Berkes, and C. Folke, "Indigenous Knowledge for Biodiversity Conservation," *Ambio* **22**, 151–156 (1993).

Gagnon, C., P. Hirsch, and R. Howitt, "Can SIA Empower Communities?" *Environmental Impact Assessment Review* **13**, 229–254 (1993).

Gardner, J., and M. Roseland, "Thinking Globally: The Role of Social Equity in Sustainable Development," *Alternatives* **16**, 26–34 (1989).

George, C., "Assessing Global Impacts at Sector and Project Levels," *Environmental Impact Assessment Review* **17**, 227–247 (1997).

George, C., "Testing for Sustainable Development Through Environmental Assessment," *Environmental Impact Assessment Review* **19**, 175–200 (1999).

George, C., R. Nafti, and J. Curran, "Capacity Building for Trade Impact Assessment: Lessons from the Development of Environmental Impact Assessment," *Impact Assessment and Project Appraisal* **19**, 311–319 (2001).

Gerrard, M. B., *Whose Background, Whose Risk*, MIT Press, Cambridge, Massachusetts, 1995.

Gibson, R. B., in E. Lykke, ed., "Respecting Ignorance and Uncertainty," *Achieving Environmental Goals: The Concept and Practice of Environmental Performance Review*, Belhaven, London, 1992.

Gibson, R. B., "Environmental Assessment Design: Lessons from the Canadian Experience," *Environmental Professional* **15**, 12–24 (1993).

Gibson, R. B., "Favouring the Higher Test: Contribution to Sustainability as the Central Criterion for Reviews and Decisions Under the Canadian Environmental Assessment Act," *Journal of Environmental Law and Practice* **10**, 39–55 (2000).

Gibson, R. B., *Specification of Sustainability-based Environmental Assessment Decision Criteria and Implications for Determining "Significance" in Environmental Assessment*, Canadian Environmental Assessment Agency, Hull, Québec, Canada, 2001.

Giere, R. N., *Science Without Limits*, University of Chicago Press, Chicago, 1999.

Gilpin, A., *Environmental Impact Assessment: Cutting Edge for the Twenty-First Century*, Cambridge University Press, Cambridge, 1995.

Glasson, J., R. Thérivel, and A. Chadwick, *Introduction to Environmental Impact Assessment*, 2nd ed., UCL Press Ltd., London, 1999.

Gleick, J., *Chaos: Making a New Science*, Viking Penguin, New York, 1988.

Goldstein, H. A., "Planning as Argumentation," *Environment and Planning: B—Planning and Design* **11**, 297–312 (1984).

Goodland, R., "Definition of Environmental Sustainability," *IAIA Newsletter* **5**, 1–2 (1993).

Gordon, W. J. J., *Synectics: The Development of Creative Capacity*, Collier Books, New York, 1961.

Gorz, A., *Ecology as Politics*, Black Rose Books, Montréal, Québec, Canada, 1980.

Government of British Columbia (GBC), *An Introduction Guide to Adaptive Management*, Forest Practice Branch, BC Forest Services, Ministry of Forests, Victoria, British Columbia, Canada, 1999.

Government of Canada, *A Canadian Perspective on the Precautionary Approach/Principle: Discussion Document*, Ottawa, Canada, September 2001.

Gower, B., *Scientific Method: An Historical and Philosophical Introduction*, Routledge, London, 1997.

Graham, J. D., "The Role of Precaution in Risk Assessment and Management: An American's View," *presentation to the Conference on the US, Europe, Precaution and Risk Management: A Comparative Case Study Analysis of the Management of Risk in a Complex World*, Administrator, Office of Information and Regulatory Affairs, OMB, Washington, D.C., 2002.

Gray, B., *Collaborating: Finding Common Ground for Multiparty Problems*, Jossey-Bass, San Francisco, 1989.

Greenall, W., "Organizing and Managing the Planning Team," *presentation to the Seminar on Integrated Approaches to Resource Planning*, Banff School of Management, Banff, Alberta, Canada, 1985.

Greene, B., *The Elegant Universe*, Vantage Press, New York, 1999.

Greer-Wooten, B., in A. J. Sinclair, ed., "Contexualizing the Environmental Impact Assessment Process," *Canadian Environmental Assessment in Transition*, Canadian Association of Geographers, Waterloo, Canada, 1997.

Grima, A. P., P. Timmerman, C. D. Fowle and P. Byer, *Risk Management and EIA: Research Needs and Opportunities*, Canadian Environmental Assessment Research Council, Ministry of Supply and Services, Ottawa, Canada, 1986.

Gullett, W., "Environmental Protection and the 'Precautionary Principle': A Response to Scientific Uncertainty in Environmental Management," *Environmental and Planning Law Journal* **14**, 52–69 (1997).

Gullett, W., "Environmental Impact Assessment and the Precautionary Principle: Legislating Caution in Environmental Protection," *Australian Journal of Environmental Management* **5**, 146–158 (1998).

Gullett, W., "The Precautionary Principle in Australian Policy, Law and Potential for Precautionary EIAs," *paper presented at the Risk Assessment and Policy Association Meeting*, Washington, DC, 1999.

Gullett, W., "Environmental Decision Making in a Transboundary Context: Principles and Challenges for the Denmark-Sweden Øresund Fixed Link," *Journal of Environmental Assessment Policy and Management* **2**, 529–559 (2000).

Gunderson, L., "Resilience, Flexibility and Adaptive Management: Antidotes for Spurious Certitude?" *Conservation Ecology* **3**, 7 (1999).

Habermas, J., *Justification and Application: Remarks on Discourse Ethics*, MIT Press, Cambridge, Massachusetts, 1993.

Hainer, R. M., in E. Glatt and M. V. Shelly, eds., "Rationalism, Pragmatism and Existentialism: Perceived But Undiscovered Multicultural Problems," *The Research Society*, Gordon and Breach, New York, 1968.

Halstead, J. H., R. A. Chase, S. H. Murdock, and F. L. Leistritz, *Socioeconomic Impact Management*, Westview Press, Boulder, Colorado, 1984.

Hance, B. J., C. Chess, and P. M. Sandman, *Industry Risk Communications Manual: Improving Dialogue with Communities*, Lewis Publishers, Chelsea, Michigan, 1990.

Haney, A., and R. L. Power, "Adaptive Management for Sound Ecosystem Management," *Environmental Management* **20**, 879–886 (1996).

Harashina, S., "Environmental Dispute Resolution Process and Information Exchange," *Environmental Impact Assessment Review* **15**, 69–80 (1995).

Hardstaff, P., "The Precautionary Principle, Trade and the WTO," *Discussion Paper for the European Commission Consultation on Trade and Sustainable Development*, Royal Society for the Protection of Birds, Bedfordshire, U.K., 2000.

Harper, T. L., and S. M. Stein, "The Centrality of Normative Ethical Theory to Contemporary Planning Theory," *Journal of Planning Education and Research* **11**, 105–116 (1992).

Harremoës, P., D. Gee, M. MacGarvins, J. Stirling, J. Keys, B. Wynee, and S. Guedes Vaz, eds., *Late Lessons from Early Warnings: The Precautionary Principle, 1896–2000*, Environmental Issues Report No. 22, European Environmental Agency, Copenhagen, Denmark, 2002.

Harrop, D. O., and J. A. Nixon, *Environmental Assessment in Practice*, Routledge, London, 1999.

Harvey, S., and A. Usher, "Communities and Natural Resource Management: Bridging the Gap," *Plan Canada* **36**, 34–38 (1996).

Healey, P., *Collaborative Planning: Shaping Places in Fragmented Societies*, UBC Press, Vancouver, British Columbia, Canada, 1997.

Health Canada, *Canadian Handbook on Health Impact Assessment*: Volume 1, The Basics; Volume 2, Decision Making in Environmental Health Impact Assessment; Volume 3, Roles for the Health Practitioner, Ottawa, Canada, 2000a.

Health Canada, *Public Involvement Framework and Guidelines*, Minister of Public Works and Government Services Canada, Ottawa, Canada, 2000b.

Hegmann, G. H., and G. A. Yarranton, *Cumulative Effects and the Energy Resources Conservation Board's Review Process*, Macleod Institute for Environmental Analysis, Working Paper No. 1, Calgary, Alberta, Canada, July 1995.

Heinman, M. K., "Science by the People: Grassroots Environmental Monitoring and the Debate over Scientific Expertise," *Journal of Planning Education and Research* **16**, 291–299 (1997).

Hendler, S., "Feminist Planning Ethics," *Journal of Planning Literature* **9**, 115–127 (1994).

Hettwer, R., "Ethical Factors in Environmental Negotiation," *Environmental Impact Assessment Review* **11**, 197–202 (1991).

Hirschhorn, L., "Scenario Writing: A Developmental Approach," *American Planning Association Journal* **46**, 172–183 (1980).

Hirschmann, D. J., "Peer Review and the Contract Between Science and Society," *Environmental Professional* **16**, 79–89 (1994).

Hoch, C., "Doing Good and Being Right," *Journal of the American Planning Association* **50**, 335–345 (1984).

Hoch, C., "Racism and Planning," *Journal of the American Planning Association* **59**, 451–460 (1993).

Hodgson, P., and R. P. White, *Relax Its Only Uncertainty*, Pearson Education Ltd., London, 2001.

Hoffmann, N., and G. Davidson, "Moving Toward Effective Decision-making," *Plan Canada* **37**, 29–32 (1997).

Hofstadter, D. R., *Metamagical Themes: Questing for the Essence of Mind and Pattern*, Basic Books, New York, 1985.

Hollick, M., "Self-organizing Systems and Environmental Management," *Environmental Management* **17**, 621–628 (1993).

Holling, C. S., ed., *Adaptive Environmental Assessment and Management*, John Wiley & Sons, Chichester, UK, 1978.

Holm, S., and J. Harris, "Precautionary Principle Stifles Discovery," *Nature* **400**, July 29, 1999.

Homer-Dixon, T., *The Ingenuity Gap*, Vintage Canada, Toronto, Canada, 2000.

Hood, J., and S. Nicholl, "The Role of Environmental Risk Management and Reporting: An Empirical Analysis," *Journal of Environmental Assessment Policy and Management* **4**, 1–29 (2002).

Hooper, B. P., G. T. McDonald, and B. Mitchell, "Facilitating Integrated Resource and Environmental Management: Australian and Canadian Perspectives," *Journal of Environmental Planning and Management* **42**, 747–766 (1999).

Howe, E., "Normative Ethics in Planning," *Journal of Planning Literature* **5**, 123–149 (1990).

Howell, R. E., M. E. Olsen, and D. Olsen, *Designing a Citizen Involvement Program: A Guidebook for Involving Citizens in the Resolution of Environmental Issues*, Oregon State University, Convallis, Oregon, 1987.

Hughes, R., in A. Donnelly, B. Dalal-Clayton, and R. Hughes, eds., "Environmental Impact Assessment and Stakeholder Involvement," *A Directory of Impact Assessment Guidelines*, 2nd ed., International Institute for Environment and Development, Russell Press, Nottingham, UK, 1998.

Hummel, R. P., *The Bureaucratic Experience*, St. Martin's Press, New York, 1977.

Hunsaker, Jr., D. B., and D. W. Lee, in B. Sadler, ed., "Environmental Impact Assessment of Abnormal Events: A Follow-up Study," *Auditing and Evaluation in Environmental Assessment and Management—Canadian and International Experience*, Enviromental Protection Services of Environment Canada and the Banff Centre, School of Management, Banff, Alberta, Canada, 1985.

Huxley, M., and O. Yiftachel, "New Paradigm or Old Myopia? Unsettling the Communicative Turn in Planning Theory," *Journal of Planning Education and Research* 19, 333–342 (2000).

Hyman, E. L., B. Stiftel, D. H. Moreau, and R. C. Nichols, *Combining Facts and Values in Environmental Impact Assessment*, Westview Press, Boulder, Colorado, 1988.

Ignatieff, M., *The Rights Revolution*, Canadian Broadcasting Corporation, Etobicoke, Ontario, Canada, 2000.

Iles, A. T., "Adaptive Management: Making Environmental Law Policy More Dynamic Experimentalist and Learning, *Environmental and Planning Law Journal* 13, 288–307 (1996).

Imperial College Consultants Ltd. (ICCL), Babtie Allott and Lomax, Austrian Institute for the Development of Environmental Assessment, ECA, CESAM, University of Ashus, *SEA and Integration of the Environment into Strategic Decision-making*, CEC Contract No. B4–3044/99/136634/MAR/B4, London, U.K., May 2001.

Innes, J. E., "Planning Theory's Emerging Paradigm: Communicative Action and Interactive Practice," *Journal of Planning Education and Research* 14, 183–190 (1995).

Innes, J. E., "Planning Through Consensus Building," *American Planning Association Journal* 62, 460–472 (1996).

Innes, J. E., "Information in Communicative Planning," *American Planning Association Journal* 64, 52–63 (1998).

Innes, J. E., and D. E. Booher, "Consensus Building and Complex Adaptive Systems," *Journal of the American Planning Association* 65, 412–423 (1999).

Institute of Public Health in Ireland (IPHI), *Health Impact Assessment: An Introductory Paper*, Dublin, Ireland, September 2001.

Interagency Ecosystem Management Task Force (IEMTF), *The Ecosystem Approach: Health Ecosystems and Sustainable Economics*: Volume I, Overview; Volume II, Implementation Issues, Washington, DC, November 1995.

International Associates, *Conflict Resolution in Organizations*, Handout from the Third National Conference in Peace-making and Conflict Resolution, Denver, Colorado, 1986.

International Association for Impact Association in cooperation with the Institute of Environmental Assessment (IAIA and IEA), *Principles of Environmental Impact Assessment Best Practice*, Fargo, N.D. and Lincoln, U.K., undated.

Interorganizational Committee on Guidelines and Principles, "Guidelines and Principles for Social Impact Assessment," *Impact Assessment* 12, 107–152 (1994).

Ison, E., and A. Miller, "The Use of LCA to Introduce Life-Cycle Thinking into Decision-making for the Purchase of Medical Devices in the NHS," *Journal of Environmental Assessment Policy and Management* 2, 453–476 (2000).

Jain, R. K., L. V. Urban, and H. E. Balbach, *Environmental Assessment*, McGraw-Hill, New York, 1993.

Jantscht, E., "Inter and Transdisciplinary University: A Systems Approach to Education and Innovation," *Ekistics* 32, 433 (1971).

Jasanoff, S., *Risk Precaution and Environmental Values*, Carnegie Council on Ethics and International Affairs, New York, undated.

Jeltes, R., and P. A. H. Hermens, "Towards a Systematic Approach of Effect Prediction: A Handbook as Guide and Aid for Environmental Prediction: A Review," *Environmental Management* **14**, 161–166 (1990).

Jepson, E. J., "Sustainability and Planning: Diverse Concepts and Close Associations," *Journal of Planning Literature* **14**, 499–510 (2001).

Jiliberto, R., "Decisional Environmental Values as the Object of Analysis by Strategic Environmental Assessment," *Impact Assessment and Project Appraisal* **20**, 61–70 (2002).

João, E. M., in A. L. Porter and J. Fittipaldi, eds., "Use of Geographical Information Systems in Impact Assessment," *Environmental Methods Review: Retooling Impact Assessment for the New Century*, Army Environmental Policy Institute and International Institute for Impact Assessment, Fargo, North Dakota, 1998.

Johannes, R. E., in J. T. Inglis, ed., "Integrating Traditional Ecological Knowledge and Management with Environmental Impact Assessment," *Traditional Ecological Knowledge: Concepts and Cases*, International Program on Traditional Ecological Knowledge and International Development Research Centre, Ottawa, Canada, 1999.

Johnson, B. L., "The Role of Adaptive Management as an Operational Approach for Resource Management Agencies," *Conservation Ecology* **3**, 8 (1999).

Johnson, S., *Emergence*, Scribner's, New York, 2001.

Jones, M. L., and L. A. Grieg in V. W. Maclaren and J. B. Whitney, eds., "Adaptive Environmental Assessment and Management: A New Approach to Impact Assessment," *New Directions in Environmental Impact Assessment*, Methuen, Toronto, Canada, 1988.

Julien, B., "Current and Future Directions for Structured Impact Assessments," *Impact Assessment* **13**, 403–432 (1995).

Kamrin, M. A., "The Need for Risk by Consensus," *Impact Assessment* **11**, 167–174 (1993).

Kasperson, R. E., P. G. Derr, and R. W. Kates in M. J. Pasqualetti and K. D. Pijawka, eds., "Confronting Equity in Radioactive Waste Management: Modest Proposals for a Socially Just and Acceptable Program," *Nuclear Power: Assessing and Managing Hazardous Technology*, Westview Press, Boulder, Colorado, 1984.

Kasperson, R. E., S. Ratick, and O. Renn, *Assessing the State/National Distributional Equity Issues Associated with the Proposed Yucca Mountain Repository: A Conceptual Approach*, State of Nevada Agency for Nuclear Projects/Nuclear Waste Project Office, Carson City, Nevada, 1988.

Kaufman, H., in R. Clark and L. Canter, eds., "The Role of NEPA in Sustainable Development," *Environmental Policy and NEPA*, St. Lucie Press, Boca Raton, Florida, 1997.

Kennedy, A. J., and W. A. Ross, "An Approach to Integrate Impact Scoping with Environmental Impact Assessment," *Environmental Management* **16**, 475–484 (1992).

Kent, R. B., and R. E. Klosterman, "GIS and Mapping: Pitfalls for Planning," *Journal of the American Planning Association* **66**, 189–198 (2000).

Keysar, E., and A. Steinemann, "Integrating Environmental Impact Assessment with Master Planning: Lessons from the US Army," *Impact Assessment Review* **22**, 583–611 (2002).

King, T. F., "Cultural Resources in an Environmental Assessment Under NEPA," *Environmental Practice* **4**, 137–144 (2002).

Koestler. A., *The Act of Creation*, Pan Piper, London, 1964.

Kørnøv, L., "Strategic Environmental Assessment and the Limits of Rationality in Decision Making Processes," *Presentation to the International Association for Impact Assessment*, Christchurch, New Zealand, 1998.

Kørnøv, L., and W. A. H. Thissen, "Rationality in Decision-making and Policy-making: Implications for Strategic Environmental Assessment," *Impact Assessment and Project Appraisal* **18**, 191–200 (2000).

Kozlowski, J., "Toward Ecological Orientation of the Planning Process: A Planner's Perspective," *Impact Assessment Bulletin* **8**, 47–68 (1990).

Kreske, D. L., *Environmental Impact Statements: A Practical Guide for Agencies, Citizens and Consultants*, John Wiley & Sons, New York, 1996.

Krieger, M. H., *Advice and Planning*, Temple University Press, Philadelphia, 1981.

Lahlou, M., and L. W. Canter, "Alternatives Evaluation and Selection in Development and Environmental Remediation Projects," *Environmental Impact Assessment Review* **13**, 37–62 (1993).

Lal, P., H. Lim-Applegate, and M. Scoccimarro, "The Adaptive Decision-making Process as a Tool for Integrated Natural Resource Management: Focus, Attitudes and Approach, *Conservation Ecology* **5** (2001).

Lauria, M., and M. J. Soll, "Communicative Action, Power and Misinformation in a Site Selection Process," *Journal of Planning Education and Research* **15**, 199–211 (1996).

Lawrence, D. P., "Designing and Adapting the EIA Planning Process," *Environmental Professional* **16**, 2–21 (1994).

Lawrence, D. P., "Quality and Effectiveness of Environmental Impact Assessment: Lessons and Insights from Ten Assessments in Canada," *Project Appraisal* **12**, 219–232 (1997a).

Lawrence, D. P., "Integrating Sustainability and Environmental Impact Assessment," *Environmental Management* **21**, 23–42 (1997b).

Lawrence, D. P., "The Need for EIA Theory-Building," *Environmental Impact Assessment Review* **17**, 79–107 (1997c).

Lawrence, D. P., "Choices for EIA Process Design and Management," *Journal of Environmental Assessment Policy Management* **3**, 437–464 (2001).

Lawrence, R. L., S. E. Daniels, and G. H. Stankey, "Procedural Justice and Public Involvement in Natural Resources Decision Making," *Society and Natural Resources* **10**, 577–589 (1997).

Laws, D., "The Practice of Fairness," *Environmental Impact Assessment Review* **16**, 65–70 (1996).

Lee, B., L. Hawath, and C. Brunk, "Value and Science in Impact Assessment," *Environments* **23**, 93–100 (1992).

Lee, K. N., "Appraising Adaptive Management," *Conservation Ecology* **3**, 2 (1999).

Lee, N., in N. Lee and C. George, eds., "Reviewing the Quality of Environmental Assessments," *Environmental Impact Assessment in Developing and Transitional Countries: Principles, Methods and Practice*, John Wiley & Sons, Chichester, UK, 2000.

Lehto, J., and A. Ritsatakis, "Health Impact Assessment as a Tool for Intersectoral Health Policy," *Discussion Paper for the Conference on Health Impact Assessment: From Theory to Practice*, European Centre for Health Policy, WHO Regional Office for Europe, Gothenburg, Sweden, 1999.

Lein, J. K., "Expressing Environmental Risk Using Fuzzy Variables: A Preliminary Examination," *Environmental Professional* **14**, 257–267 (1992).

Leith B., "The Social Cost of Sustainability," *Alternatives* **21**, 18–24 (1995).

Lemons, J., and D. A. Brown, "The Role of Science in the Decision to Site a High-Level Nuclear Waste Repository at Yucca Mountain, Nevada, USA," *Environmentalist* **10**, 3–24 (1990).

Lindblom, C., *The Intelligence of Democracy*, Free Press, New York, 1965.

Liu, F., "Dynamics and Causation of Environmental Equity, Locally Unwanted Land Uses, and Neighborhood Changes," *Environmental Management* **21**, 643–656 (1997).

Lockie, S., "SIA in Review: Setting the Agenda for Impact Assessment in the 21st Century," *Impact Assessment and Project Appraisal* **19**, 277–287 (2001).

Lou, D. K., and E. J. Rykiel, Jr., "Integrated Resource Management Systems: Coupling Expert Systems with Database Management and Geographic Information Systems," *Environmental Management* **16**, 167–177 (1992).

Lowry, K., P. Adler, and N. Milner, "Participating the Public: Group Process, Politics and Planning," *Journal of Planning Education and Research* **16**, 177–187 (1997).

Lynn, F. M., "Community–Scientist Collaboration in Environmental Research," *American Behavioral Scientist* **44**, 648–662 (2000).

Lynn, S., and P. Wathern, "Intervenor Funding in Environmental Assessment Processes in Canada," *Project Appraisal* **6**, 169 (1991).

MacNiven, C. D., *The Moral Question: Ethical Theory*, Ontario Educational Communications Authority, Toronto, Canada, 1982.

Malik, M., and R. Bartlett, "Formal Guidance for the Use of Science in EIA: Analysis of Agency Procedures for Implementing NEPA," *Environmental Professional* **15**, 34–45 (1993).

Mandelbaum, S. J., "Telling Stories," *Journal of Planning Education and Research* **10**, 209–214 (1991).

Manitoba Conservation, *Building a Sustainable Future: Proposed Changes to Manitoba's Environment Act: A Discussion Paper*, Winnipeg, Manitoba, Canada, September 2001.

Manring, N., P. C. West, and P. Bidol, "Social Impact Assessment and Environmental Conflict Management: Potential for Integration and Application," *Environmental Impact Assessment Review* **10**, 247–252 (1990).

March, F., *NEPA Effectiveness: Measuring the Process*, Government Institutes, Inc., Rockville, Maryland, 1998.

Margerum, R. D., "Integrated Approaches to Environmental Planning and Management," *Journal of Planning Literature* **11**, 459–475 (1997).

Marriott, B., *Environmental Impact Assessment: A Practical Guide*, McGraw-Hill, New York, 1997.

Marris, P., *The Politics of Uncertainty: Attachment in Public and Private Life*, Routledge, London, 1996.

Maser, C., *Resolving Environmental Conflict*, St. Lucie Press, Delray Beach, Florida, 1996.

Mayda, J., "Reform Impact Assessment: Issues, Premises and Elements," *Impact Assessment* **14**, 87–96 (1996).

Maynes, C., *Public Consultation: A Citizens Handbook*, Ontario Environmental Network, Toronto, Canada, 1989.

Mazmanion, D. A., and D. Morell in N. J. Vig and M. E. Kraft, eds., "The 'NIMBY' Syndrome: Facility Siting and the Failure of Democratic Discourse," *Environmental Policy in the 1990s*, 2nd ed., Congressional Quarterly, Inc., Washington, DC, 1994.

McClendon, B. W., "The Paradigm of Empowerment," *Journal of the American Planning Association* **59**, 145–147 (1993).

McDonald, G. T., and A. L. Brown, "Planning and Management Processes and Environmental Assessment," *Impact Assessment Bulletin* **8**, 261–274 (1990).

McGlennon, J. A. S., and L. Susskind, "Responsibility, Accountability and Liability in the Conduct of Environmental Negotiations," *The Place of Negotiations in Environmental Assessment*, Canadian Environmental Assessment Research Council, Hull, Québec, undated.

McIntyre, L., and M. Petticrew, *Methods of Health Impact Assessment: A Literature Review*, Occasional Paper, Medical Research Council Social and Public Health Services Unit, University of Glasgow, Glasgow, Scotland, 1999.

McLain, R. J., and R. G. Lee, "Adaptive Management: Promises and Pitfalls," *Environmental Management* **20**, 437–448 (1996).

McLoughlin, J. B., *Urban and Regional Planning: A Systems Approach*, Faber & Faber, London, 1969.

McNeil, D., and P. Freiberger, *Fuzzy Logic*, Simon and Schuster, New York, 1994.

Mebratu, D., "Sustainability and Sustainable Development: Historical and Conceptual Review," *Environmental Impact Assessment Review* **18**, 493–520 (1998).

Menard, L., *Pragmatism: A Reader*, Vintage Books, New York, 1997.

Metzger, J. T., "The Theory and Practice of Equity Planning: An Annotated Bibliography," *Journal of Planning Literature* **11**, 112–126 (1996).

Mezirow, J., "Understanding Transformation Theory," *Adult Education Quarterly* **44**, 222–232 (1994).

Michael, D. N., "Forecasting and Planning in an Incoherent Context," *Technological Forecasting and Social Change* **36**, 79–87 (1989).

Miller, A., "The Role of Analytical Science in Natural Resource Decision Making," *Environmental Management* **17**, 563–574 (1993).

Mintzberg, H., *The Rise and Fall of Strategic Planning*, Free Press, New York, 1994.

Mittelstaedt, G., T. Williams, and K. Ordon in R. Clark and L. Canter, eds., "NEPA and Tribal Matters," *Environmental Policy and NEPA: Past, Present and Future*, St. Lucie Press, Boca Raton, Florida, 1997.

Moore, C. W., *Negotiating Bargaining and Conflict Management*, prepared by CDR Associates for U.S. Army Corps of Engineers, Huntsville, Alabama, 1986.

Moore, C. W., in J. L. Creighton, C. M. Dunning, J. D. Priscoli, and J. D. Ayres, eds., "Negotiations," *Public Involvement and Dispute Resolution: A Reader on the Second Decade of Experience at the Institute for Water Resources*, Institute for Water Resources, U.S. Army Corps of Engineers, Alexandria, Virginia, 1999.

Moore, S. N., "A View of the History of the National Environmental Policy Act," *NAEP Newsletter* **17**, 8–9 (1992).

Moore, C. W., and P. Woodrow, in L. Susskind, S. McKearnen, and J. Thomas-Lamar, eds., "Visioning," *The Consensus Building Handbook–A Comprehensive Guide to Reaching Agreements*, Sage, Thousand Oaks, California, 1997.

Morell, D., "Siting and the Politics of Equity," *Hazardous Waste* **1**, 555–571 (1984).

Morgan, R. K., *Environmental Impact Assessment: A Methodological Perspective*, Kluwer Academic Press, Boston, 1998.

Morris, P., and R. Thérivel, eds., *Methods of Environmental Impact Assessment*, UBC Press, Vancouver, British Columbia, Canada, 1995.

Morrison-Saunders, A., J. Arts, J. Baker, and P. Caldwell, "Roles and Stakes in Environmental Impact Assessment Follow-up," *Impact Assessment and Project Appraisal* **19**, 289–296 (2001).

Morrone, M., "Walking a Tight Rope: Planning and the Environmental Movement," *Journal of Planning Literature* **7**, 100–107 (1992).

Mosley, D., C. Moore, M. Stagle, and D. Burns, in J. L. Creighton, C. M. Dunning, J. D. Priscoli, and D. B. Ayres, eds., "Partnering: Win-Win Strategic Planning in Action," *Public Involvement and Dispute Resolution—A Reader on the Second Decade of Experience at the Institute of Water Resources*, Institute of Water Resources, U.S. Army Corps of Engineers, Alexandria, Virginia, 1999.

Mostert, E., "Subjective Environmental Impact Assessment: Causes, Problems and Solutions," *Impact Assessment* **14**, 191–214 (1996).

Muvihill, P. R., and R. F. Keith, "Institutional Requirements for Adaptive EIS: The Kativik Environmental Quality Commission," *Environmental Impact Assessment Review* **9**, 399–412, 1989.

Myers, D., and A. Kituse, "Constructing the Future in Planning: A Survey of Theories and Tools," *Journal of Planning Education and Research* **19**, 221–231 (2001).

Nagel, J. H., *Participation*, Prentice Hall, Englewood Cliffs, New Jersey, 1987.

Neuman, M., "Communicate This! Does Consensus Lead to Advocacy and Pluralism?" *Journal of Planning Education and Research* **19**, 343–350 (2000).

Newman, W. L., *Social Research Methods*, 3rd ed., Allyn and Bacon, Boston, 1997.

Nicolas, G., and I. Prigogine, *Exploring Complexity: An Introduction*, W.H. Freeman and Co., New York, 1989.

Nilsson, M., and H. Dalkmann, "Decision Making and Strategic Environmental Assessment," *Journal of Environmental Assessment Policy and Management* **3**, 305–328 (2001).

Nitz, T., and A. L. Brown, "SEA Must Learn How Policy Making Works," *Journal of Environmental Assessment Policy and Management* **3**, 329–342 (2001).

Noble, B. F., "Strategic Environmental Assessment: What Is It and What Makes It Strategic?" *Journal of Environmental Assessment Policy and Management* **2**, 203–224 (2000a).

Noble, B. F., "Strengthening EIA Through Adaptive Management: A Systems Perspective," *Environmental Impact Assessment Review* **20**, 97–112 (2000b).

Noorbakhsh, F., and S. Ranjan, "A Model of Sustainable Development: Integrating Environmental Assessment and Project Planning," *Impact Assessment and Project Appraisal* **17**, 283–294 (1999).

Nooteboom, S., "Environmental Assessments of Strategic Decisions and Project Decisions: Integration and Benefits," *Impact Assessment and Project Appraisal* **18**, 151–160 (2000).

Nooteboom, S., and K. Wieringa, "Comparing Strategic Environmental Assessment and Integrated Environmental Assessment," *Journal of Environmental Assessment Policy and Management* **1**, 441–457 (1999).

Northern and Yorkshire Public Health Observatory (NYPHO), "An Overview of Health Impact Assessment," *Occasional Paper No. 1*, NYPHO, Stockton on Tees, UK, May 2001.

Novek, J., "Environmental Impact Assessment and Sustainable Development: Case Studies of Environmental Conflict," *Society and Natural Resources* **8**, 145–159 (1995).

Offringa, P., in R. Clark and L. Canter, eds., "Creating a User-Friendly NEPA," *Environmental Policy and NEPA: Past, Present and Future*, St. Lucie Press, Boca Raton, FL, 1997.

O'Riordan, T. ed., *Environmental Science for Environmental Management*, John Wiley & Sons, New York, 1995.

O'Riordan, T., and J. Cameron, *Interpreting the Precautionary Principle*, Earthscan Publishers Ltd., London, 1994.

Ortolano, L., "Controls on Project Proponents and Environmental Impact Assessment Effectiveness," *Environmental Professional* **15**, 352–363 (1993).

Ortolano, L., *Environmental Regulation and Impact Assessment*, John Wiley & Sons, New York, 1997.

Ortolano, L., and A. Shepherd in F. Vanclay and D. A. Bronstein, eds., "Environmental Impact Assessment," *Environmental Impact Assessments*, John Wiley & Sons, New York, 1995.

Ostrom, E., "A Behavioral Approach to the Rational Choice Theory of Collective Action: Presidential Address, American Political Science Association," *American Political Science Review* **92**, 1–22 (1998).

Ozawa, C. P., *Recasting Science: Consensual Procedures in Public Policy Making*, Westview Policy Making, Boulder, Colorado, 1991.

Paci, C., A. Tobin, and P. Robb, "Reconsidering the Canadian Environmental Impact Assessment Act: A Place for Traditional Environmental Knowledge," *Environmental Impact Assessment Review* **22**, 118–128 (2002).

Parashar, A., R. Paliwai, and P. Rambabu, "Utility of Fuzzy Cross-Impact Simulation in Environmental Assessment," *Environmental Impact Assessment Review* **17**, 427–447 (1997).

Parenteau, R., *Public Participation in Environmental Decision-making*, Federal Environmental Assessment Review Office, Hull, Québec, Canada, 1988.

Parkin, J. B., "When Science Is Not Enough: A Case Study in Social Impact Mitigation," *Impact Assessment* **14**, 321–327 (1996).

Parry, J., and A. Stevens, "Prospective Health Impact Assessment: Pitfalls, Problems and Possible Ways Forward," *BMJ* **323**, 1177–1182 (2001).

Partidário, M. R., "Strategic Environmental Assessment: Key Issues Emerging From Recent Practice," *Environmental Impact Assessment Review* **16**, 31–55 (1996).

Pateman, C., *Participation and Democratic Theory*, Cambridge University Press, London, 1970.

Patterson, M. E., and D. R. Williams, "Paradigms and Problems: The Practice of Social Science in Natural Resource Management," *Society and Natural Resources* **11**, 279–295 (1998).

Patton, B. R., K. Giffin, and E. N. Patton, *Decision-making Group Interaction*, HarperCollins, New York, 1989.

Patton, C. V., and D. S. Sawicki, *Basic Methods of Policy Analysis and Planning*, Prentice Hall, Upper Saddle River, New Jersey, 1993.

Petts, J., ed., *Handbook of Environmental Impact Assessment: Process, Methods and Potential*, Blackwell Science, Oxford, 1999.

Pickering, A., *The Mangle of Practice: Time, Agency and Science*, University of Chicago Press, Chicago, 1995.

Pickvance, C., "The Rise and Fall of Urban Movements and the Role of Comparative Analysis," *Environment and Planning D: Society and Space* **3**, 31–53 (1985).

Porritt, J., *Playing Safe: Science and the Environment*, Thames and Hudson, New York, 2000.

Porter, A. L., "Technology Assessment," *Impact Assessment* **13**, 135–152 (1995).

Poulton, M. C., "Rationalism Versus Radicalism: The Debate Revisited," *Plan Canada* **30**, 37–40 (1990).

Powell, D., *The Role of Risk Perception and the Public in Decision-making on Risk*, Solid and Hazardous Waste Management Series, Publication WM-84-12, University of Toronto, Toronto, Canada, 1984.

Power, M., G. Power, and D. G. Dixon, "Decision-making in Environmental Effects Monitoring," *Environmental Monitoring* **19**, 629–639 (1995).

Power, M., and S. M. Adams, "Perspectives on the Scientific Community on the Status of Ecological Risk Assessment," *Environmental Management* **21**, 803–830 (1997).

Praxis, *Public Involvement: Planning and Implementing Public Involvement Programs*, prepared for the Canadian Federal Environmental Assessment Review Office, Calgary, Alberta, Canada, 1988.

Presidential/Congressional Commission on Risk Assessment and Risk Management (PCCRARM), *Framework for Environmental Health Risk Management*, Final Report, Volume 1, Washington, DC, 1997a.

Presidential/Congressional Commission on Risk Assessment and Risk Management (PCCRARM), *Risk Assessment and Risk Management in Regulatory Decision-making*, Final Report, Volume 2, Washington, DC, 1997b.

Presidential Council on Sustainable Development (PCSD), *Lessons Learned from Collaborative Approaches*, Final Draft, New National Opportunities Task Force, Washington, D.C., April 1997.

Pressman, J. L., and A. B. Wildavsky, *Implementation*, University of California Press, Berkeley, California, 1973.

Priscoli, J. D., "The Enduring Myths of Public Involvement," *Citizen Participation* **3**, 5–7 (1982).

Priscoli, J. D., in J. L. Creighton, C. M. Dunning, J. D. Priscoli, and D. B. Ayres, eds., "Public Involvement; Conflict Management; and Dispute Resolution in Water Resources and Environmental Decision-making," *Public Involvement and Dispute Resolution: A Reader on the Second Decade of Experience at the Institute for Water Resources*, Institute for Water Resources, U.S. Army Corps of Engineers, Alexandria, Virginia, 1999.

Priscoli, J. D., and P. Homenuck in R. Lang, ed., "Consulting the Publics," *Integrated Approaches to Resource Planning and Management*, Banff Centre School of Management, Banff, Alberta, Canada, 1986.

Rabe, B. G., *Beyond NIMBY: Hazardous Waste Siting in Canada and the United States*, Brookings Institution, Washington, DC, 1994.

Radford, K. J., in B. H. Massam, ed., "Complex Decision Situations," *Complex Locational Problems: Interdisciplinary Approaches*, Institute for Social Research, North York, Canada, 1988.

Raffensperger, C., and P. deFur, "A Paradigm Shift: Rethinking Environmental Decision Making and Risk Assessment," a paper presented to the Risk Assessment and Policy Association Meeting, Virginia, March 6, 1997.

Rahm-Crites, L., in A. L. Porter and J. J. Fittipaldi, eds., "Risk Communication in Environmental Assessment," *Environmental Methods Review: Retooling Impact Assessment for the New Century*, Army Environmental Policy Institute and International Association for Impact Assessment, Fargo, North Dakota, 1998.

Raiffa E., "Science and Policy: Their Separation and Integration in Risk Analysis," *The American Statistician* **36**, 225–233 (1982).

Ratanachai, C., "Environmental Impact Assessment as a Tool for Risk Management," *Toxicology and Industrial Health* **7**, 379–391 (1991).

Ravetz, J., "Integrated Assessment Models: From Global to Local," *Impact Assessment and Project Appraisal* **16**, 147–154 (1998).

Rawls, J., *A Theory of Justice*, Harvard University Press, Cambridge, Massachusetts, 1971.

Rawls, J., *Justice as Fairness: A Restatement*, Harvard University Press, Cambridge, Massachusetts, 2001.

Reckhow, K. E., "Importance of Scientific Uncertainty in Decision Making," *Environmental Management* **18**, 161–166 (1994).

Reinke, D. C., and L. L. Swartz, *The NEPA Reference Guide*, Battelle Press, Columbia, Ohio, 1999.

Rickson, R. E., R. B. Burdge, T. Hundloe, and G. T. McDonald, "Institutional Constraints to Adoption of Social Impact Assessment as a Decision–making and Planning Tool," *Environmental Impact Assessment Review* **10**, 233–243 (1990a).

Rickson, R. E., J. S. Western, and R. J. Burdge, "Social Impact Assessment: Knowledge and Development," *Environmental Impact Assessment Review* **10**, 1–10 (1990b).

Ridgeway, B., "The Project Cycle and the Role of EIA and EMS," *Journal of Environmental Assessment and Management* **3**, 93–106 (1999).

Rittel, H. W. J., and M. W. Webber, "Dilemmas in a General Theory of Planning," *Policy Sciences* **4**, 155–169 (1973).

Ritzdorf, M., in S. Campbell and S. Fainstein, eds., "Feminist Thoughts in the Theory and Practice of Planning," *Readings in Planning Theory*, Blackwell Publishers Inc., Cambridge, Massachusetts, 1996.

Ritzer, G., *Modern Social Theory*, 4th ed., McGraw-Hill, New York, 1996.

Robinson, J., G. Francis, R. Legge, and S. Lerner, "Defining a Sustainable Society," *Alternatives* **17**, 36–46 (1990).

Rocha, E. M., "A Ladder of Empowerment," *Journal of Planning Education and Research* **17**, 31–44 (1997).

Rodwin, M. A., "Can Bargaining and Negotiation Change the Administrative Process?" *Environmental Impact Assessment Review* **3**, 373–386 (1982).

Rorty, R., *Objectivity, Relativism and Truth*, Cambridge University Press, Cambridge, 1991.

Ross, H., "Community Social Impact Assessment: A Framework for Indigenous People," *Environmental Impact Assessment Review* **10**, 185–194 (1990).

Rothblatt, D. N., "Multiple Advocacy," *Journal of the American Institute of Planners* **44**, 193–199 (1978).

Rothman, T., and G. Sudarshan, *Doubt and Certainty*, Perseus Books, Reading, Massachusetts, 1998.

Rowe, P. G., *Design Thinking*, MIT Press, Cambridge, Massachusetts, 1991.

Rowe, W. D., "Understanding Uncertainty," *Risk Analysis* **14**, 41–50 (1994).

Rowson, J., in J. A. Sinclair, ed., "The Appraisal of Environmental Impact Assessment Processes: Towards a 'Situated' Approach," *Canadian Environmental Assessment in Transition*, University of Waterloo, Waterloo, Ontario, Canada, 1997.

Saarikoski, H., "Environmental Impact Assessment as a Collaborative Learning Process," *Environmental Impact Assessment Review* **20**, 681–700 (2000).

Sadler, B., "Environmental Assessment in a Changing World; Evaluating Practice to Improve Performance (final report)," *International Study of the Effectiveness of Environmental Assessment*, Canadian Environmental Assessment Agency and International Association for Impact Assessment, Hull, Québec, Canada, 1996.

Sadler, B., "Building Mediation into the Federal Environmental Assessment Process," *The Place of Negotiations in Environmental Assessment*, Canadian Environmental Assessment Agency, Hull, Québec, Canada, undated.

Sager, T., *Communicative Planning Theory*, Avebury, Brookfield, Vermont, 1994.

Sallenave, J., "Giving Traditional Ecological Knowledge Its Rightful Place in Environmental Impact Assessment," *CARC—Northern Perspectives* **22**, 1–7 (1994).

Sánchez, L. E., and T. Hacking, "An Approach to Linking Environmental Impact Assessment and Environmental Management Systems," *Impact Assessment and Project Appraisal* **20**, 25–38 (2002).

Sandman, P. M., *Responding to Community Outrage: Strategies for Effective Risk Communications*, American Industrial Hygiene Association, Fairfax, Virginia, 1992.

Saskatchewan Environment and Resource Management (SERM), *Public Involvement Policy Framework and Guidelines*, Saskatchewan Environment and Resource Management, Public Involvement Working Group, Regino, Saskatchewan, Canada, undated.

Saskatchewan Indian Federated College (SIFC), *Co-managing Natural Resources with First Nations*, Guidelines to Reaching Agreements and Making Them Work, Department of Indian Affairs and Northern Development, Regino, Saskatchewan, Canada, April 1996.

Saul, J. R., *Voltaire's Bastards*, Penguin Books, London, 1992.

Saunders, R., and A. Stephens, "Mount Sterling: Political and Environmental Convergence for Sustainable Development," *Environmental Impact Assessment Review* **19**, 319–332 (1999).

Science and Environmental Health Network (SEHN), *Risk Assessment and Management*, Ames, Indiana, undated.

Schön, D. A., *The Reflective Practitioner*, Basic Books, New York, 1983.

Schön, D. A., and M. Rein, *Frame Reflection: Toward the Resolution of Intractable Policy Disputes*, Basic Books, New York, 1994.

Scottish Needs Assessment Programme, *Health Impact Assessment: Piloting the Process in Scotland*, Scottish Needs Assessment Programme, Glasgow, Scotland, 2000.

Scott-Samuel, A., M. Birley, and K. Ardern, *The Merseyside Guidelines for Health Impact Assessment*, Merseyside Health Impact Assessment Steering Group, Liverpool, UK, 1998.

Seley, J. E., *The Politics of Public-Facility Planning*, Lexington Books, Lexington, Massachusetts, 1983.

Serafin, R., G. Nelson, and R. Butler, "Post Hoc Assessment in Resource Management and Environmental Planning: A Typology and Three Case Studies," *Environmental Impact Assessment Review* **12**, 271–294 (1992).

Shearman, R., "The Meaning and Ethics of Sustainability," *Environmental Management* **14**, 1–8 (1990).

Shepherd, A., and C. Bowler, "Beyond the Requirements: Improving Public Participation in EIA," *Journal of Environmental Planning and Management* **40**, 739–750 (1997).

Shoemaker, D. J., *Cumulative Environmental Assessment*, Waterloo, Ontario, Department of Geography Publications Series No. 42, University of Waterloo, Waterloo, Ontario, Canada, 1994.

Simon, H., *Administrative Behavior: A Study of Decision-Making Processes in Administrative Organizations*, Third Edition, Free Press, New York, 1976.

Sinclair, A. J., and A. P. Diduck, "Public Involvement in EA in Canada: A Transformative Learning Perspective," *Environmental Impact Assessment Review* **21**, 113–136 (2001).

Sipe, N. G., "An Empirical Analysis of Environmental Mediation," *Journal of the American Planning Association* **64**, 275–285 (1998).

Sipe, N. G., and B. Stiftel, "Mediating Environmental Enforcement Disputes: How Well Does It Work?" *Environmental Impact Assessment Review* **15**, 139–156 (1995).

Siting Process Task Force (SPTF), *Options for Co-operation: Report of the Siting Process Task Force on Low-Level Radioactive Waste Disposal*, Energy, Mines and Resources Canada, Ottawa, Canada, undated.

Skea, J., "Scenarios as Frameworks for Thinking: The Foresight Process," *Environmental Assessment*, June (1999).

Sköllerhorn, E., "Habermas and Nature: The Nature of Communicative Action for Studying Environmental Policy," *Journal of Environmental Planning and Management* **41**, 555–573 (1998).

Slocombe, D. S., "Environmental Planning, Ecosystem Science and Ecosystem Approaches for Integrating Environment and Development," *Environmental Management* **17**, 289–303 (1993).

Slovic, P., "Perception of Risk," *Science*, 280–285 (1987).

Smit, B., and H. Spaling, "Methods for Cumulative Effects Assessment," *Environmental Impact Assessment Review* **15**, 81–106 (1995).

Smith, L. G., *Impact Assessment and Sustainable Resource Management*, John Wiley & Sons, New York, 1993.

Smith, S. P., and W. R. Sheate, "Sustainability Appraisal of English Regional Plans: Incorporating the Requirements of the EU Strategic Environmental Assessment Directive," *Impact Assessment and Project Appraisal* **19**, 263–276 (2001).

Smith, L. G., C. Y. Nell, and M. V. Prystupa, "The Converging Dynamics of Interest Representation in Resources Management," *Environmental Management* **21**, 139–146 (1997).

Soloman, R. M., S. Yonts-Shepard, and W. T. Supulski II in R. Clark and L. Canter, eds., "Public Involvement Under NEPA: Trends and Opportunities," *Environmental Policy and NEPA: Past, Present and Future*, St. Lucie Press, Boca Raton, Florida, 1997.

Sorenson, A. D., and M. L. Auster, "Fatal Remedies: The Sources of Ineffectiveness," *Town Planning Review* **60**, 29–43 (1989).

Spaling, H., B. Smit, and R. Kreutzwiser, "Evaluating Environmental Impact Assessment: Approaches, Lessons and Prospects," *Environments* **22**, 63–74 (1993).

Spooner, B., in A. Donnelly, B. Dalal-Clayton, and R. Hughes, eds., "Review of the Quality of EIA Guidelines, Their Use and Circumnavigation," *A Directory of Impact Assessment Guidelines*, 2nd ed., International Institute for Environment and Development, Russell Press, Nottingham, UK, 1998.

Stackelberg, K. V., and D. E. Burmaster, "A Discussion of the Use of Probabilistic Risk Assessment in Human Health Risk Assessment," *Environmental Impact Assessment Review* **14**, 385–401 (1994).

Starr, G., A. Langley, and A. Taylor, *Environmental Health Risk Perception in Australia*, A Research Report to the Commonwealth Department of Health and Aged Care, Centre for Population Studies in Epidemiology, South Australian Department of Human Services, Adelaide, Australia, 2000.

Stec, S., ed., *Handbook on Access to Justice Under the Aarhus Convention*, Regional Environmental Center for Central and Eastern Europe, Szentendre, Hungary, 2003.

Steinemann, A., "Rethinking Human Health Impact Assessment," *Environmental Impact Assessment Review* **20**, 627–645 (2000).

Steinemann, A., "Improving Alternatives for Environmental Impact Assessment," *Environmental Impact Assessment Review* **21**, 13–22 (2001).

Stern, P. C., and H. V. Fineberg, *Understanding Risk: Informing Decisions in a Democratic Society*, National Research Council, National Academic Press, Washington, DC, 1996.

Susskind, L., in L. Susskind, S. McKearnen, and J. Thomas-Lamar, eds., "Part One: A Short Guide to Consensus Building," *The Consensus Building Handbook: A Comprehensive Guide to Reaching Agreement*, Sage, Thousand Oaks, California, 1999.

Susskind, L., and J. Cruikshank, *Breaking the Impasse: Consensual Approaches to Resolving Public Disputes*, Basic Books, New York, 1987.

Susskind, L., and D. Madigan, "New Approaches to Resolving Disputes in the Public Sector," *Justice Systems Journal* **9**, 179–203 (1984).

Susskind, L., S. McKearnen, and J. Thomas-Larmar, *The Consensus Building Handbook: A Comprehensive Guide to Reaching Agreement*, Sage, Thousand Oaks, California, 1999.

Suter, G. W. II., *Ecological Risk Assessment*, Lewis Publishers, Chelsea, UK, 1993.

Taylor, N., "Mistaken Interests and the Discourse Model of Planning," *Journal of the American Planning Association* **64**, 64–76 (1998).

Taylor, P. W., *Respect for Nature: A Theory of Environmental Ethics*, Princeton University Press, Princeton, New Jersey, 1986.

Teller, J., and A. Bond, "Review of Present Environmental Policies and Legislation Involving Cultural Heritage," *Environmental Impact Assessment Review* **22**, 611–632 (2002).

Thérivel, R., "Systems of Strategic Environmental Impact Assessment," *Environmental Impact Assessment Review* **13**, 145–168 (1993).

Thérivel, R., and D. Minas, "Ensuring Effective Sustainability Appraisal," *Impact Assessment and Project Appraisal* **29**, 81–91 (2002).

Thérivel, R., E. Wilson, S. Thompson, D. Heaney, and D. Prichard, *Strategic Environmental Assessment*, Earthscan Publications, London, 1992.

Thompson, J. G., and G. Williams, "Vertical Linkage and Competition for Local Political Power: A Case of Natural Resource Development and Federal Land Policy," *Impact Assessment Bulletin* **10**, 33–58 (1992).

Throgmorton, J. A., "On the Virtues of Skillful Meandering," *American Planning Association Journal* **66**, 367–383 (2000).

Tickner, J., "A Commonsense Framework for Operationalizing the Precautionary Principle," *paper presented at the wingspread Conference, Strategies for Implementing the Precautionary Principle*, Racine, Washington, January, 23–25 1998.

Tickner, J., and C. Raffensperger, *The Precautionary Principle in Action: A Handbook*, Science and Environmental Health Network, Ames, Indiana, 1998.

Todd, S., "Building Consensus on Divisive Issues: A Case Study of the Yukon Wolf Management Team," *Environmental Impact Assessment Review* **22**, 655–684 (2002).

Tomlinson, P., and S. F. Atkinson, "Environmental Audits: Proposed Terminology," *Environmental Monitoring and Assessment* **8**, 187–198 (1987).

Tonn, B. E., "Environmental Decision Making in the Face of Uncertainty," *Environmental Practice* **2**, 188–202 (2000).

Torgerson, D., *Industrialization and Assessment*, York University Publications in Northern Studies, Toronto, Canada, 1980.

Torgerson, D., in F. J. Tester and W. Mykes, eds., "SIA as a Social Phenomenon: The Problem of Contextuality," *Social Impact Assessment*, Detselig Enterprises, Calgary, Alberta, Canada, 1981.

Treasury Board of Canada Secretariat, *Integrated Risk Management Framework*, Ottawa, Ontario, Canada, March 2000.

Treweek, J., "Ecological Impact Assessment," *Impact Assessment* **13**, 289–315 (1995).

Treweek, J., *Ecological Impact Assessment*, Blackwell Science, Oxford, 1999.

Treweek, J., and P. Handkard, in A. L. Porter and J. Fittipaldi, eds., "Ecological Impact Assessment," *Enviromental Methods Review: Retooling Impact Assessment for the New Century*, Press Club, Fargo, North Dakota, 1998.

Trist, E., "The Environment and System-Response Capability," *Futures* **12**, 113–127 (1980).

Tywoniuk, N., "Integrating Environmental Factors into Both Planning and Impact Assessment Functions," *Impact Assessment Bulletin* **8**, 275–288 (1990).

Ugoretz, S. M., in S. G. Hildebrand and S. G. Cannon, eds., "Risk Assessment and Environmental Impact Assessment," *Environmental Analysis: The NEPA Experience*, Lewis Publishers, Boca Raton, Florida, 1993.

United Nations Economic Commission for Europe (UNECE), *Convention on Access to Information, Public Participation in Decision-making and Access to Justice in Environmental Matters*, Fourth Ministerial Conference—Environment for Europe, Arthus, Denmark, 1998.

United Nations Environment Program (UNEP), *Environmental Impact Assessment Training Manual*, prepared for the UNEP by Environment Australia, Canberra, Australia, 1997.

United National Environment Program (UNEP), *Guidelines for Incorporating Environmental Impact Assessment Legislation and/or Process and in Strategic Environmental Assessment*, Secretariat of the Convention on Biological Diversity, The Hague, the Netherlands, 2002.

U.S. Army Corps of Engineers (US ACE), *Guidelines for Risk and Uncertainty Analysis in Water Resources Planning*, Volume 1, Principles with Technical Appendices, Water Resources Support Center, Institute for Water Resources, IWR Report 92-R-1, Alexandria, Virginia, March 1992.

U.S. Council on Environmental Quality (US CEQ), *Incorporating Biodiversity Considerations into Environmental Impact Analysis Under the National Environmental Policy Act*, CEQ, Executive Office of the President, Washington, DC, 1993.

U.S. Council on Environmental Quality (US CEQ), *The National Environmental Policy Act—A Study of its Effectiveness After Twenty-five Years*, CEQ, Executive Office of the President, Washington, DC, January 1997a.

U.S. Council on Environmental Quality (US CEQ), *Considering Cumulative Effects Under the National Environmental Policy Act*, CEQ, Executive Office of the President, Washington, DC, January 1997b.

U.S. Department of Energy (US DOE), *NEPA Environmental Impact Statement Checklist*, US DOE, Office of Environment, Safety and Health, Office of NEPA Policy and Assistance, Washington, DC, November 1997.

U.S. Department of Energy (US DOE), *Effective Public Participation Under the National Environmental Policy Act*, Environment, Safety and Health Office of NEPA Policy and Assistance, Washington, DC, August 1998.

U.S. Department of Energy (US DOE), *Directory of Potential Stakeholders for Department of Energy Actions under the National Environmental Policy Act*, 17th ed., Washington, DC, January 2002.

U.S. Environmental Protection Agency (US EPA), *Final Guidance for Incorporating Environmental Justice Concerns in EPA's NEPA Compliance Analysis*, Washington, DC, April 1998a.

U.S. Environmental Protection Agency (US EPA), *Principles of Environmental Impact Assessment: An International Training Course*, Washington, DC, April 1998b.

U.S. Environmental Protection Agency (US EPA), *Guidelines for Ecological Risk Assessment*, EPA/630/R-95/002F, Risk Assessment Forum, Federal Register **63(93)**: 26845–26924, Washington, DC, April 1998c.

U.S. Environmental Protection Agency (US EPA), *Guidelines for Carcinogen Risk Assessment*, Risk Assessment Forum, NCEA-F-0644, Review Draft, Washington, DC, July 1999.

U.S. Environmental Protection Agency (US EPA), *The Model Plan for Public Participation*, developed by the Public Participation and Accountability Subcommittee of the National Environmental Justice Advisory Council, Washington, DC, February 2000a.

U.S. Environmental Protection Agency (US EPA), *Resource Guide: Resolving Environmental Conflicts in Communities*, EPA 360-F-00-001, Washington, DC, May 2000b.

U.S. Environmental Protection Agency (US EPA), *Stakeholder Involvement and Public Participation at the US EPA: Lessons Learned, Barriers and Innovative Approaches*, Washington, DC, January 2001a.

U.S. Environmental Protection Agency (US EPA), *Public Involvement in EPA Decisions*, A National Dialogue Convened by the U.S. Environmental Protection Agency, Washington, DC, July 2001b.

U.S. National Research Council (US NRC), *Risk Assessment in the Federal Government: Managing the Process*, National Academy Press, Washington, DC, 1983.

U.S. National Research Council (US NRC), *Science and Judgment in Risk Assessment*, National Academy Press, Washington, DC, 1994.

U.S. National Research Council (US NRC), *Understanding Risk: Informing Decisions in a Democratic Society*, National Academy Press, Washington, DC, 1997.

Vanclay, F., "Conceptualizing Social Impacts," *Environmental Impact Assessment Review* **22**, 183–212 (2002).

Vanclay, F., and S. A. Bronstein, *Environmental and Social Impact Assessment*, John Wiley & Sons, New York, 1995.

Van Der Vorst, R., A. Grafé-Buckens, and W. R. Sheate, "A Systematic Framework for Environmental Decision-making," *Journal of Environmental Assessment Policy and Management* **1**, 1–27 (1999).

VanGundy, A. B., *Techniques of Structured Problem Solving*, 2nd ed., Van Nostrand Reinhold Company, New York, 1988.

Verloo, M., and C. Roggeband, "Gender Impact Assessment: The Development of a New Instrument in the Netherlands," *Impact Assessment* **14**, 3–20 (1996).

Verma, N., *Similarities, Connections and Systems: The Search for a New Rationality for Planning and Management*, Lexington Books, Lanham, Maryland, 1998.

Verma, V. K., *Organizing Projects for Success*, The Human Aspects of Project Management, Project Management Institute, Upper Darby, Pennsylvania, 1995.

Wachs, M., "Forecasting Versus Envisioning," *American Planning Association Journal* **67**, 367–372 (2001).

Wackernagel, M., and W. Rees, *Our Ecological Footprint*, New Society Publishers, Gabriola Island, British Columbia, Canada, 1996.

Walker, L. J., and J. Johnston, *Guidelines for the Assessment of Indirect and Cumulative Impacts as Well as Impact Interactions*, EC DG XI, Environment, Nuclear Safety and Civil Protection, Hyder, European Commission, Brussels, Belgium, May 1999.

Walters, C., *Adaptive Management of Renewable Resources*, Macmillan, New York, 1986.

Walters, C. J., "Challenges in Adaptive Management of Riparian and Coastal Ecosystems," *Conservation Ecology* **1** (1997).

Wandesforde-Smith, G., in R. V. Bartlett, ed., "Environmental Impact Assessment, Entrepreneurship and Policy Change," *Policy Through Impact Assessment*, Greenwood Press, New York, 1989.

Warner, M., "Integration of Regional Land Use Planning with Environmental Impact Assessment: Practical Land Suitability Approach," *Impact Assessment* **14**, 155–190 (1996).

Weaver, C., and A. M. Cunningham, "Social Theory, Impact Assessment and Northern Native Communities," *Berkeley Planning Journal* **2**, 6–29 (1985).

Webber, M. M., "The Myth of Rationality: Development Planning Reconsidered," *Environment and Planning B: Planning and Design* **10**, 89–99 (1983).

Weber, E. P., "Successful Collaboration: Negotiating Effective Regulations," *Environment* **40**, 10–37 (1998).

Webler, T., H. Kastenholz, and O. Renn, "Public Participation in Impact Assessment: A Social Learning Perspective," *Environmental Impact Assessment* **15**, 443–464 (1995).

Webster, R., in R. Clark and L. Canter, eds., "Increasing the Efficiency and Effectiveness of NEPA Through the Use of Technology," *Environmental Policy and NEPA: Past, Present and Future*, St. Lucie Press, Boca Raton, Florida, 1997.

Weinberg, A. S., "The Environmental Justice Debate: New Agendas for a Third Generation of Research," *Society and Natural Resources* **11**, 605–614 (1998).

Weiner, K. S., in R. Clark and L. Canter, eds., "Basic Purpose and Policies of the NEPA Regulations," *Environmental Policy and NEPA: Past, Present and Future*, St. Lucie Press, Boca Raton, Florida, 1997.

Weiss, E. H., "An Unreadable EIS Is An Environmental Hazard," *Environmental Professional* **11**, 236–240 (1989).

Welles, H., in R. Clark and L. Canter, eds., "The CEQ NEPA Effectiveness Study: Learning from Our Past and Shaping Our Future," *Environmental Policy and NEPA: Past, Present and Future*, St. Lucie Press, Boca Raton, Florida, 1997.

Wende, W., "Evaluation of the Effectiveness and Quality of Environmental Impact Assessment in the Federal Republic of Germany," *Impact Assessment and Project Appraisal* **29**, 93–100 (2002).

Westman, W. E., *Ecology, Impact Assessment and Environmental Planning*, John Wiley & Sons, New York, 1985.

Whelen, E. M., "Our 'Stolen Future' and the Precautionary Principle," *Priorities* **8**, 3 (1996).

Whitney, J. B. R., in V. W. Maclaren and J. B. Whitney, eds., "Integrated Economic Models in Environmental Impact Assessment," *New Directions in Environmental Impact Assessment in Canada*, Methuen, Toronto, Canada, 1986.

Whyte, A. V., and I. Burton, eds., *Environmental Risk Assessment*, SCOPE 15, John Wiley & Sons, New York, 1980.

Wiener, J. B., and M. D. Rogers, *Comparing Precaution in the United States and Europe*, WP 2002-01, Duke Center for Environmental Solutions, Duke University, Durham, North Carolina, 2002.

Wieringa, M. J., and A. G. Morton, "Hydropower, Adaptive Management and Biodiversity," *Environmental Management* **20**, 831–840 (1996).

Wiesner, D., *E.I.A.: The Environmental Impact Assessment Process—What Is It and How to Do One?* Prism Press, Dorset, UK, 1995.

Wildavsky, A., *But Is It True? A Citizen's Guide to Environmental Health and Safety Issues*, Harvard University Press, Cambridge, Massachusetts, 1995.

Wiles, A., J. McEwen, and M. H. Sadar, "Use of Traditional Ecological Knowledge in Environmental Assessment of Uranium Mining in Saskatchewan," *Impact Assessment and Project Appraisal* **17**, 107–114 (1999).

Wilkinson, C., in A. Porter and J. Fittipaldi, eds., "Environmental Justice Impact Assessment, Key Components and Emerging Issues," *Environmental Methods Review: Retooling Impact Assessment for the New Century*, International Association for Impact Assessment and Army Environmental Policy Institute, Fargo, North Dakota, 1998.

Wilson, E. O., *Consilience: The Unity of Knowledge*, Alfred A. Knopf, New York, 1998.

Wilson, J., in R. Boardman, ed., "Green Lobbies: Pressure Groups and Environmental Policy," *Canadian Environmental Policy: Ecosystems, Politics and Process*, Oxford University Press, Toronto, Canada, 1992.

Winters, L., and A. Scott-Samuel, *Health Impact Assessment of Community Safety Projects*, Huyton SRB Area, Observatory Report Series No. 38, 1997.

Witty, D., "The Practice Behind the Theory: Co-management as a Community Development Tool," *Plan Canada* **36**, 22–27 (1994).

Wolfe, A. K., N. Kerchner, and T. Wilbanks, "Public Involvement on a Regional Scale," *Environmental Impact Assessment Review* **21**, 431–448 (2001).

Wolfe, L. D. S., *Methods of Scoping Environmental Impact Assessments: A Review of Literature and Experience*, Federal Environmental Assessment Review Office, Vancouver, British Columbia, Canada, January 1987.

Wood, C., *Environmental Impact Assessment: A Comparative Review*, Longman Scientific and Technical, Harlow, Essex, UK, 1995.

Wood, C., in N. Lee and C. George, eds., "Screening and Scoping," *Environmental Impact Assessment in Developing and Transitional Countries: Principles, Methods and Practice*, John Wiley & Sons, Chichester, UK, 2000.

Wood, C., and M. Dejeddour, "Strategic Environmental Assessment: EA of Policies, Plans and Programmes," *Impact Assessment Bulletin* **10**, 3–22 (1992).

Wood, C., A. Barker, C. Jones, and J. Hughes, *Evaluation of the Performance of the EIA Process: Final Report, Volume 1: Main Report*, EIA Centre, University of Manchester, Manchester, UK, October 1996.

World Bank, *The Impact of Environmental Assessment: A Review of World Bank Experience*, Environmental Department, World Bank, Washington, DC, 1997.

World Commission on Environment and Development (WCED), *Our Common Future*, Oxford University Press, Oxford, 1987.

World Health Organization (WHO), "The Constitution of the World Health Organization," *World Health Organization Chronicles* **1**, 29 (1967).

World Health Organization (WHO), *Health and Safety Component of Environmental Impact Assessment*, Environmental Health Series No. 15, WHO Regional Office for Europe, Copenhagen, Denmark, 1987.

World Health Organization Regional Office for Europe (WHOROE), *Gothenburg Consensus Paper Health Impact Assessment: Main Concepts and Suggested Approach*, European Centre for Health Policy, Brussels, Belgium, 1999.

World Health Organization Regional Office for Europe (WHOROE), *HIA in Strategic Environmental Assessment Brochure*, Copenhagen, Denmark, 2001a.

World Health Organization Regional Office for Europe (WHOROE), *HIA in Strategic Environmental Assessment: Current Policy Issues for Debate and Action by the Health Sector*, Copenhagen, Denmark, 2001b.

World Health Organization Regional Office for Europe (WHOROE), *Precautionary Policies and Health Protection: Principles and Applications*, Report on a WHO Workshop, Rome, Italy, 2001c.

Yiftachel, O., "Towards a New Typology of Urban Planning Theories," *Environment and Planning B. Planning and Design* **16**, 23–39 (1989).

Yoe, C. H., *An Introduction to Risk and Uncertainty in the Evaluation of Environmental Investments*, Evaluation of Environmental Investments Research Program, U.S. Army Corps of Engineers, IWR Report 96-R-8, Alexandria, Virginia and Vicksburg, Mississippi, March 1996.

Yoe, C. H., and L. Skaggs, *Risk and Uncertainty Analysis Procedures for the Evaluation of Environmental Outputs*, Evaluation of Environmental Investments Research Program, U.S. Army Corps of Engineers, IWR Report 97-R-7, Alexandria, Virginia and Vicksburg, Mississippi, August 1997.

INDEX

Environmental Impact Assessment: Practical Solutions to Recurrent Problems, By David P. Lawrence
ISBN 0-471-45722-1 Copyright © 2003 John Wiley & Sons, Inc.

DATE DUE
Fecha Para Retornar

THE LIBRARY STORE #47-0100